W9-BVQ-474

PERGAMON INTERNATIONAL LIBRARY
of Science, Technology, Engineering and Social Studies

The 1000-volume original paperback library in aid of education,
industrial training and the enjoyment of leisure

Publisher: Robert Maxwell, M.C.

COURSE OF THEORETICAL PHYSICS

Volume 5

STATISTICAL PHYSICS

Part 1

Third edition, revised and enlarged

THE PERGAMON TEXTBOOK
INSPECTION COPY SERVICE

An inspection copy of any book published in the Pergamon International Library
will gladly be sent to academic staff without obligation for their consideration
for course adoption or recommendation. Copies may be retained for a period
of 60 days from receipt and returned if not suitable. When a particular title
is adopted or recommended for adoption for class use and the recommendation
results in a sale of 12 or more copies, the inspection copy may be retained
with our compliments. The Publishers will be pleased to receive suggestions for
revised editions and new titles to be published in this important International
Library.

Other Titles in the Course of Theoretical Physics

Vol. 1. MECHANICS, 3rd edition

Vol. 2. THE CLASSICAL THEORY OF FIELDS, 4th edition

Vol. 3. QUANTUM MECHANICS (NON-RELATIVISTIC THEORY),
3rd edition

Vol. 4. RELATIVISTIC QUANTUM THEORY (published in two parts)

Vol. 6. FLUID MECHANICS

Vol. 7. THEORY OF ELASTICITY, 2nd edition

Vol. 8. ELECTRODYNAMICS OF CONTINUOUS MEDIA

Vol. 9. STATISTICAL PHYSICS, Part 2

Vol. 10. PHYSICAL KINETICS

Also of interest:

A Shorter Course of Theoretical Physics
(Based on the Course of Theoretical Physics)

Vol. 1. Mechanics and Electrodynamics

Vol. 2. Quantum Mechanics

STATISTICAL PHYSICS

by

L. D. LANDAU AND E. M. LIFSHITZ

INSTITUTE OF PHYSICAL PROBLEMS,
U.S.S.R. ACADEMY OF SCIENCES

Volume 5 of *Course of Theoretical Physics*

PART 1
THIRD EDITION, REVISED AND ENLARGED
by E. M. LIFSHITZ and L. P. PITAEVSKII

Translated from the Russian by
J. B. SYKES AND M. J. KEARSLEY

PERGAMON PRESS

OXFORD · NEW YORK · TORONTO · SYDNEY · PARIS · FRANKFURT

U.K.	Pergamon Press Ltd., Headington Hill Hall, Oxford OX3 0BW, England
U.S.A.	Pergamon Press Inc., Maxwell House, Fairview Park, Elmsford, New York 10523, U.S.A.
CANADA	Pergamon Press Canada Ltd., Suite 104, 150 Consumers Rd., Willowdale, Ontario M2J 1P9, Canada
AUSTRALIA	Pergamon Press (Aust.) Pty. Ltd., P.O. Box 544, Potts Point, N.S.W. 2011, Australia
FRANCE	Pergamon Press SARL, 24 rue des Ecoles, 75240 Paris, Cedex 05, France
FEDERAL REPUBLIC OF GERMANY	Pergamon Press GmbH, 6242 Kronberg-Taunus, Hammerweg 6, Federal Republic of Germany

First published in English 1959
Second Impression 1959
Third Impression 1963
Second Revised and Enlarged Edition 1968
Third Revised and Enlarged Edition 1980
Reprinted 1982

British Library Cataloguing in Publication Data

Landau, Lev Davidovich
Statistical physics.
Part 1.—3rd ed., revised and enlarged.—
(Course of theoretical physics; vol. 5).
1. Mathematical physics 2. Mathematical statistics
I. Title   II. Lifshits, Evgenii Mikhailovich
III. Sykes, John Bradbury   IV. Kearsley, Mary Jennifer
V. Pitaevskii, Lev Petrovich
530.1'5'95   QC21.2   78-40140

ISBN 0-08-023039-3 (Hardcover)
ISBN 0-08-023038-5 (Flexicover)

Translated from *Statisticheskaya fizika,* chast' 1, Third edition, published in 1976 by Nauka, Moscow.

Printed in Great Britain by A. Wheaton & Co. Ltd., Exeter

CONTENTS

XIII. THE SYMMETRY OF CRYSTALS

XIV. PHASE TRANSITIONS OF THE SECOND KIND AND CRITICAL PHENOMENA

XV. SURFACES

CONTENTS OF PART 2

Contents of part 2

IV. GREEN'S FUNCTIONS AT NON-ZERO TEMPERATURES

V. SUPERCONDUCTIVITY

VI. ELECTRONS IN THE CRYSTAL LATTICE

VII. MAGNETISM

VIII. ELECTROMAGNETIC FLUCTUATIONS

IX. HYDRODYNAMIC FLUCTUATIONS

PREFACE TO THE THIRD RUSSIAN EDITION

In this edition the book has been considerably augmented and revised, with the assistance of L. P. Pitaevskiĭ throughout.

New sections have been added on the magnetic properties of gases, the thermodynamics of a degenerate plasma, liquid crystals, the fluctuation theory of phase transitions of the second kind, and critical phenomena. The chapters on solids and on the symmetry of crystals have been substantially enlarged, in particular by a fuller account of the theory of irreducible representations of space groups as applied to the physics of the crystal state. The sections on the fluctuation–dissipation theorem have been revised and extended.

Some sections have been removed from the book, dealing with the theory of quantum liquids and the related theory of almost ideal degenerate gases. The physics of quantum liquids, which was founded and largely developed by the pioneering experiments of P. L. Kapitza and the theoretical work of Landau himself, has now become a wide subject whose significance goes far beyond its original concern, the liquid helium isotopes. An account of the theory of quantum liquids must now occupy its rightful place in even a general course of theoretical physics, and the few sections given to it in the earlier editions of this book are insufficient.

They will appear, in a considerably expanded form, in another volume of this course, now being prepared by Pitaevskiĭ and myself, which will also give a detailed treatment of the Green's function method and the diagram technique, which have largely determined the development of statistical physics in the last 20 years. The transfer of these (and some other) topics to a separate volume is dictated not only by the fact that their inclusion in the present one would make it too large and would considerably alter its whole character. There is also the reason that such topics are essentially akin to hydrodynamics and macroscopic electrodynamics; for example, in presenting the microscopic theory of superconductivity it is convenient to make use of the known macroscopic theory of this phenomenon. For this reason, the new volume will stand as one of the course, after *Mechanics* and *Electrodynamics of Continuous Media*.

The first version of this book (which included only classical statistical physics) appeared in 1938. The reader of today may be surprised to find that the use of the general Gibbs method in statistical physics even in the 1930s called for reasoning such as is given in the extracts (reproduced below) from the preface to that book. Perhaps it was just in the development of the exposition of general principles and numerous applications of statistical physics that Landau most showed his astonishing breadth of grasp of the whole subject, his astonishing ability to discern the most direct and effective way of deriving every result of the theory, whether major or minor.

Lastly, on behalf of L. P. Pitaevskiĭ and myself, may I sincerely thank I. E. Dzyaloshinskiĭ, I. M. Lifshitz and V. L. Pokrovskiĭ for many discussions of matters arising in the revision of this book.

Moscow   E. M. LIFSHITZ
May 1975

FROM THE PREFACES TO PREVIOUS
RUSSIAN EDITIONS

IT IS a fairly widespread delusion among physicists that statistical physics is the least well-founded branch of theoretical physics. Reference is generally made to the point that some of its conclusions are not subject to rigorous mathematical proof; and it is overlooked that every other branch of theoretical physics contains just as many non-rigorous proofs, although these are not regarded as indicating an inadequate foundation for such branches.

Yet the work of Gibbs transformed the statistical physics of Clausius, Maxwell and Boltzmann into a logically connected and orderly system. Gibbs provided a general method, which is applicable in principle to all problems that can be posed in statistical physics, but which unfortunately has not been adequately taken up. The fundamental inadequacy of the majority of existing books on statistical physics is precisely that their authors, instead of taking this general method as a basis, give it only incidentally.

Statistical physics and thermodynamics together form a unit. All the concepts and quantities of thermodynamics follow most naturally, simply and rigorously from the concepts of statistical physics. Although the general statements of thermodynamics *can* be formulated non-statistically, their application to specific cases always requires the use of statistical physics.

We have tried in this book to give a systematic account of statistical physics and thermodynamics together, based on the Gibbs method. All specific problems are statistically analysed by general methods. In the proofs, our aim has been not mathematical rigour, which is not readily attainable in theoretical physics, but chiefly to emphasise the interrelation of different physical statements.

In the discussion of the foundations of classical statistical physics, we consider from the start the statistical distribution for small parts (subsystems) of systems, not for entire closed systems. This is in accordance with the fundamental problems and aims of physical statistics, and allows a complete avoidance of the problem of the ergodic and similar hypotheses, which in fact is not important as regards these aims.

An ideal gas is regarded as a particular case from the standpoint of general methods, and we have therefore not described the Boltzmann method as such. This method cannot be independently justified; in particular, the use of *a priori* probabilities is difficult to justify. The Boltzmann expression for the entropy of an ideal gas is derived from the general formulae of the Gibbs method.

L. D. LANDAU

E. M. LIFSHITZ

1937–9

NOTATION

OPERATORS are denoted by a circumflex.

Mean values of quantities are denoted by a bar over the symbol or by angle brackets (see the footnote after (1.5)).

Phase space

p, q generalised momenta and coordinates
$dp \, dq = dp_1 \, dp_2 \ldots dp_s \, dq_1 \, dq_2 \ldots dq_s$ volume element in phase space (with s degrees of freedom)
$d\Gamma = dp \, dq/(2\pi\hbar)^s$
$\int' \ldots d\Gamma$ integral over all physically different states

Thermodynamic quantities

T temperature
V volume
P pressure
E energy
S entropy
$W = E+PV$ heat function
$F = E-TS$ free energy
$\Phi = E-TS+PV$ thermodynamic potential
$\Omega = -PV$ thermodynamic potential
C_p, C_v specific heats
c_p, c_v molecular specific heats
N number of particles
μ chemical potential
α surface-tension coefficient
$\mathscr{S}$ area of interface

In all formulae the temperature is expressed in energy units; the method of converting to degrees is described in footnotes to §§ 9 and 42.

References to other volumes in the *Course of Theoretical Physics*:
Mechanics = Vol. 1 (*Mechanics*, third English edition, 1976).
Fields = Vol. 2 (*The Classical Theory of Fields*, fourth English edition, 1975)
Quantum Mechanics = Vol. 3 (*Quantum Mechanics*, third English edition, 1977).
RQT = Vol. 4 (*Relativistic Quantum Theory*, Part 1, English edition, 1971).
Elasticity = Vol. 7 (*Theory of Elasticity*, second English edition, 1970).
Electrodynamics = Vol. 8 (*Electrodynamics of Continuous Media*, English edition, 1960).

All are published by Pergamon Press.

CHAPTER I

THE FUNDAMENTAL PRINCIPLES
OF STATISTICAL PHYSICS

§ 1. Statistical distributions

Statistical physics, often called for brevity simply *statistics*, consists in the study of the special laws which govern the behaviour and properties of macroscopic bodies (that is, bodies formed of a very large number of individual particles, such as atoms and molecules). To a considerable extent the general character of these laws does not depend on the mechanics (classical or quantum) which describes the motion of the individual particles in a body, but their substantiation demands a different argument in the two cases. For convenience of exposition we shall begin by assuming that classical mechanics is everywhere valid.

In principle, we can obtain complete information concerning the motion of a mechanical system by constructing and integrating the equations of motion of the system, which are equal in number to its degrees of freedom. But if we are concerned with a system which, though it obeys the laws of classical mechanics, has a very large number of degrees of freedom, the actual application of the methods of mechanics involves the necessity of setting up and solving the same number of differential equations, which in general is impracticable. It should be emphasised that, even if we could integrate these equations in a general form, it would be completely impossible to substitute in the general solution the initial conditions for the velocities and coordinates of all the particles.

At first sight we might conclude from this that, as the number of particles increases, so also must the complexity and intricacy of the properties of the mechanical system, and that no trace of regularity can be found in the behaviour of a macroscopic body. This is not so, however, and we shall see below that, when the number of particles is very large, new types of regularity appear.

These *statistical laws* resulting from the very presence of a large number of particles forming the body cannot in any way be reduced to purely mechanical laws. One of their distinctive features is that they cease to have meaning when applied to mechanical systems with a small number of degrees of

1

freedom. Thus, although the motion of systems with a very large number of degrees of freedom obeys the same laws of mechanics as that of systems consisting of a small number of particles, the existence of many degrees of freedom results in laws of a different kind.

The importance of statistical physics in many other branches of theoretical physics is due to the fact that in Nature we continually encounter macroscopic bodies whose behaviour can not be fully described by the methods of mechanics alone, for the reasons mentioned above, and which obey statistical laws.

In proceeding to formulate the fundamental problem of classical statistics, we must first of all define the concept of *phase space*, which will be constantly used hereafter.

Let a given macroscopic mechanical system have s degrees of freedom: that is, let the position of points of the system in space be described by s coordinates, which we denote by q_i, the suffix i taking the values 1, 2, ..., s. Then the state of the system at a given instant will be defined by the values at that instant of the s coordinates q_i and the s corresponding velocities $\dot{q}_i$. In statistics it is customary to describe a system by its coordinates and momenta p_i, not velocities, since this affords a number of very important advantages. The various states of the system can be represented mathematically by points in *phase space* (which is, of course, a purely mathematical concept); the coordinates in phase space are the coordinates and momenta of the system considered. Every system has its own phase space, with a number of dimensions equal to twice the number of degrees of freedom. Any point in phase space, corresponding to particular values of the coordinates q_i and momenta p_i of the system, represents a particular state of the system. The state of the system changes with time, and consequently the point in phase space representing this state (which we shall call simply the *phase point* of the system) moves along a curve called the *phase trajectory*.

Let us now consider a macroscopic body or system of bodies, and assume that the system is closed, i.e. does not interact with any other bodies. A part of the system, which is very small compared with the whole system but still macroscopic, may be imagined to be separated from the rest; clearly, when the number of particles in the whole system is sufficiently large, the number in a small part of it may still be very large. Such relatively small but still macroscopic parts will be called *subsystems*. A subsystem is again a mechanical system, but not a closed one; on the contrary, it interacts in various ways with the other parts of the system. Because of the very large number of degrees of freedom of the other parts, these interactions will be very complex and intricate. Thus the state of the subsystem considered will vary with time in a very complex and intricate manner.

An exact solution for the behaviour of the subsystem can be obtained only by solving the mechanical problem for the entire closed system, i.e. by setting up and solving all the differential equations of motion with given initial conditions, which, as already mentioned, is an impracticable task. Fortunately, it is just this very complicated manner of variation of the state of subsystems which, though rendering the methods of mechanics inapplicable, allows a different approach to the solution of the problem.

A fundamental feature of this approach is the fact that, because of the extreme complexity of the external interactions with the other parts of the system, during a sufficiently long time the subsystem considered will be many times in every possible state. This may be more precisely formulated as follows. Let $\Delta p \, \Delta q$ denote some small "volume" of the phase space of the subsystem, corresponding to coordinates q_i and momenta p_i lying in short intervals Δq_i and Δp_i. We can say that, in a sufficiently long time T, the extremely intricate phase trajectory passes many times through each such volume of phase space. Let Δt be the part of the total time T during which the subsystem was in the given volume of phase space $\Delta p \, \Delta q$.[†] When the total time T increases indefinitely, the ratio $\Delta t/T$ tends to some limit

$$w = \lim_{T \to \infty} \Delta t/T. \tag{1.1}$$

This quantity may clearly be regarded as the probability that, if the subsystem is observed at an arbitrary instant, it will be found in the given volume of phase space $\Delta p \, \Delta q$.

On taking the limit of an infinitesimal phase volume[‡]

w – probability

$$dq \, dp = dq_1 \, dq_2 \ldots dq_s \, dp_1 \, dp_2 \ldots dp_s, \tag{1.2}$$

we can define the probability dw of states represented by points in this volume element, i.e. the probability that the coordinates q_i and momenta p_i have values in given infinitesimal intervals between q_i, p_i and $q_i + dq_i, p_i + dp_i$. This probability dw may be written

probability density

$$dw = \varrho(p_1, \ldots, p_s, q_1, \ldots, q_s) \, dp \, dq, \tag{1.3}$$

where $\varrho(p_1, \ldots, p_s, q_1, \ldots, q_s)$ is a function of all the coordinates and momenta; we shall usually write for brevity $\varrho(p, q)$ or even ϱ simply. The function ϱ, which represents the "density" of the probability distribution in phase space, is called the *statistical distribution function,* or simply the

---

[†] For brevity, we shall usually say, as is customary, that the system "is in the volume $\Delta p \, \Delta q$ of phase space", meaning that the system is in states represented by phase points in that volume.

[‡] In what follows we shall always use the conventional notation dp and dq to denote the products of the differentials of all the momenta and all the coordinates of the system respectively.

distribution function, for the body concerned. This function must obviously satisfy the *normalisation condition*

$$\int \varrho \, dp \, dq = 1 \tag{1.4}$$

(the integral being taken over all phase space), which simply expresses the fact that the sum of the probabilities of all possible states must be unity.

The following circumstance is extremely important in statistical physics. The statistical distribution of a given subsystem does not depend on the initial state of any other small part of the same system, since over a sufficiently long time the effect of this initial state will be entirely outweighed by the effect of the much larger remaining parts of the system. It is also independent of the initial state of the particular small part considered, since in time this part passes through all possible states, any of which can be taken as the initial state. Without having to solve the mechanical problem for a system (taking account of initial conditions), we can therefore find the statistical distribution for small parts of the system.

The determination of the statistical distribution for any subsystem is in fact the fundamental problem of statistical physics. In speaking of "small parts" of a closed system, we must bear in mind that the macroscopic bodies with which we have to deal are usually themselves such "small parts" of a large closed system consisting of these bodies together with the external medium which surrounds them.

If this problem is solved and the statistical distribution for a given subsystem is known, we can calculate the probabilities of various values of any physical quantities which depend on the states of the subsystem (i.e. on the values of its coordinates q and momenta p). We can also calculate the mean value of any such quantity $f(p, q)$, which is obtained by multiplying each of its possible values by the corresponding probability and integrating over all states. Denoting the averaging by a bar, we can write

$$\bar{f} = \int f(p, q) \, \varrho(p, q) \, dp \, dq, \tag{1.5}$$

from which the mean values of various quantities can be calculated by using the statistical distribution function.[†]

The averaging with respect to the distribution function (called *statistical averaging*) frees us from the necessity of following the variation with time of the actual value of the physical quantity $f(p, q)$ in order to determine its mean value. It is also obvious that, by the definition (1.1) of the probability, the statistical averaging is exactly equivalent to a time averaging. The latter

[†] In this book we shall denote averaging by a bar over the symbol or by angle brackets: $\bar{f}$ or $\langle f \rangle$, being influenced in this solely by convenience in writing the formulae. The second way is preferable for writing the mean values of lengthy expressions.

would involve following the variation of the quantity with time, establishing the function $f = f(t)$, and determining the required mean value as

$$\bar{f} = \lim_{T \to \infty} \frac{1}{T} \int_0^T f(t)\, dt.$$

The foregoing discussion shows that the deductions and predictions concerning the behaviour of macroscopic bodies which are made possible by statistical physics are probabilistic. In this respect statistical physics differs from (classical) mechanics, the deductions of which are entirely deterministic. It should be emphasised, however, that the probabilistic nature of the results of classical statistics is not an inherent property of the objects considered, but simply arises from the fact that these results are derived from much less information than would be necessary for a complete mechanical description (the initial values of the coordinates and momenta are not needed).

In practice, however, when statistical physics is applied to macroscopic bodies, its probabilistic nature is not usually apparent. The reason is that, if any macroscopic body (in external conditions independent of time) is observed over a sufficiently long period of time, it is found that all physical quantities describing the body are practically constant (and equal to their mean values) and undergo appreciable changes relatively very rarely; we mean, of course, macroscopic quantities describing the body as a whole or macroscopic parts of it, but not individual particles.[†] This result, which is fundamental to statistical physics, follows from very general considerations (to be discussed in § 2) and becomes more and more nearly valid as the body considered becomes more complex and larger. In terms of the statistical distribution, we can say that, if by means of the function $\varrho(p, q)$ we construct the probability distribution function for various values of the quantity $f(p, q)$, this function will have an extremely sharp maximum for $f = \bar{f}$, and will be appreciably different from zero only in the immediate vicinity of this point.

Thus, by enabling us to calculate the mean values of quantities describing macroscopic bodies, statistical physics enables us to make predictions which are valid to very high accuracy for by far the greater part of any time interval which is long enough for the effect of the initial state of the body to be entirely eliminated. In this sense the predictions of statistical physics become

---

[†] We may give an example to illustrate the very high degree of accuracy with which this is true. If we consider a region in a gas which contains, say, only 1/100 gram-molecule, we find that the mean relative variation of the energy of this quantity of matter from its mean value is only $\sim 10^{-11}$. The probability of finding (in a single observation) a relative deviation of the order of 10^{-6}, say, is given by a fantastically small number, $\sim 10^{-3 \times 10^{15}}$.

practically determinate and not probabilistic. (For this reason, we shall henceforward almost always omit the bar when using mean values of macroscopic quantities.)

If a closed macroscopic system is in a state such that in any macroscopic subsystem the macroscopic physical quantities are to a high degree of accuracy equal to their mean values, the system is said to be in a state of *statistical equilibrium* (or *thermodynamic* or *thermal equilibrium*). It is seen from the foregoing that, if a closed macroscopic system is observed for a sufficiently long period of time, it will be in a state of statistical equilibrium for much the greater part of this period. If, at any initial instant, a closed macroscopic system was not in a state of statistical equilibrium (if, for example, it was artificially disturbed from such a state by means of an external interaction and then left to itself, becoming again a closed system), it will necessarily enter an equilibrium state. The time within which it will reach statistical equilibrium is called the *relaxation time*. In using the term "sufficiently long" intervals of time, we have meant essentially times long compared with the relaxation time.

The theory of processes relating to the attainment of an equilibrium state is called *kinetics*. It is not part of statistical physics proper, which deals only with systems in statistical equilibrium.

§ 2. Statistical independence

The subsystems discussed in § 1 are not themselves closed systems; on the contrary, they are subject to the continuous interaction of the remaining parts of the system. But since these parts, which are small in comparison with the whole of the large system, are themselves macroscopic bodies also, we can still suppose that over not too long intervals of time they behave approximately as closed systems. For the particles which mainly take part in the interaction of a subsystem with the surrounding parts are those near the surface of the subsystem; the relative number of such particles, compared with the total number of particles in the subsystem, decreases rapidly when the size of the subsystem increases, and when the latter is sufficiently large the energy of its interaction with the surrounding parts will be small in comparison with its internal energy. Thus we may say that the subsystems are *quasi-closed*. It should be emphasised once more that this property holds only over not too long intervals of time. Over a sufficiently long interval of time, the effect of interaction of subsystems, however weak, will ultimately appear. Moreover, it is just this relatively weak interaction which leads finally to the establishment of statistical equilibrium.

The fact that different subsystems may be regarded as weakly interacting has the result that they may also be regarded as statistically independent.

By *statistical independence* we mean that the state of one subsystem does not affect the probabilities of various states of the other subsystems.

Let us consider any two subsystems, and let $dp^{(1)} dq^{(1)}$ and $dp^{(2)} dq^{(2)}$ be volume elements in their phase spaces. If we regard the two subsystems together as one composite subsystem, then the statistical independence of the subsystems signifies mathematically that the probability of the composite subsystem's being in its phase volume element $dp^{(12)} dq^{(12)} = dp^{(1)} dq^{(1)}$. $dp^{(2)} dq^{(2)}$ can be written as the product of the probabilities for the two subsystems to be respectively in $dp^{(1)} dq^{(1)}$ and $dp^{(2)} dq^{(2)}$, each of these probabilities depending only on the coordinates and momenta of the subsystem concerned. Thus we can write

$$\varrho_{12} \, dp^{(12)} \, dq^{(12)} = \varrho_1 \, dp^{(1)} \, dq^{(1)} \cdot \varrho_2 \, dp^{(2)} \, dq^{(2)},$$

or

$$\varrho_{12} = \varrho_1 \varrho_2, \tag{2.1}$$

where ϱ_{12} is the statistical distribution of the composite subsystem, and ϱ_1, ϱ_2 the distribution functions of the separate subsystems. A similar relation is valid for a group of several subsystems.[†]

The converse statement is clearly also true: if the probability distribution for a compound system is a product of factors, each of which depends only on quantities describing one part of the system, then the parts concerned are statistically independent, and each factor is proportional to the probability of the state of the corresponding part.

If f_1 and f_2 are two physical quantities relating to two different subsystems, then from (2.1) and the definition (1.5) of mean values it follows immediately that the mean value of the product $f_1 f_2$ is equal to the product of the mean values of the quantities f_1 and f_2 separately:

$$\overline{f_1 f_2} = \bar{f_1} \cdot \bar{f_2}. \tag{2.2}$$

Let us consider a quantity f relating to a macroscopic body or to a part of it. In the course of time this quantity varies, fluctuating about its mean value. We may define a quantity which represents the average range of this fluctuation. The mean value of the difference $\Delta f = f - \bar{f}$ is not suitable for this purpose, since the quantity f varies from its mean value in both directions, and the difference $f - \bar{f}$, which is alternately positive and negative, has mean value zero regardless of how often f undergoes considerable deviations from its mean value. The required characteristic may conveniently be defined as the mean square of this difference. Since $(\Delta f)^2$ is always positive, its mean value tends to zero only if $(\Delta f)^2$ itself tends to zero; that is, the mean value

---

[†] Provided, of course, that these subsystems together still form only a small part of the whole closed system.

is small only when the probability of considerable deviations of f from $\bar{f}$ is small. The quantity $\langle(\Delta f)^2\rangle^{1/2}$ is called the *root-mean-square* (r.m.s.) *fluctuation* of the quantity f. Multiplying out the square $(f-\bar{f})^2$ shows that

$$\langle(\Delta f)^2\rangle = \bar{f^2}-(\bar{f})^2, \tag{2.3}$$

i.e. the r.m.s. fluctuation is determined by the difference between the mean square of the quantity and the square of its mean.

The ratio $\langle(\Delta f)^2\rangle^{1/2}/\bar{f}$ is called the *relative fluctuation* of the quantity f. The smaller this ratio is, the more negligible is the proportion of time during which the body is in states where the deviation of f from its mean value is a considerable fraction of the mean value.

We shall show that the relative fluctuations of physical quantities decrease rapidly when the size of the bodies (that is, the number of particles) to which they relate increases. To prove this, we first note that the majority of quantities of physical interest are additive. This property is a consequence of the fact that the various parts of a body are quasi-closed systems, and signifies that the value of such a quantity for the whole body is the sum of its values for the various (macroscopic) parts of the body. For example, since the internal energies of these parts are, as shown above, large compared with their interaction energies, it is sufficiently accurate to assume that the energy of the whole body is equal to the sum of the energies of its parts.

Let f be such an additive quantity. We imagine the body concerned to be divided into a large number N of approximately equal small parts. Then

$$\bar{f} = \sum_{i=1}^{N} \bar{f_i}, \quad \propto N$$

where the quantities f_i relate to the individual parts of the body.

It is clear that, as the size of the body increases, $\bar{f}$ increases approximately in proportion to N. Let us also determine the r.m.s. fluctuation of f. We have

$$\langle(\Delta f)^2\rangle = \left\langle\left(\sum_i \Delta f_i\right)^2\right\rangle.$$

Because of the statistical independence of the different parts of the body, the mean values of the products $\Delta f_i \Delta f_k$ are

$$\overline{\Delta f_i \Delta f_k} = \overline{\Delta f_i}\cdot\overline{\Delta f_k} = 0 \quad (i \neq k),$$

since each $\overline{\Delta f_i} = 0$. Hence

$$\langle(\Delta f)^2\rangle = \sum_{i=1}^{N} \langle(\Delta f_i)^2\rangle. \quad \propto N \tag{2.4}$$

It follows that, as N increases, the mean square $\langle(\Delta f)^2\rangle$ also increases in proportion to N. The relative fluctuation is therefore inversely proportional to $\sqrt{N}$:

$$\frac{\langle (\Delta f)^2 \rangle^{1/2}}{\bar{f}} \propto \frac{1}{\sqrt{N}}. \qquad \text{for large } N. \qquad (2.5)$$

On the other hand, if we consider a homogeneous body to be divided into parts of a given small size, it is clear that the number of parts will be proportional to the total number of particles (molecules) in the body. Hence the result can also be stated by saying that the relative fluctuation of any additive quantity f decreases inversely as the square root of the number of particles in a macroscopic body, and so, when the number of these is sufficiently large, the quantity f itself may be regarded as practically constant in time and equal to its mean value. This conclusion has already been used in § 1.

§ 3. Liouville's theorem

Let us now return to a further study of the properties of the statistical distribution function, and suppose that a subsystem is observed over a very long interval of time, which we divide into a very large (in the limit, infinite) number of equal short intervals between instants t_1, t_2, ... At each of these instants the subsystem considered is represented in its phase space by a point A_1, A_2, ... The set of points thus obtained is distributed in phase space with a density which in the limit is everywhere proportional to the distribution function $\varrho(p, q)$. This follows from the significance of the latter function as determining the probabilities of various states of the subsystem.

Instead of considering points representing states of one subsystem at different instants t_1, t_2, ..., we may consider simultaneously, in a formal manner, a very large (in the limit, infinite) number of exactly identical subsystems,[†] which at some instant, say $t = 0$, are in states represented by the points A_1, A_2, ...

stat ensemble.

We now follow the subsequent movement of the phase points which represent the states of these subsystems over a not too long interval of time, such that a quasi-closed subsystem may with sufficient accuracy be regarded as closed. The movement of the phase points will then obey the equations of motion, which involve the coordinates and momenta only of the particles in the subsystem.

It is clear that at any instant t these points will be distributed in phase space according to the same distribution function $\varrho(p, q)$, in just the same way as at $t = 0$. In other words, as the phase points move about in the course of time, they remain distributed with a density which is constant at any given point and is proportional to the corresponding value of ϱ.

[†] Such an imaginary set of identical systems is usually called a *statistical ensemble.*

$S = 3N$?

This movement of phase points may be formally regarded as a steady flow of a "gas" in phase space of $2s$ dimensions, and the familiar equation of continuity may be applied, which expresses the constancy of the total number of "particles" (in this case, phase points) in the gas. The ordinary equation of continuity is

Conservation of mass

$$\partial\varrho/\partial t + \mathrm{div}\,(\varrho v) = 0,$$

$$\frac{\partial \ell}{\partial t} + \nabla(\varrho \vec{v}) = 0$$

where ϱ is the density and v the velocity of the gas. For steady flow, we have

$$\mathrm{div}\,(\varrho v) = 0.$$

$$\nabla(\varrho\vec{v}) = 0$$

Volume in phase space!

For a space of $2s$ dimensions, this will become

$$\sum_{i=1}^{2s} \frac{\partial}{\partial x_i}(\varrho v_i) = 0.$$

Note: $v_i \neq q_i$ only
$v_i = \dot{q}_i, \dot{p}_i$

In the present case the "coordinates" x_i are the coordinates q and momenta p, and the "velocities" $v_i = \dot{x}_i$ are the time derivatives $\dot{q}$ and $\dot{p}$ given by the equations of motion. Thus we have

$$\sum_{i=1}^{s}\left[\frac{\partial}{\partial q_i}(\varrho \dot{q}_i) + \frac{\partial}{\partial p_i}(\varrho \dot{p}_i)\right] = 0.$$ ✓

Expanding the derivatives gives

$$\sum_{i=1}^{s}\left[\dot{q}_i\frac{\partial\varrho}{\partial q_i} + \dot{p}_i\frac{\partial\varrho}{\partial p_i}\right] + \varrho\sum_{i=1}^{s}\left[\frac{\partial\dot{q}_i}{\partial q_i} + \frac{\partial\dot{p}_i}{\partial p_i}\right] = 0.$$ (3.1)

With the equations of motion in Hamilton's form:

$$\dot{q}_i = \partial H/\partial p_i, \quad \dot{p}_i = -\partial H/\partial q_i,$$

where $H = H(p, q)$ is the Hamiltonian for the subsystem considered, we see that

$$\partial\dot{q}_i/\partial q_i = \partial^2 H/\partial q_i\,\partial p_i = -\partial\dot{p}_i/\partial p_i.$$

The second term in (3.1) is therefore identically zero. The first term is just the total time derivative of the distribution function. Thus

$$\frac{d\varrho}{dt} = \sum_{i=1}^{s}\left(\frac{\partial\varrho}{\partial q_i}\dot{q}_i + \frac{\partial\varrho}{\partial p_i}\dot{p}_i\right) = 0.$$ (3.2)

We therefore reach the important conclusion that the distribution function is constant along the phase trajectories of the subsystem. This is *Liouville's theorem.* Since quasi-closed subsystems are under dicussion, the result is valid only for not too long intervals of time, during which the subsystem behaves as if closed, to a sufficient approximation.

§ 4. The significance of energy

It follows at once from Liouville's theorem that the distribution function must be expressible entirely in terms of combinations of the variables p and q which remain constant when the subsystem moves as a closed subsystem. These combinations are the *mechanical invariants* or *integrals of the motion*, which are the first integrals of the equations of motion. We may therefore say that the distribution function, being a function of the mechanical invariants, is itself an integral of the motion.

It proves possible to restrict very considerably the number of integrals of the motion on which the distribution function can depend. To do this, we must take into account the fact that the distribution ϱ_{12} for a combination of two subsystems is equal to the product of the distribution functions ϱ_1 and ϱ_2 of the two subsystems separately: $\varrho_{12} = \varrho_1 \varrho_2$. Hence

$$\log \varrho_{12} = \log \varrho_1 + \log \varrho_2, \tag{4.1}$$

so that the logarithm of the distribution function is an additive quantity. We therefore reach the conclusion that the logarithm of the distribution function must be not merely an integral of the motion, but an additive integral of the motion.

As we know from mechanics, there exist only seven independent additive integrals of the motion: the energy, the three components of the momentum vector and the three components of the angular momentum vector. We shall denote these quantities for the ath subsystem (as functions of the coordinates and momenta of the particles in it) by $E_a(p, q)$, $\mathbf{P}_a(p, q)$, $\mathbf{M}_a(p, q)$ respectively. The only additive combination of these quantities is a linear combination of the form

$$\log \varrho_a = \alpha_a + \beta E_a(p, q) + \boldsymbol{\gamma} \cdot \mathbf{P}_a(p, q) + \boldsymbol{\delta} \cdot \mathbf{M}_a(p, q) \tag{4.2}$$

with constant coefficients α_a, β, $\boldsymbol{\gamma}$, $\boldsymbol{\delta}$, of which β, $\boldsymbol{\gamma}$, $\boldsymbol{\delta}$ must be the same for all subsystems in a given closed system.

We shall return in Chapter III to a detailed study of the distribution (4.2); here we need note only the following points. The coefficient α_a is just the normalisation constant, given by the condition $\int \varrho_a \, dp^{(a)} \, dq^{(a)} = 1$. The constants β, $\boldsymbol{\gamma}$, $\boldsymbol{\delta}$, involving seven independent quantities altogether, may be determined from the seven constant values of the additive integrals of the motion for the whole closed system. Thus we reach a conclusion very important in statistical physics. The values of the additive integrals of the motion (energy, momentum and angular momentum) completely define the statistical properties of a closed system, i.e. the statistical distribution of any of its subsystems, and therefore the mean values of any physical quantities relating to them. These seven additive integrals of the motion replace the unimaginable multiplicity of data (initial conditions) which would be required in the approach from mechanics.

The above arguments enable us at once to set up a simple distribution function suitable for describing the statistical properties of a closed system. Since, as we have seen, the values of non-additive integrals of the motion do not affect these properties, the latter can be described by any function ϱ which depends only on the values of the additive integrals of the motion for the system and which satisfies Liouville's theorem. The simplest such function is $\varrho = $ constant for all points in phase space which correspond to given constant values of the energy (E_0), momentum $(\mathbf{P}_0)$ and angular momentum $(\mathbf{M}_0)$ of the system (regardless of the values of the non-additive integrals) and $\varrho = 0$ at all other points. It is clear that a function so defined will certainly remain constant along a phase trajectory of the system, i.e. will satisfy Liouville's theorem.

This formulation, however, is not quite exact. The reason is that the points defined by the equations

$$E(p, q) = E_0, \quad \mathbf{P}(p, q) = \mathbf{P}_0, \quad \mathbf{M}(p, q) = \mathbf{M}_0 \tag{4.3}$$

form a manifold of only $2s-7$ dimensions, not $2s$ like the phase volume. Consequently, if the integral $\int \varrho \, dp \, dq$ is to be different from zero, the function $\varrho(p, q)$ must become infinite at these points. The correct way of writing the distribution function for a closed system is

$$\varrho = \text{constant} \times \delta(E - E_0) \, \delta(\mathbf{P} - \mathbf{P}_0) \, \delta(\mathbf{M} - \mathbf{M}_0). \tag{4.4}$$

The presence of the delta functions[†] ensures that ϱ is zero at all points in phase space where one or more of the quantities E, $\mathbf{P}$, $\mathbf{M}$ is not equal to the given value E_0, $\mathbf{P}_0$ or $\mathbf{M}_0$. The integral of ϱ over the whole of a phase volume which includes all or part of the above-mentioned manifold of points is finite. The distribution (4.4) is called *microcanonical*.[‡]

The momentum and angular momentum of a closed system depend on its motion as a whole (uniform translation and uniform rotation). We can therefore say that the statistical state of a system executing a given motion depends only on its energy. In consequence, energy is of exceptional importance in statistical physics.

[†] The definition and properties of the delta function are given, for example, in *Quantum Mechanics*, § 5.

[‡] It should be emphasised once more that this distribution is not the true statistical distribution for a closed system. Regarding it as the true distribution is equivalent to asserting that, in the course of a sufficiently long time, the phase trajectory of a closed system passes arbitrarily close to every point of the manifold defined by equations (4.3). But this assertion (called the *ergodic hypothesis*) is certainly not true in general.

principle of Equal a priori Probabilities.

In order to exclude the momentum and angular momentum from the subsequent discussion we may use the following device. We imagine the system
to be enclosed in a rigid "box" and take coordinates such that the "box" is
at rest. Under these conditions the momentum and angular momentum are
not integrals of the motion, and the only remaining additive integral of the
motion is the energy. The presence of the "box", on the other hand, clearly
does not affect the statistical properties of small parts of the system (subsystems). Thus for the logarithms of the distribution functions of the subsystems,
instead of (4.2), we have the still simpler expressions

$$\log \varrho_a = \alpha_a + \beta E_a(p, q). \tag{4.5}$$

The microcanonical distribution for the whole system is

$$\varrho = \text{constant} \times \delta(E - E_0). \tag{4.6}$$

So far we have assumed that the closed system as a whole is in statistical
equilibrium; that is, we have considered it over times long compared with its
relaxation time. In practice, however, it is usually necessary to discuss a
system over times comparable with or even short relative to the relaxation
time. For large systems this can be done, owing to the existence of what are
called *partial* (or *incomplete*) *equilibria* as well as the complete statistical
equilibrium of the entire closed system. Such equilibria occur because the
relaxation time increases with the size of the system, and so the separate
small parts of the system attain the equilibrium state considerably more
quickly than equilibrium is established between these small parts. This
means that each small part of the system is described by a distribution function of the form (4.2), with the parameters β, γ, δ of the distribution having
different values for different parts. In such a case the system is said to be in
partial equilibrium. In the course of time, the partial equilibrium gradually
becomes complete, and the parameters β, γ, δ for each small part slowly
vary and finally become equal throughout the closed system.

Another kind of partial equilibrium is also of frequent occurrence, namely
that resulting from a difference in the rates of the various processes occurring
in the system, not from a considerable difference in relaxation time between
the system and its small parts. One obvious example is the partial equilibrium
in a mixture of several substances involved in a chemical reaction. Owing to
the comparative slowness of chemical reactions, equilibrium as regards the
motion of the molecules will be reached, in general, considerably more rapidly than equilibrium as regards reactions of molecules, i.e. as regards the composition of the mixture. This enables us to regard the partial equilibria of the
mixture as equilibria at a given (actually non-equilibrium) chemical composition.

The existence of partial equilibria leads to the concept of *macroscopic states* of a system. Whereas a mechanical microscopic description of the system specifies the coordinates and momenta of every particle in the system, a macroscopic description is one which specifies the mean values of the physical quantities determining a particular partial equilibrium, for instance the mean values of quantities describing separate sufficiently small but macroscopic parts of the system, each of which may be regarded as being in a separate equilibrium.

9-4

§ 5. **The statistical matrix**

Turning now to the distinctive features of quantum statistics, we may note first of all that the purely mechanical approach to the problem of determining the behaviour of a macroscopic body in quantum mechanics is of course just as hopeless as in classical mechanics. Such an approach would require the solution of Schrödinger's equation for a system consisting of all the particles in the body, a problem still more hopeless, one might even say, than the integration of the classical equations of motion. But even if it were possible in some particular case to find a general solution of Schrödinger's equation, it would be utterly impossible to select and write down the particular solution satisfying the precise conditions of the problem and specified by particular values of an enormous number of different quantum numbers. Moreover, we shall see below that for a macroscopic body the concept of stationary states itself becomes to some extent arbitrary, a fact of fundamental significance.

Let us first elucidate some purely quantum-mechanical features of macroscopic bodies as compared with systems consisting of a relatively small number of particles.

These features amount to an extremely high density of levels in the energy eigenvalue spectrum of a macroscopic body. The reason for this is easily seen if we note that, because of the very large number of particles in the body, a given quantity of energy can, roughly speaking, be "distributed" in innumerable ways among the various particles. The relation between this fact and the high density of levels becomes particularly clear if we take as an example a macroscopic body consisting of a "gas" of N particles which do not interact at all, enclosed in some volume. The energy levels of such a system are just the sums of the energies of the individual particles, and the energy of each particle can range over an infinite series of discrete values.[†] It is clear that, on choosing in all possible ways the values of the N terms in this sum, we shall

---

[†] The separations between successive energy levels of a single particle are inversely proportional to the square of the linear dimensions L of the volume enclosing it ($\sim \hbar^2/mL^2$, where m is the mass of the particle and $\hbar$ the quantum constant).

obtain a very large number of possible values of the energy of the system in any appreciable finite part of the spectrum, and these values will therefore lie very close together.

It may be shown (see (7.18)) that the number of levels in a given finite range of the energy spectrum of a macroscopic body increases exponentially with the number of particles in the body, and the separations between levels are given by numbers of the form 10^{-N} (where N is a number of the order of the number of particles in the body), whatever the units, since a change in the unit of energy has no effect on such a fantastically small number.[†]

In consequence of the extremely high density of levels, a macroscopic body in practice can never be in a strictly stationary state. First of all, it is clear that the value of the energy of the system will always be "broadened" by an amount of the order of the energy of interaction between the system and the surrounding bodies. The latter is very large in comparison with the separations between levels, not only for quasi-closed subsystems but also for systems which from any other aspect could be regarded as strictly closed. In Nature, of course, there are no completely closed systems, whose interaction with any other body is exactly zero; and whatever interaction does exist, even if it is so small that it does not affect other properties of the system, will still be very large in comparison with the infinitesimal intervals in the energy spectrum.

In addition to this, there is another fundamental reason why a macroscopic body in practice cannot be in a stationary state. It is known from quantum mechanics that the state of a system described by a wave function is the result of some process of interaction of the system with another system which obeys classical mechanics to a sufficient approximation. In this respect the occurrence of a stationary state implies particular properties of the system. Here we must distinguish between the energy E of the system before the interaction and the energy E' of the state which results from the interaction. The uncertainties ΔE and $\Delta E'$ in the quantities E and E' are related to the duration Δt of the interaction process by the formula

$$|\Delta E' - \Delta E| \sim \hbar/\Delta t;$$

see *Quantum Mechanics*, § 44. The two errors ΔE and $\Delta E'$ are in general of the same order of magnitude, and analysis shows that we cannot make

---

[†] It should be mentioned that this discussion is inapplicable to the initial part of the energy spectrum; the separations between the first few energy levels of a macroscopic body may even be independent of the size of the body. This point, however, does not affect the subsequent conclusions: when referred to a single particle, the separations between the first few levels for a macroscopic body are negligibly small, and the high density of levels mentioned in the text is reached for very small values of the energy relative to a single particle.

$\Delta E' \ll \Delta E$. We can therefore say that $\Delta E' \sim \hbar/\Delta t$. In order that the state may be regarded as stationary, the uncertainty $\Delta E'$ must certainly be small in comparison with the separations between adjoining levels. Since the latter are extremely small, we see that, in order to bring the macroscopic body into a particular stationary state, an extremely long time $\Delta t \sim \hbar/\Delta E'$ would be necessary. In other words, we again conclude that strictly stationary states of a macroscopic body cannot exist.

To describe the state of a macroscopic body by a wave function at all is impracticable, since the available data concerning the state of such a body are far short of the complete set of data necessary to establish its wave function. Here the position is somewhat similar to that which occurs in classical statistics, where the impossibility of taking account of the initial conditions for every particle in a body makes impossible an exact mechanical description of its behaviour; the analogy is imperfect, however, since the impossibility of a complete quantum-mechanical description and the lack of a wave function describing a macroscopic body may, as we have seen, possess a much more profound significance.

The quantum-mechanical description based on an incomplete set of data concerning the system is effected by means of what is called a *density matrix*; see *Quantum Mechanics*, § 14. A knowledge of this matrix enables us to calculate the mean value of any quantity describing the system, and also the probabilities of various values of such quantities. The incompleteness of the description lies in the fact that the results of various kinds of measurement which can be predicted with a certain probability from a knowledge of the density matrix might be predictable with greater or even complete certainty from a complete set of data for the system, from which its wave function could be derived.

We shall not pause to write out here the formulae of quantum mechanics relating to the density matrix in the coordinate representation, since this representation is seldom used in statistical physics, but we shall show how the density matrix may be obtained directly in the energy representation, which is necessary for statistical applications.

Let us consider some subsystem, and define its "stationary states" as the states obtained when all interactions of the subsystem with the surrounding parts of a closed system are entirely neglected. Let $\psi_n(q)$ be the normalised wave functions of these states (without the time factor), q conventionally denoting the set of all coordinates of the subsystem, and the suffix n the set of all quantum numbers which distinguish the various stationary states; the energies of these states will be denoted by E_n.

Let us assume that at some instant the subsystem is in a completely described state with wave function ψ. The latter may be expanded in terms of the functions $\psi_n(q)$, which form a complete set; we write the ex-

pansion as

$$\psi = \sum_n c_n \psi_n. \tag{5.1}$$

The mean value of any quantity f in this state can be calculated from the coefficients c_n by means of the formula

$$\bar{f} = \sum_n \sum_m c_n^* c_m f_{nm}, \tag{5.2}$$

where

$$f_{nm} = \int \psi_n^* \hat{f} \psi_m \, dq \tag{5.3}$$

are the matrix elements of the quantity f ($\hat{f}$ being the corresponding operator).

The change from the complete to the incomplete quantum-mechanical description of the subsystem may be regarded as a kind of averaging over its various ψ states. In this averaging, the products $c_n^* c_m$ give a double set (two suffixes) of quantities, which we denote by w_{mn} and which cannot be expressed as products of any quantities forming a single set. The mean value of f is now given by

$$\bar{f} = \sum_m \sum_n w_{mn} f_{nm}. \tag{5.4}$$

The set of quantities w_{mn} (which in general are functions of time) is the density matrix in the energy representation; in statistical physics it is called the *statistical matrix*.[†]

If we regard the w_{mn} as the matrix elements of some *statistical operator* $\hat{w}$, then the sum $\sum_n w_{mn} f_{nm}$ will be a diagonal matrix element of the operator product $\hat{w}\hat{f}$, and the mean value $\bar{f}$ becomes the trace (sum of diagonal elements) of this operator:

$$\bar{f} = \sum_n (\hat{w}\hat{f})_{nn} = \mathrm{tr}\,(\hat{w}\hat{f}). \tag{5.5}$$

This formula has the advantage of enabling us to calculate with any complete set of orthonormal wave functions: the trace of an operator is independent of the particular set of functions with respect to which the matrix elements are defined; see *Quantum Mechanics*, § 12.

---

† The energy representation is mentioned here, as being the one generally used in statistical physics. We have not so far, however, made direct use of the fact that the ψ_n are wave functions of stationary states. It is therefore clear that the same method could be used to define the density matrix with respect to any complete set of wave functions.

The usual coordinate density matrix $\varrho(q, q')$ (see *Quantum Mechanics*, § 14) is expressed in terms of the matrix w_{mn} by

$$\varrho(q, q') = \sum_{m,n} w_{mn} \psi_n^*(q') \, \psi_m(q).$$

The other expressions of quantum mechanics which involve the quantities c_n are similarly modified, the products $c_n^* c_m$ being everywhere replaced by the "averaged values" w_{mn}:

$$c_n^* c_m \;\rightarrow\; w_{mn}.$$

For example, the probability that the subsystem is in the nth state is equal to the corresponding diagonal element w_{nn} of the density matrix (instead of the squared modulus $c_n^* c_n$). It is evident that these elements, which we shall denote by w_n, are always positive:

$$w_n = w_{nn} > 0, \tag{5.6}$$

and satisfy the normalisation condition

$$\mathrm{tr}\,\hat{w} = \sum_n w_n = 1 \tag{5.7}$$

$\left(\text{corresponding to the condition } \sum_n |c_n|^2 = 1\right).$

It must be emphasised that the averaging over various ψ states, which we have used in order to illustrate the transition from a complete to an incomplete quantum-mechanical description, has only a very formal significance. In particular, it would be quite incorrect to suppose that the description by means of the density matrix signifies that the subsystem can be in various ψ states with various probabilities and that the averaging is over these probabilities. Such a treatment would be in conflict with the basic principles of quantum mechanics.

The states of a quantum-mechanical system that are described by wave functions are sometimes called *pure states*, as distrinct from *mixed states*, which are described by a density matrix. Care should, however, be taken not to misunderstand the latter term in the way indicated above.

The averaging by means of the statistical matrix according to (5.4) has a twofold nature. It comprises both the averaging due to the probabilistic nature of the quantum description (even when as complete as possible) and the statistical averaging necessitated by the incompleteness of our information concerning the object considered. For a pure state only the first averaging remains, but in statistical cases both types of averaging are always present. It must be borne in mind, however, that these constituents cannot be separated; the whole averaging procedure is carried out as a single operation, and cannot be represented as the result of successive averagings, one purely quantum-mechanical and the other purely statistical.

The statistical matrix in quantum statistics takes the place of the distribution function in classical statistics. The whole of the discussion in the previous sections concerning classical statistics and the, in practice, deterministic nature of its predictions applies entirely to quantum statistics also. The proof given in § 2 that the relative fluctuations of additive physical quantities tend to zero as the number of particles increases made no use of any specific properties of classical mechanics, and so remains entirely valid in the quantum case. We can therefore again assert that macroscopic quantities remain practically equal to their mean values.

In classical statistics the distribution function $\varrho(p, q)$ gives directly the probability distribution of the various values of the coordinates and momenta of the particles of the body. In quantum statistics this is no longer true; the quantities w_n give only the probabilities of finding the body in a particular quantum state, with no direct indication of the values of the coordinates and momenta of the particles.

From the very nature of quantum mechanics, the statistics based on it can deal only with the determination of the probability distribution for the coordinates and momenta separately, not together, since the coordinates and momenta of a particle cannot simultaneously have definite values. The required probability distributions must reflect both the statistical uncertainty and the uncertainty inherent in the quantum-mechanical description. To find these distributions, we repeat the arguments given above. We first assume that the body is in a pure quantum state with the wave function (5.1). The probability distribution for the coordinates is given by the squared modulus

$$|\psi|^2 = \sum_n \sum_m c_n^* c_m \psi_n^* \psi_m,$$

so that the probability that the coordinates have values in a given interval $dq = dq_1\, dq_2\, \ldots\, dq_s$ is $dw_q = |\psi|^2\, dq$. For a mixed state, the products $c_n^* c_m$ are replaced by the elements w_{mn} of the statistical matrix, and $|\psi|^2$ thus becomes

$$\sum_n \sum_m w_{mn} \psi_n^* \psi_m.$$

By the definition of the matrix elements,

$$\sum_m w_{mn} \psi_m = \hat{w} \psi_n,$$

and so

$$\sum_n \sum_m w_{mn} \psi_n^* \psi_m = \sum_n \psi_n^* \hat{w} \psi_n.$$

Thus we have the following formula for the coordinate probability distribution:

$$dw_q = \sum_n \psi_n^* \hat{w} \psi_n \cdot dq. \tag{5.8}$$

In this expression the functions ψ_n may be any complete set of normalised wave functions.

Let us next determine the momentum probability distribution. The quantum states in which all the momenta have definite values correspond to free motion of all the particles. We denote the wave functions of these states by $\psi_p(q)$, the suffix p conventionally representing the set of values of all the momenta. As we know, the diagonal elements of the density matrix are the probabilities that the system is in the corresponding quantum states. Hence, having determined the density matrix with respect to the set of functions ψ_p, we obtain the required momentum probability distribution from the formula[†]

$$dw_p = w_{pp}\, dp = dp \cdot \int \psi_p^* \hat{w} \psi_p \, dq, \tag{5.9}$$

where $dp = dp_1 \, dp_2 \ldots dp_s$.

It is interesting that both distributions (coordinate and momentum) can be obtained by integrating the same function

$$I(q, p) = \psi_p^*(q)\, \hat{w}\psi_p(q). \tag{5.10}$$

Integration of this expression with respect to q gives the momentum distribution (5.9). Integration with respect to p gives

$$dw_q = dq \int \psi_p^*(q)\, \hat{w}\psi_p(q) \, dp, \tag{5.11}$$

in agreement with the general definition (5.8). Note also that the function (5.10) can be expressed in terms of the coordinate density matrix $\varrho(q, q')$ by

$$I(q, p) = \psi_p^*(q) \int \varrho(q, q')\, \psi_p(q') \, dq'. \tag{5.12}$$

It must be emphasised, however, that this does not at all imply that the function $I(q, p)$ may be regarded as a probability distribution for coordinates and momenta simultaneously; the expression (5.10) is in any case complex, quite apart from the fact that such a treatment would altogether contradict the fundamental principles of quantum mechanics.[‡]

[†] The functions $\psi_p(q)$ are plane waves in the configuration space of the system; they are assumed normalised by the delta function of all the momenta.

[‡] Since $I(q, p)$ has no direct physical significance, the definition of the function with the properties stated is of course not unique. For example, the q and p distributions can be obtained by the same method from the function

$$I_W(q, p) = \int_{-\infty}^{\infty} \varrho(q + \tfrac{1}{2}\xi, q - \tfrac{1}{2}\xi)\, \psi_p^*(q + \tfrac{1}{2}\xi)\, \psi_p(q - \tfrac{1}{2}\xi) \, d\xi, \tag{5.10a}$$

where ξ denotes the set of auxiliary variables $\xi_1, \ldots, \xi_s$, and $d\xi = d\xi_1 \ldots d\xi_s$ (E. P. Wigner, 1932): since

$$\int \psi_p^*(q + \tfrac{1}{2}\xi)\, \psi_p(q - \tfrac{1}{2}\xi) \, dp = \delta(q + \tfrac{1}{2}\xi - q + \tfrac{1}{2}\xi) = \delta(\xi),$$

the integral $\int I_W \, dp = \varrho(q, q)$. The integral $\int I_W \, dq$, after the change of variables $q + \tfrac{1}{2}\xi \to q$, $q - \tfrac{1}{2}\xi \to q'$, is the same as $\int I \, dq$. Unlike $I(q, p)$, $I_W(q, p)$ is real (as can be easily seen by using the fact that the matrix $\varrho(q, q')$ is Hermitian), but in general it is not everywhere positive.

§ 6. Statistical distributions in quantum statistics

In quantum mechanics a theorem can be proved which is analogous to Liouville's theorem derived in § 3 on the basis of classical mechanics.

To do this, we first derive a general equation of quantum mechanics which gives the time derivative of the statistical matrix of any (closed) system.[†] Following the method used in § 5, we first assume that the system is in a pure state with a wave function represented in the form of a series (5.1). Since the system is closed, its wave function will have the same form at all subsequent instants, but the coefficients c_n will depend on the time, being proportional to factors $e^{-iE_n t/\hbar}$. We therefore have

$$\frac{\partial}{\partial t}(c_n^* c_m) = \frac{i}{\hbar}(E_n - E_m)c_n^* c_m.$$

The change to the statistical matrix in the general case of mixed states is now effected by replacing the products $c_n^* c_m$ by w_{mn}, and this gives the required equation:

$$\dot{w}_{mn} = (i/\hbar)(E_n - E_m)w_{mn}. \tag{6.1}$$

This equation can be written in a general operator form by noticing that

$$(E_n - E_m)w_{mn} = \sum_l (w_{ml}H_{ln} - H_{ml}w_{ln}),$$

where H_{mn} are the matrix elements of the Hamiltonian $\hat{H}$ of the system; this matrix is diagonal in the energy representation, which we are using. Hence

$$\dot{\hat{w}} \quad \hat{w} = (i/\hbar)(\hat{w}\hat{H} - \hat{H}\hat{w}). \tag{6.2}$$

It should be pointed out that this expression differs in sign from the usual quantum-mechanical expression for the operator of the time derivative of a quantity.

We see that, if the time derivative of the statistical matrix is zero, the operator $\hat{w}$ must commute with the Hamiltonian of the system. This result is the quantum analogue of Liouville's theorem: in classical mechanics the requirement of a stationary distribution function has the result that w is an integral of the motion, while the commutability of the operator of a quantity with the Hamiltonian is just the condition, in quantum mechanics, that that quantity is conserved.

In the energy representation used here, the condition is particularly simple: (6.1) shows that the matrix w_{mn} must be diagonal, again in accordance with

---

[†] In § 5 the density matrix of a subsystem was discussed, having regard to its fundamental applications in statistical physics, but a density matrix can of course also be used to describe a closed system in a mixed state.

the usual matrix condition that a quantity is conserved in quantum mechanics, namely that the matrix of such a quantity can be brought to diagonal form simultaneously with the Hamiltonian.

As in § 3, we can now apply the results obtained to quasi-closed subsystems, for intervals of time during which they behave to a sufficient approximation as closed systems. Since the statistical distributions (or in this case the statistical matrices) of subsystems must be stationary, by the definition of statistical equilibrium, we first of all conclude that the matrices w_{mn} are diagonal for all subsystems.[†] The problem of determining the statistical distribution therefore amounts to a calculation of the probabilities $w_n = w_{nn}$, which represent the "distribution function" in quantum statistics. Formula (5.4) for the mean value of any quantity f becomes simply

$$\bar{f} = \sum_n w_n f_{nn}, \qquad (6.3)$$

and contains only the diagonal matrix elements f_{nn}.

Next, using the facts that w must be a quantum-mechanical integral of the motion and that the subsystems are quasi-independent, we find in a similar way to the derivation of (4.5) that the logarithm of the distribution function for subsystems must be of the form

$$\log w_n^{(a)} = \alpha^{(a)} + \beta E_n^{(a)}, \qquad (6.4)$$

where the index a corresponds to the various subsystems. Thus the probabilities w_n can be expressed as a function of the energy level alone: $w_n = w(E_n)$.

Finally, the discussion in § 4 concerning the significance of additive integrals of the motion, and in particular the energy, as determining all the statistical properties of a closed system, remains entirely valid. This again enables us to set up for a closed system a simple distribution function suitable for describing its statistical properties though (as in the classical case) certainly not the true distribution function.

To formulate mathematically this "quantum microcanonical distribution" we must use the following device. The energy spectra of macroscopic bodies being "almost continuous", we make use of the concept of the number of quantum states of a closed system which "belong" to a particular infinitesimal range of values of its energy.[‡] We denote this number by $d\Gamma$; it plays a

---

[†] Since this statement involves neglecting the interactions between subsystems, it is more precise to say that the non-diagonal elements w_{mn} tend to zero as the relative importance of these interactions decreases, and therefore as the number of particles in the subsystems increases.

[‡] It will be remembered that in § 4 we agreed to ignore entirely the momentum and angular momentum of the system as a whole, for which purpose it is sufficient to consider a system enclosed in a rigid "box" with coordinates such that the box is at rest.

part analogous to that of the phase volume element $dp\,dq$ in the classical case.

If we regard a closed system as consisting of subsystems, and neglect the interaction of the latter, every state of the whole system can be described by specifying the states of the individual subsystems, and the number $d\Gamma$ is a product

$$d\Gamma = \prod_a d\Gamma_a \qquad (6.5)$$

of the numbers $d\Gamma_a$ of the quantum states of the subsystems (such that the sum of the energies of the subsystems lies in the specified interval of energy of the whole system).

We can now formulate the microcanonical distribution analogously to the classical expression (4.6), writing

$$dw = \text{constant} \times \delta(E-E_0) \prod_a d\Gamma_a \qquad (6.6)$$

for the probability dw of finding the system in any of the $d\Gamma$ states.

§ 7. Entropy

Let us consider a closed system for a period of time long compared with its relaxation time; this implies that the system is in complete statistical equilibrium.

The following discussion will be given first of all for quantum statistics. Let us divide the system into a large number of macroscopic parts (subsystems) and consider any one of them. Let w_n be the distribution function for this subsystem; to simplify the formulae we shall at present omit from w_n (and other quantities) the suffix indicating the subsystem. By means of the function w_n we can, in particular, calculate the probability distribution of the various values of the energy E of the subsystem. We have seen that w_n may be written as a function of the energy alone, $w_n = w(E_n)$. In order to obtain the probability $W(E)\,dE$ that the subsystem has an energy between E and $E+dE$, we must multiply $w(E)$ by the number of quantum states with energies in this interval; here we use the same idea of a "broadened" energy spectrum as was mentioned at the end of § 6. Let $\Gamma(E)$ denote the number of quantum states with energies less than or equal to E. Then the required number of states with energy between E and $E+dE$ can be written

$$\frac{d\Gamma(E)}{dE}\,dE,$$

and the energy probability distribution is

$$W(E) = \frac{d\Gamma(E)}{dE}\,w(E). \qquad (7.1)$$

$W(E)$ – prob in Γ space

$W(E)$ – prob in E space

The normalisation condition

$$\int W(E)\, dE = 1$$

signifies geometrically that the area under the curve $W = W(E)$ is unity.

In accordance with the general statements in § 1, the function $W(E)$ has a very sharp maximum at $E = \bar{E}$, being appreciably different from zero only in the immediate neighbourhood of this point. We may define the "width" ΔE of the curve $W = W(E)$ as the width of a rectangle whose height is equal to the value of the function $W(E)$ at the maximum and whose area is unity:

$$W(\bar{E})\, \Delta E = 1. \tag{7.2}$$

Using the expression (7.1), we can write this definition as

$$w(\bar{E})\, \Delta \Gamma = 1, \tag{7.3}$$

where

$$\Delta \Gamma = \frac{d\Gamma(\bar{E})}{dE}\, \Delta E \tag{7.4}$$

is the number of quantum states corresponding to the interval ΔE of energy. The quantity $\Delta \Gamma$ thus defined may be said to represent the "degree of broadening" of the macroscopic state of the subsystem with respect to its microscopic states. The interval ΔE is equal in order of magnitude to the mean fluctuation of energy of the subsystem.

These definitions can be immediately applied to classical statistics, except that the function $w(E)$ must be replaced by the classical distribution function ϱ, and $\Delta \Gamma$ by the volume of the part of phase space defined by the formula

$$\varrho(\bar{E})\, \Delta p\, \Delta q = 1. \tag{7.5}$$

The phase volume $\Delta p\, \Delta q$, like $\Delta \Gamma$, represents the size of the region of phase space in which the subsystem will almost always be found.

It is not difficult to establish the relation between $\Delta \Gamma$ in quantum theory and $\Delta p\, \Delta q$ in the limit of classical theory. In the quasi-classical case, a correspondence can be set up between the volume of a region of phase space and the "corresponding" number of quantum states (see *Quantum Mechanics*, § 48): we can say that a "cell" of volume $(2\pi\hbar)^s$ (where s is the number of degrees of freedom of the system) "corresponds" in phase space to each quantum state. It is therefore clear that in the quasi-classical case the number of states $\Delta \Gamma$ may be written

$$\Delta \Gamma = \Delta p\, \Delta q / (2\pi\hbar)^s, \tag{7.6}$$

where s is the number of degrees of freedom of the subsystem considered. This formula gives the required relation between $\Delta \Gamma$ and $\Delta p\, \Delta q$.

ΔΓ — number of accessible states.

The quantity $\Delta\Gamma$ is called the *statistical weight* of the macroscopic state of the subsystem, and its logarithm

$$S = \log\Delta\Gamma \qquad \left(S = k\, \ln\Delta\Gamma\right) \tag{7.7}$$

is called the *entropy* of the subsystem. In the case of classical statistics the corresponding expression is

$$S = \log\frac{\Delta p\,\Delta q}{(2\pi\hbar)^s}. \qquad \left(\times\ k\right) \tag{7.8}$$

The entropy thus defined is dimensionless, like the statistical weight itself. Since the number of states $\Delta\Gamma$ is not less than unity, the entropy cannot be negative. The concept of entropy is one of the most important in statistical physics.

It is apposite to mention that, if we adhere strictly to the standpoint of classical statistics, the concept of the "number of microscopic states" cannot be defined at all, and we should have to define the statistical weight simply as $\Delta p\,\Delta q$. But this quantity, like any volume in phase space, has the dimensions of the product of s momenta and s coordinates, i.e. the sth power of action ((erg·sec)^s). The entropy, defined as $\log\Delta p\,\Delta q$, would then have the peculiar dimensions of the logarithm of action. This means that the entropy would change by an additive constant when the unit of action changed: if the unit were changed by a factor a, $\Delta p\,\Delta q$ would become $a^s\,\Delta p\,\Delta q$, and $\log\Delta p\,\Delta q$ would become $\log\Delta p\,\Delta q + s\log a$. In purely classical statistics, therefore, the entropy is defined only to within an additive constant which depends on the choice of units, and only differences of entropy, i.e. changes of entropy in a given process, are definite quantities independent of the choice of units.

This accounts for the appearance of the quantum constant $\hbar$ in the definition (7.8) of the entropy for classical statistics. Only the concept of the number of discrete quantum states, which necessarily involves a non-zero quantum constant, enables us to define a dimensionless statistical weight and so to give an unambiguous definition of the entropy.

We may write the definition of the entropy in another form, expressing it directly in terms of the distribution function. According to (6.4), the logarithm of the distribution function of a subsystem has the form

$$\log w(E_n) = \alpha + \beta E_n.$$

Since this expression is linear in E_n, the quantity

$$\log w(\bar{E}) = \alpha + \beta\bar{E}$$

can be written as the mean value $\langle\log w(E_n)\rangle$. The entropy $S = \log\Delta\Gamma = -\log w(\bar{E})$ (from (7.3)) can therefore be written

$$S = -\langle\log w(E_n)\rangle, \tag{7.9}$$

i.e. the entropy can be defined as minus the mean logarithm of the distribution function of the subsystem. From the significance of the mean value,

$$S = -\sum_n w_n \log w_n;$$ (7.10)

this expression can be written in a general operator form independent of the choice of the set of wave functions with respect to which the statistical matrix elements are defined:[†]

$$S = -\text{tr}\,(\hat{w} \log \hat{w}).$$ (7.11)

Similarly, in classical statistics, the definition of the entropy can be written

$$S = -\langle \log [(2\pi\hbar)^s \varrho] \rangle$$
$$= -\int \varrho \log [(2\pi\hbar)^s \varrho]\,dp\,dq.$$ (7.12)

Let us now return to the closed system as a whole, and let $\Delta\Gamma_1, \Delta\Gamma_2, \ldots$ be the statistical weights of its various subsystems. If each of the subsystems can be in one of $\Delta\Gamma_a$ quantum states, this gives

$$\Delta\Gamma = \prod_a \Delta\Gamma_a$$ (7.13)

as the number of different states of the whole system. This is called the statistical weight of the closed system, and its logarithm is the entropy S of the system. Clearly

$$S = \sum_a S_a,$$ (7.14)

i.e. the entropy thus defined is additive: the entropy of a composite system is equal to the sum of the entropies of its parts.

For a clear understanding of the way in which entropy is defined, it is important to bear in mind the following point. The entropy of a closed system (whose total energy we denote by E_0) in complete statistical equilibrium can also be defined directly, without dividing the system into subsystems. To do this, we imagine that the system considered is actually only a small part of a fictitious very large system (called in this connection a *thermostat* or *heat bath*). The thermostat is assumed to be in complete equilibrium, in such a way that the mean energy of the system considered (which is now a non-closed subsystem of the thermostat) is equal to its actual energy E_0. Then we can formally assign to the system a distribution function of the same form as for any subsystem of it, and by means of this distribution determine its statis-

---

[†] In accordance with the general rules, the operator log $\hat{w}$ must be understood as an operator whose eigenvalues are equal to the logarithms of the eigenvalues of the operator $\hat{w}$, and whose eigenfunctions are the same as those of $\hat{w}$.

$$\bar{f} = \langle f \rangle = \oint \sum_m f_m w_m$$

tical weight $\Delta\Gamma$, and therefore the entropy, directly from the same formulae (7.3)–(7.12) as were used for subsystems. It is clear that the presence of the thermostat has no effect on the statistical properties of individual small parts (subsystems) of the system considered, which in any case are not closed and are in equilibrium with the remaining parts of the system. The presence of the thermostat therefore does not alter the statistical weights $\Delta\Gamma_a$ of these parts, and the statistical weight defined in the way just described will be the same as that previously defined as the product (7.13).

So far we have assumed that the closed system is in complete statistical equilibrium. We must now generalise the above definitions to systems in arbitrary macroscopic states (partial equilibria).

Let us suppose that the system is in some state of partial equilibrium, and consider it over time intervals Δt which are small compared with the relaxation time for complete equilibrium. Then the entropy must be defined as follows. We imagine the system divided into parts so small that their respective relaxation times are small compared with the intervals Δt (remembering that the relaxation times in general decrease with decreasing size of the system). During the time Δt such parts may be regarded as being in their own particular equilibrium states, described by certain distribution functions. We can therefore apply to them the previous definition of the statistical weights $\Delta\Gamma_a$, and so calculate their entropies S_a. The statistical weight $\Delta\Gamma$ of the whole system is then defined as the product (7.13), and the corresponding entropy S as the sum of the entropies S_a.

It should be emphasised, however, that the entropy of a non-equilibrium system, defined in this way as the sum of the entropies of its parts (satisfying the above condition), cannot now be calculated by means of the thermostat concept without dividing the system into parts. At the same time this definition is unambiguous in the sense that further division of the subsystems into even smaller parts does not alter the value of the entropy, since each subsystem is already in "complete" internal equilibrium.

In particular, attention should be drawn to the significance of time in the definition of entropy. The entropy is a quantity which describes the average properties of a body over some non-zero interval of time Δt. If Δt is given, to determine S we must imagine the body divided into parts so small that their relaxation times are small in comparison with Δt. Since these parts must also themselves be macroscopic, it is clear that when the intervals Δt are too short the concept of entropy becomes meaningless; in particular, we cannot speak of its instantaneous value.

Having thus given a complete definition of the entropy, let us now ascertain the most important properties and the fundamental physical significance of this quantity. To do so, we must make use of the microcanonical distribution, according to which a distribution function of the form (6.6) may be used to describe the statistical properties of a closed system:

$$dw = \text{constant} \times \delta(E-E_0) \cdot \prod_a d\Gamma_a.$$

Here $d\Gamma_a$ may be taken as the differential of the function $\Gamma_a(E_a)$, which represents the number of quantum states of a subsystem with energies less than or equal to E_a. We can write dw as

$$dw = \text{constant} \times \delta(E-E_0) \cdot \prod_a (d\Gamma_a/dE_a)\, dE_a. \qquad (7.15)$$

The statistical weight $\Delta\Gamma_a$, by definition, is a function of the mean energy $\bar{E}_a$ of the subsystem; the same applies to $S_a = S_a(\bar{E}_a)$. Let us now formally regard $\Delta\Gamma_a$ and S_a as functions of the actual energy E_a (the same functions as they really are of $\bar{E}_a$). Then we can replace the derivatives $d\Gamma_a(E_a)/dE_a$ in (7.15) by the ratios $\Delta\Gamma_a/\Delta E_a$, where $\Delta\Gamma_a$ is a function of E_a in this sense, and ΔE_a the interval of energy corresponding to $\Delta\Gamma_a$ (also a function of E_a). Finally, replacing $\Delta\Gamma_a$ by $e^{S_a(E_a)}$, we obtain

$$dw = \text{constant} \times \delta(E-E_0)e^S \prod_a dE_a/\Delta E_a, \qquad (7.16)$$

where

$$S = \sum_a S_a(E_a)$$

is the entropy of the whole closed system, regarded as a function of the exact values of the energies of its parts. The factor e^S, whose exponent is an additive quantity, is a very rapidly varying function of the energies E_a. In comparison with this function, the energy dependence of the quantity $\prod \Delta E_a$ is quite unimportant, and we can therefore replace (7.16) with very high accuracy by

$$dw = \text{constant} \times \delta(E-E_0)e^S \prod_a dE_a. \qquad (7.17)$$

But dw expressed in a form proportional to the product of all the differentials dE_a is just the probability that all the subsystems have energies in given intervals between E_a and E_a+dE_a. Thus we see that this probability is determined by the entropy of the system as a function of the energies of the subsystems; the factor $\delta(E-E_0)$ ensures that the sum $E = \Sigma E_a$ has the given value E_0 of the energy of the system. This property of the entropy, as we shall see later, is the basis of its applications in statistical physics.

We know that the most probable values of the energies E_a are their mean values $\bar{E}_a$. This means that the function $S(E_1, E_2, \ldots)$ must have its maximum possible value (for a given value of the sum $\Sigma E_a = E_0$) when $E_a = \bar{E}_a$. But the $\bar{E}_a$ are just the values of the energies of the subsystems which correspond to complete statistical equilibrium of the system. Thus we reach the important conclusion that the entropy of a closed system in a state of complete statistical equilibrium has its greatest possible value (for a given energy of the system).

Finally, we may mention another interesting interpretation of the function $S = S(E)$, the entropy of any subsystem or closed system; in the latter case it is assumed that the system is in complete equilibrium, so that its entropy may be expressed as a function of its total energy alone. The statistical weight $\Delta\Gamma = e^{S(E)}$, by definition, is the number of energy levels in the interval ΔE which describes in a certain way the width of the energy probability distribution. Dividing ΔE by $\Delta\Gamma$, we obtain the mean separation between adjoining levels in this interval (near the energy E) of the energy spectrum of the system considered. Denoting this distance by $D(E)$, we can write

$$D(E) = \Delta E \cdot e^{-S(E)}. \tag{7.18}$$

Thus the function $S(E)$ determines the density of levels in the energy spectrum of a macroscopic system. Since the entropy is additive, we can say that the mean separations between the levels of a macroscopic body decrease exponentially with increasing size of the body (i.e. with increasing number of particles in it).

§ 8. The law of increase of entropy

If a closed system is not in a state of statistical equilibrium, its macroscopic state will vary in time, until ultimately the system reaches a state of complete equilibrium. If each macroscopic state of the system is described by the distribution of energy between the various subsystems, we can say that the sequence of states successively traversed by the system corresponds to more and more probable distributions of energy. This increase in probability is in general very considerable, because it is exponential, as shown in § 7. We have seen that the probability is given by e^S, the exponent being an additive quantity, the entropy of the system. We can therefore say that the processes occurring in a non-equilibrium closed system do so in such a way that the system continually passes from states of lower to those of higher entropy until finally the entropy reaches the maximum possible value, corresponding to complete statistical equilibrium.

Thus, if a closed system is at some instant in a non-equilibrium macroscopic state, the most probable consequence at later instants is a steady increase in the entropy of the system. This is the *law of increase of entropy* or *second law of thermodynamics*, discovered by R. Clausius (1865); its statistical explanation was given by L. Boltzmann in the 1870s.

In speaking of the "most probable" consequence, we must remember that in reality the probability of transition to states of higher entropy is so enormous in comparison with that of any appreciable decrease in entropy that in practice the latter can never be observed in Nature. Ignoring decreases in entropy due to negligible fluctuations, we can therefore formulate the law of increase of entropy as follows: if at some instant the entropy of a closed system does not have its maximum value, then at subsequent instants the entropy will not decrease; it will increase or at least remain constant.

There is no doubt that the foregoing simple formulations accord with reality; they are confirmed by all our everyday observations. But when we consider more closely the problem of the physical nature and origin of these laws of behaviour, substantial difficulties arise, which to some extent have not yet been overcome.

Firstly, if we attempt to apply statistical physics to the entire Universe, regarded as a single closed system, we immediately encounter a glaring contradiction between theory and experiment. According to the results of statistics, the universe ought to be in a state of complete statistical equilibrium. More precisely, any finite region of it, however large, should have a finite relaxation time and should be in equilibrium. Everyday experience shows us, however, that the properties of Nature bear no resemblance to those of an equilibrium system; and astronomical results show that the same is true throughout the vast region of the Universe accessible to our observation.

The escape from this contradiction is to be sought in the general theory of relativity. The reason is that, when large regions of the Universe are considered, the gravitational fields present become important. These fields are just a change in the space–time metric. When the statistical properties of bodies are discussed, the metric properties of space–time may in a sense be regarded as "external conditions" to which the bodies are subject. The statement that a closed system must, over a sufficiently long time, reach a state of equilibrium, applies of course only to a system in steady external conditions. On the other hand, the general cosmological expansion of the Universe means that its metric depends essentially on time, so that the "external conditions" are by no means steady in this case. Here it is important that the gravitational field cannot itself be included in a closed system, since the conservation laws which are, as we have seen, the foundation of statistical physics would then reduce to identities. For this reason, in the general theory of relativity, the Universe as a whole must be regarded not as a closed system but as a system in a variable gravitational field. Consequently the application of the law of increase of entropy does not prove that statistical equilibrium must necessarily exist.

Thus this aspect of the problem of the Universe as a whole indicates the physical basis of the apparent contradictions. There are, however, other difficulties in understanding the physical nature of the law of increase of entropy.

Classical mechanics itself is entirely symmetrical with respect to the two directions of time. The equations of mechanics remain unaltered when the time t is replaced by $-t$; if these equations allow any particular motion, they will therefore allow the reverse motion, in which the mechanical system passes through the same configurations in the reverse order. This symmetry must naturally be preserved in a statistics based on classical mechanics. Hence, if any particular process is possible which is accompanied by an increase in the entropy of a closed macroscopic system, the reverse process must also be possible, in which the entropy of the system decreases. The formulation of the law of increase of entropy given above does not itself contradict this symmetry, since it refers only to the most probable consequence of a macroscopically described state. In other words, if some non-equilibrium macroscopic state is given, the law of increase of entropy asserts only that, out of all the microscopic states which meet the given macroscopic description, the great majority lead to an increase of entropy at subsequent instants.

A contradiction arises, however, if we look at another aspect of the problem. In formulating the law of increase of entropy, we have referred to the most probable consequence of a macroscopic state given at some instant. But this state must itself have resulted from some other states by means of processes occurring in Nature. The symmetry with respect to the two directions of time means that, in any macroscopic state arbitrarily selected at some instant $t = t_0$, we can say not only that much the most probable consequence at $t > t_0$ is an increase in entropy, but also that much the most probable origin of the state was from states of greater entropy; that is, the presence of a minimum of entropy as a function of time at the arbitrarily chosen instant $t = t_0$ is much the most probable.[†]

[†] For a better understanding of this symmetry, we may plot diagrammatically the variation of the entropy of a system that is closed during a very long interval of time (Fig. 1). Let a macroscopic state with entropy $S = S_1 < S_{max}$ be observed in

FIG. 1

such a system, arising from a (very improbable) large fluctuation. Then we can say that it will be, with very high probability, a point of type 1, at which the entropy has reached a minimum, and not one of type 2, at which the entropy will decrease further.

This assertion, of course, is not at all equivalent to the law of increase of entropy, according to which the entropy never decreases (apart from entirely negligible fluctuations) in any closed systems which actually occur in Nature. And it is precisely this general formulation of the law of increase of entropy which is confirmed by all natural phenomena. It must be emphasised that it is certainly not equivalent to the formulation given at the beginning of this section, as it might appear to be. In order to derive one formulation from the other, it would be necessary to use the concept of an observer who artificially "creates" a closed system at some instant, so that the problem of its previous behaviour does not arise. Such a dependence of the laws of physics on the nature of an observer is quite inadmissible, of course.

It is doubtful whether the law of increase of entropy thus formulated could be derived on the basis of classical mechanics. Moreover, because of the invariance of the equations of classical mechanics under time reversal, one could seek only to derive a monotonic variation of entropy. In order to obtain a law of monotonic increase, we should have to define the direction of time as that in which the entropy increases. The problem would then arise of proving that such a thermodynamic definition was identical with the quantum-mechanical definition (see below).

In quantum mechanics, the situation is substantially changed. The fundamental equation of quantum mechanics, namely Schrödinger's equation, is itself symmetrical under time reversal, provided that the wave function Ψ is also replaced by Ψ^*. This means that, if at some instant $t = t_1$ the wave function $\Psi = \Psi(t_1)$ is given, and if according to Schrödinger's equation it should become $\Psi(t_2)$ at some other instant t_2, then the change from $\Psi(t_1)$ to $\Psi(t_2)$ is reversible; in other words, if $\Psi = \Psi^*(t_2)$ at the initial instant t_1, then $\Psi = \Psi^*(t_1)$ at t_2.

However, despite this symmetry, quantum mechanics does in fact involve an important non-equivalence of the two directions of time. This appears in connection with the interaction of a quantum object with a system which with sufficient accuracy obeys the laws of classical mechanics, a process of fundamental significance in quantum mechanics. If two interactions A and B with a given quantum object occur in succession, then the statement that the probability of any particular result of process B is determined by the result of process A can be valid only if process A occurred earlier than process B; see also *Quantum Mechanics*, § 7.

Thus in quantum mechanics there is a physical non-equivalence of the two directions of time, and theoretically the law of increase of entropy might be its macroscopic expression. In that case, there must exist an inequality involving the quantum constant $\hbar$ which ensures the validity of this law and is satisfied in the real world. Up to the present, however, no such relation has been at all convincingly shown to exist.

Quasi static   system always in equillibrium.

The question of the physical foundations of the law of monotonic increase of entropy thus remains open: it may be of cosmological origin and related to the general problem of initial conditions in cosmology; the violation of symmetry under time reversal in some weak interactions between elementary particles may play some part. The answers to such questions may be achieved only in the course of further synthesis of physical theories.

Summarising, we may repeat the general formulation of the law of increase of entropy: in all closed systems which occur in Nature, the entropy never decreases; it increases, or at least remains constant. In accordance with these two possibilities, all processes involving macroscopic bodies are customarily divided into *irreversible* and *reversible* processes. The former comprise those which are accompanied by an increase of entropy of the whole closed system; the reverse processes cannot occur, since the entropy would then have to decrease. Reversible processes are those in which the entropy of the closed system remains constant,[†] and which can therefore take place in the reverse direction. A strictly reversible process is, of course, an ideal limiting case; processes actually occurring in Nature can be reversible only to within a certain degree of approximation.

closed system $E_0 = \sum_a E_a$ 8 6

prob of having $d\Gamma$ states $= d\omega = A\, S(E-E_0)\, d\Gamma$

$$= A\, S(E-E_0) \prod_a d\Gamma_a$$

Example: $d\Gamma_1 \boxed{1 \mid 2} d\Gamma_2$

$E_1 + E_2 = E_0$

$d\Gamma = d\Gamma_1 d\Gamma_2$

$d\Gamma_1 \qquad E_1$

$d\Gamma_2 \qquad E_2 = E_0 - E_1$

$d\Gamma_2$

$d\Gamma_1 \qquad E_1$

Same for E_2

S $\bar{E_1}, E_1$

$d\Gamma$ $\bar{E_1}$

∴ S peaked at $\frac{}{E_i}$

† It must be emphasised that the entropies of the individual parts of the system need not remain constant also.

flaw: If a closed system is not in equilibrium, the most probable consequence is that the entropy will increase.

CHAPTER II

THERMODYNAMIC QUANTITIES

§ 9. Temperature

Thermodynamic physical quantities are those which describe macroscopic states of bodies. They include some which have both a thermodynamic and a purely mechanical significance, such as energy and volume. There are also, however, quantities of another kind, which appear as a result of purely statistical laws and have no meaning when applied to non-macroscopic systems, for example entropy.

In what follows we shall define a number of relations between thermodynamic quantities which hold good whatever the particular bodies to which these quantities relate. These are called *thermodynamic relations*.

When thermodynamic quantities are discussed, the negligible fluctuations to which they are subject are usually of no interest. Accordingly, we shall entirely ignore such fluctuations, and regard the thermodynamic quantities as varying only with the macroscopic state of the body.[†]

Let us consider two bodies in thermal equilibrium with each other, forming a closed system. Then the entropy S of this system has its maximum value (for a given energy E of the system). The energy E is the sum of the energies E_1 and E_2 of the two bodies: $E = E_1 + E_2$. The same applies to the entropy S of the system, and the entropy of each body is a function of its energy: $S = S_1(E_1) + S_2(E_2)$. Since $E_2 = E - E_1$, E being a constant, S is really a function of one independent variable, and the necessary condition for a maximum may be written

$$\frac{\mathrm{d}S}{\mathrm{d}E_1} = \frac{\mathrm{d}S_1}{\mathrm{d}E_1} + \frac{\mathrm{d}S_2}{\mathrm{d}E_2}\frac{\mathrm{d}E_2}{\mathrm{d}E_1}$$

$$= \frac{\mathrm{d}S_1}{\mathrm{d}E_1} - \frac{\mathrm{d}S_2}{\mathrm{d}E_2} = 0,$$

whence

$$\mathrm{d}S_1/\mathrm{d}E_1 = \mathrm{d}S_2/\mathrm{d}E_2.$$

[†] Fluctuations of thermodynamic quantities will be discussed in a separate chapter (Chapter XII).

34

This conclusion can easily be generalised to any number of bodies in equilibrium with one another.

Thus, if a system is in a state of thermodynamic equilibrium, the derivative of the entropy with respect to the energy is the same for every part of it, i.e. is constant throughout the system. A quantity which is the reciprocal of the derivative of the entropy S of a body with respect to its energy E is called the *absolute temperature* T (or simply the *temperature*) of the body:

$$ dS/dE = 1/T. \qquad (9.1) $$

The temperatures of bodies in equilibrium with one another are therefore equal: $T_1 = T_2$.

Like the entropy, the temperature is seen to be a purely statistical quantity, which has meaning only for macroscopic bodies.

Let us next consider two bodies forming a closed system but not in equilibrium with each other. Their temperatures T_1 and T_2 are then different. In the course of time, equilibrium will be established between the bodies, and their temperatures will gradually become equal. During this process, their total entropy $S = S_1 + S_2$ must increase, i.e. its time derivative is positive:

$$ \frac{dS}{dt} = \frac{dS_1}{dt} + \frac{dS_2}{dt} $$

$$ = \frac{dS_1}{dE_1}\frac{dE_1}{dt} + \frac{dS_2}{dE_2}\frac{dE_2}{dt} \geqslant 0. $$

Since the total energy is conserved, $dE_1/dt + dE_2/dt = 0$, and so

$$ \frac{dS}{dt} = \left(\frac{dS_1}{dE_1} - \frac{dS_2}{dE_2}\right)\frac{dE_1}{dt} = \left(\frac{1}{T_1} - \frac{1}{T_2}\right)\frac{dE_1}{dt} > 0. $$

Let the temperature of the second body be greater than that of the first $(T_2 > T_1)$. Then $dE_1/dt > 0$, and $dE_2/dt < 0$. In other words, the energy of the second body decreases and that of the first increases. This property of the temperature may be formulated as follows: energy passes from bodies at higher temperature to bodies at lower temperature.

The entropy S is a dimensionless quantity. The definition (9.1) therefore shows that the temperature has the dimensions of energy, and so can be measured in energy units, for example ergs. In ordinary circumstances, however, the erg is too large a quantity, and in practice the temperature is customarily measured in its own units, called *degrees* Kelvin or simply *degrees*. The conversion factor between ergs and degrees, i. e. the number of ergs per degree, is called *Boltzmann's constant* and is usually denoted by

k; its value is[†]

$$k = 1.38 \times 10^{-16} \text{ erg/deg.}$$

In all subsequent formulae the temperature will be assumed measured in energy units. To convert to the temperature measured in degrees, in numerical calculations, we need only replace T by kT. The continual use of the factor k, whose only purpose is to indicate the conventional units of temperature measurement, would merely complicate the formulae.

If the temperature is in degrees, the factor k is usually included in the definition of entropy:

$$S = k \log \Delta\Gamma, \tag{9.2}$$

instead of (7.7), in order to avoid the appearance of k in the general relations of thermodynamics. Then formula (9.1) defining the temperature, and therefore all the general thermodynamic relations derived subsequently in this chapter, are unaffected by the change to degrees.

Thus the rule for conversion to degrees is to substitute in all formulae

$$T \rightarrow kT, \quad S \rightarrow S/k. \tag{9.3}$$

§ 10. Macroscopic motion

As distinct from the microscopic motion of molecules, the *macroscopic motion* is that in which the various macroscopic parts of a body participate as a whole. Let us consider the possibility of macroscopic motion in a state of thermodynamic equilibrium.

Let the body be divided into a large number of small (but macroscopic) parts, and let M_a, E_a and $\mathbf{P}_a$ denote the mass, energy and momentum of the ath part. The entropy S_a of each part is a function of its internal energy, i.e. the difference between its total energy E_a and the kinetic energy $P_a^2/2M_a$ of its macroscopic motion.[‡] The total entropy of the body can therefore be written

$$S = \sum_a S_a(E_a - P_a^2/2M_a). \tag{10.1}$$

[†] For reference, we may also give the conversion factor between degrees and electron-volts:

$$1 \text{ eV} = 11,606 \text{ deg.}$$

[‡] The fact that the entropy of a body is a function only of its internal energy follows at once from Galileo's relativity principle; the number of quantum states, and therefore the statistical weight (whose logarithm is the entropy), must be the same in all inertial frames of reference, and in particular that in which the body is at rest.

We shall assume that the body is a closed system. Then its total momentum and angular momentum are conserved, as well as its energy:

$$\sum_a \mathbf{P}_a = \text{constant}, \quad \sum_a \mathbf{r}_a \times \mathbf{P}_a = \text{constant}, \tag{10.2}$$

where $\mathbf{r}_a$ is the radius vector of the ath part. In a state of equilibrium, the total entropy S of the body as a function of the momenta $\mathbf{P}_a$ has a maximum subject to the conditions (10.2). Using the familiar Lagrange's method of undetermined multipliers, we find the necessary conditions for a maximum by equating to zero the derivatives with respect to $\mathbf{P}_a$ of the sum

$$\sum_a \{S_a + \mathbf{a} \cdot \mathbf{P}_a + \mathbf{b} \cdot \mathbf{r}_a \times \mathbf{P}_a\}, \tag{10.3}$$

where $\mathbf{a}$ and $\mathbf{b}$ are constant vectors. Differentiation of S_a with respect to $\mathbf{P}_a$ gives,[†] by the definition of the temperature,

$$\frac{\partial}{\partial \mathbf{P}_a} S_a\left(E_a - \frac{P_a^2}{2M_a}\right) = -\frac{\mathbf{P}_a}{M_a T} = -\frac{\mathbf{v}_a}{T},$$

where $\mathbf{v}_a = \mathbf{P}_a/M_a$ is the velocity of the ath part of the body. Differentiation of (10.3) therefore gives

$$-\mathbf{v}_a/T + \mathbf{a} + \mathbf{b} \times \mathbf{r}_a = 0,$$

or

$$\mathbf{v}_a = \mathbf{u} + \boldsymbol{\Omega} \times \mathbf{r}_a, \tag{10.4}$$

where $\mathbf{u} = T\mathbf{a}$ and $\boldsymbol{\Omega} = T\mathbf{b}$ are constant vectors.

This result has a simple physical significance. If the velocities of all the parts of a body are given by formula (10.4) with the same $\mathbf{u}$ and $\boldsymbol{\Omega}$, this means that we have a translational motion of the body as a whole with constant velocity $\mathbf{u}$ and a rotation of the body as a whole with constant angular velocity $\boldsymbol{\Omega}$. Thus we arrive at the important result that in thermodynamic equilibrium a closed system can execute only a uniform translational and rotational motion as a whole. No internal macroscopic motion is possible in a state of equilibrium.[‡]

[†] The derivative with respect to a vector is to be understood as another vector whose components are equal to the derivatives with respect to the components of the first vector.

[‡] To avoid misunderstanding, an important exception to this rule should be noted: superfluid liquid helium cannot rotate as a whole. This effect will be discussed elsewhere in the *Course of Theoretical Physics*. Here we shall only mention that the proof given above is invalid in that case, because the velocity distribution is subject to a further condition (the superfluid flow is a potential flow) and the entropy maximum is to be sought under this condition.

In what follows we shall usually consider bodies at rest, and the energy E will accordingly be the internal energy of the body.

So far we have made use only of the necessary condition for a maximum of entropy as a function of the momenta, but not of the sufficient condition to be imposed on the second derivatives. It is easy to see that the latter condition leads to the very important result that the temperature must be positive: $T > 0$.[†] To deduce this, it is not in fact necessary to calculate the second derivatives; instead, we can argue as follows.

Let us consider a body forming a closed system, at rest as a whole. If the temperature were negative, the entropy would increase with decreasing argument. Since the entropy tends to increase, the body would spontaneously seek to break up into dispersing parts (with total momentum $\Sigma \mathbf{P}_a = 0$), so that the argument of each S_a in the sum (10.1) should take its least possible value. In other words, bodies in equilibrium could not exist with $T < 0$.

The following point should be noted, however. Although the temperature of a body or any part of it can never be negative, there may exist partial equilibria in which the temperature corresponding to a particular group of degrees of freedom of the body is negative. This is further discussed in § 73.

§ 11. Adiabatic processes

Among the various kinds of external interactions to which a body is subject, those which consist in a change in the external conditions form a special group. By "external conditions" we mean in a wide sense various external fields. In practice the external conditions are most often determined by the fact that the body must have a prescribed volume. In one sense this case may also be regarded as a particular type of external field, since the walls which limit the volume are equivalent in effect to a potential barrier which prevents the molecules in the body from escaping.

If the body is subject to no interactions other than changes in external conditions, it is said to be *thermally isolated.* It must be emphasised that, although a thermally isolated body does not interact directly with any other bodies, it is not in general a closed system, and its energy may vary with time.

In a purely mechanical way, a thermally isolated body differs from a closed system only in that its Hamiltonian (the energy) depends explicitly on the time: $E = E(p, q, t)$, because of the variable external field. If the body also interacted directly with other bodies, it would have no Hamiltonian of its

[†] The temperature $T = 0$ (absolute zero) corresponds to $-273.15°C$.

own, since the interaction would depend not only on the coordinates of the molecules of the body in question but also on those of the molecules in the other bodies.

This leads to the result that the law of increase of entropy is valid not only for closed systems but also for a thermally isolated body, since here we regard the external field as a completely specified function of coordinates and time, and in particular neglect the reaction of the body on the field. That is, the field is a purely mechanical and not a statistical object, whose entropy can in this sense be taken as zero. This proves the foregoing statement.

Let us suppose that a body is thermally isolated, and is subject to external conditions which vary sufficiently slowly. Such a process is said to be *adiabatic*. We shall show that, in an adiabatic process, the entropy of the body remains unchanged, i.e. the process is reversible.

We shall describe the external conditions by certain parameters which are given functions of time. For example, suppose that there is only one such parameter, which we denote by λ. The time derivative dS/dt of the entropy will depend in some manner on the rate of variation $d\lambda/dt$ of the parameter λ. Since $d\lambda/dt$ is small, we can expand dS/dt in powers of $d\lambda/dt$. The zero-order term in this expansion, which does not involve $d\lambda/dt$, is zero, since if $d\lambda/dt = 0$ then $dS/dt = 0$ also, because the entropy of a closed system in thermodynamic equilibrium must remain constant under constant external conditions. The first-order term, which is proportional to $d\lambda/dt$, must also be zero, since this term changes sign with $d\lambda/dt$, whereas dS/dt is always positive, according to the law of increase of entropy. Hence it follows that the expansion of dS/dt begins with the second-order term, i.e. for small $d\lambda/dt$ we have

$$\frac{dS}{d\lambda}\frac{d\lambda}{dt} = dS/dt = A(d\lambda/dt)^2,$$

or

$$dS/d\lambda = A\, d\lambda/dt.$$

Thus, when $d\lambda/dt$ tends to zero, so does $dS/d\lambda$, which proves that the adiabatic process is reversible.

It must be emphasised that, although an adiabatic process is reversible, not every reversible process is adiabatic. The condition for a process to be reversible requires only that the total entropy of the whole of a closed system be constant, while the entropies of its individual parts may either increase or decrease. In an adiabatic process, a stronger condition holds: the entropy of a body which is only a part of a closed system also remains constant.

We have defined an adiabatic process as one which is sufficiently slow. More precisely, we can say that the external conditions must change so

(handwritten margin notes) λ's – E,B fields; Volume; Ex: particle in box $\varepsilon = \frac{\hbar^2 n^2}{8m V^{2/3}}$; $\frac{dS}{dt} = f\left(\frac{d\lambda}{dt}\right)$; $\frac{dS}{dt} = a + b\left(\frac{d\lambda}{dt}\right) + c\left(\frac{d\lambda}{dt}\right)^2$; $a=0$; $b=0$; $\frac{dS}{dt} > 0$; $-$ ext variable; $dE_s = -\lambda\, d\lambda$ ↳ generalized force; $\frac{dS}{d\lambda}\frac{d\lambda}{dt} = c\frac{d\lambda}{dt}$; $\frac{d\lambda}{dt}$; $\frac{dS}{d\lambda} = c\lambda$; $\lim_{\lambda \to 0}\frac{dS}{d\lambda} = 0$; quasi static; S constant; $\Delta\Gamma$ const. W_n do not change.

slowly that at any instant the body may be regarded as being in a state of equilibrium corresponding to the prevailing external conditions. That is, the process must be slow in comparison with the processes leading to the establishment of equilibrium in the body concerned.[†]

We may derive a formula to calculate by a purely thermodynamic method various mean values. To do so, we assume that a body undergoes an adiabatic process, and determine the time derivative dE/dt of its energy. By definition, the thermodynamic energy is

$$E = \overline{E(p, q; \lambda)},$$

where $E(p, q; \lambda)$ is the Hamiltonian of the body, depending on λ as a parameter. We know from mechanics that the total time derivative of the Hamiltonian is equal to its partial time derivative (see *Mechanics*, § 40):

$$\frac{dE(p, q; \lambda)}{dt} = \frac{\partial E(p, q; \lambda)}{\partial t}.$$

In the present case $E(p, q; \lambda)$ depends explicitly on the time through $\lambda(t)$, and we can therefore write

$$\frac{dE(p, q; \lambda)}{dt} = \frac{\partial E(p, q; \lambda)}{\partial \lambda}\frac{d\lambda}{dt}.$$

Since the operations of averaging over the statistical distribution and differentiating with respect to time can clearly be interchanged, we have

$$\frac{dE}{dt} = \frac{\overline{dE(p, q; \lambda)}}{dt} = \frac{\overline{\partial E(p, q; \lambda)}}{\partial \lambda}\frac{d\lambda}{dt}; \qquad (11.1)$$

the derivative $d\lambda/dt$ is a given function of time, and can be taken outside the averaging.

It is very important that, since the process is adiabatic, the mean value of the derivative $\partial E(p, q; \lambda)/\partial \lambda$ in (11.1) can be taken as the mean value over

---

[†] In practice this may be a very weak condition, so that the "slow" adiabatic process may be quite a "fast" one. For example, in the expansion of a gas, say in a cylinder with a piston moving outwards, the speed of the piston need be small only compared with the velocity of sound in the gas, i.e. it may in practice be very large.

In general textbooks on physics an adiabatic expansion (or compression) is often defined as one which is "sufficiently rapid". This refers to a different aspect of the problem: the process must occur so rapidly that the body cannot exchange heat with the surrounding medium. Thus the condition in question is one which will in practice ensure that the body is thermally isolated, and the condition of slowness compared with processes leading to the establishment of equilibrium is tacitly assumed satisfied.

the statistical distribution corresponding to equilibrium for a given value of the parameter λ, i.e. for the external conditions prevailing at a given instant.

The derivative dE/dt can also be written in another form by regarding the thermodynamic quantity E as a function of the entropy S of the body and the external parameters λ. Since, in an adiabatic process, the entropy S remains constant, we have

$$\frac{dE}{dt} = \left(\frac{\partial E}{\partial \lambda}\right)_S \frac{d\lambda}{dt}, \tag{11.2}$$

where the subscript to the parenthesis indicates that the derivative is taken for constant S.

Comparison of (11.1) and (11.2) shows that

$$\frac{\overline{\partial E(p, q; \lambda)}}{\partial \lambda} = \left(\frac{\partial E}{\partial \lambda}\right)_S. \tag{11.3}$$

This is the required formula. It enables us to calculate thermodynamically the mean values (over the equilibrium statistical distribution) of quantities of the form $\partial E(p, q; \lambda)/\partial \lambda$. Such quantities are continually encountered when studying the properties of macroscopic bodies, and in consequence formula (11.3) is of great importance in statistical physics. It appears in the calculation of various forces acting on a body (the parameters λ being the coordinates of a particular part of the body; see § 12), the calculation of the magnetic or electric moment of bodies (the parameters λ being the magnetic or electric field strengths), and so on.

The arguments given here for classical mechanics are entirely applicable to the quantum theory, except that the energy $E(p, q; \lambda)$ must be everywhere replaced by the Hamiltonian operator $\hat{H}$. Then formula (11.3) becomes

$$\frac{\overline{\partial \hat{H}}}{\partial \lambda} = \left(\frac{\partial E}{\partial \lambda}\right)_S, \tag{11.4}$$

the bar denoting complete statistical averaging (which automatically includes the quantum averaging).

§ 12. Pressure

The energy E of a body, as a thermodynamic quantity, has the property of being additive: the energy of the body is equal to the sum of the energies of its individual (macroscopic) parts.[†] Another fundamental thermodynamic quantity, the entropy, also has this property.

[†] In so far as we neglect the energy of interaction of these parts; this is not permissible if we are interested in effects arising from the presence of interfaces between different bodies. Chapter XV deals with this topic.

The additivity of the energy and the entropy leads to the following important result. If a body is in thermal equilibrium, we can say that, for a
given energy, the entropy depends only on the volume of the body, and not
on its shape; the same is true of the energy for a given entropy.[†] For a
change in the shape of the body can be regarded as a rearrangement of its
individual parts, and so the entropy and energy, being additive, will remain
unchanged. Here, of course, it is assumed that the body is not in an external
field of force, so that the motion of the parts of the body in space does not
involve a change in their energy.

Thus the macroscopic state of a body at rest in equilibrium is entirely
determined by only two quantities, for example the volume and the energy.
All other thermodynamic quantities can be expressed as functions of these
two. Of course, because of this mutual dependence of the various thermodynamic quantities, any other pair could be regarded as the independent
variables.

Let us now calculate the force exerted by a body on the surface bounding
its volume. According to the formulae of mechanics, the force acting on a
surface element ds is

$$F = -\partial E(p, q; \mathbf{r})/\partial \mathbf{r},$$

where $E(p, q; \mathbf{r})$ is the energy of the body as a function of the coordinates
and momenta of its particles and of the radius vector of the surface element
considered, which here acts as an external parameter. Averaging this equation and using formula (11.3), we obtain

$$\bar{\mathbf{F}} = -\frac{\overline{\partial E(p, q; \mathbf{r})}}{\partial \mathbf{r}} = -\left(\frac{\partial E}{\partial \mathbf{r}}\right)_S = -\left(\frac{\partial E}{\partial V}\right)_S \frac{\partial V}{\partial \mathbf{r}},$$

where V is the volume. Since the change in volume is ds·dr, we have $\partial V/\partial \mathbf{r} =$
ds, the surface element, and so

$$\bar{\mathbf{F}} = -(\partial E/\partial V)_S \, ds.$$

Hence we see that the mean force on a surface element is normal to the element and proportional to its area (*Pascal's law*). The magnitude of the
force per unit area is

$$P = -(\partial E/\partial V)_S. \qquad (12.1)$$

This quantity is called the *pressure*.

[†] These statements are applicable in practice to liquids and gases but not to
solids. A change in shape (deformation) of a solid involves the doing of work, so
that the energy of the body is changed. This is because the deformed state of the
solid is, strictly speaking, an incomplete thermodynamic equilibrium (but the
relaxation time for the establishment of complete equilibrium is so long that in
many respects the deformed body behaves as if in equilibrium).

In the top margin (handwritten):

$$\frac{1}{T} = \frac{dS}{dE} \implies \boxed{\frac{dE}{dS} = T}$$

$$E = E \text{ (energy levels, occupation)}$$
$$= E(V, S)$$

$$dE = \left(\frac{\partial E}{\partial S}\right)_V dS + \left(\frac{\partial E}{\partial V}\right)_S dV$$

(handwritten, "general external variable")

$$= \left(\frac{\partial E}{\partial S}\right)_V dS$$
$$- P dV$$
$$= T dS - P dV$$

In defining the temperature by formula (9.1) we were essentially considering a body which is not in direct contact with any other bodies, and in particular is not surrounded by any external medium. Under these conditions it was possible to speak of the change in energy and entropy of the body without making more specific the nature of the process. In the general case of a body in an external medium, or surrounded by the walls of a vessel, formula (9.1) must be made more precise. For if during the process the volume of the body changes, this will necessarily affect the state of the bodies in contact with it, and in order to define the temperature we should have to take into consideration at the same time all the bodies in contact (for example, both the body in question and the vessel containing it). If it is desired to define the temperature in terms of thermodynamic quantities for the given body only, its volume must be regarded as constant. In other words, the temperature is defined as the derivative of the energy of the body with respect to its entropy, taken at constant volume:

$$T = (\partial E/\partial S)_V. \tag{12.2}$$

The equations (12.1), (12.2) can also be written together as a relation between differentials:

(handwritten: First law)

$$\boxed{dE = T\,dS - P\,dV.}$$

(handwritten: $+ \sum_i \mu_i dN_i$) $\tag{12.3}$

(handwritten: Quasi-static processes only)

This is one of the most important relations in thermodynamics.

The pressures of bodies in equilibrium with one another are equal. This follows immediately from the fact that thermal equilibrium necessarily presupposes mechanical equilibrium; in other words, the forces exerted on each other by any two of these bodies at their surface of contact must be equal in magnitude and opposite in direction, and thus balance.

The equality of pressures in equilibrium can also be derived from the condition of maximum entropy, in the same way as the equality of temperatures was shown in § 9. To do this, we consider two parts, in contact, of a closed system in equilibrium. One necessary condition for the entropy to be a maximum is that it should be a maximum with respect to a change in the volumes V_1 and V_2 of these two parts when the states of the other parts undergo no change (this means, in particular, that $V_1 + V_2$ remains constant). If the entropies of the two parts are S_1 and S_2, we have

$$\frac{\partial S}{\partial V_1} = \frac{\partial S_1}{\partial V_1} + \frac{\partial S_2}{\partial V_2}\frac{\partial V_2}{\partial V_1} = \frac{\partial S_1}{\partial V_1} - \frac{\partial S_2}{\partial V_2} = 0.$$

From the relation (12.3) in the form

(handwritten left)
$$T = \left(\frac{dE}{dS}\right)_V$$

$$dS = \frac{1}{T}\,dE + \frac{P}{T}\,dV$$

(handwritten right)
$$\frac{1}{T} = \left(\frac{\partial S}{\partial E}\right)_V$$

(handwritten left)
$$P = -\left(\frac{dE}{dV}\right)_S$$

$$S = S(E, V)$$

(handwritten)
$$\frac{P}{T} = \left(\frac{\partial S}{\partial V}\right)_E$$

(handwritten: E - thermodynamic potential)

$$P = T\left(\frac{\partial S}{\partial V}\right)_E$$

$E_o = E_1 + E_2$

$S_o = S_1 + S_2$

$V = V_1 + V_2$

$S_1 = S_1(V_1, E_1)$

$S_2 = S_2(V_2, E_2)$

$ds = \left(\frac{\partial S_1}{\partial E_1}\right)_{V_1} dE_1$

$+\left(\frac{\partial S_1}{\partial V_1}\right)_{E_1} dV_1$

$+\left(\frac{\partial S_2}{\partial E_2}\right)_{V_2} dE_2$

$+\left(\frac{\partial S_2}{\partial V_2}\right)_{E_2} dE_2$

$dE_1 = -dE_2$

$dV_1 = -dV_2$

$\frac{\partial S_1}{\partial E_1} = \frac{1}{T_1}$

$\frac{\partial S_2}{\partial E_2} = \frac{1}{T_2}$

$\frac{\partial S_1}{\partial V_1} = \frac{P_1}{T_1}$

$ds = \left(-\frac{1}{T_1} + \frac{1}{T_2}\right) dE_2$

$= \left(\frac{P_1}{T_1} - \frac{P_2}{T_2}\right) dV_2$

$= 0$ in Eq.

$\boxed{T_1 = T_2}$

$\boxed{P_1 = P_2}$

44

Thermodynamic Quantities

it is seen that $\partial S/\partial V = P/T$, and so $P_1/T_1 = P_2/T_2$. Since the temperatures T_1 and T_2 are the same in equilibrium, we therefore find that the pressures are equal, $P_1 = P_2$.

It must be remembered that, when thermal equilibrium is established, the equality of pressures (i.e. mechanical equilibrium) is reached much more rapidly than that of temperatures, and so cases are often met with in which the pressure is constant throughout a body but the temperature is not. The reason is that the non-constancy of pressure is due to the presence of uncompensated forces; these bring about macroscopic motion so as to equalise the pressure much more rapidly than the equalisation of temperature, which does not involve macroscopic motion.

It is easy to see that the pressure must be positive in any equilibrium state: when $P > 0$ we have $(\partial S/\partial V)_E > 0$, and the entropy could increase only by an expansion of the body, which is prevented by the surrounding bodies. If $P < 0$, however, then we should have $(\partial S/\partial V)_E < 0$, and the body would spontaneously contract so as to increase its entropy.

There is, however, an important difference between the requirements of positive temperature and positive pressure. Bodies of negative temperature would be completely unstable and cannot exist in Nature. States (non-equilibrium) of negative pressure can exist in Nature with restricted stability. The reason is that the spontaneous contraction of the body involves "detaching" it from the walls of the vessel or the formation of cavities within it, that is, the formation of a new surface, and this leads to the possibility of the existence of negative pressures in what are called *metastable states*.[†]

§ 13. Work and quantity of heat

The external forces applied to a body can do *work* on it, which is determined, according to the general rules of mechanics, by the products of these forces and the displacements which they cause. This work may serve to bring the body into a state of macroscopic motion (or in general to change its kinetic energy), or to move the body in an external field (for instance, to raise it against gravity). We shall, however, be mainly interested in cases where the volume of a body is changed as a result of work done on it (i.e. the external forces compress the body but leave it at rest as a whole).

We shall everywhere regard as positive an amount of work R done on a given body by external forces. Negative work $(R < 0)$ will correspondingly mean that the body itself does work (equal to $|R|$) on some external objects (for example, in expanding).

[†] These are defined in § 21. Negative pressures are further discussed in the footnote following (83.1).

Bearing in mind that the force per unit area of the surface of the body is the pressure, and that the product of the area of a surface element and its displacement is the volume swept out by it, we find that the work done on the body per unit time when its volume changes is

$$dR/dt = -P \, dV/dt; \tag{13.1}$$

in compression, $dV/dt < 0$, so that $dR/dt > 0$. This formula is applicable to both reversible and irreversible processes; only one condition need be satisfied, namely that throughout the process the body must be in a state of mechanical equilibrium, i.e. at each instant the pressure must be constant throughout the body.

If the body is thermally isolated, the whole of the change in its energy is due to the work done on it. In the general case of a body not thermally isolated, in addition to the work done, the body gains or loses energy by direct transfer from or to other bodies in contact with it. This part of the change in energy is called the quantity of *heat* Q gained or lost by the body. Thus the change in the energy of the body per unit time may be written

$$\frac{dE}{dt} = \frac{dR}{dt} + \frac{dQ}{dt}. \tag{13.2}$$

Like the work, the heat will be regarded as positive if gained by the body from external sources.

The energy E in (13.2) must, in general, be understood as the total energy of the body, including the kinetic energy of its macroscopic motion. We shall, however, usually consider the work corresponding to the change in volume of a body at rest, in which case the energy reduces to the internal energy of the body.

Under conditions where the work is defined by formula (13.1), we have for the quantity of heat

$$\frac{dQ}{dt} = \frac{dE}{dt} + P\frac{dV}{dt}. \tag{13.3}$$

Let us assume that at every instant throughout the process the body may be regarded as being in a state of thermal equilibrium corresponding to its energy and volume at that instant; it must be emphasised that this does not mean that the process is necessarily reversible, since the body may not be in equilibrium with surrounding bodies. Then, from the relation (12.3), which gives the differential of the function $E(S, V)$, the energy of the body in the equilibrium state, we can put

$$\frac{dE}{dt} = T\frac{dS}{dt} - P\frac{dV}{dt}.$$

$\dfrac{dQ}{dt} = T \dfrac{dS}{dt}$

$\dfrac{dS}{dt} = \dfrac{d}{dt}\; \{\upsilon_m \ell_m \dot{\upsilon}_m$ changes

Quasi-static

Comparison with (13.3) shows that

$$dQ/dt = T\,dS/dt. \tag{13.4}$$

The work dR and the quantity of heat dQ gained by the body in an infinitesimal change of state are not the total differentials of any quantities.[†] Only the sum $dQ + dR$, i.e. the change in energy dE, is a total differential. We can therefore speak of the energy E in a given state, but not, for example, of the quantity of heat which a body possesses in a given state. In other words, the energy of the body cannot be divided into thermal and mechanical parts; this is possible only when considering the change in energy. The change in energy when a body goes from one state to another can be divided into the quantity of heat gained or lost by the body and the work done on it or by it. This division is not uniquely determined by the initial and final states of the body, but depends also on the nature of the process itself. That is, the work and the quantity of heat are functions of the process undergone by the body and not only of its initial and final states. This is seen particularly when the body undergoes a cyclic process, starting and finishing in the same state. The change in energy is then zero, but the body may gain or lose a quantity of heat or work. Mathematically this corresponds to the fact that the integral of the total differential dE around a closed circuit is zero, but the integral of dQ or dR, which are not total differentials, is not zero.

The quantity of heat which must be gained in order to raise the temperature of the body by one unit is called its *specific heat*. This clearly depends on the conditions under which the heating takes place. A distinction is usually made between the specific heat at constant volume C_v and that at constant pressure C_p. Clearly

$$\left(\dfrac{\partial Q}{\partial T}\right)_V = C_v = T(\partial S/\partial T)_V, \tag{13.5}$$

$$\left(\dfrac{\partial Q}{\partial T}\right)_P = C_p = T(\partial S/\partial T)_P. \tag{13.6}$$

Let us consider cases where formula (13.4) for the quantity of heat is inapplicable, but at the same time it is possible to establish certain inequalities for this quantity. There exist processes in which the body is not in thermal equilibrium although the temperature (and pressure) are constant throughout the body; for example, chemical reactions in a homogeneous mixture of reactants. Owing to the irreversible process (the chemical reaction) occurring in the body, its entropy increases independently of the heat gained, and so we can say that the inequality

$$dQ/dt < T\,dS/dt \tag{13.7}$$

holds.

[†] In this sense the notation dR and dQ is not quite precise, and we therefore avoid it as far as possible.

Another case where a similar inequality can be stated is an irreversible process in which the body goes from one equilibrium state to another neighbouring one but is not in equilibrium during the process.[†] Then the inequality

$$\delta Q < T \, \delta S \tag{13.8}$$

holds between the quantity of heat δQ gained by the body in this process and its entropy change δS.

[handwritten: $E = E(S, P, V, T)$]

[handwritten: $dE = T \, dS - P \, dV$]

§ 14. The heat function

[handwritten: $+ \sum_i \lambda_i d\lambda_i$]

If the volume of a body remains constant during a process, then $dQ = dE$, i.e. the quantity of heat gained by the body is equal to the change in its energy. If the process occurs at constant pressure, the quantity of heat can be written as the differential

[handwritten: 4 potentials]
[handwritten: E - energy S, V]
[handwritten: F - free energy T, V]

$$dQ = d(E+PV) = dW \tag{14.1}$$

of a quantity

$$W = E+PV, \tag{14.2}$$

[handwritten: ϕ - Thermo-potential (Gibbs)]
[handwritten: T, P]
[handwritten: W - heat fn. Enthalpy S, P]

called the *heat function* of the body.[‡] The change in the heat function in processes occurring at constant pressure is therefore equal to the quantity of heat gained by the body.

It is easy to find an expression for the total differential of the heat function. Putting $dE = T \, dS - P \, dV$ and $dW = dE + P \, dV + V \, dP$, we have

$$dW = T \, dS + V \, dP. \tag{14.3}$$

From this it follows that

$$T = (\partial W/\partial S)_P, \quad V = (\partial W/\partial P)_S. \tag{14.4}$$

If the body is thermally isolated (which, it will be remembered, does not imply that it is a closed system), $dQ = 0$, and (14.1) shows that, in processes occurring at constant pressure and involving a thermally isolated body,

$$W = \text{constant}, \tag{14.5}$$

i.e. the heat function is conserved.

The specific heat C_v can be written, using the relation $dE = T \, dS - P \, dV$, as

$$C_v = (\partial E/\partial T)_V. \tag{14.6}$$

[†] An example is the *Joule–Thomson process* (see § 18) with a small change in pressure.

[‡] Also called the *enthalpy* or *heat content*.

SP - C

[Handwritten left margin notes:]

Energy $E(s,v)$

E

$dE = Tds - PdV$

$E(s,v)$

$T = \left(\frac{\partial E}{\partial s}\right)_v$

$P = -\left(\frac{\partial E}{\partial v}\right)_s$

$\left(\frac{\partial T}{\partial v}\right)_s = \left(\frac{\partial P}{\partial s}\right)_v$

s - constant
$dE = $ work (dR)

v - constant
$dE = dQ$

$dQ = dE + PdV$
always

$dQ \leq Tds$

const.
$dE \leq 0$

Similarly, we have for the specific heat C_p

$$C_p = (\partial W/\partial T)_P.$$

We see that at constant pressure the heat function has properties similar to those of the energy at constant volume.

§ 15. The free energy and the thermodynamic potential

The work done on a body in an infinitesimal isothermal reversible change of state can be written as a differential:

$$dR = dE - dQ = dE - T\,dS$$
$$= d(E - TS)$$

or

$$dR = dF, \tag{15.1}$$

where

$$F = E - TS \tag{15.2}$$

is another function of the state of the body, called the *free energy*. Thus the work done on the body in a reversible isothermal process is equal to the change in its free energy.

Let us find the differential of the free energy. Substituting $dE = T\,dS - P\,dV$ and $dF = dE - T\,dS - S\,dT$, we have

$$dF = -S\,dT - P\,dV. \tag{15.3}$$

Hence it is evident that

$$S = -(\partial F/\partial T)_V, \quad P = -(\partial F/\partial V)_T. \tag{15.4}$$

Using the relation $E = F + TS$, we can express the energy in terms of the free energy as

$$E = F - T(\partial F/\partial T)_V$$
$$= -T^2\left(\frac{\partial}{\partial T}\frac{F}{T}\right)_V. \tag{15.5}$$

Formulae (12.1), (12.2), (14.4) and (15.4) show that, if we know any of the quantities E, W and F as a function of the corresponding two variables and take its partial derivatives, we can determine all the remaining thermodynamic quantities. For this reason E, W and F are called *thermodynamic potentials* (by analogy with the mechanical potential) or *characteristic functions*: the energy E with respect to the variables S, V; the heat function W with respect to S, P; the free energy F with respect to V, T.

We still lack a thermodynamic potential with respect to the variables P, T
To derive this we substitute in (15.3) $P\,dV = d(PV) - V\,dP$, take $d(PV)$ to
the left-hand side of the equation, and obtain

$$d\Phi = -S\,dT + V\,dP,\tag{15.6}$$

with a new quantity

$$\Phi = E - TS + PV$$

$$= F + PV$$

$$= W - TS,\tag{15.7}$$

called the *thermodynamic potential* (in a restricted sense of the term).[†]
From (15.6) we clearly have

$$S = -(\partial\Phi/\partial T)_P, \quad V = (\partial\Phi/\partial P)_T.\tag{15.8}$$

The heat function is expressed in terms of Φ in the same way as E in terms
of F:

$$W = \Phi - T(\partial\Phi/\partial T)_P$$

$$= -T^2\left(\frac{\partial}{\partial T}\frac{\Phi}{T}\right)_P.\tag{15.9}$$

If there are other parameters λ_i besides the volume which define the state
of the system, the expression for the differential of the energy must be augmented by terms proportional to the differentials $d\lambda_i$:

$$dE = T\,dS - P\,dV + \sum_i \Lambda_i\,d\lambda_i,\tag{15.10}$$

where the Λ_i are some functions of the state of the body. Since the transformation to other potentials does not affect the variables λ_i, it is clear that
similar terms will be added to the differentials of F, Φ, W:

$$dF = -S\,dT - P\,dV + \sum_i \Lambda_i\,d\lambda_i,$$

etc. Hence the quantities Λ_i can be obtained by differentiation with respect to
λ_i of any of these potentials (it must be remembered which other variables
are treated as constant in the differentiation). Using also formula (11.3), we
can write down the analogous relation

$$\frac{\partial E(p, q; \lambda)}{\partial\lambda} = \left(\frac{\partial F}{\partial\lambda}\right)_{T,\,V},\tag{15.11}$$

[†] In Western literature, the functions F and Φ are often called respectively the
Helmholtz free energy and the *Gibbs free energy*.

Enthalpy $W(S, P)$

$$W = E + PV$$

$$dW = T\,dS + V\,dP$$

$$W(S, P)$$

$$T = \left(\frac{\partial W}{\partial S}\right)_P$$

$$V = \left(\frac{\partial W}{\partial P}\right)_S$$

P - constant
$$dW = dQ$$

const.
$$dW \le 0$$

Free Energy $F(T, v)$

$F = E - TS$

$dF = -S dT - P dV$

$F(T, v)$

$S = -\left(\dfrac{\partial F}{\partial T}\right)_V$

$P = -\left(\dfrac{\partial F}{\partial V}\right)_T$

const

$dF \le 0$

$C_V = \left(\dfrac{\partial Q}{\partial T}\right)_V$

$= T\left(\dfrac{\partial S}{\partial T}\right)_V$

$C_P = \left(\dfrac{\partial Q}{\partial V}\right)_T$

$= T\left(\dfrac{\partial S}{\partial T}\right)_P$

$F = -T \ln Z$

which expresses the mean value of the derivative of the Hamiltonian with respect to any parameter as the derivative of the free energy with respect to that parameter (and similar relations involving the derivatives of Φ and W).

The following point may be noted. If the values of the parameters λ_i change slightly, the quantities E, F, W and Φ will also undergo small changes. It is evident that these changes will be equal if each is considered for the appropriate pair of constant quantities:

$$(\delta E)_{S, V} = (\delta F)_{T, V} = (\delta W)_{S, P} = (\delta \Phi)_{T, P}. \tag{15.12}$$

This statement, which we call the *theorem of small increments*, will be used several times below.

The free energy and the thermodynamic potential have an important property which determines the direction in which they change in various irreversible processes. From the inequality (13.7), substituting dQ/dt from (13.3), we obtain

$$\frac{dE}{dt} + P\frac{dV}{dt} < T\frac{dS}{dt}. \tag{15.13}$$

Let us assume that the process is isothermal and occurs at constant volume ($T = $ constant, $V = $ constant). Then this inequality may be written

$$\frac{d(E - TS)}{dt} = \frac{dF}{dt} < 0. \tag{15.14}$$

Thus irreversible processes occurring at constant temperature and constant volume are accompanied by a decrease in the free energy of the body.

Similarly, for $P = $ constant and $T = $ constant the inequality (15.13) becomes

$$d\Phi/dt < 0; \tag{15.15}$$

that is, irreversible processes occurring at constant temperature and constant pressure are accompanied by a decrease in the thermodynamic potential.[†]

Correspondingly, in a state of thermal equilibrium the free energy and the thermodynamic potential have minimum values, the former with respect to all changes of state with T and V constant, and the latter with respect to changes of state with T and P constant.

PROBLEM

How can the mean kinetic energy of the particles in a body be calculated if the formula for its free energy is known?

† It should be remembered that in both cases the processes in question are those (such as chemical reactions) for which the body is not in equilibrium, so that its state is not uniquely defined by the temperature and the volume (or pressure).

SOLUTION. The Hamiltonian function (or, in the quantum case, the Hamiltonian operator) may be written in the form $E(p, q) = U(q) + K(p)$, where $U(q)$ is the potential energy of interaction of the particles in the body, and $K(p)$ their kinetic energy. The latter is a quadratic function of the momenta, inversely proportional to the particle mass m (for a body consisting of identical particles). Regarding m as a parameter, we can therefore write

$$\frac{\partial E(p, q; m)}{\partial m} = -\frac{1}{m}K(p).$$

Then, applying formula (15.11), we obtain the mean kinetic energy $K = \overline{K(p)}$:

$$K = -m(\partial F/\partial m)_{T, V}.$$

§ 16. Relations between the derivatives of thermodynamic quantities

In practice the most convenient, and the most widely used, pairs of thermodynamic variables are T, V and T, P. It is therefore necessary to transform various derivatives of the thermodynamic quantities with respect to one another to different variables, both dependent and independent.

If V and T are used as independent variables, the results of the transformation can be conveniently expressed in terms of the pressure P and the specific heat C_v (as functions of V and T). The equation which relates the pressure, volume and temperature is called the *equation of state* for a given body. Thus the purpose of the formulae in this case is to make it possible to calculate various derivatives of thermodynamic quantities from the equation of state and the specific heat C_v.

Similarly, when P and T are taken as the basic variables the results of the transformation should be expressed in terms of V and C_p (as functions of P and T).

Here it must be remembered that the dependence of C_v on V or of C_p on P (but not on the temperature) can itself be determined from the equation of state. It is easily seen that the derivative $(\partial C_v/\partial V)_T$ can be transformed so that it is defined in terms of the function $P(V, T)$. Using the fact that $S = -(\partial F/\partial T)_V$, we have

$$\left(\frac{\partial C_v}{\partial V}\right)_T = T\frac{\partial^2 S}{\partial V \, \partial T} = -T\frac{\partial^3 F}{\partial V \, \partial T^2}$$

$$= -T\frac{\partial^2}{\partial T^2}\left(\frac{\partial F}{\partial V}\right)_T,$$

and since $(\partial F/\partial V)_T = -P$, we have the required formula

$$(\partial C_v/\partial V)_T = T(\partial^2 P/\partial T^2)_V. \tag{16.1}$$

Thermo. Pot $\phi(T,P)$

$\phi = E - TS + PV$

$d\phi = -SdT + VdP$

$\phi(T, P)$

$S = -\left(\dfrac{\partial\phi}{\partial T}\right)_S$

$V = \left(\dfrac{\partial\phi}{\partial P}\right)_T$

const

$d\phi \leq 0$

Measurables

$\alpha = \frac{1}{V}\left(\frac{\partial V}{\partial T}\right)_P$

vol. coeff of exp.

$K = \frac{1}{V}\left(\frac{\partial V}{\partial P}\right)_T$

isothermal
compressibility

$\frac{\alpha}{K} = \left(\frac{dP}{dT}\right)_V$

Similarly we find

$$(\partial C_p/\partial P)_T = -T(\partial^2 V/\partial T^2)_P, \qquad (16.2)$$

formulae (15.8) being used in the calculation.

We shall show how some of the thermodynamic derivatives most often encountered may be transformed.

The derivatives of the entropy with respect to volume or pressure can be calculated from the equation of state by means of the following formulae, which are a direct consequence of the expressions for the differentials of the thermodynamic quantities. We have

$$\left(\frac{\partial S}{\partial V}\right)_T = -\frac{\partial}{\partial V}\left(\frac{\partial F}{\partial T}\right)_V = -\frac{\partial}{\partial T}\left(\frac{\partial F}{\partial V}\right)_T$$

or

$$(\partial S/\partial V)_T = (\partial P/\partial T)_V. \qquad (16.3)$$

Similarly

$$\left(\frac{\partial S}{\partial P}\right)_T = -\frac{\partial}{\partial P}\left(\frac{\partial \Phi}{\partial T}\right)_P = -\frac{\partial}{\partial T}\left(\frac{\partial \Phi}{\partial P}\right)_T$$

or

$$(\partial S/\partial P)_T = -(\partial V/\partial T)_P. \qquad (16.4)$$

The derivative $(\partial E/\partial V)_T$ is calculated from the equation

$$dE = T\,dS - P\,dV$$

as

$$\left(\frac{\partial E}{\partial V}\right)_T = T\left(\frac{\partial S}{\partial V}\right)_T - P$$

or, substituting (16.3),

$$\left(\frac{\partial E}{\partial V}\right)_T = T\left(\frac{\partial P}{\partial T}\right)_V - P. \qquad (16.5)$$

Similarly we can derive

$$\left(\frac{\partial E}{\partial P}\right)_T = -T\left(\frac{\partial V}{\partial T}\right)_P - P\left(\frac{\partial V}{\partial P}\right)_T, \qquad (16.6)$$

$$\left(\frac{\partial W}{\partial V}\right)_T = T\left(\frac{\partial P}{\partial T}\right)_V + V\left(\frac{\partial P}{\partial V}\right)_T, \qquad \left(\frac{\partial W}{\partial P}\right)_T = V - T\left(\frac{\partial V}{\partial T}\right)_P, \qquad (16.7)$$

$$\left(\frac{\partial E}{\partial T}\right)_P = C_p - P\left(\frac{\partial V}{\partial T}\right)_P, \qquad \left(\frac{\partial W}{\partial T}\right)_V = C_v + V\left(\frac{\partial P}{\partial T}\right)_V. \qquad (16.8)$$

Finally, we shall show how the specific heat C_v may be calculated from the specific heat C_p and the equation of state, using T and P as the basic variables. Since $C_v = T(\partial S/\partial T)_V$, we have to transform the derivative

$(\partial S/\partial T)_V$ to different independent variables. A transformation of this type is most simply effected by the use of Jacobians.[†] We write

$$C_v = T(\partial S/\partial T)_V$$

$$= T \partial(S, V)/\partial(T, V)$$

$$= T \frac{\partial(S, V)/\partial(T, P)}{\partial(T, V)/\partial(T, P)}$$

$$= T \frac{(\partial S/\partial T)_P (\partial V/\partial P)_T - (\partial S/\partial P)_T (\partial V/\partial T)_P}{(\partial V/\partial P)_T}$$

$$= C_p - T \frac{(\partial S/\partial P)_T (\partial V/\partial T)_P}{(\partial V/\partial P)_T}.$$

Substituting (16.4), we obtain the required formula:

$$C_p - C_v = -T[(\partial V/\partial T)_P]^2/(\partial V/\partial P)_T. \tag{16.9}$$

Similarly, transforming $C_p = T(\partial S/\partial T)_P$ to the variables T, V, we can derive the formula

$$C_p - C_v = -T[(\partial P/\partial T)_V]^2/(\partial P/\partial V)_T. \tag{16.10}$$

The derivative $(\partial P/\partial V)_T$ is negative: in an isothermal expansion of a body, its pressure always decreases. This will be rigorously proved in § 21. It therefore follows from (16.10) that for all bodies

$$C_p > C_v. \tag{16.11}$$

[†] The Jacobian $\partial(u, v)/\partial(x, y)$ is defined as the determinant

$$\frac{\partial(u, v)}{\partial(x, y)} = \begin{vmatrix} \partial u/\partial x & \partial u/\partial y \\ \partial v/\partial x & \partial v/\partial y \end{vmatrix}. \tag{I}$$

It clearly has the following properties:

$$\frac{\partial(v, u)}{\partial(x, y)} = -\frac{\partial(u, v)}{\partial(x, y)}, \tag{II}$$

$$\frac{\partial(u, y)}{\partial(x, y)} = \left(\frac{\partial u}{\partial x}\right)_y. \tag{III}$$

The following relations also hold:

$$\frac{\partial(u, v)}{\partial(x, y)} = \frac{\partial(u, v)}{\partial(t, s)} \cdot \frac{\partial(t, s)}{\partial(x, y)}, \tag{IV}$$

$$\frac{d}{dt} \frac{\partial(u, v)}{\partial(x, y)} = \frac{\partial(du/dt, v)}{\partial(x, y)} + \frac{\partial(u, dv/dt)}{\partial(x, y)}. \tag{V}$$

Thermodynamic Quantities

In adiabatic expansion (or contraction) of a body its entropy remains constant. The relation between the temperature, volume and pressure of the body in an adiabatic process is therefore determined by various derivatives taken at constant entropy. We shall derive formulae whereby these derivatives may be calculated from the equation of state of the body and its specific heat.

For the derivative of the temperature with respect to volume we have, changing to independent variables V, T,

$$\left(\frac{\partial T}{\partial V}\right)_S = \frac{\partial(T, S)}{\partial(V, S)} = \frac{\partial(T, S)/\partial(V, T)}{\partial(V, S)/\partial(V, T)}$$

$$= -\frac{(\partial S/\partial V)_T}{(\partial S/\partial T)_V}$$

$$= -\frac{T}{C_v}\left(\frac{\partial S}{\partial V}\right)_T,$$

or, substituting (16.3),

$$\left(\frac{\partial T}{\partial V}\right)_S = -\frac{T}{C_v}\left(\frac{\partial P}{\partial T}\right)_V. \qquad (16.12)$$

Similarly we find

$$\left(\frac{\partial T}{\partial P}\right)_S = \frac{T}{C_p}\left(\frac{\partial V}{\partial T}\right)_P. \qquad (16.13)$$

These formulae show that, according as the thermal expansion coefficient $(\partial V/\partial T)_P$ is positive or negative, the temperature of the body falls or rises in an adiabatic expansion.[†]

Let us next calculate the adiabatic compressibility $(\partial V/\partial P)_S$ of the body, writing

$$\left(\frac{\partial V}{\partial P}\right)_S = \frac{\partial(V, S)}{\partial(P, S)} = \frac{\partial(V, S)/\partial(V, T)}{\partial(P, S)/\partial(P, T)} \cdot \frac{\partial(V, T)}{\partial(P, T)} = \frac{(\partial S/\partial T)_V}{(\partial S/\partial T)_P} \cdot \left(\frac{\partial V}{\partial P}\right)_T$$

or

$$\left(\frac{\partial V}{\partial P}\right)_S = \frac{C_v}{C_p}\left(\frac{\partial V}{\partial P}\right)_T. \qquad (16.14)$$

The inequality $C_p > C_v$ therefore implies that the adiabatic compressibility is always smaller in absolute value than the isothermal compressibility.

[†] In § 21 it will be shown rigorously that C_v is always positive, and therefore so is C_p.

Using formulae (16.9) and (16.10), we can derive from (16.14) the relations

$$\left(\frac{\partial V}{\partial P}\right)_S = \left(\frac{\partial V}{\partial P}\right)_T + \frac{T}{C_p}\left[\left(\frac{\partial V}{\partial T}\right)_P\right]^2, \qquad (16.15)$$

$$\left(\frac{\partial P}{\partial V}\right)_S = \left(\frac{\partial P}{\partial V}\right)_T - \frac{T}{C_v}\left[\left(\frac{\partial P}{\partial T}\right)_V\right]^2. \qquad (16.16)$$

§ 17. The thermodynamic scale of temperature

We shall show how a thermodynamic scale of temperature may be construct-ed, at least in principle, using for this purpose an arbitrary body whose equation of state is not assumed known *a priori*. The problem is thus to establish by means of this body the relation $T = T(\tau)$ between the absolute scale of temperature T and some purely arbitrary scale τ defined by an arbitrarily calibrated "thermometer".

To do this, we start from the following relation (in which all quantities refer to the body in question):

$$(\partial Q/\partial P)_T = T(\partial S/\partial P)_T = -T(\partial V/\partial T)_P,$$

where (16.4) has been used. Since τ and T are in one-to-one relation, it does not matter whether the derivative is written for constant T or constant τ. The derivative $(\partial V/\partial T)_P$ may be written as

$$\left(\frac{\partial V}{\partial T}\right)_P = \left(\frac{\partial V}{\partial \tau}\right)_P \frac{d\tau}{dT}.$$

Then

$$\left(\frac{\partial Q}{\partial P}\right)_\tau = -T\left(\frac{\partial V}{\partial \tau}\right)_P \frac{d\tau}{dT},$$

or

$$\frac{d \log T}{d\tau} = -\frac{(\partial V/\partial \tau)_P}{(\partial Q/\partial P)_\tau}. \qquad (17.1)$$

The right-hand side involves quantities which can be measured directly as functions of the arbitrary temperature τ: $(\partial Q/\partial P)_\tau$ is the quantity of heat which must be supplied to the body in order to maintain its temperature constant during expansion, and the derivative $(\partial V/\partial \tau)_P$ is determined by the change in volume of the body on heating. Thus formula (17.1) gives the solution of the problem and can be used to determine the required relation $T = T(\tau)$.

Here it must be remembered that the integration of (17.1) determines $\log T$ only to within an additive constant. The temperature T is therefore determined only to within an arbitrary constant factor. This is as it should be, of course: the choice of the units of measurement of the absolute temperature remains arbitrary, which is equivalent to the presence of an arbitrary factor in the function $T = T(\tau)$.

§ 18. The Joule–Thomson process

Let us consider a process which consists in a gas (or liquid) at pressure P_1 being steadily transferred to a vessel where its pressure is P_2. By "steadily" we mean that the pressures P_1 and P_2 remain constant throughout the process. Such a process may be diagrammatically represented as a passage of the gas through a porous partition (a in Fig. 2), the constancy of pressure on either

FIG. 2

side of the partition being maintained by pistons moving inward and outward in an appropriate manner. If the holes in the partition are sufficiently small, the macroscopic flow velocity of the gas may be taken as zero. We shall also assume that the gas is thermally isolated from the external medium.

This process is called a *Joule–Thomson process*. It must be emphasised that it is an irreversible process, as may be seen simply from the presence of the partition with very small holes, which creates a large amount of friction and destroys the velocity of the gas.

Let a quantity of gas, occupying a volume V_1 at pressure P_1, pass (thermally isolated) into the volume V_2, the pressure becoming equal to P_2. The change in energy $E_2 - E_1$ of this gas is equal to the work $P_1 V_1$ done on the gas to move it out of the volume V_1, minus the work $P_2 V_2$ done by the gas in occupying the volume V_2 at pressure P_2. Thus $E_2 - E_1 = P_1 V_1 - P_2 V_2$, or

$$E_1 + P_1 V_1 = E_2 + P_2 V_2,$$

that is,

$$W_1 = W_2. \tag{18.1}$$

Thus the heat function of the gas is conserved in a Joule–Thomson process.

The change in temperature caused by a small change of pressure in a Joule–Thomson process is given by the derivative $\partial T/\partial P$ taken with the heat function constant. We may transform this derivative to independent variables P and T:

$$\left(\frac{\partial T}{\partial P}\right)_W = \frac{\partial(T, W)}{\partial(P, W)} = \frac{\partial(T, W)/\partial(P, T)}{\partial(P, W)/\partial(P, T)} = -\frac{(\partial W/\partial P)_T}{(\partial W/\partial T)_P},$$

whence, by means of formulae (14.7) and (16.7), we obtain

$$\left(\frac{\partial T}{\partial P}\right)_W = \frac{1}{C_p}\left[T\left(\frac{\partial V}{\partial T}\right)_P - V\right]. \tag{18.2}$$

The change in entropy is given by the derivative $(\partial S/\partial P)_W$. From the relation $dW = T\,dS + V\,dP$, written in the form $dS = dW/T - V\,dP/T$, we have

$$(\partial S/\partial P)_W = -V/T. \tag{18.3}$$

This quantity is always negative, as it should be: the change of a gas to a lower pressure by an irreversible Joule–Thomson process results in an increase in entropy.

We may add a few words concerning a process in which a gas originally in one of two communicating vessels expands into the other vessel; this process, of course, is not a steady one, the pressures in the two vessels varying until they become equal. When a gas expands into a vacuum in this way, its energy E is conserved. If, as a result of the expansion, the total volume is changed only slightly, the change in temperature is given by the derivative $(\partial T/\partial V)_E$. On converting this derivative to independent variables V, T, we obtain the formula

$$\left(\frac{\partial T}{\partial V}\right)_E = \frac{1}{C_v}\left[P - T\left(\frac{\partial P}{\partial T}\right)_V\right]. \tag{18.4}$$

The change in entropy is given by

$$(\partial S/\partial V)_E = P/T. \tag{18.5}$$

The entropy increases on expansion, as it should.

§ 19. Maximum work

Let us consider a thermally isolated system consisting of several bodies not in thermal equilibrium with one another. While equilibrium is being established, the system may do work on some external objects. The transition to equilibrium may, however, occur in different ways, and the final equilibrium states of the system will also be different; in particular, its energy and entropy will be different.

Accordingly, the total work which can be got from a non-equilibrium system will depend on the manner in which equilibrium is established, and we may ask how the equilibrium state must be reached in order that the system should do the maximum possible amount of work. Here we are concerned with the work done because the system is not in equilibrium; that is, we must exclude any work done by a general expansion of the system, since this work could also be done by a system in equilibrium. We shall therefore assume that the total volume of the system is unchanged by the process (although it may vary during the process).

Let the original energy of the system be E_0, and the energy in the equilibrium state, as a function of the entropy of the system in that state, be $E(S)$. Since the system is thermally isolated, the work which it does is just the change in energy:

$$|R| = E_0 - E(S);$$

we write $|R|$, since $R < 0$ in accordance with convention if work is done by the system.

Differentiating $|R|$ with respect to the entropy S of the final state, we have

$$\partial|R|/\partial S = -(\partial E/\partial S)_V = -T,$$

where T is the temperature of the final state; the derivative is taken with the volume of the system in its final state constant (the same as in the initial state). We see that this derivative is negative, i.e. $|R|$ decreases with increasing S. The entropy of a thermally isolated system cannot decrease, and the greatest possible value of $|R|$ therefore occurs if S remains constant throughout the process.

Thus we conclude that the system does maximum work when its entropy remains constant, i.e. when the process of reaching equilibrium is reversible.

Let us determine the maximum work which can be done when a small quantity of energy is transferred between two bodies at different temperatures T_1 and T_2, with $T_2 > T_1$. First of all, it must be emphasised that, if the energy transfer occurred directly between the bodies on contact, no work would be done. The process would be irreversible, the entropy of the two bodies increasing by $\delta E(1/T_1 - 1/T_2)$, where δE is the amount of energy transferred.

Consequently, in order to achieve a reversible transfer of energy and so maximise the work, some further body (the *working medium*) must be brought into the system and caused to execute a reversible cyclic process. This process must be carried out in such a way that the bodies between which direct transfer of energy occurs are at the same temperature. The working medium at temperature T_2 is brought into contact with the body at that temperature

---

Handwritten marginal notes:

— Isolation

T_H
body 1 T_L
body 2 δb_2
δR

Thermal Ins

δS_H T_H δQ_H

δS_L

$\delta R \rightarrow$

δQ_L

$\delta Q_H = \delta R + \delta Q_L$

$\delta S_H < \delta S_L$

$eff \equiv \dfrac{\delta R}{\delta Q_H}$

CASE I
$\delta Q_H = \delta Q_L \ (\delta R = 0)$

$T_H \delta S_H = T_L \delta S_L$

$\delta S_L = \dfrac{T_H}{T_L} \delta S_H$

Hot and cold bodies together

CASE II (IDEAL)
$\delta Q_H = \delta R \ (\delta Q_L = 0)$

$T_H \delta Q_H = \delta S_H$
$\dfrac{}{T_H}$ (i.e. $\delta Q_H = T_H \delta S_H$)

$\delta S_L - \delta S_H < 0$

CASE III
$\delta S_L = \delta S_H$

$\dfrac{\delta Q_L}{T_L} = \dfrac{\delta Q_H}{T_H}$

$\delta Q_L = \dfrac{T_L}{T_H} \delta Q_H$

$\delta R = \delta Q_H - \delta Q_L$

$= \left(1 - \dfrac{T_L}{T_H}\right) \delta Q_L$

η eff.   ideal engine

and receives a certain amount of energy from it isothermally. It is then adiabatically cooled to T_1, releases energy at this temperature to the body at T_1,
and finally is adiabatically returned to its original state. In the expansions
involved in this process the working medium does work on external objects.
The cyclic process just described is called a *Carnot cycle*.

To calculate the resulting maximum work, we first note that the working
medium may be ignored, since it is returned to its initial state at the end of
the process. Let the hotter body 2 lose an amount of energy $-\delta E_2 = -T_2\,\delta S_2$
and body 1 gain energy $\delta E_1 = T_1\,\delta S_1$. Since the process is reversible, the
sum of the entropies of the two bodies remains constant, i.e. $\delta S_1 = -\delta S_2$.
The work done is equal to the decrease in the total energy of the two bodies,
i.e.

$$|\delta R|_{max} = -\delta E_1 - \delta E_2 = -T_1\,\delta S_1 - T_2\,\delta S_2$$
$$= -(T_2 - T_1)\,\delta S_2,$$

or

$$|\delta R|_{max} = \frac{T_2 - T_1}{T_2}\,|\delta E_2|. \tag{19.1}$$

The ratio of the work done to the amount of energy expended is called
the *efficiency* η. The maximum efficiency when energy is transferred from a
hotter to a cooler body is, from (19.1),

$$\eta_{max} = (T_2 - T_1)/T_2. \tag{19.2}$$

A more convenient quantity is the *utilisation coefficient* n, defined as the ratio
of the work done to the maximum work which can be obtained in given
conditions. Clearly

$$n = \eta/\eta_{max}. \tag{19.3}$$

§ 20. Maximum work done by a body in an external medium

Let us now consider a different formulation of the maximum-work problem. Let a body be in an external medium whose temperature T_0 and pressure
P_0 differ from the temperature T and pressure P of the body. The body can do
work on some object, assumed thermally isolated both from the medium and
from the body. The medium, together with the body in it and the object on
which work is done, forms a closed system. The volume and energy of the
medium are so large that the change in these quantities due to processes involving the body does not lead to any appreciable change in the temperature
and pressure of the medium, which may therefore be regarded as constant.

If the medium were absent, the work done by the body on the thermally
isolated object, for a given change in state of the body (i.e. for given initial
and final states) would be completely defined, and equal to the change in the

energy of the body. The presence of the medium which also takes part in the process makes the result indefinite, and the question arises of the maximum work which the body can do for a given change in its state.

If a body does work on an external object in a transition from one state to another, then in the reverse transition from the second state to the first some external source of work must do work on the body. A transition in which the body does the maximum work $|R|_{max}$ corresponds to a reverse transition which requires the external source to do the minimum work R_{min}. These must obviously be the same, so that the calculation of the one is equivalent to that of the other, and we shall speak below of the work done on the body by a thermally isolated external source of work.

During the process, the body may exchange heat and work with the medium. The work done on the body by the medium must of course be subtracted from the total work done on the body, since we are concerned only with the work done by the external source. Thus the total change ΔE in the energy of the body in some (not necessarily small) change in its state consists of three parts: the work R done on the body by the external source, the work done by the medium, and the heat gained from the medium. As already mentioned, owing to the large size of the medium its temperature and pressure may be taken as constant, and the work done by it on the body is therefore $P_0 \Delta V_0$, while the heat given up by it is $-T_0 \Delta S_0$ (the suffix zero indicates quantities pertaining to the medium, while those for the body have no suffix). Thus

$$\Delta E = R + P_0 \Delta V_0 - T_0 \Delta S_0.$$

Since the total volume of the medium and the body remains constant, $\Delta V_0 = -\Delta V$, and the law of increase of entropy shows that $\Delta S + \Delta S_0 \geqslant 0$; the entropy of the thermally isolated source of work does not vary. Thus $\Delta S_0 \geqslant -\Delta S$. From $R = \Delta E - P_0 \Delta V_0 + T_0 \Delta S_0$ we therefore find

$$R \geqslant \Delta E - T_0 \Delta S + P_0 \Delta V. \tag{20.1}$$

The equality occurs for a reversible process. Thus we again conclude that the change occurs with minimum expenditure of work, and the reverse change with maximum work, if it occurs reversibly. The value of the minimum work is

$$R_{min} = \Delta(E - T_0 S + P_0 V) \tag{20.2}$$

(T_0 and P_0, being constants, can be placed after Δ), i.e. this work is equal to the change in the quantity $E - T_0 S + P_0 V$. For maximum work the formula must be written with the opposite sign:

$$|R|_{max} = -\Delta(E - T_0 S + P_0 V), \tag{20.3}$$

since the initial and final states are interchanged.

(margin notes, handwritten)

$T_0 \; P_0$

Medium

$\Delta S_0 \rightarrow T \; P$

$\Delta V_0 \rightarrow$

body $\leftarrow R$

object does work

Body has E, V, S

$\Delta E = R + P_0 \Delta V_0 - T_0 \Delta S_0$

$V_{TOT} = V_0 + V$

$S_{tot} = S_0 + S$

$\Delta S_{TOT} \geqslant 0$

$\Delta S_0 \geqslant -\Delta S$

Minimum work into body

If the body is in an equilibrium state at every instant during the process (but not, of course, in equilibrium with the medium), then for an infinitesimal change in its state formula (20.2) may be written differently. Substituting $dE = T \, dS - P \, dV$ in $dR_{min} = dE - T_0 \, dS + P_0 \, dV$, we find

$$dR_{min} = (T - T_0) \, dS - (P - P_0) \, dV. \tag{20.4}$$

Two important particular cases may be noted. If the volume and temperature of the body remain constant, the latter being equal to the temperature of the medium, (20.2) gives $R_{min} = \Delta(E - TS)$, or

$$R_{min} = \Delta F, \tag{20.5}$$

i.e. the minimum work is equal to the change in the free energy of the body. Secondly, if the temperature and pressure of the body are constant and equal to T_0 and P_0, we have

$$R_{min} = \Delta\Phi, \qquad = \Delta(E + PV - TS) \tag{20.6}$$

i.e. the work done by the external source is equal to the change in the thermodynamic potential of the body.

It should be emphasised that in both these particular cases the body concerned must be one not in equilibrium, so that its state is not defined by T and V (or P) alone; otherwise, the constancy of these quantities would mean that no process could occur at all. We must consider, for example, a chemical reaction in a mixture of reacting substances, a process of dissolution, or the like.

Let us now assume that a body in an external medium is left to itself and no work is done on it. Spontaneous irreversible processes will occur in the body and bring it into equilibrium. In the inequality (20.1) we must now put $R = 0$, and so

$$\Delta(E - T_0 S + P_0 V) \leqslant 0. \tag{20.7}$$

This means that the processes occurring in the body will cause the quantity $E - T_0 S + P_0 V$ to decrease, and it will reach a minimum at equilibrium.

In particular, for spontaneous processes at constant temperature $T = T_0$ and constant pressure $P = P_0$, the thermodynamic potential Φ of the body decreases, and for processes at constant temperature $T = T_0$ and constant volume of the body its free energy F decreases. These results have already been derived by a different approach in § 15. It may be noted that the derivation given here does not essentially assume that the temperature and volume (or pressure) of the body remain constant throughout the process: we may say that the thermodynamic potential (or free energy) of a body decreases as a result of any process for which the initial and final temperature and pressure (or volume) are the same (and equal to the temperature and pressure of the medium), even if they vary during the process.

Another thermodynamic significance may also be ascribed to the minimum work. Let S_t be the total entropy of the body and the medium. If the body is in equilibrium with the medium, S_t is a function of their total energy E_t:

$$S_t = S_t(E_t).$$

If the body is not in equilibrium with the medium, their total entropy differs from $S_t(E_t)$ for the same value of the total energy E_t by some amount $\Delta S_t < 0$. In Fig. 3 the continuous line shows the function $S_t(E_t)$ and the vertical

FIG. 3

segment ab is $-\Delta S_t$. The horizontal segment bc is the change in the total energy when the body goes reversibly from the state of equilibrium with the medium to the state corresponding to the point b. In other words, this segment represents the minimum work which must be done by some external source to bring the body from the state of equilibrium with the medium to the state considered; the equilibrium state in question (the point c in Fig. 3) is, of course, not the same as that corresponding to the given value of E_t (point a).

Since the body is a very small part of the whole system, the processes involving it cause only a negligible relative change in the total energy and entropy. Figure 3 therefore shows that

$$\Delta S_t = -\frac{dS_t(E_t)}{dE_t}\, R_{\min}.$$

But the derivative dE_t/dS_t is the equilibrium temperature of the system, i.e. the temperature T_0 of the medium. Thus

$$\Delta S_t = -\frac{R_{\min}}{T_0} = -\frac{1}{T_0}(\Delta E - T_0\,\Delta S + P_0\,\Delta V). \qquad (20.8)$$

This formula determines the amount by which the entropy of a closed system (body + medium) differs from its greatest possible value if the body is not in equilibrium with the medium; ΔE, ΔS and ΔV are here the differences between the energy, entropy and volume of the body and their values in a state of complete equilibrium.

§ 21. Thermodynamic inequalities

In deriving the conditions of thermal equilibrium from that of maximum entropy, we have so far considered only the first derivatives. By equating to zero the derivatives with respect to energy and volume, we have deduced in §§ 9 and 12 the equality of temperature and pressure in all parts of the body as the conditions of equilibrium. But the vanishing of the first derivatives is only a necessary condition for an extremum and does not ensure that the entropy is in fact a maximum. The determination of the sufficient conditions for a maximum involves, of course, an examination of the second derivative of the function.

Such an examination is, however, more conveniently carried out not from the condition of maximum entropy of a closed system but from another equivalent condition.[†] Let us consider some small but macroscopic part of the body concerned. With respect to this part, the remainder of the body may be regarded as an external medium. Then, as shown in § 20, we can state that in equilibrium the quantity

$$E - T_0 S + P_0 V$$

has a minimum, where E, S and V are the energy, entropy and volume of the part considered, and T_0, P_0 the temperature and pressure of the medium, i.e. of the remainder of the body. Clearly T_0 and P_0 are also the temperature and pressure of the part considered when in equilibrium.

Thus in any small deviation from equilibrium the change in the quantity $E - T_0 S + P_0 V$ must be positive, i.e.

$$\delta E - T_0 \, \delta S + P_0 \, \delta V > 0. \tag{21.1}$$

In other words, the minimum work which must be done to bring this part of the body from equilibrium to any neighbouring state is positive.

In what follows the equilibrium values will be implied for any coefficients appearing in the deviations of thermodynamic quantities from their equilibrium values, and the zero suffixes will therefore be omitted.

[†] As regards the dependence of the entropy on the momenta of macroscopic motion, we have already investigated the conditions to be imposed on both the first and the second derivatives (§ 10), obtaining in this way the conditions that internal macroscopic motions in the body be absent and that the temperature be positive.

Expanding δE as a series (regarding E as a function of S and V), we have as far as the second-order terms

$$\delta E = \frac{\partial E}{\partial S}\,\delta S + \frac{\partial E}{\partial V}\,\delta V + \frac{1}{2}\left[\frac{\partial^2 E}{\partial S^2}(\delta S)^2 + 2\frac{\partial^2 E}{\partial S\,\partial V}\,\delta S\,\delta V + \frac{\partial^2 E}{\partial V^2}(\delta V)^2\right].$$

But $\partial E/\partial S = T$, $\partial E/\partial V = -P$, so that the first-order terms are $T\,\delta S - P\,\delta V$, and cancel when δE is substituted in (21.1). Thus we obtain the condition

$$\frac{\partial^2 E}{\partial S^2}(\delta S)^2 + 2\frac{\partial^2 E}{\partial S\,\partial V}\,\delta S\,\delta V + \frac{\partial^2 E}{\partial V^2}(\delta V)^2 > 0. \qquad (21.2)$$

If such an inequality holds for arbitrary δS and δV, two conditions must be satisfied:[†]

$$\partial^2 E/\partial S^2 > 0, \qquad (21.3)$$

$$\frac{\partial^2 E}{\partial S^2}\frac{\partial^2 E}{\partial V^2} - \left(\frac{\partial^2 E}{\partial S\,\partial V}\right)^2 > 0. \qquad (21.4)$$

Since

$$\partial^2 E/\partial S^2 = (\partial T/\partial S)_V = T/C_v,$$

the condition (21.3) becomes $T/C_v > 0$, or

$$\boxed{C_v > 0,} \qquad (21.5)$$

i.e. the specific heat at constant volume is always positive.

The condition (21.4) may be written in terms of the Jacobian

$$\frac{\partial[(\partial E/\partial S)_V, (\partial E/\partial V)_S]}{\partial(S, V)} = -\partial(T, P)/\partial(S, V) > 0.$$

Changing to the variables T and V, we have

$$\frac{\partial(T, P)}{\partial(S, V)} = \frac{\partial(T, P)/\partial(T, V)}{\partial(S, V)/\partial(T, V)} = \frac{(\partial P/\partial V)_T}{(\partial S/\partial T)_V} = \frac{T}{C_v}\left(\frac{\partial P}{\partial V}\right)_T < 0.$$

Since $C_v > 0$, this is equivalent to the condition

$$(\partial P/\partial V)_T < 0, \qquad (21.6)$$

i.e. an increase in volume at constant temperature is always accompanied by a decrease in pressure.

The conditions (21.5) and (21.6) are called *thermodynamic inequalities*. States in which these conditions are not satisfied are unstable and cannot exist in Nature.

[†] The special case where the equality sign holds in (21.4) will be discussed in § 152.

It has already been noted in § 16 that from the inequality (21.6) and formula (16.10) we always have $C_p > C_v$. From (21.5) we can therefore conclude that

$$C_p > 0 \tag{21.7}$$

always.

The fact that C_v and C_p are positive means that the energy is a monotonically increasing function of temperature at constant volume, and the heat function behaves similarly at constant pressure. The entropy increases monotonically with temperature at either constant volume or constant pressure.

The conditions (21.5), (21.6), which have been derived for an arbitrary small part of a body, are of course valid for the whole body also, since in equilibrium the temperatures and pressures of all parts of the body are the same. Here it is assumed that the body is homogeneous (only such bodies have been considered so far). It must be emphasised that the fulfilment of the conditions (21.5), (21.6) depends on the homogeneity of the body. We can, for example, consider a body whose particles are held together by gravitational forces. Such a body will clearly be inhomogeneous, having a higher density towards the centre, and the specific heat of the body as a whole may be less than zero, so that its temperature rises as its energy decreases. We may note that this does not contradict the result that the specific heat is positive for every small part of the body, since in these conditions the energy of the whole body is not equal to the sum of the energies of its parts; there is also the energy of the gravitational interaction between these parts.

The inequalities derived above are conditions of equilibrium, but their fulfilment is not sufficient for the equilibrium to be completely stable. There can exist states such that the entropy decreases for an infinitesimal deviation from the state and the body then returns to its initial state, whereas for a finite deviation the entropy may be greater than in the original state. After such a finite deviation the body does not return to its original state, but will tend to pass to some other equilibrium state corresponding to a maximum entropy greater than that in the original state. Accordingly, we must distinguish between *metastable* and *stable* equilibrium states. A body in a metastable state may not return to it after a sufficient deviation. Although a metastable state is stable within certain limits, the body will always leave it sooner or later for another state which is stable, corresponding to the greatest of the possible maxima of entropy. A body which is displaced from this state will always eventually return to it.

§ 22. Le Chatelier's principle

Let us consider a closed system consisting of a body and a medium surrounding it. Let S be the total entropy of the system, and y a quantity pertaining to the body, such that the condition for S to be a maximum relative to y, i.e. $\partial S/\partial y = 0$, signifies that the body itself is in equilibrium, though it is not necessarily in equilibrium with the medium. Also, let x be another thermodynamic quantity pertaining to the same body, such that if both $\partial S/\partial y = 0$ and $\partial S/\partial x = 0$ the body is not only in internal equilibrium but also in equilibrium with the medium.

We shall use the notation

$$X = -\partial S/\partial x, \quad Y = -\partial S/\partial y. \tag{22.1}$$

In complete thermodynamic equilibrium the entropy S must be a maximum. For this, besides the conditions

$$X = 0, \quad Y = 0, \tag{22.2}$$

the conditions

$$(\partial X/\partial x)_y > 0, \quad (\partial Y/\partial y)_x > 0 \tag{22.3}$$

and

$$\left(\frac{\partial X}{\partial x}\right)_y \left(\frac{\partial Y}{\partial y}\right)_x - \left[\left(\frac{\partial X}{\partial y}\right)_x\right]^2 > 0 \tag{22.4}$$

must be satisfied.

Let us now assume that the equilibrium of the body with the medium is destroyed by some small external interaction, the quantity x being somewhat changed and the condition $X = 0$ no longer satisfied; we assume that y is not directly affected by the interaction in question. Let the change in x be Δx. Then the change in X at the instant of interaction is

$$(\Delta X)_y = (\partial X/\partial x)_y \, \Delta x.$$

The change in x at constant y leads, of course, to a violation of the condition $Y = 0$ also, i.e. of internal equilibrium of the body. When equilibrium is again restored, the quantity $X \equiv \Delta X$ will be

$$(\Delta X)_{Y=0} = (\partial X/\partial x)_{Y=0} \, \Delta x,$$

where the derivative is taken at constant $Y(= 0)$.

To compare the two values of ΔX, using the properties of Jacobians, we have

$$\left(\frac{\partial X}{\partial x}\right)_{Y=0} = \frac{\partial(X, Y)}{\partial(x, Y)} = \frac{\partial(X, Y)/\partial(x, y)}{\partial(x, Y)/\partial(x, y)} = \left(\frac{\partial X}{\partial x}\right)_y - \frac{[(\partial X/\partial y)_x]^2}{(\partial Y/\partial y)_x}.$$

The denominator of the second term in this expression is positive by the condition (22.3); using also (22.4), we find that

$$(\partial X/\partial x)_y > (\partial X/\partial x)_{Y=0} > 0, \tag{22.5}$$

or

$$|(\Delta X)_y| > |(\Delta X)_{Y=0}|. \tag{22.6}$$

The inequality (22.5) or (22.6) forms the content of what is called *Le Chatelier's principle.*

We shall regard the change Δx of the quantity x as a measure of the external interaction acting on the body, and ΔX as a measure of the change in properties of the body resulting from this interaction. The inequality (22.6) shows that, when the internal equilibrium of the body is restored after the external interaction which disturbed it, the value of ΔX is reduced. Thus Le Chatelier's principle may be formulated as follows: an external interaction which disturbs the equilibrium brings about processes in the body which tend to reduce the effects of this interaction.

The above may be illustrated by some examples.

First of all, it is convenient to modify somewhat the definition of the quantities X and Y by using formula (20.8), according to which the change in entropy of the system (medium + body) is $-R_{min}/T_0$, where T_0 is the temperature of the medium and R_{min} the minimum work needed to bring the body from a state of equilibrium with the medium to the state in question. We can therefore write

$$X = \frac{1}{T_0} \frac{\partial R_{min}}{\partial x}, \quad Y = \frac{1}{T_0} \frac{\partial R_{min}}{\partial y}. \tag{22.7}$$

For an infinitesimal change in the state of the body we have (see (20.4))

$$dR_{min} = (T-T_0)\, dS - (P-P_0)\, dV;$$

here and below all quantities without suffix relate to the body, and those with suffix 0 to the medium.

Let x be the entropy S of the body. Then $X = (T-T_0)/T_0$. The equilibrium condition $X = 0$ gives $T = T_0$, i.e. the temperatures of the body and the medium are equal. The inequalities (22.5) and (22.6) become

$$(\partial T/\partial S)_y > (\partial T/\partial S)_{Y=0} > 0, \tag{22.8}$$

$$|(\Delta T)_y| > |(\Delta T)_{Y=0}|. \tag{22.9}$$

The significance of these inequalities is as follows. The change in x (the entropy of the body) means that a quantity of heat is given to or taken from the body. This destroys the equilibrium of the body itself and, in particular, changes its temperature by $(\Delta T)_y$. The restoration of equilibrium in the body has the result that the absolute value of the change in temperature decreases, becoming $(\Delta T)_{Y=0}$, i.e. it is as if the result of the interaction which brings the body out of equilibrium were reduced. We can say that heating or cooling a body brings about processes in it which tend to lower or raise the temperature respectively.

Now let x be the volume V of a body. Then $X = -(P-P_0)/T_0$. In equilibrium $X = 0$, i.e. $P = P_0$. The inequalities (22.5) and (22.6) give

$$(\partial P/\partial V)_y < (\partial P/\partial V)_{Y=0} < 0, \tag{22.10}$$

$$|(\Delta P)_y| > |(\Delta P)_{Y=0}|. \tag{22.11}$$

If the body is disturbed from equilibrium by a change in its volume at constant temperature, then, in particular, its pressure is changed; the restoration of equilibrium in the body leads to a decrease in the absolute value of the change in pressure. Since a decrease in the volume of the body causes an increase in its pressure, and vice versa, we can say that decreasing or increasing the volume of a body brings about processes in it which tend to lower or raise the pressure respectively.

Later we shall meet with numerous applications of these results (to solutions, chemical reactions and so on).

It may also be noted that, if y in the inequalities (22.8) is taken to be the volume of the body, we have

$$(\partial T/\partial S)_y = (\partial T/\partial S)_V = T/C_v,$$

$$(\partial T/\partial S)_{Y=0} = (\partial T/\partial S)_P = T/C_p,$$

since the condition $Y = 0$ then denotes $P = P_0$, i.e. constant pressure. Thus we again obtain the already familiar inequalities $C_p > C_v > 0$. Similarly, if in (22.10) y is taken as the entropy of a body, the condition $Y = 0$ implies that the temperature is constant, $T = T_0$, and we find

$$(\partial P/\partial V)_S < (\partial P/\partial V)_T < 0,$$

another result already known.

§ 23. Nernst's theorem

The fact that the specific heat C_v is positive means that the energy is a monotonically increasing function of the temperature. Conversely, when the temperature falls the energy decreases monotonically, and therefore, when the temperature has its least possible value, i.e. at absolute zero, a body must be in the state of least possible energy. If we regard the energy of a body as the sum of the energies of the parts into which it may be imagined to be divided, we can say that each of these parts will also be in the state of least energy; it is clear that the minimum value of the sum must correspond to the minimum value of each term.

Thus at absolute zero any part of the body must be in a particular quantum state, the ground state. In other words, the tsatistical weights of these parts are equal to unity, and therefore so is their product, i.e. the statistical weight of the macroscopic state of the body as a whole. The entropy of the body, being the logarithm of its statistical weight, is therefore zero.

We consequently reach the important result that the entropy of any body vanishes at the absolute zero of temperature. This is called *Nernst's theorem* (W. Nernst, 1906).[†]

It should be emphasised that this theorem is a deduction from quantum statistics, in which the concept of discrete quantum states is of essential importance. The theorem cannot be proved in purely classical statistics, where the entropy is determined only to within an arbitrary additive constant (see § 7).

Nernst's theorem enables us to draw conclusions also concerning the behaviour of certain other thermodynamic quantities as $T \to 0$.

For instance, it is easy to see that for $T = 0$ the specific heats C_p and C_v both vanish:

$$C_p = C_v = 0 \quad \text{for} \quad T = 0. \tag{23.1}$$

This follows immediately from the definition of the specific heat in the form

$$C = T \, \partial S / \partial T$$
$$= \partial S / \partial \log T.$$

When $T \to 0$, $\log T \to -\infty$, and since S tends to a finite limit, namely zero, it is clear that the derivative tends to zero.

The thermal expansion coefficient also tends to zero:

$$(\partial V / \partial T)_P = 0 \quad \text{for} \quad T = 0. \tag{23.2}$$

For this derivative is equal to the derivative $-(\partial S / \partial P)_T$ (see (16.4)), which vanishes for $T = 0$, since $S = 0$ for $T = 0$ and any pressure.

Similarly, we can see that

$$(\partial P / \partial T)_V = 0 \quad \text{for} \quad T = 0. \tag{23.3}$$

The entropy usually vanishes, for $T \to 0$, according to a power law, i.e. as $S = aT^n$, where a is a function of pressure or volume. In this case, clearly, the specific heats and $(\partial V / \partial T)_P$, $(\partial P / \partial T)_V$ will tend to zero in the same way (with the same value of n).

[†] To avoid misunderstandings we should emphasise that this refers to the temperature tending to zero with other conditions remaining unchanged—say at constant volume, or at constant pressure. If, on the other hand, the temperature of a gas tends to zero while its density decreases without limit, for example, the entropy need not tend to zero.

Finally, it may be seen that the difference $C_p - C_v$ tends to zero more rapidly than the specific heats themselves, i.e.

$$(C_p - C_v)/C_p = 0 \quad \text{for} \quad T = 0. \tag{23.4}$$

For let the entropy tend to zero as $S \propto T^n$ for $T \to 0$. From formula (16.9) we then see that $C_p - C_v \propto T^{2n+1}$, so that $(C_p - C_v)/C_p \propto T^{n+1}$; it should be borne in mind that the compressibility $(\partial V / \partial P)_T$ is in general finite and not zero when $T = 0$.

If the specific heat of a body is known for all temperatures, the entropy can be calculated by integration, and Nernst's theorem gives the value of the constant of integration. For example, the dependence of the entropy on temperature for a given pressure is determined by

$$S = \int_0^T (C_p / T) \, dT. \tag{23.5}$$

The corresponding formula for the heat function is

$$W = W_0 + \int_0^T C_p \, dT, \tag{23.6}$$

where W_0 is the value of the heat function for $T = 0$. Similarly, for the thermodynamic potential $\Phi = W - TS$ we have

$$\Phi = W_0 + \int_0^T C_p \, dT - T \int_0^T \frac{C_p}{T} \, dT. \tag{23.7}$$

§ 24. The dependence of the thermodynamic quantities on the number of particles

As well as the energy and entropy, such thermodynamic quantities as F, Φ and W also have the property of additivity, as follows directly from their definitions if we bear in mind that the pressure and temperature are constant throughout a body in equilibrium. From this property we can draw certain conclusions concerning the manner in which each of these quantities depends on the number of particles in the body. Here we shall consider bodies consisting of identical particles (molecules); all the results can be immediately generalised to mixtures of different particles (see § 85).

The additivity of a quantity signifies that, when the amount of matter (and therefore the number N of particles) is changed by a given factor, the quantity is changed by the same factor. In other words, we can say that an additive thermodynamic quantity must be a homogeneous function of the first order with respect to the additive variables.

$$f(tx, ty) = t^N f(x, y)$$

Let us express the energy of the body as a function of the entropy, volume, and number of particles. Since S and V are themselves additive, this function must be of the form

$$E = Nf(S/N, V/N), \tag{24.1}$$

the most general homogeneous function of the first order in N, S and V.

The free energy F is a function of N, T and V. Since the temperature is constant throughout the body, and the volume is additive, a similar argument gives

$$F = Nf(V/N, T). \tag{24.2}$$

In exactly the same way we have for the heat function W, expressed as a function of N, S and the pressure P,

$$W = Nf(S/N, P). \tag{24.3}$$

Finally, the thermodynamic potential as a function of N, P and T is

$$\Phi = Nf(P, T). \tag{24.4}$$

In the foregoing discussion we have essentially regarded the number of particles as a parameter which has a given constant value for each body. We shall now formally consider N as a further independent variable. Then the expressions for the differentials of the thermodynamic potentials must include terms proportional to dN. For example, the total differential of the energy will be written

$$dE = T \, dS - P \, dV + \mu \, dN, \tag{24.5}$$

where μ denotes the partial derivative

define

$$\mu = (\partial E/\partial N)_{S, V}. \tag{24.6}$$

The quantity μ is called the *chemical potential* of the body. Similarly we have

$$dW = T \, dS + V \, dP + \mu \, dN, \tag{24.7}$$

$$dF = -S \, dT - P \, dV + \mu \, dN, \tag{24.8}$$

$$d\Phi = -S \, dT + V \, dP + \mu \, dN, \tag{24.9}$$

with the same μ. These formulae show that

$$\mu = (\partial W/\partial N)_{S, P} = (\partial F/\partial N)_{T, V} = (\partial \Phi/\partial N)_{P, T}, \tag{24.10}$$

i.e. the chemical potential can be obtained by differentiating any of the quantities E, W, F and Φ with respect to the number of particles, but the result is expressed in terms of different variables in each case.

Differentiating Φ in the form (24.4), we find that $\mu = \partial\Phi/\partial N = f(P, T)$, i.e.

$$\Phi = N\mu. \tag{24.11}$$

Thus the chemical potential of a body (consisting of identical particles) is just its thermodynamic potential per molecule. When expressed as a function of P and T, the chemical potential is independent of N. Thus we can immediately write down for the differential of the chemical potential

$$d\mu = -s\,dT + v\,dP, \tag{24.12}$$

where s and v are the entropy and volume per molecule.

If we consider (as we have usually done hitherto) a definite amount of matter, the number of particles in it is a given constant, while the volume is variable. Let us now take a certain volume within the body, and consider the matter enclosed therein; the number of particles N will now be variable, and the volume V constant. Then, for example, equation (24.8) reduces to

$$dF = -S\,dT + \mu\,dN.$$

Here the independent variables are T and N. We may define a thermodynamic potential such that the second independent variable is μ, not N. To do so, we substitute $\mu\,dN = d(\mu N) - N\,d\mu$, obtaining

$$d(F - \mu N) = -S\,dT - N\,d\mu.$$

But $\mu N = \Phi$, and $F - \Phi = -PV$. Thus the new thermodynamic potential (denoted by Ω) is just

$$\Omega = -PV, \tag{24.13}$$

and

$$d\Omega = -S\,dT - N\,d\mu. \tag{24.14}$$

The number of particles is obtained by differentiating Ω with respect to the chemical potential at constant temperature and volume:

$$N = -(\partial\Omega/\partial\mu)_{T,V} = V(\partial P/\partial\mu)_{T,V}. \tag{24.15}$$

In the same way as we proved the equality of small changes in E, W, F and Φ (with the appropriate pairs of quantities constant), we can easily show that the change $(\delta\Omega)_{T,\mu,V}$ at constant T, μ, V has the same property:

$$(\delta E)_{S,V,N} = (\delta F)_{T,V,N} = (\delta\Phi)_{T,P,N} = (\delta W)_{S,P,N} = (\delta\Omega)_{T,V,\mu}. \tag{24.16}$$

These equations refine and extend the theorem of small increments (15.12).

Finally, as in §§ 15 and 20 for the free energy and the thermodynamic potential, we may show that the work in a reversible process occurring at constant T, V and μ is equal to the change in the potential Ω. In a state of thermal equilibrium the potential Ω is a minimum with respect to any change of state at constant T, V, μ.

PROBLEM

Derive an expression for the specific heat C_v in terms of the variables T, μ, V.

SOLUTION. We transform the derivative $C_v = T(\partial S/\partial T)_{V,N}$ to the variables T, V, μ, writing (with V regarded as a constant throughout)

$$\left(\frac{\partial S}{\partial T}\right)_N = \frac{\partial(S,N)}{\partial(T,N)} = \frac{\partial(S,N)/\partial(T,\mu)}{\partial(T,N)/\partial(T,\mu)} = \left(\frac{\partial S}{\partial T}\right)_\mu - \frac{(\partial S/\partial\mu)_T (\partial N/\partial T)_\mu}{(\partial N/\partial\mu)_T}.$$

But $(\partial S/\partial\mu)_T = -\partial^2\Omega/\partial T\,\partial\mu = (\partial N/\partial T)_\mu$, and therefore

$$C_v = T\left\{\left(\frac{\partial S}{\partial T}\right)_\mu - \frac{[(\partial N/\partial T)_\mu]^2}{(\partial N/\partial\mu)_T}\right\}.$$

§ 25. Equilibrium of a body in an external field

Let us consider a body in an external field which is constant in time. The different parts of the body are in different conditions, and the body will therefore be inhomogeneous. One of the conditions of equilibrium of such a body is again that the temperature should be constant throughout it, but the pressure will now vary from point to point.

To derive the second condition of equilibrium, let us consider two adjoining volumes in the body and maximise their entropy $S = S_1 + S_2$ when the remainder of the body is in a fixed state. One necessary condition for a maximum is that the derivative $\partial S/\partial N_1$ should be zero. Since the total number of particles $N_1 + N_2$ in these two parts of the body is regarded as constant, we have

$$\frac{\partial S}{\partial N_1} = \frac{\partial S_1}{\partial N_1} + \frac{\partial S_2}{\partial N_2}\frac{\partial N_2}{\partial N_1} = \frac{\partial S_1}{\partial N_1} - \frac{\partial S_2}{\partial N_2} = 0.$$

The equation $dE = T\,dS + \mu\,dN$, written in the form

$$dS = \frac{dE}{T} - \frac{\mu}{T}\,dN,$$

shows that the derivative $\partial S/\partial N$ for constant E and V is $-\mu/T$. Thus $\mu_1/T_1 = \mu_2/T_2$. But in equilibrium $T_1 = T_2$, so that $\mu_1 = \mu_2$. We therefore conclude that in equilibrium in an external field, in addition to the constancy of temperature, we must have

$$\mu = \text{constant}, \tag{25.1}$$

i.e. the chemical potential of every part of the body must be the same. The chemical potential of each part is a function of its temperature and pressure, as well as of the parameters which define the external field. If there is no field, the constancy of μ and T necessarily implies that of the pressure.

(handwritten margin notes)

S_1	S_2
E_1	$dE \to E_2$
V_1	$dV \to V_2$
N_1	$dN \to N_2$

$S = S_1 + S_2$

$V_1 + V_2 = V$

$N_1 + N_2 = 0$

$dS(E V N) \geq 0$

$\frac{\partial S}{\partial E}\,dE + \frac{\partial S}{\partial V}\,dV + \frac{\partial S}{\partial N}\,dN \geq 0$

$\Rightarrow T_1 > T_2$

$P_1 > P_2$

$\mu_1 > \mu_2$

In a gravitational field the potential energy u of a molecule is a function only of the coordinates x, y, z of its centre of gravity (and not of the arrangement of the atoms within the molecule). In this case the change in the thermodynamic quantities for the body amounts to adding to its energy the potential energy of the molecules in the field. In particular, the chemical potential (the thermodynamic potential per molecule) has the form $\mu = \mu_0 + u(x, y, z)$, where $\mu_0(P, T)$ is the chemical potential in the absence of the field. Thus the condition of equilibrium in a gravitational field may be written

$$\mu_0(P, T) + u(x, y, z) = \text{constant.} \qquad (25.2)$$

In particular, in a uniform gravitational field $u = mgz$ (where m is the mass of a molecule, g the acceleration due to gravity, and z the vertical coordinate). Differentiating equation (25.2) with respect to the coordinate z at constant temperature, we have $v\, dP = -mg\, dz$, where $v = (\partial\mu_0/\partial P)_T$ is the specific volume. For small changes in pressure, v may be regarded as constant. Substituting the density $\varrho = m/v$ and integrating, we obtain

$$P = \text{constant} - \varrho gz,$$

the customary formula for the hydrostatic pressure in an incompressible fluid.

§ 26. Rotating bodies

In a state of thermal equilibrium, as we have seen in § 10, only a uniform translational motion and a uniform rotation of a body as a whole are possible. The uniform translational motion needs no special treatment, since by Galileo's relativity principle it has no effect on the mechanical properties of the body, nor therefore on its thermodynamic properties, and the thermodynamic quantities are unchanged except that the energy of the body is increased by its kinetic energy.

Let us consider a body in uniform rotation round a fixed axis with angular velocity Ω. Let $E(p, q)$ be the energy of the body in a fixed coordinate system and $E'(p, q)$ the energy in a coordinate system rotating with the body. We know from mechanics that these quantities are related by

$$E'(p, q) = E(p, q) - \Omega \cdot M(p, q), \qquad (26.1)$$

where $M(p, q)$ is the angular momentum of the body.[†]

[†] See *Mechanics*, § 39. Although the derivation of formula (39.13) is based on classical mechanics, in quantum theory exactly the same relations apply to the operators of the corresponding quantities. Hence all the thermodynamic relations derived below are independent of which mechanics describes the motion of the particles in the body.

[Handwritten marginal notes:]

$E_B = E_A + Mg z$

$\mu = \left(\frac{\partial E}{\partial N}\right)_{V,S}$

$\mu_B = \mu_A + mgz$

$PV = NT$

$\mu(P,T,z) =$

$\mu(P,T) + mgz$

$d\mu(PTz)$

$= \frac{\partial \mu}{\partial P} dP$

$+ \frac{\partial \mu}{\partial T} dT$

$+ mg\, dz$

$dT = 0$

$d\mu = 0$

$\left(\frac{\partial \mu}{\partial P}\right) = \frac{\partial \phi/N}{\partial P}$

$= \frac{1}{N}\frac{\partial \phi}{\partial P} = \frac{V}{N}$

$\frac{V}{N} = \frac{I}{P}$

$\frac{I}{P} dP + mg\, dz = 0$

$dP = -\frac{mg}{T} dz$

$\ln \frac{P}{P_0} = -\frac{mg}{T} z$

$P = \frac{NT}{V}$

$= MT$

Thus the energy $E'(p, q)$ depends on the angular velocity Ω as a parameter, and

$$\partial E'(p, q)/\partial\Omega = -M(p, q).$$

Averaging this equation over the statistical distribution and using formula (11.3), we obtain

$$(\partial E'/\partial\Omega)_S = -\mathbf{M}, \tag{26.2}$$

where $E' = \overline{E'(p, q)}$, $\mathbf{M} = \overline{\mathbf{M}(p, q)}$ are the mean (thermodynamic) energy and angular momentum of the body.

From this relation we can write down the differential of the energy of a rotating body of given volume:

$$dE' = T\,dS - \mathbf{M}\cdot d\mathbf{\Omega}. \tag{26.3}$$

Similarly, for the free energy $F' = E' - TS$ (in the rotating coordinate system) we have

$$dF' = -S\,dT - \mathbf{M}\cdot d\mathbf{\Omega}. \tag{26.4}$$

Averaging equation (26.1) gives

$$E' = E - \mathbf{M}\cdot\mathbf{\Omega}. \tag{26.5}$$

Differentiating this equation and substituting (26.3), we obtain the differential of the energy in the fixed coordinate system:

$$dE = T\,dS + \mathbf{\Omega}\cdot d\mathbf{M}. \tag{26.6}$$

Correspondingly, for the free energy $F = E - TS$

$$dF = -S\,dT + \mathbf{\Omega}\cdot d\mathbf{M}. \tag{26.7}$$

Thus in these relations the independent variable is not the angular velocity but the angular momentum, and

$$\mathbf{\Omega} = (\partial E/\partial\mathbf{M})_S = (\partial F/\partial\mathbf{M})_T. \tag{26.8}$$

As we know from mechanics, a uniform rotation is in a certain sense equivalent to the presence of two fields of force, centrifugal and Coriolis. The centrifugal forces are proportional to the size of the body, as they involve the distance from the axis of rotation; the Coriolis forces are independent of the size of the body. For this reason the effect of the Coriolis forces on the thermodynamic properties of a rotating macroscopic body is entirely negligible in comparison with that of the centrifugal forces, and the former can usually be neglected.[†] The condition of thermal equilibrium of a rotating body is therefore obtained by simply substituting for $u(x, y, z)$ in (25.2) the centrifugal energy of the particles:

$$\mu_0(P, T) - \tfrac{1}{2}m\Omega^2 r^2 = \text{constant}, \tag{26.9}$$

[†] It may be shown that in classical statistics the Coriolis forces do not affect the statistical properties of the body; see § 34.

where μ_0 is the chemical potential of the body at rest, m the mass of a molecule, and r the distance from the axis of rotation.

For the same reason, the total energy E of a rotating body may be written
as the sum of its internal energy (here denoted by E_{in}) and its kinetic energy
of rotation:

$$E = E_{in} + M^2/2I, \tag{26.10}$$

where I is the moment of inertia of the body with respect to the axis of rotation. It should be remembered that rotation in general changes the distribution of mass in the body, and so the moment of inertia and internal energy
of the body are themselves in general functions of Ω (or of M). They may be
regarded as constants independent of Ω only when the rotation is sufficiently
slow.

Let us consider an isolated uniformly rotating solid with a given mass
distribution. Since the entropy of a body is a function of its internal energy,
we have in this case $S = S(E - M^2/2I)$. Because the body is a closed system,
its total energy and angular momentum are conserved, and the entropy must
have the maximum value possible for the given M and E. We therefore conclude that the equilibrium rotation of the body takes place about the axis
with respect to which the moment of inertia has the greatest possible value.
This assumes that the axis of rotation is necessarily a principal axis of inertia
of the body, but the latter result is evident: if the body rotates about an axis
other than a principal axis of inertia, then, as we know from mechanics, the
axis of rotation will itself precess in space, and the rotation will be non-
uniform, and therefore not an equilibrium rotation.

§ 27. Thermodynamic relations in the relativistic region

Relativistic mechanics leads to a number of changes in the usual thermodynamic relations. Here we shall discuss the most interesting of these changes.

If the microscopic motion of the particles forming a body becomes relativistic, the general thermodynamic relations are unchanged, but there is an
important inequality between the pressure and energy of the body:

$$P < E/3V, \tag{27.1}$$

where E is the energy of the body including the rest energy of the particles
in it.[†]

[†] See *Fields*, § 35. It should be mentioned, however, that there is as yet no general
proof of this inequality valid for all types of interaction (not only electromagnetic)
which exist between particles in Nature.

The changes caused by the general theory of relativity in the conditions of thermal equilibrium, taking account of the gravitational field of the body itself, are of fundamental importance. Let us consider a macroscopic body at rest; its gravitational field is, of course, constant. In a constant gravitational field we must distinguish the conserved energy E_0 of any small part of the body from the energy E measured by an observer situated at a given point. These two quantities are related by

$$E_0 = E\sqrt{g_{00}},$$

where g_{00} is the time component of the metric tensor; see *Fields*, § 88, formula (88.9) with $v = 0$, $mc^2 = E$. But, from the sense of the proof given in § 9 that the temperature is constant throughout a body in equilibrium, it is clear that the quantity obtained by differentiating the entropy with respect to the conserved energy E_0 must be constant. The temperature T measured by an observer situated at a given point in space is, however, obtained by differentiating the entropy with respect to the energy E, and will therefore be different at different points in the body.

To derive a quantitative relation, we note that the entropy, by definition, depends only on the internal state of the body and so is unchanged by the presence of a gravitational field (provided that this field does not affect the internal properties of the body, a condition which is always satisfied in practice). The derivative with respect to entropy of the conserved energy E_0 is therefore $T\sqrt{g_{00}}$, and so one of the conditions of thermal equilibrium is that

$$T\sqrt{g_{00}} = \text{constant} \tag{27.2}$$

throughout the body.

A similar change occurs in the second condition of equilibrium, the constancy of the chemical potential. The latter is defined as the derivative of the energy with respect to the number of particles. Since this number is of course unaffected by a gravitational field, we have for the chemical potential measured at any given point a relation of the same kind as for the temperature:

$$\mu\sqrt{g_{00}} = \text{constant}. \tag{27.3}$$

We may note that the relations (27.2), (27.3) may be written

$$
\begin{aligned}
T &= \text{constant} \times dx^0/ds, \\
\mu &= \text{constant} \times dx^0/ds,
\end{aligned}
\tag{27.4}
$$

which enable us to consider the body not only in the frame of reference in which it is at rest but also in those where it is moving (rotating as a whole). The derivative dx^0/ds must be taken along the world line described by the point considered in the body.

In a weak (Newtonian) gravitational field, $g_{00} = 1 + 2\phi/c^2$, where ϕ is the gravitational potential (see *Fields*, § 87). Substituting this expression in (27.2) and taking the square root, we have to the same approximation

$$T = \text{constant} \times (1 - \phi/c^2).\qquad(27.5)$$

Since $\phi < 0$, this shows that in equilibrium the temperature is higher at points in the body where $|\phi|$ is greater, i.e. within the body. In the limit of non-relativistic mechanics $(c \to \infty)$, (27.5) becomes $T = \text{constant}$, as it should.

We can similarly transform the condition (27.3), bearing in mind that the relativistic chemical potential, in the limit of classical mechanics, does not directly become the ordinary (non-relativistic) expression for the chemical potential in the absence of a field, which we now denote by μ_0, but $\mu_0 + mc^2$, where mc^2 is the rest energy of a particle of the body. Thus we have

$$\mu\sqrt{g_{00}} \cong (\mu_0 + mc^2)(1 + \phi/c^2)$$
$$\cong \mu_0 + mc^2 + m\phi,$$

so that the condition (27.3) becomes

$$\mu_0 + m\phi = \text{constant};$$

this agrees with (25.2), as it should.

Finally, we may mention a useful relation which follows immediately from the conditions (27.2) and (27.3). Dividing one by the other, we find that $\mu/T = \text{constant}$, and hence

$$d\mu/\mu = dT/T.$$

From (24.12), at constant volume (equal to unity) we have

$$dP = S\,dT + N\,d\mu,$$

where S and N are the entropy and number of particles in unit volume of the body. Substituting $dT = (T/\mu)\,d\mu$ and noting that $\mu N + ST = \Phi + ST = \varepsilon + P$, where ε is the energy per unit volume, gives[†]

$$d\mu/\mu = dP/(\varepsilon + P).\qquad(27.6)$$

[†] In the non-relativistic case, putting $\mu \cong mc^2$, $\varepsilon \cong \varrho c^2 \gg P$ (where ϱ is the density), we get $d\mu = v\,dP$ with $v = m/\varrho$ the volume per particle; this is as it should be for $T = \text{constant}$.

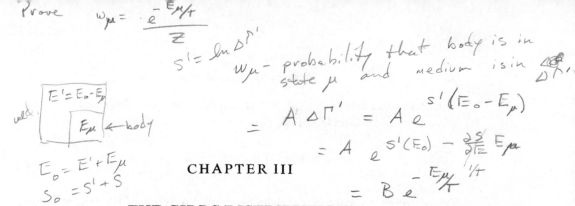

Prove $\quad w_\mu = \dfrac{e^{-E_\mu/T}}{Z}$

$s' = \ln \Delta \Gamma'$

w_μ — probability that body is in state μ and medium is in $\Delta \Gamma'$.

$\boxed{E' = E_0 - E_\mu}$

$E_\mu \leftarrow$ body

$$= A \,\Delta\Gamma' = A\, e^{s'(E_0 - E_\mu)}$$

$$= A\, e^{s'(E_0) - \frac{\partial S}{\partial E}\big|_E E_\mu}$$

$E_0 = E' + E_\mu$

$S_0 = s' + S$

$$= B\, e^{-E_\mu/T}$$

$$s' = s'(E_0 - E_\mu)$$

$$Z = \sum e^{-E/T}$$

$\sum w_\mu = 1$

$\Rightarrow B = \dfrac{1}{Z}$

CHAPTER III

THE GIBBS DISTRIBUTION

§ 28. The Gibbs distribution

LET us now turn to the problem stated in Chapter I of finding the distribution function for a subsystem, i.e. any macroscopic body which is a small part of some large closed system. The most convenient and general method of approaching the solution of this problem is based on the application of the microcanonical distribution to the whole system.

Distinguishing the body in question from the rest of the closed system, we may consider the system as consisting of these two parts. The rest of the system will be called the "medium" in relation to the body.

The microcanonical distribution (6.6) can be written in the form

$$dw = \text{constant} \times \delta(E + E' - E^{(0)})\, d\Gamma\, d\Gamma', \qquad (28.1)$$

where E, $d\Gamma$ and E', $d\Gamma'$ relate to the body and the medium respectively, and $E^{(0)}$ is the given value of the energy of the closed system, which must be equal to the sum $E + E'$ of the energies of the body and the medium.

Our object is to find the probability w_n of a state of the whole system such that the body concerned is in some definite quantum state (with energy E_n), i.e. a microscopically defined state. The microscopic state of the medium is of no interest, so that we shall suppose this to be in a macroscopically defined state. Let $\Delta\Gamma'$ be the statistical weight of the macroscopic state of the medium and let $\Delta E'$ be the range of values of the energy of the medium corresponding to the range $\Delta\Gamma'$ of quantum states in the sense discussed in § 7.

The required probability w_n can be found by taking $d\Gamma = 1$ in (28.1), putting $E = E_n$ and integrating with respect to Γ':

$$w_n = \text{constant} \times \int \delta(E_n + E' - E^{(0)})\, d\Gamma'.$$

Let $\Gamma'(E')$ be the total number of quantum states of the medium with energy not exceeding E'. Since the integrand depends only on E', we can change to integration with respect to E', putting $d\Gamma' = (d\Gamma'(E')/dE')dE'$. The derivative $d\Gamma'/dE'$ is replaced (cf. § 7) by

$$d\Gamma'/dE' = e^{S'(E')}/\Delta E',$$

where $S'(E')$ is the entropy of the medium as a function of its energy; $\Delta E'$ is, of course, also a function of E'. Thus we have

$$w_n = \text{constant} \times \int \frac{e^{S'}}{\Delta E'} \, \delta(E' + E_n - E^{(0)}) \, dE'.$$

Owing to the presence of the delta function, the result of the integration is simply to replace E' by $E^{(0)} - E_n$:

$$w_n = \text{constant} \times \left(\frac{e^{S'}}{\Delta E'} \right)_{E' = E^{(0)} - E_n}. \qquad (28.2)$$

We now use the fact that, since the body is small, its energy E_n is small in comparison with $E^{(0)}$. The quantity $\Delta E'$ undergoes only a very small relative change when E' varies slightly, and so in $\Delta E'$ we can simply put $E' = E^{(0)}$; it then becomes a constant independent of E_n. In the exponential factor $e^{S'}$, we must expand $S'(E^{(0)} - E_n)$ in powers of E_n as far as the linear term:

$$S'(E^{(0)} - E_n) = S'(E^{(0)}) - E_n \, dS'(E^{(0)})/dE^{(0)}.$$

The derivative of the entropy S' with respect to energy is just $1/T$, where T is the temperature of the system; the temperatures of the body and the medium are the same, since the system is assumed to be in equilibrium.

Thus we have finally for w_n the expression

$$w_n = A e^{-E_n/T}, \qquad (28.3)$$

where A is a normalisation constant independent of E_n. This is one of the most important formulae in statistical physics. It gives the statistical distribution of any macroscopic body which is a comparatively small part of a large closed system. The distribution (28.3) is called the *Gibbs distribution* or *canonical distribution*; it was discovered by J. W. Gibbs for classical statistics in 1901.

The normalisation constant A is given by the condition $\Sigma w_n = 1$, whence

$$\frac{1}{A} = \sum_n e^{-E_n/T}. \qquad (28.4)$$

The mean value of any physical quantity f pertaining to the body can be calculated by means of the Gibbs distribution, using the formula

$$\bar{f} = \sum_n w_n f_{nn}$$

$$= \sum_n f_{nn} e^{-E_n/T} \Big/ \sum_n e^{-E_n/T}. \qquad (28.5)$$

In classical statistics an expression exactly corresponding to (28.3) is obtained for the distribution function in phase space:

$$\varrho(p, q) = Ae^{-E(p, q)/T}, \tag{28.6}$$

where $E(p, q)$ is the energy of the body as a function of its coordinates and momenta.[†] The normalisation constant A is given by the condition

$$\int \varrho \, dp \, dq = A \int e^{-E(p, q)/T} \, dp \, dq = 1. \tag{28.7}$$

In practice, cases are frequently encountered where it is not the entire microscopic motion of the particles which is quasi-classical, but only the motion corresponding to some of the degrees of freedom, whereas the motion with respect to the remaining degrees of freedom is quantised (for example, the translational motion of the molecules may be quasi-classical while the motion of the atoms within the molecules is quantised). Then the energy levels of the body may be written as functions of the quasi-classical coordinates and momenta: $E_n = E_n(p, q)$, where n denotes the set of quantum numbers defining the "quantised part" of the motion, for which p and q are parameters. The Gibbs distribution formula then becomes

$$dw_n(p, q) = Ae^{-E_n(p, q)/T} \, dp_{cl} \, dq_{cl}, \tag{28.8}$$

where $dp_{cl} \, dq_{cl}$ is the product of differentials of the "quasi-classical" coordinates and momenta.

Finally, the following comment is necessary concerning the group of problems which may be solved by means of the Gibbs distribution. We have spoken of the latter throughout as the statistical distribution for a subsystem, as in fact it is. It is very important to note, however, that this same distribution can quite succesfully be used also to determine the fundamental statistical properties of bodies forming closed systems, since such properties of a body as the values of the thermodynamic quantities or the probability distributions for the coordinates and velocities of its individual particles are clearly independent of whether we regard the body as a closed system or as being placed in an imaginary thermostat (§ 7). But in the latter case the body becomes a "subsystem" and the Gibbs distribution is immediately applicable to it. The difference between bodies forming closed and non-closed systems when the Gibbs distribution is used appears essentially only in the treatment of the fairly unimportant problem of fluctuations in the total energy of the body. The Gibbs distribution gives for the mean fluctuation of this quantity a non-

---

[†] To avoid misunderstanding, let us mention once more that the w_n (or ϱ) are monotonic functions of energy and need not have maxima for $E = \bar{E}$. It is the distribution function with respect to energy, obtained by multiplying w_n by $d\Gamma(E)/dE$, wich has a sharp maximum at $E = \bar{E}$.

$$W(E) \, \Delta E = 1 \quad \mu \ \text{canonical}$$

$$W(E)$$

$$W_n = \text{constant}$$
$$E = E_0 \pm \Delta E$$
$$= 0$$
$$E \neq E_0$$

canonical
$$W_n = \frac{e^{-E_n/T}}{z}$$

$$W_n$$

degen.

$$E_n$$

$$W(E)$$
$$= W_m \, \Delta \Gamma$$

zero value, which is meaningful for a body in a medium but is entirely spurious for a closed system, since the energy of such a body is by definition constant and does not fluctuate.

The possibility of applying the Gibbs distribution (in the manner described) to closed systems is also seen from the fact that this distribution hardly differs from the microcanonical distribution, while being very much more convenient for practical calculations. For the microcanonical distribution is, roughly speaking, equivalent to regarding as equally probable all microstates of the body which correspond to a given value of its energy. The canonical distribution is "spread" over a certain range of energy values, but the width of this range (of the order of the mean fluctuation of energy) is negligible for a macroscopic body.

§ 29. The Maxwellian distribution

The energy $E(p, q)$ in the Gibbs distribution formula of classical statistics can always be written as the sum of two parts: the kinetic energy and the potential energy. The first of these is a quadratic function of the momenta of the atoms,[†] and the second is a function of their coordinates, the form of which depends on the law of interaction between the particles within the body (and on the external field, if any). If the kinetic and potential energies are denoted by $K(p)$ and $U(q)$ respectively, then $E(p, q) = K(p) + U(q)$, and the probability $dw = \varrho(p, q)\, dp\, dq$ becomes

$$dw = Ae^{-U(q)/T}e^{-K(p)/T}\, dp\, dq,$$

i.e. is the product of two factors, one of which depends only on the coordinates and the other only on the momenta. This means that the probabilities for momenta and coordinates are independent, in the sense that any particular values of the momenta do not influence the probabilities of the various values of the coordinates, and vice versa. Thus the probability of the various values of the momenta can be written

$$dw_p = ae^{-K(p)/T}\, dp, \tag{29.1}$$

and the probability distribution for the coordinates is

$$dw_q = be^{-U(q)/T}\, dq. \tag{29.2}$$

Since the sum of the probabilities of all possible values of the momenta must be unity (and the same applies to the coordinates), each of the probabilities dw_p and dw_q must be normalised, i.e. their integrals over all possible

---

† It is assumed that Cartesian coordinates are used.

values of the momenta and coordinates respectively for the body concerned must be equal to unity. From these conditions we can determine the constants a and b in (29.1) and (29.2).

Let us consider the probability distribution for the momenta, and once again emphasise the very important fact that in classical statistics this distribution does not depend on the nature of the interaction of particles within the system or on the nature of the external field, and so can be expressed in a form applicable to all bodies.[†]

The kinetic energy of the whole body is equal to the sum of the kinetic energies of each of the atoms composing it, and the probability again falls into a product of factors, each depending on the momenta of only one atom. This again shows that the momentum probabilities of different atoms are independent, i.e. the momentum of one does not affect the probabilities of various momenta of any other. We can therefore write the probability distribution for the momenta of each atom separately.

For an atom of mass m the kinetic energy is $(p_x^2 + p_y^2 + p_z^2)/2m$, where p_x, p_y, p_z are the Cartesian components of its momentum, and the probability distribution is

$$dw_p = ae^{-(p_x^2 + p_y^2 + p_z^2)/2mT} \, dp_x \, dp_y \, dp_z.$$

The constant a is given by the normalisation condition. The integrations over p_x, p_y and p_z are separate; by means of the formula

$$\int_{-\infty}^{\infty} e^{-\alpha x^2} \, dx = \sqrt{(\pi/\alpha)}$$

we find $a = (2\pi mT)^{-3/2}$, and the momentum probability distribution takes the final form

$$dw_p = \frac{1}{(2\pi mT)^{3/2}} e^{-(p_x^2 + p_y^2 + p_z^2)/2mT} \, dp_x \, dp_y \, dp_z. \tag{29.3}$$

Changing from momenta to velocities ($\mathbf{p} = m\mathbf{v}$), we can write the corresponding velocity distribution as

$$dw_v = \left(\frac{m}{2\pi T}\right)^{3/2} e^{-m(v_x^2 + v_y^2 + v_z^2)/2T} \, dv_x \, dv_y \, dv_z. \tag{29.4}$$

This is the *Maxwellian distribution* (J. C. Maxwell, 1860). It again consists of a product of three independent factors

$$dw_{v_z} = \sqrt{\frac{m}{2\pi T}} e^{-mv_z^2/2T} \, dv_x, \ \ldots, \tag{29.5}$$

† In quantum statistics this statement is not true in general.

$$= \frac{4\pi}{(2\pi m T)^{3/2}} \, p^2 \, e^{-p^2/2mT} \, dp$$

$$\boxed{d^3p = 4\pi p^2 \, dp}$$

each of which gives the probability distribution for a single velocity component.

If the body consists of molecules (e.g. a polyatomic gas), then together with the Maxwellian distribution for the individual atoms there is a similar distribution for the translational motion of each molecule as a whole: from the kinetic energy of the molecule we can separate a term which gives the energy of the translational motion, and so the required distribution separates in the form (29.4), where m must now be taken as the total mass of the molecule, and v_x, v_y, v_z as the velocity components of its centre of mass. It should be emphasised that the Maxwellian distribution for the translational motion of molecules can be valid quite independently of the nature of the motion of the atoms within the molecule (and the rotation of the molecule), and in particular when a quantised description of the latter is necessary.[†]

The expression (29.4) is written in terms of Cartesian coordinates in "velocity space". If we change from Cartesian to spherical polar coordinates, the result is

$$dw_{\mathbf{v}} = \left(\frac{m}{2\pi T}\right)^{3/2} e^{-mv^2/2T}v^2 \sin\theta \, d\theta \, d\phi \, dv, \tag{29.6}$$

where v is the absolute magnitude of the velocity, and θ and ϕ the polar angle and azimuthal angle which determine the direction of the velocity. Integration with respect to angle gives the probability distribution for the absolute magnitude of the velocity:

$$dw_v = 4\pi \left(\frac{m}{2\pi T}\right)^{3/2} e^{-mv^2/2T}v^2 \, dv. \tag{29.7}$$

It is sometimes convenient to use cylindrical coordinates in velocity space. Then

$$dw_{\mathbf{v}} = \left(\frac{m}{2\pi T}\right)^{3/2} e^{-m(v_z^2+v_r^2)/2T}v_r \, dv_r \, dv_z \, d\phi, \tag{29.8}$$

where v_z is the velocity component along the z-axis, v_r the component perpendicular to that axis, and ϕ the angle which gives the direction of this component.

Let us calculate the mean kinetic energy of an atom. According to the definition of the mean, and using (29.5), we find for any Cartesian velocity com-

---

† The Maxwellian distribution clearly applies also to the Brownian motion of particles suspended in a liquid.

ponent[†]

$$\overline{v_x^2} = \sqrt{\frac{m}{2\pi T}} \int_{-\infty}^{\infty} v_x^2 e^{-mv_x^2/2T}\, dv_x$$

$$= T/m. \tag{29.9}$$

The mean value of the kinetic energy of the atom is therefore $3T/2$. We can thus say that the mean kinetic energy of all the particles in the body in classical statistics is always $3NT/2$, where N is the total number of particles.

PROBLEMS

PROBLEM. 1. Find the mean value of the nth power of the absolute magnitude of the velocity.

SOLUTION. Using (29.7), we find

$$\overline{v^n} = 4\pi \left(\frac{m}{2\pi T}\right)^{3/2} \int_0^{\infty} e^{-mv^2/2T} v^{n+2}\, dv$$

$$= \frac{2}{\sqrt{\pi}} \left(\frac{2T}{m}\right)^{n/2} \Gamma\left(\frac{n+3}{2}\right).$$

[†] For reference we shall give the values of the integrals of the form

$$I_n = \int_0^{\infty} e^{-\alpha x^2} x^n\, dx,$$

which often occur in applications of the Maxwellian distribution. The substitution $\alpha x^2 = y$ gives

$$I_n = \tfrac{1}{2}\alpha^{-(n+1)/2} \int_0^{\infty} e^{-y} y^{(n-1)/2}\, dy$$

$$= \tfrac{1}{2}\alpha^{-(n+1)/2} \Gamma(\tfrac{1}{2}n + \tfrac{1}{2}),$$

where $\Gamma(x)$ is the gamma function. In particular, if $n = 2r$ with $r > 0$, then

$$I_{2r} = \frac{(2r-1)!!}{2^{r+1}} \sqrt{\frac{\pi}{\alpha^{2r+1}}},$$

where $(2r-1)!! = 1 \cdot 3 \cdot 5 \cdot \ldots (2r-1)$. If $r = 0$, then

$$I_0 = \tfrac{1}{2}\sqrt{(\pi/\alpha)}.$$

If $n = 2r+1$, then

$$I_{2r+1} = r!/2\alpha^{r+1}.$$

The same integral from $-\infty$ to $+\infty$ is zero if $n = 2r+1$ and twice the integral from 0 to ∞ if $n = 2r$.

In particular, if n is even ($= 2r$), then

$$\overline{v^{2r}} = (T/m)^r (2r+1)!!;$$

if $n = 2r+1$, then

$$\overline{v^{2r+1}} = \frac{2}{\sqrt{\pi}} \left(\frac{2T}{m}\right)^{(2r+1)/2} (r+1)!.$$

PROBLEM 2. Find the mean square fluctuation of the velocity.

SOLUTION. The result of Problem 1 with $n = 1$ and $n = 2$ gives

$$\langle(\varDelta v)^2\rangle = \overline{v^2} - \overline{v}^2 = (T/m)(3 - 8/\pi).$$

PROBLEM 3. Find the mean energy, the mean square energy, and the mean square fluctuation of the kinetic energy of an atom.

SOLUTION. From the results of Problem 1 we find

$$\bar{\varepsilon} = \tfrac{1}{2} m\overline{v^2} = 3T/2,$$

$$\overline{\varepsilon^2} = 15T^2/4,$$

$$\langle(\varDelta\varepsilon)^2\rangle = 3T^2/2.$$

PROBLEM 4. Find the probability distribution for the kinetic energy of an atom.

SOLUTION.

$$dw_\varepsilon = \frac{2}{\sqrt{(\pi T^3)}} e^{-\varepsilon/T} \sqrt{\varepsilon}\, d\varepsilon.$$

PROBLEM 5. Find the probability distribution for the angular velocities of rotation of molecules.

SOLUTION. Just as for translational motion, we can write the probability distribution for the rotation of each molecule separately (in classical statistics). The kinetic energy of rotation of a molecule regarded as a rigid body (which is permissible, owing to the smallness of the atomic vibrations within the molecule) is

$$\varepsilon_{rot} = \frac{1}{2}(I_1\Omega_1^2 + I_2\Omega_2^2 + I_3\Omega_3^2) = \frac{1}{2}\left(\frac{M_1^2}{I_1} + \frac{M_2^2}{I_2} + \frac{M_3^2}{I_3}\right),$$

where I_1, I_2, I_3 are the principal moments of inertia, $\Omega_1, \Omega_2, \Omega_3$ are the components of the angular velocity along the principal axes of inertia, and $M_1 = I_1\Omega_1$, $M_2 = I_2\Omega_2$, $M_3 = I_3\Omega_3$ are the components of the angular momentum, which act as generalised momenta with respect to the velocities Ω_1, Ω_2, Ω_3. The normalised probability distribution for the angular-momentum components is

$$dw_M = (2\pi T)^{-3/2}(I_1 I_2 I_3)^{-1/2} \exp\left[-\frac{1}{2T}\left(\frac{M_1^2}{I_1} + \frac{M_2^2}{I_2} + \frac{M_3^2}{I_3}\right)\right] dM_1\, dM_2\, dM_3$$

and for the angular velocity

$$dw_\Omega = (2\pi T)^{-3/2} (I_1 I_2 I_3)^{1/2} \exp\left[-\frac{1}{2T}(I_1\Omega_1^2 + I_2\Omega_2^2 + I_3\Omega_3^2)\right] d\Omega_1\, d\Omega_2\, d\Omega_3.$$

PROBLEM 6. Find the mean squares of the absolute magnitudes of the angular velocity and angular momentum of a molecule.

SOLUTION. The above distributions give

$$\overline{\Omega^2} = T\left(\frac{1}{I_1}+\frac{1}{I_2}+\frac{1}{I_3}\right),$$

$$\overline{M^2} = T(I_1+I_2+I_3).$$

§ 30. The probability distribution for an oscillator

Let us consider a body whose atoms are executing small oscillations about some equilibrium positions. They may be atoms in a crystal or in a gas molecule; in the latter case the motion of the molecule as a whole does not affect the oscillations of the atoms within it and so does not influence the results.

As we know from mechanics, the Hamiltonian (the energy) of a system consisting of an arbitrary number of particles executing small oscillations can be written as a sum:

$$E(p, q) = \tfrac{1}{2} \sum_{\alpha} (p_\alpha^2 + \omega_\alpha^2 q_\alpha^2),$$

where q_α are what are called the *normal coordinates* of the oscillations (equal to zero at points of equilibrium), $p_\alpha = \dot{q}_\alpha$ are the corresponding generalised momenta, and ω_α are the oscillation frequencies. In other words, $E(p, q)$ is a sum of independent terms, each corresponding to a separate normal oscillation (or, as we say, to an *oscillator*). In quantum mechanics the same is true of the Hamiltonian operator of the system, so that each oscillator is independently quantised and the energy levels of the system are given by the sums

$$\sum_{\alpha} \hbar\omega_\alpha(n_\alpha + \tfrac{1}{2}),$$

the n_α being integers.

As a result of these facts the Gibbs distribution for the whole system is a product of independent factors, each giving the statistical distribution for a separate oscillator. In consequence we shall consider a single oscillator in what follows.

Let us determine the probability distribution for the coordinate q of an oscillator;[†] the suffix α which gives the number of the oscillator will be omitted henceforward. In classical statistics the solution to this problem would be very simple: since the potential energy of the oscillator is $\frac{1}{2}\omega^2 q^2$, the probability distribution is

$$dw_q = Ae^{-\omega^2 q^2/2T} dq,$$

† The normal coordinate has the dimensions $cm \cdot g^{1/2}$.

or, determining A from the normalisation condition,

$$dw_q = \frac{\omega}{\sqrt{(2\pi T)}} e^{-\omega^2 q^2/2T} \, dq; \qquad (30.1)$$

the integration with respect to q may be taken from $-\infty$ to $+\infty$, since the integral is rapidly convergent.

Let us now consider the solution of this problem in the quantum case. Let $\psi_n(q)$ be the wave functions of the stationary states of the oscillator, corresponding to the energy levels

$$\varepsilon_n = \hbar\omega(n+\tfrac{1}{2}).$$

If the oscillator is in the nth state, the quantum probability distribution for its coordinate is given by ψ_n^2 (in the present case the functions ψ_n are real, and so we write simply ψ_n^2 instead of the squared modulus $|\psi_n|^2$). The required statistical probability distribution is obtained by multiplying ψ_n^2 by the probability w_n of finding the oscillator in the nth state, and then summing over all possible states.

According to the Gibbs distribution,

$$w_n = a e^{-\varepsilon_n/T},$$

where a is a constant. Thus we have the formula

$$dw_q = a \, dq \sum_{n=0}^{\infty} e^{-\varepsilon_n/T} \psi_n^2, \qquad (30.2)$$

which is, of course, entirely in agreement with the general formula (5.8).

To calculate the sum, we can proceed as follows. With the notation $dw_q = \varrho_q \, dq$, we form the derivative

$$\frac{d\varrho_q}{dq} = 2a \sum_{n=0}^{\infty} e^{-\varepsilon_n/T} \psi_n \frac{d\psi_n}{dq}.$$

Using the momentum operator $\hat{p} = i\hbar \, d/dq$ and the fact that the oscillator momentum has non-zero matrix elements (see *Quantum Mechanics*, § 23) only for transitions with $n \to n\pm1$, we can write

$$\frac{d\psi_n}{dq} = \frac{i}{\hbar}\hat{p}\psi_n$$

$$= \frac{i}{\hbar}(p_{n-1,\,n}\psi_{n-1} + p_{n+1,\,n}\psi_{n+1})$$

$$= \frac{\omega}{\hbar}(q_{n-1,\,n}\psi_{n-1} - q_{n+1,\,n}\psi_{n+1}).$$

Here we have used the relations

$$p_{n-1,n} = -i\omega q_{n-1,n}, \qquad p_{n+1,n} = i\omega q_{n+1,n}$$

between the momentum and coordinate matrix elements. Thus

$$\frac{d\varrho_q}{dq} = \frac{2a\omega}{\hbar}\left\{ \sum_{n=0}^{\infty} q_{n-1,n}\psi_n\psi_{n-1}e^{-\varepsilon_n/T} - \sum_{n=0}^{\infty} q_{n+1,n}\psi_n\psi_{n+1}e^{-\varepsilon_n/T} \right\}.$$

In the first sum we change the summation suffix from n to $n+1$ and use the relations

$$\varepsilon_{n+1} = \varepsilon_n + \hbar\omega, \qquad q_{n+1,n} = q_{n,n+1}, \qquad q_{-1,0} = 0,$$

obtaining

$$\frac{d\varrho_q}{dq} = -\frac{2a\omega}{\hbar}(1 - e^{-\hbar\omega/T}) \sum_{n=0}^{\infty} q_{n,n+1}\psi_n\psi_{n+1}e^{-\varepsilon_n/T}.$$

In an exactly similar manner we can prove that

$$q\varrho_q = a(1 + e^{-\hbar\omega/T}) \sum_{n=0}^{\infty} q_{n,n+1}\psi_n\psi_{n+1}e^{-\varepsilon_n/T}.$$

A comparison of the two equations gives

$$\frac{d\varrho_q}{dq} = -\left(\frac{2\omega}{\hbar} \tanh \frac{\hbar\omega}{2T}\right)q\varrho_q,$$

whence

$$\varrho_q = \text{constant} \times \exp\left\{-q^2 \frac{\omega}{\hbar} \tanh \frac{\hbar\omega}{2T}\right\}.$$

Determining the constant from the normalisation condition, we finally obtain the formula

$$dw_q = \left(\frac{\omega}{\pi\hbar} \tanh \frac{\hbar\omega}{2T}\right)^{1/2} \exp\left\{-q^2 \frac{\omega}{\hbar} \tanh \frac{\hbar\omega}{2T}\right\} dq \qquad (30.3)$$

(F. Bloch, 1932). Thus in the quantum case also the probabilities of various values of the coordinate of an oscillator are distributed according to a law of the form $\exp(-\alpha q^2)$, but the coefficient α differs from that in the classical case. In the limit $\hbar\omega \ll T$, where the quantisation is unimportant, formula (30.3) becomes (30.1), as we should expect.

In the opposite limiting case $\hbar\omega \gg T$, formula (30.3) becomes

$$dw_q = \sqrt{\frac{\omega}{\pi\hbar}} \exp(-q^2\omega/\hbar) dq,$$

i.e. the purely quantum probability distribution for the coordinate in the ground state of the oscillator.[†] This corresponds to the fact that when $T \ll \hbar\omega$ the oscillations are hardly excited at all.

The probability distribution for the momentum of the oscillator can be written by analogy with (30.3) without repeating the calculation. The reason is that the problem of quantisation of the oscillator is completely symmetrical as regards the coordinate and the momentum, and the oscillator wave functions in the p representation are the same as its ordinary coordinate wave functions (q being replaced by p/ω; see *Quantum Mechanics*, § 23, Problem 1. The required distribution is therefore

$$dw_p = \left(\frac{1}{\pi\hbar\omega} \tanh \frac{\hbar\omega}{2T} \right)^{1/2} \exp \left\{ -\frac{p^2}{\hbar\omega} \tanh \frac{\hbar\omega}{2T} \right\} dp. \qquad (30.4)$$

In the limit of classical mechanics ($\hbar\omega \ll T$) this becomes the usual Maxwellian distribution:

$$dw_p = (2\pi T)^{-1/2} e^{-p^2/2T} \, dp. \qquad (30.5)$$

PROBLEM

Determine the coordinate density matrix for a harmonic oscillator.

SOLUTION. The coordinate density matrix for an oscillator in statistical equilibrium is

$$\varrho(q, q') = a \sum_{n=0}^{\infty} e^{-\varepsilon_n/T} \psi_n(q') \, \psi_n(q);$$

see the footnote following (5.4). We put $q = r+s$, $q' = r-s$, and calculate the derivative $(\partial\varrho/\partial s)_r$. As in the similar calculation above, we find

$$\frac{\partial\varrho}{\partial s} = \frac{\partial\varrho}{\partial q} - \frac{\partial\varrho}{\partial q'}$$

$$= -\frac{a\omega}{\hbar}(1+e^{-\hbar\omega/T}) \sum_{n=0}^{\infty} q_{n,\,n+1}[\psi_{n+1}(q) \, \psi_n(q') - \psi_n(q) \, \psi_{n+1}(q')].$$

On calculating $s\varrho = \frac{1}{2}(q-q')\varrho$ in the same way and comparing it with the derivative, we obtain

$$(\partial\varrho/\partial s)_r = -s\varrho(2\omega/\hbar) \coth (\hbar\omega/2T),$$

whence

$$\varrho(q, q') = A(r) \exp \left(-s^2 \frac{\omega}{\hbar} \coth \frac{\hbar\omega}{2T} \right).$$

[†] This is the squared modulus of the wave function of the ground state of the oscillator.

The function $A(r)$ is determined by the condition that for $s = 0$, i.e. $q = q' = r$, the "diagonal elements" $\varrho(q, q)$ of the density matrix agree with (30.3). The final result is

$$\varrho(q, q') = \left(\frac{\omega}{\pi\hbar} \tanh \frac{\hbar\omega}{2T}\right)^{1/2} \exp\left\{-\frac{\omega(q+q')^2}{4\hbar} \tanh \frac{\hbar\omega}{2T} - \frac{\omega(q-q')^2}{4\hbar} \coth \frac{\hbar\omega}{2T}\right\}.$$

§ 31. The free energy in the Gibbs distribution

According to formula (7.9) the entropy of a body can be calculated as the mean logarithm of its distribution function:

$$S = -\langle \log w_n \rangle.$$

Substituting the Gibbs distribution (28.3) gives

$$S = -\log A + \bar{E}/T,$$

whence $\log A = (\bar{E} - TS)/T$. But the mean energy $\bar{E}$ is just what is meant by the term "energy" in thermodynamics; hence $\bar{E} - TS = F$ and $\log A = F/T$, i.e. the normalisation constant of the distribution is directly related to the free energy of the body.

Thus the Gibbs distribution may be written in the form

$$w_n = e^{(F-E_n)/T}, \qquad (31.1)$$

and this is the form most frequently used. The same method gives in the classical case, using (7.12), the expression

$$\varrho = (2\pi\hbar)^{-s} e^{[F-E(p,q)]/T}. \qquad (31.2)$$

The normalisation condition for the distribution (31.1) is

$$\sum_n w_n = e^{F/T} \sum_n e^{-E_n/T} = 1,$$

or

$$F = -T \log \sum_n e^{-E_n/T}. \qquad = -T \ln z \qquad (31.3)$$

This formula is fundamental in thermodynamic applications of the Gibbs distribution. It affords, in principle, the possibility of calculating the thermodynamic functions for any body whose energy spectrum is known.

The sum in the logarithm in (31.3) is usually called the *partition function* (or *sum over states*). It is just the trace of the operator $\exp(-\hat{H}/T)$, where $\hat{H}$ is the Hamiltonian of the body:[†]

$$Z \equiv \sum_n e^{-E_n/T} = \text{tr} \exp(-\hat{H}/T). \qquad (31.4)$$

[†] In accordance with the general rules, $\exp(-\hat{H}/T)$ denotes an operator whose eigenfunctions are the same as those of the operator $\hat{H}$ and whose eigenvalues are $e^{-E_n/T}$.

$$S = \ln \delta\Gamma = -\ln w(E)$$

$$= -\ln\left(e^{-E/T}/z\right) = \frac{E}{T} + \ln z$$

$$\bar{E} - TS = -T \ln z = F$$

This notation has the advantage that any complete set of wave functions may be used in order to calculate the trace.

A similar formula in classical statistics is obtained from the normalisation condition for the distribution (31.2). First of all, however, we must take account of the following fact, which was unimportant so long as we were discussing the distribution function as such and not relating the normalisation coefficient to a particular quantitative property of the body, viz. its free energy. If, for example, two identical atoms change places, then afterwards the microstate of the body is represented by a different phase point, obtained from the original one by replacing the coordinates and momenta of one atom by those of the other. On the other hand, since the interchanged atoms are identical, the two states of the body are physically identical. Thus a number of points in phase space correspond to one physical microstate of the body. In integrating the distribution (31.2), however, each state must of course be taken only once.[†] In other words, we must integrate only over those regions of phase space which correspond to physically different states of the body. This will be denoted by a prime to the integral sign.

Thus we have the formula

$$F = -T \log \int{}' e^{-E(p,q)/T} \, d\Gamma; \tag{31.5}$$

here and in all similar cases below, $d\Gamma$ denotes the volume element in phase space divided by $(2\pi\hbar)^s$:

$$d\Gamma = dp \, dq/(2\pi\hbar)^s. \tag{31.6}$$

Thus the partition function in the quantum formula (31.3) becomes an *integral over states*. As already mentioned in § 29, the classical energy $E(p, q)$ can always be written as the sum of the kinetic energy $K(p)$ and the potential energy $U(q)$. The kinetic energy is a quadratic function of the momenta, and the integration with respect to the latter can be effected in a general form. The problem of calculating the partition function therefore actually reduces to that of integrating the function $e^{-U(q)/T}$ with respect to the coordinates.

In the practical calculation of the partition function it is usually convenient to extend the region of integration and include an appropriate correction

[†] This becomes particularly evident if we consider the classical partition function (integral over states) as the limit of the quantum partition function. In the latter the summation is over all the different quantum states, and there is no problem (remembering that, because of the principle of symmetry of wave functions in quantum mechanics, the quantum state is unaffected by interchanges of identical particles).

From the purely classical viewpoint the need for this interpretation of the statistical integration arises because otherwise the statistical weight would no longer be multiplicative, and so the entropy and the other thermodynamic quantities would no longer be additive.

factor. For example, let us consider a gas of N identical atoms. Then we can integrate with respect to the coordinates of each atom separately, extending the integration over the whole volume occupied by the gas; but the result must be divided by the number of possible permutations of N atoms, which is $N!$. In other words, the integral $\int'$ can be replaced by the integral over all phase space, divided by $N!$:

$$\int' \ldots d\Gamma = \frac{1}{N!} \int \ldots d\Gamma. \tag{31.7}$$

Similarly, it is convenient to extend the region of integration for a gas consisting of N identical molecules: the integration with respect to the coordinates of each molecule as a whole (i.e. the coordinates of its centre of mass) is carried out independently over the whole volume, whilst that with respect to the coordinates of the atoms within the molecule is carried out over the "volume" belonging to each molecule (i.e. over a small region in which there is an appreciable probability of finding the atoms forming the molecule). Then the integral must again be divided by $N!$.

PROBLEMS

PROBLEM 1. The potential energy of the interaction between the particles in a body is a homogeneous function of degree n in their coordinates. Using similarity arguments, determine the form of the free energy of such a body in classical statistics.

SOLUTION. In the partition function

$$Z = \int' e^{-[K(p)+U(q)]/T}\, d\Gamma,$$

we replace each q by λq and each p by $\lambda^{n/2}p$, where λ is an arbitrary constant. If at the same time we replace T by $\lambda^n T$, the integrand is unchanged, but the limits of integration with respect to the coordinates are altered: the linear size of the region of integration is multiplied by $1/\lambda$, and so the volume is multiplied by $1/\lambda^3$. In order to restore the limits of integration, we must therefore at the same time replace V by $\lambda^3 V$. The result of these changes is to multiply the integral by $\lambda^{3N(1+n/2)}$ because of the change of variables in $d\Gamma$ ($s = 3N$ coordinates and the same number of momenta, N being the number of particles in the body). Thus we conclude that the substitutions $V \to \lambda^3 V$, $T \to \lambda^n T$ give

$$Z \to \lambda^{3N(1+n/2)} Z.$$

The most general form of function $Z(V, T)$ having this property is

$$Z = T^{3N(1/2+1/n)} f(V T^{-3/n}),$$

where f is an arbitrary function of one variable.

Hence we find for the free energy an expression of the form

$$F = -3(\tfrac{1}{2}+1/n)NT\log T + NT\phi(VT^{-3/n}/N),\qquad(1)$$

which involves only one unknown function of one variable; the number N is includ-ed in the second term in (1) so that F shall have the necessary property of additiv-ity.

PROBLEM 2. Derive the virial theorem for a macroscopic body for which the potential energy of interaction of the particles is a homogeneous function of degree n in their coordinates.

SOLUTION. Following the derivation of the virial theorem in mechanics (see *Mechanics*, § 10), we calculate the time derivative of the sum $\Sigma \mathbf{r}\cdot\mathbf{p}$, where $\mathbf{r}$ and $\mathbf{p}$ are the radius vectors and momenta of the particles in the body. Since $\dot{\mathbf{r}} = \partial K(p)/\partial \mathbf{p}$ and $K(p)$ is a homogeneous function of degree two in the momenta, we have

$$\frac{d}{dt}\Sigma \mathbf{r}\cdot\mathbf{p} = \Sigma \mathbf{p}\cdot\frac{\partial K(p)}{\partial \mathbf{p}}+\Sigma \mathbf{r}\cdot\dot{\mathbf{p}} = 2K(p)+\Sigma \mathbf{r}\cdot\dot{\mathbf{p}}.$$

The particles in the body execute a motion in a finite region of space with velocities which do not become infinite. The quantity $\Sigma \mathbf{r}\cdot\mathbf{p}$ is therefore bounded and the mean value of its time derivative is zero, so that

$$2K+\langle\Sigma \mathbf{r}\cdot\dot{\mathbf{p}}\rangle = 0,$$

where $K = \overline{K(p)}$. The derivatives $\dot{\mathbf{p}}$ are determined by the forces acting on the particles in the body. In summing over all particles we must take into account not only the forces of interaction between the particles but also the forces exerted on the surface of the body by surrounding bodies:

$$\langle\Sigma \mathbf{r}\cdot\dot{\mathbf{p}}\rangle = -\left\langle \Sigma \mathbf{r}\cdot\frac{\partial U(q)}{\partial \mathbf{r}}\right\rangle-P\oint \mathbf{r}\cdot d\mathbf{f} = -nU-3PV;$$

the surface integral is transformed to a volume integral and we use the fact that div $\mathbf{r} = 3$. Thus we have $2K-nU-3PV = 0$ or, in terms of the total energy $E = U+K$,

$$(n+2)K = nE+3PV.\qquad(2)$$

This is the required theorem. It is valid in both classical and quantum theory. In the classical case, the mean kinetic energy $K = 3NT/2$, and (2) gives

$$E+(3/n)PV = 3(\tfrac{1}{2}+1/n)NT.\qquad(3)$$

This formula could also be derived from the expression (1) for the free energy (Problem 1).

When the particles interact by Coulomb's law $(n=-1)$, we have from (2)

$$K = -E+3PV.$$

This is the limiting case of the relativistic relation

$$E-3PV = \Sigma mc^2 \sqrt{(1-v^2/c^2)},$$

in which the energy E includes the rest energy of the particles in the body; see *Fields*, § 35.

§ 32. Thermodynamic perturbation theory

In the actual calculation of thermodynamic quantities there occur cases where the energy $E(p, q)$ of a body contains relatively small terms which may be neglected to a first approximation. These may be, for instance, the potential energy of the particles of the body in an external field. The conditions under which such terms may be regarded as small are discussed below.

In these cases a kind of "perturbation theory" may be employed for the calculation of the thermodynamic quantities (R. E. Peierls, 1932). We shall first show how this is to be done when the classical Gibbs distribution is applicable.

We write the energy $E(p, q)$ in the form

$$E(p, q) = E_0(p, q) + V(p, q), \qquad (32.1)$$

where V represents the small terms. To calculate the free energy of the body, we put

$$e^{-F/T} = \int' e^{-[E_0(p, q) + V(p, q)]/T} \, d\Gamma$$

$$\cong \int' e^{-E_0/T} \left(1 - \frac{V}{T} + \frac{V^2}{2T^2}\right) d\Gamma; \qquad (32.2)$$

in the expansion in powers of V we shall always omit terms above the second order, in order to calculate the corrections only to the first and second orders of approximation. Taking logarithms and again expanding in series, we have to the same accuracy

$$F = F_0 + \int' \left(V - \frac{V^2}{2T}\right) e^{[F_0 - E_0(p, q)]/T} \, d\Gamma$$

$$+ \frac{1}{2T} \left[\int' V e^{[F_0 - E_0(p, q)]/T} \, d\Gamma\right]^2,$$

where F_0 denotes the "unperturbed" free energy, calculated for $V = 0$.

The resulting integrals are the mean values of the corresponding quantities over the "unperturbed" Gibbs distribution. Taking the averaging in this sense and noticing that $\overline{V^2} - \overline{V}^2 = \langle (V - \overline{V})^2 \rangle$, we have finally

$$F = F_0 + \overline{V} - \frac{1}{2T} \langle (V - \overline{V})^2 \rangle. \qquad (32.3)$$

Thus the first-order correction to the free energy is just the mean value of the energy perturbation V. The second-order correction is always negative, and is determined by the mean square of the deviation of V from its mean value. In particular, if the mean value $\overline{V}$ is zero, the perturbation reduces the free energy.

A comparison of the terms of the second and first orders in (32.3) enables us to ascertain the condition for this perturbation method to be applicable. Here it must be remembered that both the mean value $\bar{V}$ and the mean square $\langle (V-\bar{V})^2 \rangle$ are roughly proportional to the number of particles; cf. the discussion in § 2 concerning r.m.s. fluctuations of the thermodynamic quantities for macroscopic bodies. We can therefore formulate the desired condition by requiring that the perturbation energy per particle should be small in comparison with T.[†]

Let us now carry out the corresponding calculations for the quantum case. Instead of (32.1) we must now use the analogous expression for the Hamiltonian operator:

$$\hat{H} = \hat{H}_0 + \hat{V}.$$

According to the quantum perturbation theory (see *Quantum Mechanics*, § 38), the energy levels of the perturbed system are given, correct to the second-order terms, by

$$E_n = E_n^{(0)} + V_{nn} + {\sum_m}' \frac{|V_{nm}|^2}{E_n^{(0)} - E_m^{(0)}}, \qquad (32.4)$$

where the $E_n^{(0)}$ are the unperturbed energy levels (assumed non-degenerate); the prime to the sum signifies that the term with $m = n$ must be omitted.

This expression is to be substituted in the formula

$$e^{-F/T} = \sum_n e^{-E_n/T}$$

and expanded in the same way as above. We thus easily obtain

$$F = F_0 + \sum_n V_{nn} w_n + {\sum_n}{\sum_m}' \frac{|V_{nm}|^2 w_n}{E_n^{(0)} - E_m^{(0)}}$$
$$- \frac{1}{2T} \sum_n V_{nn}^2 w_n + \frac{1}{2T} \left(\sum_n V_{nn} w_n \right)^2, \qquad (32.5)$$

where $w_n = \exp[(F_0 - E_n^{(0)})/T]$ is the unperturbed Gibbs distribution.

The diagonal matrix element V_{nn} is just the mean value of the perturbation energy V in the given (nth) quantum state. The sum

$$\sum_n V_{nn} w_n \equiv \bar{V}_{nn}$$

is therefore the value of V averaged both over the quantum state of the body and over the (unperturbed) statistical distribution with respect to the vari-

[†] In expanding the integrand in (32.2) we have, strictly speaking, expanded in terms of a quantity V/T, which is proportional to the number of particles and is therefore certainly not small, but the further expansion of the logarithm causes the large terms to cancel, and so a series in powers of a small quantity is obtained.

ous quantum states. This value gives the correction to the free energy in the first-order approximation, formally the same as the classical result above.

Formula (32.5) may be rewritten as

$$F = F_0 + V_{nn} - \frac{1}{2} \sum_n \sum_m{}' \frac{|V_{nm}|^2 (w_m - w_n)}{E_n^{(0)} - E_m^{(0)}} - \frac{1}{2T} \langle (V_{nn} - \bar{V}_{nn})^2 \rangle. \quad (32.6)$$

All the second-order terms in this expression are negative, since $w_m - w_n$ has the same sign as $E_n^{(0)} - E_m^{(0)}$. Thus the correction to the free energy in the second-order approximation is negative in the quantum case also.

As in the classical case, the condition for this method to be applicable is that the perturbation energy per particle should be small compared with T. On the other hand, the condition for the applicability of the ordinary quantum perturbation theory (leading to the expression (32.4) for E_n) is, as we know, that the matrix elements of the perturbation should be small compared with the separations of the corresponding energy levels; roughly speaking, the perturbation energy must be small compared with the separations of the energy levels between which allowed transitions can take place.[†]

These two conditions are not the same, since the temperature is unrelated to the energy levels of the body. It may happen that the perturbation energy is small compared with T, but is not small, or indeed is even large, compared with the significant separations between energy levels. In such cases the "perturbation theory" for thermodynamic quantities, i.e. formula (32.6), will be applicable while the perturbation theory for the energy levels themselves, i.e. formula (32.4), is not; that is, the limits of convergence of the expansion represented by formula (32.6) may be wider than those of (32.4), from which the former expansion has been derived.

The converse case is, of course, also possible (at sufficiently low temperatures).

Formula (32.6) is considerably simplified if not only the perturbation energy but also the differences between energy levels are small in comparison with T. Expanding the difference $w_m - w_n$ in (32.6) in powers of $(E_n^{(0)} - E_m^{(0)})/T$, we find in this case

$$F = F_0 + V_{nn} - \frac{1}{2T} \left\{ \sum_m{}' \langle |V_{nm}|^2 \rangle + \langle (V_{nn} - \bar{V}_{nn})^2 \rangle \right\}.$$

The rule of matrix multiplication gives

$$\sum_m{}' |V_{nm}|^2 + V_{nn}^2 = \sum_m |V_{nm}|^2 = \sum_m V_{nm} V_{mn} = (V^2)_{nn},$$

[†] These are in general the transitions in which the states of only a small number of particles in the body are changed.

and we obtain an expression which is formally exactly the same as formula (32.3). Thus in this case the quantum formula is in formal agreement with the classical formula.[†]

§ 33. Expansion in powers of $\hbar$

Formula (31.5) is essentially the first and principal term in an expansion of the quantum formula (31.3) for the free energy in powers of $\hbar$ in the quasi-classical case. It is of considerable interest to derive the next non-vanishing term in this expansion (E. P. Wigner, G. E. Uhlenbeck and L. Gropper, 1932).

The problem of calculating the free energy amounts to that of calculating the partition function. For this purpose we use the fact that the latter is the trace of the operator $\exp(-\beta\hat{H})$ (see (31.4)), with the notation $\beta = 1/T$ in order to simplify the writing of the involved expressions. The trace of an operator may be calculated by means of any complete set of orthonormal wave functions. For these it is convenient to use the wave functions of free motion of a system of N non-interacting particles in a large but finite volume V. These functions are

$$\psi_p = \frac{1}{\sqrt{V^N}} \exp\left[(i/\hbar)\sum_i p_i q_i\right], \tag{33.1}$$

where the q_i are the Cartesian coordinates of the particles and the p_i the corresponding momenta, labelled by the suffix i, which takes the values $1, 2, \ldots, s$, where $s = 3N$ is the number of degrees of freedom of the system of N particles.

The subsequent calculations apply equally to systems containing identical particles (atoms) and to those where the particles are different. In order to allow in a general manner for a possible difference between the particles, we shall add to the particle mass a suffix indicating the degree of freedom: m_i. Of course the three m_i corresponding to any one particle are always equal.

The existence of identical particles in a body means that, in the quantum theory, exchange effects must be taken into account. This means, first of all, that the wave functions (33.1) must be made symmetrical or antisymmetrical in the particle coordinates, depending on the statistics obeyed by the particles. It is found, however, that this effect leads only to exponentially small terms in the free energy, and so is of no interest. Secondly, the identity of particles in quantum mechanics affects the manner in which the summation

---

[†] The more powerful methods of the "diagram technique", allowing the whole perturbation theory series to be considered for thermodynamic quantities, will be described in Part 2.

over different values of the particle momenta must be carried out. We shall meet this later, for example in calculating partition functions for an ideal quantum gas. The effect produces a term of the third order in $\hbar$ in the free energy (as shown later) and so again does not affect the terms of order $\hbar^2$ which we shall calculate here. Thus the exchange effects can be ignored in the calculation.

In each of the wave functions (33.1) the momenta p_i have definite constant values. The possible values of each p_i form a dense discrete set (the distances between neighbouring values being inversely proportional to the linear dimensions of the volume occupied by the system). The summation of the matrix elements $\exp(-\beta\hat{H})_{pp}$ with respect to all possible values of the momenta may therefore be replaced by integration with respect to p ($dp = dp_1\, dp_2 \ldots dp_s$), bearing in mind that the number of quantum states "belonging" to the volume $V^N\, dp$ of phase space (all values of the coordinates of each particle in the volume V and values of the momenta in dp) is

$$V^N\, dp/(2\pi\hbar)^s.$$

We shall use the notation

$$I = \exp\left[-(i/\hbar)\sum_i p_i q_i\right] \exp(-\beta\hat{H}) \exp\left[(i/\hbar)\sum_i p_i q_i\right]. \qquad (33.2)$$

The required matrix elements are obtained by integrating with respect to all the coordinates:

$$\exp(-\beta\hat{H})_{pp} = \frac{1}{V^N}\int I\, dq. \qquad (33.3)$$

The partition function is then obtained by integration with respect to the momenta.

Altogether, therefore, we must integrate I over all phase space, or more precisely over those of its regions which correspond to physically different states of the body, as explained in § 31. This is again denoted by a prime to the integral sign:

$$Z \equiv \sum_n e^{-\beta E_n} = \int' I\, d\Gamma. \qquad (33.4)$$

Let us first calculate I by means of the following procedure. We take the derivative

$$\frac{\partial I}{\partial \beta} = -\exp\left[-(i/\hbar)\Sigma p_i q_i\right] \hat{H}\{\exp[(i/\hbar)\Sigma p_i q_i]I\},$$

the operator $\hat{H}$ acting on all factors to its right, and expand the right-hand side, using the explicit expression for the Hamiltonian of the body:

$$\hat{H} = \sum_i \frac{\hat{p}_i^2}{2m_i} + U = -\frac{1}{2}\hbar^2 \sum_i \frac{1}{m_i}\frac{\partial^2}{\partial q_i^2} + U, \tag{33.5}$$

where $U = U(q_1, q_2, \ldots, q_s)$ is the potential energy of interaction between all particles in the body. By means of (33.5) we obtain after a straightforward calculation the following equation for I:

$$\frac{\partial I}{\partial \beta} = -E(p, q)I + \sum_i \frac{\hbar^2}{2m_i}\left(\frac{2i}{\hbar}p_i\frac{\partial I}{\partial q_i} + \frac{\partial^2 I}{\partial q_i^2}\right),$$

where

$$E(p, q) = \sum_i \frac{p_i^2}{2m_i} + U \tag{33.6}$$

is the usual classical expression for the energy of the body.

This equation is to be solved with the obvious condition that $I = 1$ when $\beta = 0$. The substitution

$$I = e^{-\beta E(p, q)}\chi \tag{33.7}$$

gives

$$\frac{\partial \chi}{\partial \beta} = \sum_i \frac{\hbar^2}{2m_i}\left[-\frac{2i\beta p_i}{\hbar}\frac{\partial U}{\partial q_i}\chi + \frac{2ip_i}{\hbar}\frac{\partial \chi}{\partial q_i} - \beta\chi\frac{\partial^2 U}{\partial q_i^2} + \beta^2\chi\left(\frac{\partial U}{\partial q_i}\right)^2\right.$$
$$\left. -2\beta\frac{\partial \chi}{\partial q_i}\frac{\partial U}{\partial q_i} + \frac{\partial^2 \chi}{\partial q_i^2}\right] \tag{33.8}$$

with the boundary condition $\chi = 1$ for $\beta = 0$.

In order to obtain an expansion in powers of $\hbar$, we solve equation (33.8) by successive approximations, putting

$$\chi = 1 + \hbar\chi_1 + \hbar^2\chi_2 + \ldots, \tag{33.9}$$

with $\chi_1 = 0$, $\chi_2 = 0$, $\ldots$ for $\beta = 0$. Substituting this expansion in equation (33.8) and separating terms in different powers of $\hbar$, we obtain the equations

$$\frac{\partial \chi_1}{\partial \beta} = -i\beta \sum_i \frac{p_i}{m_i}\frac{\partial U}{\partial q_i},$$

$$\frac{\partial \chi_2}{\partial \beta} = \sum_i \frac{1}{2m_i}\left[-2i\beta p_i\frac{\partial U}{\partial q_i}\chi_1 + 2ip_i\frac{\partial \chi_1}{\partial q_i} - \beta\frac{\partial^2 U}{\partial q_i^2} + \beta^2\left(\frac{\partial U}{\partial q_i}\right)^2\right].$$

The first equation gives χ_1, and then the second equation gives χ_2. A simple calculation leads to the results

$$\chi_1 = -\frac{1}{2} i\beta^2 \sum_i \frac{p_i}{m_i} \frac{\partial U}{\partial q_i},$$

$$\chi_2 = -\frac{1}{8} \beta^4 \left(\sum_i \frac{p_i}{m_i} \frac{\partial U}{\partial q_i}\right)^2 + \frac{1}{6}\beta^3 \sum_i \sum_k \frac{p_i}{m_i} \frac{p_k}{m_k} \frac{\partial^2 U}{\partial q_i \partial q_k} \qquad (33.10)$$

$$+ \frac{1}{6}\beta^3 \sum_i \frac{1}{m_i} \left(\frac{\partial U}{\partial q_i}\right)^2 - \frac{1}{4}\beta^2 \sum_i \frac{1}{m_i} \frac{\partial^2 U}{\partial q_i^2}.$$

The required partition function (33.4) is

$$Z = \int' (1 + \hbar\chi_1 + \hbar^2\chi_2)e^{-\beta E(p, q)} \, d\Gamma. \qquad (33.11)$$

The term of the first order in $\hbar$ in this integral is easily seen to be zero, since the integrand $\chi_1 e^{-\beta E(p, q)}$ in that term is an odd function of the momenta ($E(p, q)$ being quadratic in the momenta and χ_1, by (33.10), linear), and so the result on integrating with respect to momenta is zero. Thus we can write (33.11) as

$$Z = (1 + \hbar^2\overline{\chi_2}) \int' e^{-\beta E(p, q)} \, d\Gamma,$$

where $\overline{\chi_2}$ is the value of χ_2 averaged over the classical Gibbs distribution:

$$\overline{\chi_2} = \frac{\int' \chi_2 e^{-\beta E(p, q)} \, d\Gamma}{\int' e^{-\beta E(p, q)} \, d\Gamma}.$$

Substituting this expression for the partition function in formula (31.3), we have for the free energy

$$F = F_{cl} - \frac{1}{\beta} \log (1 + \hbar^2\overline{\chi_2}),$$

or, to the same accuracy,

$$F = F_{cl} - \hbar^2\overline{\chi_2}/\beta. \qquad (33.12)$$

Here F_{cl} is the free energy in classical statistics (formula (31.5)).

Thus the next term after the classical expression in the expansion of the free energy is of the second order in $\hbar$. This is not accidental: in equation (33.8), solved here by the method of successive approximations, the quantum constant appears only as $i\hbar$, and so the resulting expansion is one in powers of $i\hbar$; but the free energy, being a real quantity, can contain only powers of $i\hbar$ which are real. Thus this expansion of the free energy (ignoring exchange effects) is an expansion in even powers of $\hbar$.

It remains to calculate the mean value $\overline{\chi_2}$. We have seen in § 29 that in classical statistics the probability distributions for the coordinates and momenta are independent. The averaging over momenta and over coordinates can therefore be made separately.

The mean value of the product of two different momenta is clearly zero. The mean value of the square p_i^2 is m_i/β. We can therefore write

$$\langle p_i p_k \rangle = (m_i/\beta)\delta_{ik},$$

where $\delta_{ik} = 1$ for $i = k$ and 0 for $i \neq k$. Having averaged with respect to momenta by means of this formula, we obtain

$$\overline{\chi_2} = \frac{\beta^3}{24} \sum_i \frac{1}{m_i} \overline{\left(\frac{\partial U}{\partial q_i}\right)^2} - \frac{\beta^2}{12} \sum_i \frac{1}{m_i} \overline{\frac{\partial^2 U}{\partial q_i^2}}. \tag{33.13}$$

The two terms here may be combined, since the mean values are related by the formula

$$\overline{\frac{\partial^2 U}{\partial q_i^2}} = \beta \overline{\left(\frac{\partial U}{\partial q_i}\right)^2}. \tag{33.14}$$

This is easily seen by noticing that

$$\int \frac{\partial^2 U}{\partial q_i^2} e^{-\beta U} \, dq_i = \frac{\partial U}{\partial q_i} e^{-\beta U} + \beta \int \left(\frac{\partial U}{\partial q_i}\right)^2 e^{-\beta U} \, dq_i.$$

The first term on the right-hand side gives only a surface effect, and since the body is macroscopic this effect may be neglected in comparison with the second term.

Substituting the resulting expression for $\overline{\chi_2}$ in formula (33.12), and replacing β by $1/T$, we find the following final expression for the free energy:

$$F = F_{cl} + \frac{\hbar^2}{24T^2} \sum_i \frac{1}{m_i} \overline{\left(\frac{\partial U}{\partial q_i}\right)^2}. \tag{33.15}$$

We see that the correction to the classical value is always positive, and is determined by the mean squares of the forces acting on the particles. This correction decreases with increasing particle mass and increasing temperature.

According to the above discussion, the next term in the expansion given here would be of the fourth order. This enables us to calculate quite independently the term of order $\hbar^3$ which occurs in the free energy because of the peculiarities of the summation over momenta resulting from the identity of particles in quantum mechanics. The term in question is formally the same as the correction term which appears in a similar calculation for an ideal gas, and is given by formula (56.14):

$$F^{(3)} = \pm \frac{\pi^{3/2}}{2g} \frac{N^2 \hbar^3}{VT^{1/2}m^{3/2}} \tag{33.16}$$

for a body consisting of N identical particles. The upper sign applies for Fermi statistics and the lower sign for Bose statistics; g is the total degree of degeneracy with respect to the directions of the electron and nuclear angular momenta.

From these formulae we can also obtain the correction terms in the probability distribution functions for the coordinates and momenta of the atoms of the body. According to the general results in § 5, the momentum probability distribution is given by the integral of I with respect to q (see (5.10)):

$$dw_p = \text{constant} \times dp \int I \, dq.$$

The term $\chi_1 e^{-\beta E(p,\,q)}$ in I contains a total derivative with respect to the coordinates, and the integral of it gives a surface effect which can be neglected. Thus we have

$$dw_p = \text{constant} \times \exp\left(-\beta \sum_i p_i^2/2m_i\right) dp \int (1 + \hbar^2 \chi_2) e^{-\beta U} \, dq.$$

The third and fourth terms in the expression (33.10) for χ_2 give a small constant (not involving the momenta) on integration, and this can be neglected in the same approximation. Taking out also the factor $\int e^{-\beta U} \, dq$ and including it in the constant coefficient, we have

$$dw_p = \text{constant} \times \exp\left(-\beta \sum_i p_i^2/2m_i\right)\left[1 - \hbar^2 \frac{\beta^4}{8} \sum_i \sum_k \frac{p_i p_k}{m_i m_k} \overline{\frac{\partial U}{\partial q_i} \frac{\partial U}{\partial q_k}}\right.$$

$$\left. + \hbar^2 \frac{\beta^3}{6} \sum_i \sum_k \frac{p_i p_k}{m_i m_k} \overline{\frac{\partial^2 U}{\partial q_i \partial q_k}}\right] dp.$$

The mean values which appear here are related by

$$\overline{\frac{\partial^2 U}{\partial q_i \partial q_k}} = \beta \overline{\frac{\partial U}{\partial q_i} \frac{\partial U}{\partial q_k}},$$

similarly to (33.14). Hence

$$dw_p = \text{constant} \times \exp\left(-\beta \sum_i p_i^2/2m_i\right)\left[1 + \frac{\hbar^2 \beta^4}{24} \sum_i \sum_k \frac{p_i p_k}{m_i m_k} \overline{\frac{\partial U}{\partial q_i} \frac{\partial U}{\partial q_k}}\right] dp.$$

$$(33.17)$$

This expression can be conveniently rewritten in the following final form:

$$dw_p = \text{constant} \times \exp\left\{-\frac{1}{T}\left[\sum_i \frac{p_i^2}{2m_i} - \frac{\hbar^2}{24T^3} \sum_i \sum_k \frac{p_i p_k}{m_i m_k} \overline{\frac{\partial U}{\partial q_i} \frac{\partial U}{\partial q_k}}\right]\right\} dp,$$

$$(33.18)$$

the bracket in (33.17) being replaced by an exponential function to the same degree of accuracy.

Thus we see that the correction to the classical distribution function for the momenta is equivalent to adding to the kinetic energy in the exponent an expression quadratic in the momenta, with coefficients depending on the law of interaction between the particles in the body.

If it is desired to find the probability distribution for any one momentum p_i, then (33.17) must be integrated with respect to all the other momenta. All the terms involving the squares p_k^2 ($k \neq i$) will then give constants negligible compared with unity, while the terms containing products of different momenta give zero. The result is, again in exponential form,

$$dw_{p_i} = \text{constant} \times \exp\left\{-\frac{p_i^2}{2m_iT}\left[1-\frac{\hbar^2}{12T^3m_i}\overline{\left(\frac{\partial U}{\partial q_i}\right)^2}\right]\right\} dp_i. \quad (33.19)$$

Thus the distribution obtained differs from the Maxwellian only in that the true temperature T is replaced by a somewhat higher "effective temperature":

$$T_{\text{eff}} = T + \frac{\hbar^2}{12T^2m_i}\overline{\left(\frac{\partial U}{\partial q_i}\right)^2}.$$

Similarly we can calculate the corrected coordinate distribution function by integrating I with respect to the momenta:

$$dw_q = \text{constant} \times dq \int I \, dp.$$

The same calculations as led to (33.13) give

$$dw_q = \text{constant} \times \exp\left\{-\frac{1}{T}\left[U-\frac{\hbar^2}{24T^2}\sum_i \frac{1}{m_i}\left(\frac{\partial U}{\partial q_i}\right)^2+\frac{\hbar^2}{12T}\sum_i \frac{1}{m_i}\frac{\partial^2 U}{\partial q_i^2}\right]\right\} dq.$$

$$(33.20)$$

§ 34. The Gibbs distribution for rotating bodies

The problem of the thermodynamic relations for rotating bodies has already been considered in § 26. Let us now see how the Gibbs distribution is to be formulated for rotating bodies. This will complete the investigation of their statistical properties. As regards the uniform translational motion, Galileo's relativity principle shows that, as already mentioned in § 26, this motion has only a trivial effect on the statistical properties and so needs no special consideration.

In a system of coordinates rotating with the body, the usual Gibbs distribution is valid; in classical statistics,

$$\varrho = (2\pi\hbar)^{-s}\, e^{[F'-E'(p,\, q)]/T}, \quad (34.1)$$

where $E'(p, q)$ is the energy of the body in this system, as a function of the coordinates and momenta of its particles, and F' the free energy in the same system (which, of course, is not the same as the free energy of the body when at rest). The energy $E'(p, q)$ is related to the energy $E(p, q)$ in a fixed system by

$$E'(p, q) = E(p, q) - \mathbf{\Omega} \cdot \mathbf{M}(p, q),\tag{34.2}$$

where $\mathbf{\Omega}$ is the angular velocity of rotation and $\mathbf{M}(p, q)$ the angular momentum of the body (see § 26). Substituting (34.2) in (34.1), we find the Gibbs distribution for a rotating body in the form[†]

$$\varrho = (2\pi\hbar)^{-s} e^{[F' - E(p, q) + \mathbf{\Omega} \cdot \mathbf{M}(p, q)]/T}.\tag{34.3}$$

In classical statistics the Gibbs distribution for a rotating body can also be represented in another form. To obtain this, we use the following expression for the energy of the body in the rotating coordinate system:

$$E' = \Sigma \tfrac{1}{2}mv'^2 - \tfrac{1}{2}\Sigma m(\mathbf{\Omega} \times \mathbf{r})^2 + U,\tag{34.4}$$

where the $\mathbf{v}'$ are the velocities of the particles relative to the rotating system, and the $\mathbf{r}$ their radius vectors (see *Mechanics*, § 39). Denoting by

$$E_0(\mathbf{v}', \mathbf{r}) = \Sigma \tfrac{1}{2}mv'^2 + U\tag{34.5}$$

the part of the energy which is independent of $\mathbf{\Omega}$, we obtain the Gibbs distribution in the form

$$\varrho = (2\pi\hbar)^{-s} \exp\left\{\frac{1}{T}\left[F' - E_0(\mathbf{v}', \mathbf{r}) + \frac{1}{2}\Sigma m(\mathbf{\Omega} \times \mathbf{r})^2\right]\right\}.$$

The function ϱ determines the probability corresponding to the element of phase space $dx_1\, dy_1\, dz_1 \ldots dp'_{1x}\, dp'_{1y}\, dp'_{1z} \ldots$, where $\mathbf{p}' = m\mathbf{v}' + m\mathbf{\Omega} \times \mathbf{r}$ (see *Mechanics*, § 39). Since, in obtaining the differentials of the momenta, we must regard the coordinates as constant, $d\mathbf{p}' = m\, d\mathbf{v}'$, and the probability distribution expressed in terms of the coordinates and velocities of the particle is

$$dw = C \exp\left\{\frac{F'}{T} - \frac{1}{T}\left[E_0(\mathbf{v}', \mathbf{r}) - \Sigma \frac{1}{2}m(\mathbf{\Omega} \times \mathbf{r})^2\right]\right\}$$

$$\times dx_1\, dy_1\, dz_1 \ldots dv'_{1x}\, dv'_{1y}\, dv'_{1z} \ldots,\tag{34.6}$$

where C denotes for brevity the factor $(2\pi\hbar)^{-s}$ together with the product of the particle masses which appears when we go from the momentum differentials to the velocity differentials.

[†] The distribution (34.3), like the ordinary Gibbs distribution, is fully in agreement with the result (4.2) derived in § 4 from Liouville's theorem: the logarithm of the distribution function is a linear function of the energy and angular momentum of the body.

For a body at rest we have

$$dw = Ce^{[F-E_0(\mathbf{v},\, \mathbf{r})]/T}\, dx_1\, dy_1\, dz_1 \ldots dv_{1x}\, dv_{1y}\, dv_{1z} \ldots, \qquad (34.7)$$

with the same expression (34.5) for $E_0(\mathbf{v}, \mathbf{r})$, now a function of the velocities in the fixed coordinate system. Thus we see that the Gibbs distribution for the coordinates and velocities for a rotating body differs from that for a body at rest only by the additional potential energy $-\frac{1}{2}\Sigma m(\boldsymbol{\Omega}\times\mathbf{r})^2$. In other words, as regards the statistical properties of the body, the rotation is equivalent to the existence of an external field corresponding to the centrifugal force. The statistical properties are not affected by the Coriolis force.

It should be emphasised, however, that this last result applies only to classical statistics. In the quantum case the expression

$$\hat{w} = \exp\left[(F' - \hat{H} + \boldsymbol{\Omega}\cdot\hat{\mathbf{M}})/T\right] \qquad (34.8)$$

gives the statistical operator corresponding to (34.3) for a rotating body. Formally we can reduce this operator to a form analogous to (34.6), the velocities $\mathbf{v}'$ being replaced by the operators $\hat{\mathbf{v}}' = \hat{\mathbf{p}}'/m - \boldsymbol{\Omega}\times\mathbf{r}$, but the components of this vector operator do not commute, unlike those of the operator $\hat{\mathbf{v}}$ in the fixed system. The statistical operators corresponding to the expressions (34.6) and (34.7) will therefore in general be markedly different from each other, quite apart from the fact that one of them contains the centrifugal energy.

§ 35. The Gibbs distribution for a variable number of particles

So far we have always tacitly assumed that the number of particles in a body is some given constant, and have deliberately passed over the fact that in reality particles may be exchanged between different subsystems. In other words, the number N of particles in a subsystem will necessarily fluctuate about its mean value. In order to formulate precisely what we mean by the number of particles, we shall use the term *subsystem* to refer to a part of the system which is enclosed in a fixed volume. Then N will denote the number of particles within that volume.[†]

Thus the problem arises of generalising the Gibbs distribution to bodies with a variable number of particles. Here we shall write the formulae for bodies consisting of identical particles; the further generalisation to systems containing different particles is obvious (§ 85).

[†] In deriving the Gibbs distribution in § 28 we have in essence already understood subsystems in this sense; in going from (28.2) to (28.3) we differentiated the entropy whilst regarding the volume of the body (and therefore of the medium) as constant.

The distribution function now depends not only on the energy of the quantum state but also on the number N of particles in the body, and the energy levels E_{nN} are of course themselves different for different N (as indicated by the suffix N). The probability that the body contains N particles and is in the nth state will be denoted by w_{nN}.

The form of this function can be determined in exactly the same way as the function w_n in § 28. The only difference is that the entropy of the medium is now a function not only of its energy E' but also of the number N' of particles in it: $S' = S'(E', N')$. Writing $E' = E^{(0)} - E_{nN}$ and $N' = N^{(0)} - N$ (where N is the number of particles in the body, and $N^{(0)}$ the given total number of particles in the entire closed system, which is large compared with N), we have in accordance with (28.2)

$$w_{nN} = \text{constant} \times \exp \{S'(E^{(0)} - E_{nN}, N^{(0)} - N)\};$$

the quantity $\Delta E'$ is regarded as constant, as in § 28.

Next, we expand S' in powers of E_{nN} and N, again taking only the linear terms. Equation (24.5), in the form

$$dS = \frac{dE}{T} + \frac{P}{T} dV - \frac{\mu}{T} dN,$$

shows that $(\partial S/\partial E)_{V,N} = 1/T$, $(\partial S/\partial N)_{E,V} = -\mu/T$. Hence

$$S'(E^{(0)} - E_{nN}, N^{(0)} - N) \cong S'(E^{(0)}, N^{(0)}) - \frac{E_{nN}}{T} + \frac{\mu N}{T},$$

the chemical potential μ (and the temperature) being the same for the body and the medium, from the conditions of equilibrium.

Thus we obtain for the distribution function the expression

$$w_{nN} = A e^{(\mu N - E_{nN})/T}. \tag{35.1}$$

The normalisation constant A can be expressed in terms of the thermodynamic quantities in the same way as in § 31. The entropy of the body is

$$S = -\langle \log w_{nN} \rangle = -\log A - \frac{\mu \bar{N}}{T} + \frac{\bar{E}}{T},$$

and so

$$T \log A = \bar{E} - TS - \mu \bar{N}.$$

But $\bar{E} - TS = F$, and the difference $F - \mu \bar{N}$ is the thermodynamic potential Ω. Thus $T \log A = \Omega$, and (35.1) may be rewritten as

$$w_{nN} = e^{(\Omega + \mu N - E_{nN})/T}. \tag{35.2}$$

This is the final formula for the Gibbs distribution for a variable number of particles.[†]

The normalisation condition for the distribution (35.2) requires that the result of summing the w_{nN} first over all quantum states (for a given N) and then over all values of N should be equal to unity:

$$\sum_N \sum_n w_{nN} = e^{\Omega/T} \sum_N \left(e^{\mu N/T} \sum_n e^{-E_{nN}/T} \right) = 1.$$

Hence we obtain the following expression for the thermodynamic potential Ω:

$$\Omega = -T \log \sum_N \left[e^{\mu N/T} \sum_n e^{-E_{nN}/T} \right]. \qquad (35.3)$$

$$\Omega = -T \ln 3$$
$$S = -\left(\frac{\partial \Omega}{\partial T}\right)_{\mu, V}$$

$$P = -\left(\frac{\partial \Omega}{\partial V}\right)_{T, \mu}$$

$$N = -\left(\frac{\partial \Omega}{\partial \mu}\right)_{T, V}$$

This formula together with (31.3) can be used to calculate the thermodynamic quantities for specific bodies. Formula (31.3) gives the free energy of the body as a function of T, N and V, and (35.3) gives the potential Ω as a function of T, μ and V.

In classical statistics the probability distribution has the form

$$dw_N = \varrho_N \, dp^{(N)} \, dq^{(N)},$$

where

$$\varrho_N = (2\pi\hbar)^{-s} e^{[\Omega + \mu N - E_N(p, q)]/T}. \qquad (35.4)$$

The variable N is written as a subscript to the distribution function, and the same letter is written as a superscript to the element of phase volume in order to emphasise that a different phase space (of $2s$ dimensions) corresponds to each value of N. The formula for Ω correspondingly becomes

$$\Omega = -T \log \left\{ \sum_N e^{\mu N/T} \int' e^{-E_N(p, q)/T} \, d\Gamma_N \right\}. \qquad (35.5)$$

Finally, we may say a few words concerning the relation between the Gibbs distribution (35.2) for a variable number of particles derived here and the previous distribution (31.1). First of all, it is clear that, for the determination of all the statistical properties of the body except the fluctuations of the total number of particles in it, these two distributions are entirely equivalent. On neglecting the fluctuations of the number N, we obtain $\Omega + \mu N = F$, and the distribution (35.2) is identical with (31.1).

The relation between the distributions (31.1) and (35.2) is to a certain extent analogous to that between the microcanonical and canonical distributions. The description of a subsystem by means of the microcanonical distribution is equivalent to neglecting the fluctuations of its total energy; the canonical distribution in its usual form (31.1) takes into account these fluc-

[†] This distribution is sometimes called the *grand canonical ensemble*.

tuations. The latter form in turn neglects the fluctuations in the number of particles, and may be said to be "microcanonical with respect to the number of particles"; the distribution (35.2) is "canonical" with respect to both the energy and the number of particles.

Thus all three distributions, the microcanonical and the two forms of the Gibbs distribution, are in principle suitable for determining the thermo-dynamic properties of the body. The only difference from this point of view lies in the degree of mathematical convenience. In practice the microcanon-ical distribution is the least convenient and is never used for this purpose. The Gibbs distribution for a variable number of particles is usually the most convenient.

§ 36. The derivation of the thermodynamic relations from the Gibbs distribution

The Gibbs distribution plays a fundamental part throughout statistical phys-ics. We shall therefore give here another justification of it. This distribution has essentially been derived in §§ 4 and 6 directly from Liouville's theorem. We have seen that the application of Liouville's theorem (together with considerations of the multiplicativity of distribution functions for subsys-tems) enables us to deduce that the logarithm of the distribution function of a subsystem must be a linear function of its energy:

$$\log w_n = \alpha + \beta E_n, \tag{36.1}$$

the coefficients β being the same for all subsystems in a given closed system (see (6.4)), and the corresponding relation (4.5) for the classical case). Hence

$$w_n = e^{\alpha + \beta E_n};$$

using the purely formal notation $\beta = -1/T$, $\alpha = F/T$, we have an expression of the same form as the Gibbs distribution (31.1). It remains to show that the fundamental thermodynamic relations can be derived from the Gibbs distri-bution itself, i.e. in a purely statistical manner.

We have already seen that the quantity β, and therefore T, must be the same for all parts of a system in equilibrium. It is also evident that $\beta < 0$, i.e. $T > 0$, since otherwise the normalisation sum Σw_n must diverge: owing to the presence of the kinetic energy of the particles, the energy E_n can take arbitrarily large values. All these properties agree with the fundamental properties of the thermodynamic temperature.

To derive a quantitative relation, we start from the normalisation condi-tion

$$\sum_n e^{(F-E_n)/T} = 1.$$

[handwritten annotations: μ-canonical / fix $\bar{E}$, N / w-const for states w/ $\bar{E}$, N] [Canonical / N fixed, E free] [grand canonical / N, E free]

We differentiate this equation, regarding the left-hand side as a function of T and of various quantities $\lambda_1, \lambda_2, \ldots$ which represent the external conditions to which the body considered is subject; these quantities may, for example, determine the shape and size of the volume occupied by the body. The energy levels E_n depend on $\lambda_1, \lambda_2, \ldots$ as parameters.

Differentiation gives

$$\sum_n \frac{w_n}{T} \left[dF - \frac{\partial E_n}{\partial \lambda} \, d\lambda - \frac{F - E_n}{T} \, dT \right] = 0,$$

where for simplicity only one external parameter is used. Hence

$$dF \sum_n w_n = d\lambda \sum_n w_n \frac{\partial E_n}{\partial \lambda} + \frac{dT}{T} \left(F - \sum_n w_n E_n \right).$$

On the left-hand side $\Sigma w_n = 1$, and on the right-hand side

$$\sum_n w_n E_n = \bar{E}, \quad \sum_n w_n \frac{\partial E_n}{\partial \lambda} = \overline{\frac{\partial E_n}{\partial \lambda}}.$$

Using also the formulae $F - \bar{E} = -TS$ and[†]

$$\overline{\partial E_n / \partial \lambda} = \overline{\partial \hat{H} / \partial \lambda}, \tag{36.2}$$

we have finally

$$dF = -S \, dT + \overline{\partial \hat{H} / \partial \lambda} \cdot d\lambda.$$

This is the general form for the differential of the free energy.

In the same way we can derive the Gibbs distribution for a variable number of particles. If the number of particles is regarded as a dynamical variable, it is clear that it will be an "integral of the motion", and additive, for a closed system. We must therefore write

$$\log w_{nN} = \alpha + \beta E_n + \gamma N, \tag{36.3}$$

where λ, like β, must be the same for all parts of a system in equilibrium. Putting $\alpha = \Omega/T$, $\beta = -1/T$, $\gamma = \mu/T$, we obtain a distribution of the form (35.2), and then by the same method as above we can deduce an expression for the differential of the potential Ω.

[†] If the Hamiltonian $\hat{H}$ (and therefore its eigenvalues E_n) depends on a parameter λ, then

$$\partial E_n / \partial \lambda = (\partial \hat{H} / \partial \lambda)_{nn};$$

see *Quantum Mechanics*, (11.16). On statistical averaging this gives (36.2).

CHAPTER IV

IDEAL GASES — no interactions
well defined energies.

§ 37. The Boltzmann distribution

ONE of the most important subjects of study in statistical physics is an
ideal gas. By this is meant a gas in which the interaction between the particles (molecules) is so weak as to be negligible. Physically, this approximation may be allowable either because the interaction of the particles is small
whatever the distances between them or because the gas is sufficiently rarefied. In the latter case, which is the more important, the rarefaction of the gas
results in its molecules' being almost always at considerable distances apart,
such that the interaction forces are quite small.

The absence of interaction between the molecules enables the quantum-
mechanics problem of determining the energy levels E_n of the gas as a whole
to be reduced to that of determining the energy levels of a single molecule.
These levels will be denoted by ε_k, the suffix k representing the set of quantum
numbers which define the state of the molecule. The energies E_n are then given
by the sums of the energies of the various molecules.

It must be remembered, however, that, even when there is no direct force
interaction, quantum mechanics gives a peculiar mutual effect of particles
that are in the same quantum state (called the *exchange effect*). For example,
if the particles obey Fermi statistics, this effect has the result that no more
than one particle can be in each quantum state at one time;[†] a similar effect
but in a different form occurs for particles which obey Bose statistics.

Let n_k be the number of particles in a gas which are in the kth quantum
state; the numbers n_k are called the *occupation numbers* of the various
quantum states. Let us consider the problem of calculating the mean values
$\overline{n_k}$ of these numbers, and take in particular the extremely important case
where for all k

$$\overline{n_k} \ll 1. \tag{37.1}$$

[†] It should be emphasised that, when speaking of the quantum state of an individual particle, we shall always refer to states which are fully determined by a set
of values of all the quantum numbers (including the orientation of the angular
momentum of the particle, if any). These should not be confused with the quantum
energy levels; several different quantum states correspond to a given energy level
if the latter is degenerate.

111

Physically this case corresponds to a sufficiently rarefied gas. We shall later establish a criterion which ensures the fulfilment of this condition, but it may be mentioned immediately that it is in practice satisfied for all ordinary molecular or atomic gases. The condition would be violated only at such high densities that the matter concerned certainly could not be regarded as an ideal gas.

The condition $\overline{n_k} \ll 1$ for the mean occupation numbers signifies that in fact not more than one particle is in each quantum state at any instant. Consequently, we may neglect not only the direct forces of interaction of the particles but also their indirect quantum interactions mentioned above. This in turn enables us to apply the Gibbs distribution formula to the individual molecules. For the Gibbs distribution has been derived for bodies which are relatively small, but at the same time macroscopic, parts of large closed systems. The macroscopic nature of these bodies made it possible to regard them as quasi-closed, i.e. to neglect to some extent their interaction with other parts of the system. In the case under consideration the separate molecules of the gas are quasi-closed, although they are certainly not macroscopic bodies.

Applying the Gibbs distribution formula to the gas molecules, we can say that the probability that a molecule is in the kth state is proportional to $e^{-\varepsilon_k/T}$, and therefore so is the mean number $\overline{n_k}$ of molecules in that state, i.e.

$$\overline{n_k} = ae^{-\varepsilon_k/T}, \tag{37.2}$$

where a is a constant given by the normalisation condition

$$a = \frac{N}{\sum e^{-\varepsilon_k/T}} \qquad\qquad \sum_k \overline{n_k} = N \tag{37.3}$$

(N being the total number of particles in the gas). The distribution of molecules of an ideal gas among the various states that is given by formula (37.2) is called the *Boltzmann distribution*; it was discovered by Boltzmann for classical statistics in 1877.

The constant coefficient in (37.2) can be expressed in terms of the thermodynamic quantities for the gas. To do this we shall give another derivation of the formula, based on the application of the Gibbs distribution to the assembly of all particles in the gas that are in a given quantum state. We are able to do this (even if the numbers n_k are not small) since there is no direct force of interaction between these particles and the remainder (or between any of the particles in an ideal gas), and the quantum exchange effects occur only for particles in the same state. Putting $E = n_k \varepsilon_k$, $N = n_k$ and adding the suffix k to Ω in the general formula for the Gibbs distribution for a variable number of particles (35.2), we find the probability distribution for various values of n_k as

$$w_{n_k} = e^{[\Omega_k + n_k(\mu - \varepsilon_k)]/T}. \tag{37.4}$$

In particular, $w_0 = e^{\Omega_k/T}$ is the probability that there are no particles in the state concerned. In the case of interest here, for which $\bar{n}_k \ll 1$, the probability w_0 is almost unity, and so in the expression $w_1 = e^{(\Omega_k + \mu - \varepsilon_k)/T}$ for the probability of finding one particle in the kth state we can put $e^{\Omega_k/T} = 1$ to within terms of a higher order of smallness. Then $w_1 = e^{(\mu - \varepsilon_k)/T}$. The probabilities of values $n_k > 1$ must be taken as zero in the same approximation. Hence

$$\bar{n}_k = \sum_{n_k} w_{n_k} n_k = w_1 \cdot 1,$$

and we have the Boltzmann distribution in the form

$$\bar{n}_k = e^{(\mu - \varepsilon_k)/T}. \tag{37.5}$$

Thus the coefficient in (37.2) is expressed in terms of the chemical potential of the gas.

§ 38. The Boltzmann distribution in classical statistics

If the motion of gas molecules (and of the atoms in them) were subject to classical mechanics, we could use, instead of the distribution over quantum states, the distribution of molecules in phase space, i.e. over momenta and coordinates. Let dN be the mean number of molecules "contained" in a volume element of phase space of the molecule, $dp\, dq = dp_1 \dots dp_r\, dq_1 \dots dq_r$ (r being the number of degrees of freedom of the molecule). We may write this as

$$dN = n(p, q)\, d\tau, \quad d\tau = dp\, dq/(2\pi\hbar)^r \tag{38.1}$$

and call $n(p, q)$ the density in phase space (although $d\tau$ differs by a factor $(2\pi\hbar)^{-r}$ from the volume element in phase space). We then have, instead of (37.5),

$$n(p, q) = e^{[\mu - \varepsilon(p, q)]/T}, \tag{38.2}$$

where $\varepsilon(p, q)$ is the energy of the molecule as a function of the coordinates and momenta of its atoms.

Usually, however, it is not the entire motion of the molecule which is quasi-classical, but only the motion corresponding to some of its degrees of freedom. In particular, in a gas which is not in an external field, the translational motion of molecules is always quasi-classical. The kinetic energy of the translational motion then appears in the energy ε_k of the molecule as an independent term, while the remaining part of the energy does not involve the coordinates x, y, z and momenta p_x, p_y, p_z of the centre of mass of the molecule. This enables us to separate from the general formula for the Boltzmann distribution a factor which gives the distribution of the gas molecules with respect to these variables. The distribution of the molecules in the volume occupied by the gas is clearly just a uniform distribution, and we obtain for the number of molecules per unit volume with momenta (of the translational motion) in given intervals dp_x, dp_y, dp_z the Maxwellian distribution:

$$dN_p = \frac{N}{V(2\pi mT)^{3/2}} \exp\left[-(p_x^2 + p_y^2 + p_z^2)/2mT\right] dp_x \, dp_y \, dp_z, \quad (38.3)$$

$$dN_v = \frac{N}{V} \frac{m^{3/2}}{(2\pi T)^{3/2}} \exp\left[-m(v_x^2 + v_y^2 + v_z^2)/2T\right] dv_x \, dv_y \, dv_z \quad (38.4)$$

(m being the mass of a molecule), normalised to N/V particles per unit volume.

Let us next consider a gas in an external field, in which the potential energy of a molecule depends only on the coordinates of its centre of mass: $u = u(x, y, z)$ (for example, a gravitational field). If, as always occurs in practice, the translational motion in this field is quasi-classical, then $u(x, y, z)$ appears in the energy of the molecule as an independent term. The Maxwellian distribution for the velocities of the molecules remains unchanged, of course, while the distribution for the coordinates of the centre of mass is given by the formula

$$dN_r = n_0 e^{-u(x, y, z)/T} \, dV. \quad (38.5)$$

This formula gives the number of molecules in an element of volume $dV = dx \, dy \, dz$; the quantity

$$n(\mathbf{r}) = n_0 e^{-u(x, y, z)/T} \quad (38.6)$$

is the number density of the particles. The constant n_0 is the density at points where $u = 0$. Formula (38.6) is called *Boltzmann's formula*.

In particular, in a uniform gravitational field along the z-axis, $u = mgz$, and the gas density distribution is given by the *barometric formula*

$$n(z) = n_0 e^{-mgz/T}, \quad (38.7)$$

where n_0 is the density at the level $z = 0$.

At large distances from the Earth, its gravitational field must be described by the exact Newtonian expression, the potential energy u vanishing at infinity. According to formula (38.6) the gas density should remain finite and not zero at infinity, but a finite quantity of gas cannot be distributed in an infinite volume with a density which is nowhere zero. This means that in a gravitational field a gas (such as the atmosphere) cannot be in equilibrium and must be continuously dissipated into space.

PROBLEMS

PROBLEM 1. Find the density of gas in a cylinder of radius R and length l rotating about its axis with angular velocity Ω, there being a total of N molecules in the cylinder.

SOLUTION. It has been mentioned in § 34 that the rotation of a body as a whole is equivalent to the presence of an external field with potential energy $-\frac{1}{2}m\Omega^2r^2$ (where r is the distance from the axis of rotation). The gas density is therefore

$$n(r) = Ae^{m\Omega^2r^2/2T}.$$

Normalisation gives

$$n(r) = \frac{Nm\Omega^2e^{m\Omega^2r^2/2T}}{2\pi Tl(e^{m\Omega^2r^2/2T}-1)}.$$

PROBLEM 2. Find the momentum distribution of particles for a relativistic ideal gas.

SOLUTION. The energy of a relativistic particle is given in terms of its momentum by $\varepsilon = c\sqrt{(m^2c^2+p^2)}$, where c is the velocity of light. The normalised momentum distribution is

$$dN_\mathbf{p} = \frac{N}{V}\frac{\exp\{-c\sqrt{(m^2c^2+p^2)}/T\}}{2(T/mc^2)^2K_1(mc^2/T)+(T/mc^2)K_0(mc^2/T)}\frac{dp_x\,dp_y\,dp_z}{4\pi(mc)^3},$$

where K_0 and K_1 are Macdonald functions (Hankel functions of imaginary argument). In the calculation of the normalisation integral the following formulae are used:

$$\int_0^\infty e^{-z\cosh t}\sinh^2 t\,dt = K_1(z)/z,$$

$$K_1'(z) = -K_1(z)/z-K_0(z).$$

§ 39. Molecular collisions

The molecules of a gas enclosed in a vessel collide with its walls as they move. Let us calculate the mean number of impacts between the molecules of a gas and a unit area of the wall per unit time.

We take an element of surface area of the vessel wall and define a coordinate system with the z-axis perpendicular to this element, which may then be written as $dx\,dy$. Of the molecules in the gas, those which reach the vessel wall in unit time, i.e. collide with it, are just those whose z coordinate does not exceed the component v_z of their velocity along that axis (which, of course, must also be directed towards the wall, not away from it).

The number dv_v of collisions of molecules per unit time (and per unit area of the wall surface), in which the velocity components are in given intervals dv_x, dv_y, dv_z is therefore obtained by multiplying the distribution (38.4) by the volume of a cylinder of unit base area and height v_z:

$$d\nu_{\mathbf{v}} = \frac{N}{V}\left(\frac{m}{2\pi T}\right)^{3/2} \exp\left[-m(v_x^2 + v_y^2 + v_z^2)/2T\right] \times v_z \, dv_x \, dv_y \, dv_z. \quad (39.1)$$

From this we easily find the total number ν of impacts of gas molecules on unit area of the vessel wall per unit time. To do so, we integrate (39.1) over all velocities v_z from 0 to ∞ and over v_x and v_y from $-\infty$ to ∞; integration over v_z from $-\infty$ to 0 is not required, since when $v_z < 0$ the molecule is travelling away from the wall, and so does not collide with it. Hence

$$\nu = \frac{N}{V}\sqrt{\frac{T}{2\pi m}} = \frac{P}{\sqrt{(2\pi mT)}}; \quad (39.2)$$

here we have expressed the density of the gas in terms of its pressure by means of Clapeyron's equation.

Formula (39.1) may be written in spherical polar coordinates in velocity space, using instead of v_x, v_y, v_z the absolute magnitude of the velocity and the polar angles θ and ϕ which define its direction. Taking the polar axis along the z-axis, we have $v_z = v \cos\theta$ and

$$d\nu_{\mathbf{v}} = \frac{N}{V}\left(\frac{m}{2\pi T}\right)^{3/2} e^{-mv^2/2T} v^3 \sin\theta \cos\theta \, d\theta \, d\phi \, dv. \quad (39.3)$$

Let us now consider collisions between gas molecules. To do this, we must first find the velocity distribution of the molecules (the term velocity everywhere referring to the velocity of the centre of mass) relative to one another. We take any one gas molecule and consider the motion of all the other molecules relative to it, i.e. consider for each molecule not its absolute velocity v (relative to the walls of the vessel) but its velocity v' relative to some other molecule. That is, instead of dealing with individual molecules, we always consider the relative motion of a pair of molecules, ignoring the motion of their common centre of mass.

We know from mechanics that the energy of the relative motion of two particles of masses m_1 and m_2 is $\frac{1}{2}m'v'^2$, where $m' = m_1 m_2/(m_1 + m_2)$ is their "reduced mass" and v' their relative velocity. The relative-velocity distribution of the molecules of an ideal gas therefore has the same form as the absolute-velocity distribution, except that m is replaced by the reduced mass m'. Since all the molecules are alike, $m' = \frac{1}{2}m$, and the number of molecules per unit volume with a velocity relative to the selected molecule between v' and $v' + dv'$ is

$$dN_{v'} = \frac{N}{V}\frac{\pi}{2}\left(\frac{m}{\pi T}\right)^{3/2} e^{-mv'^2/4T} v'^2 \, dv'. \quad (39.4)$$

A collision between molecules may be accompanied by various processes: deflection (scattering) through a certain angle, dissociation into atoms, and so on. The processes which occur in collisions are usually described by their *cross-sections*. The cross-section for a particular process which occurs in collisions between a given particle and others is the ratio of the probability of such a collision per unit time to the particle flux density (the latter being the number of such particles per unit volume multiplied by their velocity). The number of collisions (per unit time) between this and other particles which are accompanied by a certain process with cross-section σ is therefore

$$\nu' = \frac{N}{V} \frac{\pi}{2} \left(\frac{m}{\pi T}\right)^{3/2} \int_0^\infty e^{-mv'^2/4T} \sigma v'^3 \, dv'. \tag{39.5}$$

The total number of such collisions per unit time throughout the volume of the gas is obviously $\nu'N/2$.

PROBLEMS

PROBLEM 1. Find the number of impacts of gas molecules on unit area of the wall per unit time for which the angle between the direction of the velocity of the molecule and the normal to the surface lies between θ and $\theta+d\theta$.

SOLUTION.

$$d\nu_\theta = \frac{N}{V}\left(\frac{2T}{m\pi}\right)^{1/2} \sin\theta \cos\theta \, d\theta.$$

PROBLEM 2. Find the number of impacts of gas molecules on unit area of the wall per unit time for which the absolute magnitude of the velocity lies between v and $v+dv$.

SOLUTION.

$$d\nu_v = \frac{N}{V}\pi\left(\frac{m}{2\pi T}\right)^{3/2} e^{-mv^2/2T} v^3 \, dv.$$

PROBLEM 3. Find the total kinetic energy E_{inc} of the gas molecules striking unit area of the wall per unit time.

SOLUTION.

$$E_{inc} = \frac{N}{V}\sqrt{\frac{2T^3}{m\pi}} = P\sqrt{\frac{2T}{m\pi}}.$$

PROBLEM 4. Find the number of collisions between one molecule and the rest per unit time, assuming the molecules to be rigid spheres of radius r.

SOLUTION. The cross-section for collisions between molecules is then $\sigma = \pi(2r)^2 = 4\pi r^2$ (since a collision occurs whenever two molecules pass at a distance less than $2r$). Substitution in (39.5) gives

$$\nu = 16r^2\sqrt{\frac{\pi T}{m}}\frac{N}{V} = 16r^2\sqrt{\frac{\pi}{mT}}P.$$

§ 40. Ideal gases not in equilibrium

The Boltzmann distribution can also be derived, in a quite different manner, directly from the condition of maximum entropy of the gas as a whole, regarded as a closed system. This derivation is of considerable interest in itself, since it is based on a method whereby the entropy of gas in any non-equilibrium macroscopic state may be calculated.

Any macroscopic state of an ideal gas may be described as follows. Let us distribute all the quantum states of an individual particle of the gas among groups each containing neighbouring states (which, in particular, have neighbouring energy values), both the number of states in each group and the number of particles in these states being still very large. Let the groups of states be numbered $j = 1, 2, \ldots$, and let G_j be the number of states in group j, and N_j the number of particles in these states. Then the set of numbers N_j will completely describe the macroscopic state of the gas.

The problem of calculating the entropy of the gas reduces to that of determining the statistical weight $\Delta\Gamma$ of a given macroscopic state, i.e. the number of microscopic ways in which this state can be realised. Regarding each group of N_j particles as an independent system and denoting its statistical weight by $\Delta\Gamma_j$, we can write

$$\Delta\Gamma = \prod_j \Delta\Gamma_j. \tag{40.1}$$

Thus the problem reduces to that of calculating the $\Delta\Gamma_j$.

In Boltzmann statistics the mean occupation numbers of all quantum states are small in comparison with unity. This means that the numbers of particles N_j must be small compared with the numbers of states G_j ($N_j \ll G_j$), but of course themselves still large. As has been explained in § 37, the smallness of the mean occupation numbers enables us to suppose that all the particles are entirely independently distributed among the various states. Placing each of the N_j particles in one of the G_j states, we obtain altogether $G_j^{N_j}$ possible distributions, but among these the distributions which differ only by a permutation of particles are identical, since the particles themselves are identical. The number of permutations of N_j particles is $N_j!$, and so the statistical weight of the distribution of N_j particles among G_j states is

$$\Delta\Gamma_j = G_j^{N_j}/N_j!. \tag{40.2}$$

The entropy of the gas is calculated as the logarithm of the statistical weight:

$$S = \log \Delta\Gamma = \Sigma \log \Delta\Gamma_j.$$

Substitution of (40.2) gives

$$S = \sum_j (N_j \log G_j - \log N_j!).$$

Since the numbers N_j are large, we can use the approximate formula[†]

$$\log N! \cong N \log (N/e),$$ (40.3)

obtaining

$$S = \sum_j N_j \log (eG_j/N_j).$$ (40.4)

This formula gives the solution of the problem, determining the entropy of an ideal gas in any macroscopic state defined by the set of numbers N_j. It may be rewritten by using the mean numbers $\bar{n}_j$ of particles in each of the quantum states in group j: $\bar{n}_j = N_j/G_j$. Then

$$S = \sum_j G_j \bar{n}_j \log (e/\bar{n}_j).$$ (40.5)

If the motion of the particles is quasi-classical, then in this formula we can change to the particle distribution in phase space. Let the phase space of a particle be divided into regions $\Delta p^{(j)} \Delta q^{(j)}$, each of which is small but nevertheless contains a large number of particles. The numbers of quantum states belonging to these regions are

$$G_j = \Delta p^{(j)} \Delta q^{(j)}/(2\pi\hbar)^r = \Delta\tau^{(j)},$$ (40.6)

where r is the number of degrees of freedom of the particle; the numbers of particles in these states may be written as $N_j = n(p, q) \Delta\tau^{(j)}$, where $n(p, q)$ is the particle density distribution in phase space. We substitute these expressions in (40.5), and use the fact that the regions $\Delta\tau^{(j)}$ are small in size and large in number to replace the summation over j by integration over the whole phase space of the particle:

$$S = \int n \log (e/n) \, d\tau.$$ (40.7)

In a state of equilibrium, the entropy must be a maximum (as applied to the ideal gas, this statement is sometimes called *Boltzmann's H theorem*). We shall show how this condition may be used to find the distribution function for the gas particles in a state of statistical equilibrium. The problem is to find $\bar{n}_j$ such that the sum (40.5) has the maximum value possible under the subsidiary conditions

$$\sum_j N_j = \sum_j G_j \bar{n}_j = N,$$

$$\sum_j \varepsilon_j N_j = \sum_j \varepsilon_j G_j \bar{n}_j = E,$$

[†] When N is large, the sum $\log N! = \log 1 + \log 2 + \dots + \log N$ may be approximately replaced by the integral

$$\int_0^N \log x \cdot dx,$$

which leads to (40.3).

which express the constancy of the total number of particles N and of the total energy E of the gas. Following the usual method of Lagrange's undetermined multipliers, we have to equate to zero the derivatives

$$\partial(S+\alpha N+\beta E)/\partial \bar{n}_j = 0, \tag{40.8}$$

where α and β are constants. Effecting the differentiation, we find

$$G_j(-\log \bar{n}_j + \alpha + \beta \varepsilon_j) = 0,$$

whence $\log \bar{n}_j = \alpha + \beta \varepsilon_j$, or

$$\bar{n}_j = e^{\alpha + \beta \varepsilon_j}.$$

This is just the Boltzmann distribution, the constants α and β being given in terms of T and μ by $\alpha = \mu/T$, $\beta = -1/T$.[†]

§ 41. The free energy of an ideal Boltzmann gas

We may use the general formula (31.3)

$$F = -T \log \sum_n e^{-E_n/T} \tag{41.1}$$

to calculate the free energy of an ideal gas which obeys Boltzmann statistics.

   Writing the energy E_n as a sum of energies ε_k, we can reduce the summation over all states of a gas to summation over all states of an individual molecule. Each state of the gas is defined by a set of N values of ε_k (where N is the number of molecules in the gas), which in the Boltzmann case may be regarded as all different (there being not more than one molecule in each molecular state). Writing $e^{-E_n/T}$ as a product of factors $e^{-\varepsilon_k/T}$ for each molecule and summing independently over all states of each molecule, we should obtain

$$\left(\sum_k e^{-\varepsilon_k/T}\right)^N. \tag{41.2}$$

The set of possible values of ε_k is the same for each molecule of the gas, and so the sums $\sum e^{-\varepsilon_k/T}$ are also the same.

   The following point must be borne in mind, however. Each set of N different values ε_k which differs only in the distribution of the identical gas molecules over the levels ε_k corresponds to the same quantum state of the gas. But in the partition function in formula (41.1) each state must be included

---

† These values of α and β could have been foreseen: equations (40.8) can be written as a relation between differentials, $dS+\alpha\, dN+\beta\, dE = 0$, which must be the same as the differential of the internal energy at constant volume, $dE = T\, dS+\mu\, dN$.

only once.[†] We must therefore again divide the expression (41.2) by the number of possible permutations of N molecules, i.e. by $N!$.[‡] Thus

indistinguishable particles

$$\sum_n e^{-E_n/T} = \frac{1}{N!} \left(\sum_k e^{-\varepsilon_k/T} \right)^N. \tag{41.3}$$

Substitution of this expression in (41.1) gives

$$F = -TN \log \sum_k e^{-\varepsilon_k/T} + T \log N!.$$

Since N is very large, we can use the formula (40.3) for $\log N!$. This gives the formula

$$F = -NT \log \left[(e/N) \sum_k e^{-\varepsilon_k/T} \right], \tag{41.4}$$

which enables us to calculate the free energy of any gas consisting of identical particles obeying Boltzmann statistics.

In classical statistics, formula (41.4) must be written in the form

$$F = -NT \log \left[(e/N) \int e^{-\varepsilon(p, q)/T} \, d\tau \right]; \tag{41.5}$$

the integration is taken over the phase space of the molecule, and $d\tau$ is defined by (38.1).

§ 42. The equation of state of an ideal gas

It has already been mentioned in § 38 that the translational motion of the molecules in a gas is always quasi-classical; the energy of a molecule may be written in the form

trans.   *rotation etc.*

$$\varepsilon_k(p_x, p_y, p_z) = (p_x^2 + p_y^2 + p_z^2)/2m + \varepsilon_k', \tag{42.1}$$

where the first term is the kinetic energy of the translational motion, and ε_k' denotes the energy levels corresponding to the rotation and internal state of the molecule; ε_k' is independent of the velocities and coordinates of the centre of mass of the molecule (assuming that there is no external field).

The partition function in the logarithm in formula (41.4) must now be replaced by the expression

$$\sum_k \frac{1}{(2\pi\hbar)^3} \iint \exp\left[-\varepsilon_k(p)/T \right] d^3p \, dV = V(mT/2\pi\hbar^2)^{3/2} \sum_k e^{-\varepsilon_k'/T}; \tag{42.2}$$

$$\to \ 1/v_Q \quad - \quad \text{quantum volume}$$

[†] See the second footnote to § 31.
[‡] Here it is important that in Boltzmann statistics the terms containing the same ε_k in (41.2) are of negligible significance.

$$Z = \frac{1}{N!} \left(\frac{V}{v_Q} z' \right)^N$$

$$F = -NT \ln \frac{V}{v_Q} + T \ln N! - NT \ln z'$$
$$= -NT \ln\left[\frac{eV}{N v_Q} \right] - NT \ln z'$$

the integration over V ($dV = dx \, dy \, dz$) is over the whole volume of the gas. For the free energy we obtain

$$F = -NT \log \left[\frac{eV}{N} \left(\frac{mT}{2\pi\hbar^2} \right)^{3/2} \sum_k e^{-\varepsilon_k'/T} \right].$$

(42.3)

The sum in (42.3) cannot, of course, be calculated in a general form without any assumptions as to the properties of the molecules, but an important fact is that it depends only on the temperature. The dependence of the free energy on the volume is therefore entirely determined by formula (42.3), and so we can derive from it various important general results concerning the properties of an ideal gas (which is not in an external field).

Separating in (42.3) the term containing the volume, we may write this formula as

Put V_Q into $f(T)$

$$F = -NT \log (eV/N) + Nf(T),$$

(42.4)

where $f(T)$ is some function of the temperature. Hence the pressure of the gas is $P = -\partial F/\partial V = NT/V$, or

$$PV = NT.$$

(42.5)

Thus we have the familiar equation of state of an ideal gas (Clapeyron's equation). If the temperature is measured in degrees, then[†]

$$PV = NkT.$$

(42.5a)

Knowing F, we can find the other thermodynamic quantities also. For example, the thermodynamic potential is

$$\Phi = -NT \log (eV/N) + Nf(T) + PV.$$

Substituting $V = NT/P$ according to (42.5) (since Φ must be expressed as a function of P and T) and using a new function of temperature $\chi(T) = f(T) - T \log T$, we obtain

$$\Phi = NT \log P + N\chi(T).$$

(42.6)

The entropy is defined as

$$S = -\partial F/\partial T = N \log (eV/N) - Nf'(T),$$

(42.7)

or, as a function of P and T,

$$S = -\partial \Phi/\partial T = -N \log P - N\chi'(T).$$

(42.8)

[†] For a gram-molecule of gas ($N = 6.023 \times 10^{23}$ = Avogadro's number), the product $R = Nk$ is called the *gas constant*: $R = 8.314 \times 10^7$ erg/deg.

$$\mu = \left(\frac{\partial F}{\partial N} \right)$$

Finally, the energy is

$$E = F + TS = Nf(T) - NTf'(T). \tag{42.9}$$

We see that the energy is a function only of the temperature of the gas (and the same is true of the heat function $W = E + PV = E + NT$). This is evident *a priori*: since the molecules of an ideal gas are assumed not to interact, the change in their mean distance apart when the total volume of the gas varies cannot affect its energy.

As well as E and W, the specific heats $C_v = (\partial E/\partial T)_V$ and $C_p = (\partial W/\partial T)_P$ are functions only of the temperature. In what follows it will be convenient to use the specific heats per molecule, denoted by lower-case letter c:

$$C_v = Nc_v, \quad C_p = Nc_p. \tag{42.10}$$

Since for an ideal gas $W - E = NT$, the difference $c_p - c_v$ has a fixed value:[†]

$$c_p - c_v = 1. \tag{42.11}$$

PROBLEMS

PROBLEM 1. Find the work done on an ideal gas in an isothermal change of volume from V_1 to V_2 (or of pressure from P_1 to P_2).

SOLUTION. The required work R is equal to the change in the free energy of the gas, and from (42.4) we have

$$R = F_2 - F_1 = NT \log (V_1/V_2) = NT \log (P_2/P_1).$$

The quantity of heat absorbed in this process is

$$Q = T(S_2 - S_1) = NT \log (V_2/V_1).$$

The latter result also follows directly from the fact that $R + Q$ is the change of energy and is equal to zero for an isothermal process in an ideal gas.

PROBLEM 2. Two vessels contain two identical ideal gases at the same temperature T and with equal numbers of particles N but at different pressures P_1 and P_2. The vessels are then connected. Find the change in entropy.

SOLUTION. Before the vessels are connected, the entropy of the two gases is equal to the sum of their entropies, $S_0 = -N \log (P_1 P_2) - 2N\chi'(T)$. After the connection, the temperature of the gases remains the same (as follows from the conservation of energy for the two gases). The pressure is given by the relation

$$\frac{1}{P} = \frac{V_1 + V_2}{2NT} = \frac{1}{2} \left(\frac{1}{P_1} + \frac{1}{P_2} \right).$$

[†] Since the specific heat is the derivative of the energy (quantity of heat) with respect to temperature, C must be replaced by C/k in the formulae when ordinary units (degrees) are used. For example, formula (42.11) in ordinary units becomes $c_p - c_v = k$.

$$E = F + TS = -T^2 \frac{\partial}{\partial T}\left(\frac{F}{T}\right) = \frac{3}{2}NT + E'$$

The entropy is now

$$S = 2N \log \frac{P_1 + P_2}{2P_1 P_2} - 2N\chi'(T).$$

The change in entropy is therefore

$$\Delta S = N \log \frac{(P_1 + P_2)^2}{4P_1 P_2}.$$

PROBLEM 3. Find the energy of an ideal gas in a cylindrical vessel of radius R and length l rotating about its axis with angular velocity Ω.

SOLUTION. According to § 34, the rotation is equivalent to the presence of an external "centrifugal" field with potential energy $u = -\frac{1}{2}m\Omega^2 r^2$ (r being the distance of a particle from the axis of rotation).

When an external field is present, the integrand in (42.2) contains an extra factor $e^{-u/T}$, and so in the argument of the logarithm in (42.3) the volume V is replaced by the integral $\int e^{-u/T} dV$. Thus

$$F = F_0 - NT \log \frac{1}{V} \int e^{-u/T} dV,$$

where F_0 is the free energy of the gas in the absence of the external field.

In the present case this formula for the free energy becomes (in a rotating coordinate system)

$$F' = F_0 - NT \log \frac{1}{\pi R^2 l} \int_0^l \int_0^R e^{m\Omega^2 r^2/2T} 2\pi r \, dr \, dz$$

$$= F_0 - NT \log \left[\frac{2T}{m\Omega^2 R^2} (e^{m\Omega^2 R^2/2T} - 1) \right].$$

The angular momentum of the gas is

$$M = -\partial F'/\partial \Omega$$

$$= -\frac{2NT}{\Omega} + \frac{NmR^2\Omega}{1 - e^{-m\Omega^2 R^2/2T}}.$$

The energy in a system rotating with the body is

$$E' = F' - T \partial F'/\partial T = E_0 - \frac{Nm\Omega^2 R^2}{2(1 - e^{-m\Omega^2 R^2/2T})} + NT,$$

and in a fixed system of coordinates (see (26.5))

$$E = E' + M\Omega = E_0 + \frac{Nm\Omega^2 R^2}{2(1 - e^{-m\Omega^2 R^2/2T})} - NT,$$

where E_0 is the energy of the gas at rest.

§ 43. Ideal gases with constant specific heat

We shall see later that in many important cases the specific heat of a gas is constant, independent of temperature, over a greater or smaller temperature interval. For this reason we shall now calculate in a general form the thermodynamic quantities for such a gas.

On differentiating the expression (42.9) for the energy, we find that the function $f(T)$ is related to the specific heat c_v by $-Tf''(T) = c_v$. Integration of this relation gives

$$f(T) = -c_v T \log T - \zeta T + \varepsilon_0,$$

where ζ and ε_0 are constants. Substitution in (42.4) gives for the free energy the final expression

$$F = N\varepsilon_0 - NT \log (eV/N) - Nc_v T \log T - N\zeta T. \tag{43.1}$$

The constant ζ is called the *chemical constant* of the gas. For the energy we have

$$E = N\varepsilon_0 + Nc_v T, \tag{43.2}$$

a linear function of the temperature.

The thermodynamic potential Φ of the gas is obtained by adding to (43.1) the quantity $PV = NT$, the volume of the gas being expressed in terms of the pressure and the temperature. The result is

$$\Phi = N\varepsilon_0 + NT \log P - Nc_p T \log T - N\zeta T. \tag{43.3}$$

The heat function $W = E + PV$ is

$$W = N\varepsilon_0 + Nc_p T. \tag{43.4}$$

Finally, differentiating (43.1) and (43.3) with respect to temperature, we obtain the entropy in terms of T and V and of T and P respectively:

$$S = N \log (eV/N) + Nc_v \log T + (\zeta + c_v)N, \tag{43.5}$$
$$S = -N \log P + Nc_p \log T + (\zeta + c_p)N. \tag{43.6}$$

From these expressions for the entropy we can, in particular, derive immediately a relation (called the *Poisson adiabatic*) between the volume, temperature and pressure of an ideal gas (of constant specific heat) undergoing adiabatic expansion or compression. Since the entropy remains constant in an adiabatic process, we have from (43.6) $-N \log P + Nc_p \log T = $ constant, whence $T^{c_p}/P = $ constant or, using (42.11),

$$T^\gamma P^{1-\gamma} = \text{constant}, \tag{43.7}$$

where γ denotes the constant ratio

$$\gamma = c_p/c_v. \tag{43.8}$$

Using also the equation of state $PV = NT$, we obtain relations between T and V, and P and V:

$$TV^{\gamma-1} = \text{constant}, \quad PV^\gamma = \text{constant}. \tag{43.9}$$

PROBLEMS

PROBLEM 1. Two identical ideal gases at the same pressure P and containing the same number of particles N but at different temperatures T_1 and T_2 are in vessels with volumes V_1 and V_2. The vessels are then connected. Find the change in entropy.

SOLUTION. Before the vessels are connected, the entropy of the two gases, equal to the sum of their entropies, is by (43.6) $S_0 = -2N \log P + Nc_p \log (T_1 T_2)$.[†] After the connection, the temperatures of the gases become equal. The sum of the energies of the two gases remains constant. Using the expression (43.2) for the energy, we find $T = \frac{1}{2}(T_1+T_2)$, where T is the final temperature.

After the connection, the gas contains $2N$ particles and occupies a volume $V_1+V_2 = N(T_1+T_2)/P$. Its pressure is then $2NT/(V_1+V_2) = P$, i.e. the same as before. The entropy is

$$S = -2N \log P + 2Nc_p \log \left(\tfrac{1}{2}T_1 + \tfrac{1}{2}T_2\right),$$

and the change in entropy is

$$\Delta S = S - S_0$$
$$= Nc_p \log \frac{(T_1+T_2)^2}{4T_1 T_2}.$$

PROBLEM 2. Find the work done on an ideal gas in adiabatic compression.

SOLUTION. In an adiabatic process the quantity of heat $Q = 0$, and so $R = E_2-E_1$, where E_2-E_1 is the change in energy during the process. According to (43.2) $R = Nc_v(T_2-T_1)$, where T_1 and T_2 are the gas temperatures before and after the process. R can be expressed in terms of the initial and final volumes V_1 and V_2 by means of the relation (43.9):

$$R = Nc_v T_1[(V_1/V_2)^{\gamma-1}-1]$$
$$= Nc_v T_2[1-(V_2/V_1)^{\gamma-1}].$$

PROBLEM 3. Find the quantity of heat gained by a gas in an *isochoric* process, i.e. one which occurs at constant volume.

SOLUTION. Since in this case the work $R = 0$, we have

$$Q = E_2-E_1 = Nc_v(T_2-T_1).$$

[†] We everywhere omit the constant terms in the entropy and energy which are unimportant in the solution of problems.

PROBLEM 4. Find the work done and quantity of heat gained in an *isobaric* process, i.e. one which occurs at constant pressure.

SOLUTION. At constant pressure

$$R = -P(V_2 - V_1), \quad Q = W_2 - W_1,$$

whence

$$R = N(T_1 - T_2), \quad Q = Nc_p(T_2 - T_1).$$

PROBLEM 5. Find the work done on a gas and the quantity of heat which it gains in compression from volume V_1 to V_2 in accordance with the equation $PV^n = a$ (a *polytropic* process).

SOLUTION. The work is

$$R = -\int_{V_1}^{V_2} P \, dV = \frac{a}{n-1} (V_2^{1-n} - V_1^{1-n}).$$

Since the sum of the quantity of heat gained and the work done is equal to the total change in energy, we have $Q = Nc_v(T_2 - T_1) - R$, and since $T = PV/N = (a/N)V^{1-n}$,

$$Q = a \left(c_v + \frac{1}{1-n} \right) (V_2^{1-n} - V_1^{1-n}).$$

PROBLEM 6. Find the work done on an ideal gas and the quantity of heat which it gains on going through a cyclic process (i.e. one in which it returns to its initial state at the end of the process), consisting of two isochoric and two isobaric processes: the gas goes from a state with pressure and volume P_1, V_1 to states P_1, V_2; P_2, V_2; P_2, V_1; P_1, V_1 again.

SOLUTION. The change in energy in a cyclic process is zero, since the initial and final states are the same. The work done and the quantity of heat gained in such a process are therefore the same with opposite signs ($R = -Q$). In order to find R in the present case, we note that in isochoric processes the work done is zero, and for the two isobaric processes it is respectively $-P_1(V_2 - V_1)$ and $-P_2(V_1 - V_2)$. Thus $R = (V_2 - V_1)(P_2 - P_1)$.

PROBLEM 7. The same as Problem 6, but for a cyclic process consisting of two isochoric and two isothermal processes, the successive volumes and temperatures of the gas being $V_1, T_1; V_1, T_2; V_2, T_2; V_2, T_1; V_1, T_1$.

SOLUTION.

$$R = (T_2 - T_1) N \log (V_1/V_2).$$

PROBLEM 8. The same as Problem 6, but for a cyclic process consisting of two isothermal and two adiabatic processes, the successive entropies, temperatures and pressures being $S_1, T_1, P_1; S_1, T_2; S_2, T_2, P_2; S_2, T_1; S_1, T_1, P_1$.

SOLUTION.

$$Q = (T_2 - T_1)(S_2 - S_1)$$
$$= (T_2 - T_1) [N \log (P_1/P_2) + Nc_p \log (T_2/T_1)].$$

PROBLEM 9. The same as Problem 6, but for a cyclic process consisting of two isobaric and two isothermal processes, the successive states being P_1, T_1; P_1, T_2; P_2, T_2; P_2, T_1; P_1, T_1.

SOLUTION. The work done on the gas in the isobaric processes is (see Problem 4) $N(T_1-T_2)$ and $N(T_2-T_1)$, and that in the isothermal processes is $NT_2 \log(P_2/P_1)$ and $NT_1 \log(P_1/P_2)$. The sum of these is $R = N(T_2-T_1)\log(P_2/P_1)$.

PROBLEM 10. The same as Problem 6, but for a cyclic process consisting of two isobaric and two adiabatic processes, the successive states being P_1, S_1, T_1; P_1, S_2; P_2, S_2, T_2; P_2, S_1; P_1, S_1, T_1.

SOLUTION. The temperature in the second state is $T_2(P_2/P_1)^{(1-\gamma)/\gamma}$, and in the fourth state $T_1(P_1/P_2)^{(1-\gamma)/\gamma}$; these are obtained from T_1 and T_2 by means of (43.7). The quantity of heat gained by the gas in adiabatic processes is zero, and in the isobaric processes it is (see Problem 4)

$$Nc_p[T_2(P_2/P_1)^{(1-\gamma)/\gamma}-T_1] \quad \text{and}$$
$$Nc_p[T_1(P_1/P_2)^{(1-\gamma)/\gamma}-T_2].$$

Hence

$$Q = Nc_pT_1[(P_1/P_2)^{(1-\gamma)/\gamma}-1]+Nc_pT_2[(P_2/P_1)^{(1-\gamma)/\gamma}-1].$$

PROBLEM 11. The same as Problem 6, but for a cyclic process consisting of two isochoric and two adiabatic processes, the successive states being V_1, S_1, T_1; V_1, S_2; V_2, S_2, T_2; V_2, S_1; V_1, S_1, T_1.

SOLUTION. Using the result of Problem 2, we find

$$R = Nc_vT_2[1-(V_2/V_1)^{\gamma-1}]+Nc_vT_1[1-(V_1/V_2)^{\gamma-1}].$$

PROBLEM 12. Determine the maximum work that can be obtained by connecting vessels containing two identical ideal gases at the same temperature T_0 and with equal numbers of particles N but having different volumes V_1 and V_2.

SOLUTION. The maximum work is done if the process occurs reversibly (i.e. if the entropy remains constant), and is equal to the difference between the energies before and after the process (§ 19). Before the connection of the vessels, the entropy of the two gases is equal to the sum of their entropies, i.e. by (43.5)

$$S_0 = N\log(e^2V_1V_2/N^2)+2Nc_v\log T_0.$$

After the connection we have a gas consisting of $2N$ particles occupying a volume V_1+V_2 at some temperature T. Its entropy is

$$S = 2N\log[e(V_1+V_2)/2N]+2Nc_v\log T.$$

Equating S_0 and S gives the temperature T:

$$T = T_0\left[\frac{4V_1V_2}{(V_1+V_2)^2}\right]^{(\gamma-1)/2}.$$

The energy of the two gases before and after the connection is $E_0 = 2Nc_vT_0$ and $E = 2Nc_vT$ respectively. The maximum work is therefore

$$R_{max} = E_0-E = 2Nc_v(T_0-T) = 2Nc_vT_0\left[1-\left(\frac{4V_1V_2}{(V_1+V_2)^2}\right)^{(\gamma-1)/2}\right].$$

PROBLEM 13. The same as Problem 12, but for gases with the same pressure P_0 and different temperatures T_1 and T_2 before the connection of the vessels.

SOLUTION. We have similarly

$$R_{max} = Nc_v \left\{ T_1 + T_2 - 2^\gamma \sqrt{(T_1 T_2)} \left[\frac{T_1 T_2}{(T_1 + T_2)^2} \right]^{(\gamma-1)/2} \right\}.$$

PROBLEM 14. Find the minimum work that must be done on an ideal gas in order to compress it from pressure P_1 to P_2 at a constant temperature equal to that of the surrounding medium ($T = T_0$).

SOLUTION. According to (20.2) the minimum work is $R_{min} = (E_2 - E_1) - T_0(S_2 - S_1) + P_0(V_2 - V_1)$, where the suffixes 1 and 2 refer to the gas before and after compression. In the present case the energy E is unchanged (since the temperature is constant), i.e. $E_2 - E_1 = 0$. Using (43.6), we find the change of entropy corresponding to the change of pressure from P_1 to P_2: $S_2 - S_1 = N \log (P_1/P_2)$, and the change of volume is $V_2 - V_1 = NT_0(1/P_2 - 1/P_1)$. Hence

$$R_{min} = NT_0 \left[\log \frac{P_2}{P_1} + P_0 \left(\frac{1}{P_2} - \frac{1}{P_1} \right) \right].$$

PROBLEM 15. Determine the maximum work which can be obtained from an ideal gas cooled from temperature T to the temperature of the medium T_0 at constant volume.

SOLUTION. From the general formula (20.3),

$$R_{max} = Nc_v(T - T_0) + Nc_v T_0 \log (T_0/T).$$

PROBLEM 16. The same as Problem 15, but for a gas cooled from temperature T to the temperature of the medium T_0 and at the same time expanding from pressure P to the pressure of the medium P_0.

SOLUTION.

$$R_{max} = Nc_v(T - T_0) + NT_0 \log (P/P_0) + Nc_p T_0 \log (T_0/T) + N(TP_0/P - T_0).$$

PROBLEM 17. Gas at temperature T_0 flows from a large thermally isolated reservoir into an empty thermally isolated vessel, the gas pressure in the reservoir remaining constant. Find the change in the gas temperature.

SOLUTION. The energy E of the gas in the vessel consists of the energy E_0 which it had in the reservoir and the work done on it to "expel" it from the reservoir. Since the state of the gas in the reservoir may be regarded as steady, we have the condition $W_0 = E$ (cf. § 18). The gas temperature in the vessel is therefore $T = \gamma T_0$.

§ 44. The law of equipartition

Before going on to calculate in detail the thermodynamic quantities for gases, making allowance for the various quantum effects, it is useful to consider the same problem from the point of view of purely classical statistics.

[handwritten margin notes: If T >> energy spacing, each independent quadratic term in E contributes ½T to the energy]

We shall see later when and to what extent the results obtained are applicable to actual gases.

A molecule is a configuration of atoms executing small oscillations about certain equilibrium positions which correspond to minimum potential energy of their interaction. The potential energy is then of the form

$$u = \varepsilon_0 + \sum_{i,\,k=1}^{r_{vib}} a_{ik} q_i q_k,$$

where ε_0 is the potential energy of interaction of the atoms when they are all in their equilibrium positions; the second term is a quadratic function of the coordinates which give the deviations of the atoms from their equilibrium positions. The number r_{vib} of coordinates in this function is the number of vibrational degrees of freedom of the molecule.

This number can be determined from the number n of atoms in the molecule. A molecule containing n atoms has a total of $3n$ degrees of freedom. Three of these correspond to the translational motion of the molecule as a whole, and three to its rotation as a whole. If all the atoms are collinear (and in particular for a diatomic molecule) there are only two rotational degrees of freedom. Thus a non-linear molecule of n atoms has $3n-6$ vibrational degrees of freedom, and a linear one has $3n-5$. For $n = 1$ there are, of course, no vibrational degrees of freedom, since all three degrees of freedom of an atom correspond to translational motion.

The total energy ε of the molecule is the sum of the potential and kinetic energies. The latter is a quadratic function of all the momenta, and these are equal in number to the full $3n$ degrees of freedom of the molecule. The energy ε is therefore of the form $\varepsilon = \varepsilon_0 + f_{II}(p, q)$, where $f_{II}(p, q)$ is a quadratic function of the momenta and coordinates; the total number of variables in this function is $l = 6n-6$ (for a non-linear molecule) or $6n-5$ (for a linear molecule); in a monatomic gas, $l = 3$, since the coordinates do not appear at all in the expression for the energy.

Substituting this expression for the energy in (41.5) we have

$$F = -NT \log \frac{e \cdot e^{-\varepsilon_0/T}}{N} \int e^{-f_{II}(p,\,q)/T} \, d\tau.$$

In order to find the dependence on temperature of the integral in this formula we substitute $p = p' \sqrt{T}, q = q' \sqrt{T}$ for all the l variables on which the function $f_{II}(p, q)$ depends. Since this function is quadratic, $f_{II}(p, q) = T f_{II}(p', q')$, and T cancels in the exponent of the integrand. The transformation of the differentials of these variables in $d\tau$ gives a factor $T^{l/2}$, which can be taken outside the integral. The integration over the vibrational coordinates q is taken over the range of values corresponding to vibrations in which the atoms remain within the molecule. However, since the integrand diminishes rapidly with increasing q, the integration may be extended to the whole range from $-\infty$ to ∞, as well as over all the momenta. The above-mentioned change of variables then leaves the limits of integration unaltered, and the whole integral is a constant independent of temperature. Using also the fact that the integration with respect to the coordinates of the centre of mass of the molecule gives the volume V occupied by the gas, we obtain for the free energy an expression of the form

$$F = -NT \log{(AVe^{-\varepsilon_0/T}T^{l/2}/N)},$$

where A is a constant. Expanding the logarithm, we have an expression of exactly the type (43.1) with a constant specific heat equal to

$$c_v = \tfrac{1}{2}l. \tag{44.1}$$

The specific heat $c_p = c_v + 1$ is accordingly

$$c_p = \tfrac{1}{2}(l+2). \tag{44.2}$$

Thus we see that a purely classical ideal gas must have a constant specific heat. Formula (44.1) enables us to state the following rule. Each variable in the energy $\varepsilon(p, q)$ of the molecule gives an equal contribution of $\tfrac{1}{2}$ to the specific heat c_v of the gas $\left(\tfrac{1}{2}k \text{ in ordinary units}\right)$ or, what is the same thing, an equal contribution of $\tfrac{1}{2}T$ to its energy. This is called the *law of equipartition*.

Since for the translational and rotational degrees of freedom the energy $\varepsilon(p, q)$ contains only the corresponding momenta, we can say that each of these degrees of freedom gives a contribution of $\tfrac{1}{2}$ to the specific heat. Each vibrational degree of freedom corresponds to two variables (coordinate and momentum) in the energy $\varepsilon(p, q)$, and its contribution to the specific heat is 1.

For the model considered here it is easy to find a general formula for the energy distribution of the gas molecules. For convenience we shall measure the energy of a molecule from the value ε_0, i.e. omit this constant from the expression for $\varepsilon(p, q)$. Let us consider the volume in the phase space of the molecule whose points correspond to values of $\varepsilon(p, q)$ not exceeding a given value ε, i.e. determine the integral $\tau(\varepsilon) = \int d\tau$ taken over the region $\varepsilon(p, q) \leqslant \varepsilon$. According to the foregoing discussion, $\varepsilon(p, q)$ is a quadratic function of l variables. We replace those l quantities p, q on which the energy $\varepsilon(p, q)$ depends by new variables $p' = p/\sqrt{\varepsilon}, q' = q/\sqrt{\varepsilon}$. Then the condition $\varepsilon(p, q) \leqslant \varepsilon$ becomes $\varepsilon(p', q') \leqslant 1$, and $\int d\tau$ becomes $\varepsilon^{l/2} \int d\tau'$. The integral $\int d\tau'$ is clearly independent of ε, and so $\tau = \text{constant} \times \varepsilon^{l/2}$, whence $d\tau(\varepsilon) = \text{constant} \times \varepsilon^{l/2-1} d\varepsilon$ and the energy probability distribution is

$$dw_\varepsilon = Ae^{-\varepsilon/T}\varepsilon^{l/2-1} d\varepsilon.$$

$\bar{\varepsilon}_h =$
$\sum{(a_i p_i^2 + b_j q_j^2)}$
$\bar{\varepsilon} = \int \varepsilon_h e^{-\frac{\varepsilon_h}{T}} dq \, dp$
$\int e^{-\frac{\varepsilon_h}{T}} dp \, dq$

concerning p's

$= \dfrac{\displaystyle\int(\sum a_j p_j^2)\, e^{-\frac{\sum a_j p_j^2}{T}} dp_j}{z}$

$= \left[\displaystyle\int e^{-b_j q_j^2} dq_j \displaystyle\int a_i p_i^2 e^{-\frac{a_i p_i^2}{T}} dp_i\right]\cdots$

$= \bar{\varepsilon}_1 + \bar{\varepsilon}_2 + \cdots$

$\bar{\varepsilon}_1 = \dfrac{\displaystyle\int a_i p_i^2 e^{-\frac{a_i p_i^2}{T}} dp_i}{\displaystyle\int e^{-\frac{a_i p_i^2}{T}} dp_i}$

$= \dfrac{1}{2} T$

Determining A from the normalisation condition, we find

$$dw_\varepsilon = \frac{1}{T^{1/2}\Gamma(\frac{1}{2}l)} e^{-\varepsilon/T}\varepsilon^{\frac{1}{2}l-1}\, d\varepsilon. \tag{44.3}$$

PROBLEM

Find the specific heat of an ideal gas in the extreme relativistic case, where the energy of a particle is related to its momentum by $\varepsilon = cp$, c being the velocity of light.

SOLUTION. According to (41.5) we have

$$F = -NT \log \frac{eV}{N(2\pi\hbar)^3} \int_0^\infty e^{-cp/T}\cdot 4\pi p^2\, dp$$

or, after carrying out the integration,

$$F = -NT \log (AVT^3/N),$$

where A is a constant. The specific heat is therefore $c_v = 3$, which is twice the value for a non-relativistic monatomic gas.

§ 45. Monatomic ideal gases

The complete calculation of the free energy (and therefore of the other thermodynamic quantities) for an ideal gas requires a calculation of the specific form of the partition function in the logarithm in (42.3),

$$Z = \sum_k e^{-\varepsilon_k'/T}.$$

Here ε_k' are the energy levels of the atom or molecule (the kinetic energy of the translational motion of the particle being excluded). If the summation is taken only over all the different energy levels, it must be remembered that a level may be degenerate, and in this case the corresponding term must appear in the sum over all states as many times as the degree of degeneracy. Let this be g_k. The degree of degeneracy of the level is often called in this connection its *statistical weight*. Omitting for brevity the prime in ε_k', we can write the partition function concerned in the form

$$Z = \sum_k g_k e^{-\varepsilon_k/T}. \tag{45.1}$$

The free energy of the gas is

$$F = -NT \log \left[\frac{eV}{N}\left(\frac{mT}{2\pi\hbar^2}\right)^{3/2} Z\right]. \tag{45.2}$$

$$E = \sum_{i=1}^{N} \tfrac{1}{2}m\, v_{xi}^2 + \tfrac{1}{2}m\, v_{yi}^2 + \tfrac{1}{2}m\, v_{zi}^2$$

$$= N\left(\tfrac{3}{2}T\right) = \tfrac{3}{2}NT$$

Z - individual molecules.

Turning now to the consideration of monatomic gases, we must first of all make the following important comment. As the gas temperature increases, so does the number of atoms in excited states, including the states of the continuous spectrum, which correspond to ionisation of the atom. When the temperature is not too high, the relative number of ionised atoms in the gas is negligible, but the gas is almost completely ionised at temperatures T of the order of the ionisation energy I_{ion}, and not only for $T \gg I_{ion}$ (see § 104). Thus a non-ionised gas can reasonably be considered only at temperatures such that $T \ll I_{ion}$.[†]

The atomic terms (neglecting their fine structure) are so situated that the separation between the ground state and the first excited level is comparable with the ionisation energy. At temperatures $T \ll I_{ion}$, the gas will therefore be practically free not only of ionised atoms but also of excited atoms, and so all the atoms may be regarded as being in the ground state.

Let us first consider the simplest case, that of atoms which in their ground state have neither orbital angular momentum nor spin ($L = S = 0$), such as the atoms of the inert gases. The ground state is not degenerate, and the partition function reduces to a single term, $Z = e^{-\varepsilon_0/T}$. For monatomic gases it is customary to put $\varepsilon_0 = 0$, i.e. to measure the energy from the ground state of the atom, so that $Z = 1$. Expanding the logarithm in (45.2) as a sum of logarithms, we obtain for the free energy an expression of the type (43.1), with constant specific heat

$$c_v = \tfrac{3}{2} \qquad\qquad (45.3)$$

and chemical constant

$$\zeta = \frac{3}{2}\log\frac{m}{2\pi\hbar^2} \qquad\qquad (45.4)$$

(O. Sackur and H. Tetrode, 1912).

This value of the specific heat is due entirely to the translational degrees of freedom of the atom ($\tfrac{1}{2}$ for each degree of freedom); it will be remembered that the translational motion of the gas particles is always quasi-classical. The "electronic degrees of freedom" under these conditions (no excited atoms in the gas) have, of course, no effect on the thermodynamic quantities.[‡]

[†] For different atoms the temperature I_{ion}/k lies between 5×10^4 degrees (alkali metal atoms) and 28×10^4 degrees (helium).

[‡] The electronic part of the thermodynamic quantities, naturally, can never be treated classically. In this connection we may note the fact (which in essence has been tacitly assumed already) that in classical statistics the atoms must be regarded as particles without internal structure. The impossibility of applying to effects within the atom a statistics based on classical mechanics is further shown by the absurd result obtained on substituting the interaction energy between the electrons

These expressions enable us to deduce a criterion for the validity of Boltz-mann statistics. In this statistics it is assumed that

$$\bar{n}_k = e^{(\mu - \varepsilon_k)/T} \ll 1$$

(see (37.1)). It is clearly sufficient to require the fulfilment of the condition

$$e^{\mu/T} \ll 1.$$

For the chemical potential $\mu = \Phi/N$ we have from (43.3), with c_v and ζ given by (45.3) and (45.4),

$V_Q = \left(\dfrac{2\pi \hbar}{mT}\right)^{3/2}$ Quantum Volume

$$\mu = T \log\left[\frac{P}{T^{5/2}}\left(\frac{2\pi\hbar^2}{m}\right)^{3/2}\right]$$

$$= T \log\left[\frac{N}{V}\left(\frac{2\pi\hbar^2}{mT}\right)^{3/2}\right]. \tag{45.5}$$

Thus we obtain the condition

$$(N/V)(\hbar^2/mT)^{3/2} \ll 1. \tag{45.6}$$

For a given temperature, this condition requires that the gas should be sufficiently rarefied. Substitution of numerical values shows that in practice, for any atomic (or molecular) gas, this condition can be violated only at densities where the interaction of the particles becomes important and the gas can in any case no longer be regarded as ideal.

It is useful to note the following intuitive interpretation of the above con-dition. Since the majority of atoms have energies of the order of T, and therefore momenta of the order of $\sqrt{(mT)}$, we can say that all the atoms oc-cupy in phase space a volume of the order of $V(mT)^{3/2}$, corresponding to $\sim V(mT)^{3/2}/\hbar^3$ quantum states. In the Boltzmann case this number must be large compared with the number N of particles, and hence we have (45.6).

Finally, we may make the following comment. The formulae derived in this section appear at first sight to contradict Nernst's theorem, since neither the entropy nor the specific heat is zero at $T = 0$. However, it must be remembered that, under the conditions for which Nernst's theorem is stated, all actual gases condense at sufficiently low temperatures. For Nernst's theorem requires that the entropy of a body should tend to zero at $T = 0$ for a fixed value of its volume. But as $T \to 0$ the saturated vapour pressure of all

---

and the atomic nucleus in the classical distribution formulae. This energy is of the form $-a/r$, where r is the distance of the electron from the nucleus and a is a con-stant. The substitution would give a factor $e^{a/rT}$ in the distribution, which becomes infinite for $r = 0$. This would mean that all the electrons would have to "fall" into the nucleus in thermal equilibrium.

substances becomes arbitrarily small, so that a fixed finite quantity of matter in a fixed finite volume cannot remain gaseous as $T \to 0$.

If we consider a model of a gas, possible in principle, which consists of mutually repulsive particles, then, although such a gas will never condense, at sufficiently low temperatures Boltzmann statistics ceases to be valid, and the application of Fermi or Bose statistics leads, as we shall see later, to expressions which are in agreement with Nernst's theorem.

§ 46. Monatomic gases. The effect of the electronic angular momentum

If only one of the angular momenta L and S is non-zero in the ground state of the atom, then this state again has no fine structure. In practice the absence of fine structure of the ground state is always due to a zero orbital angular momentum; the spin S is sometimes not zero (for example, atoms in the vapour of alkali metals).

A level with spin S is $(2S+1)$-fold degenerate. The only difference as compared with the case discussed in § 45 is that the partition function Z is now $2S+1$ instead of 1, and so the chemical constant (45.4) is increased by the quantity[†]

$$\zeta_S = \log(2S+1). \tag{46.1}$$

If the ground term of an atom has a fine structure, it must be remembered that the intervals in this structure may generally be comparable with T, and so all the components of the fine structure of the ground state must be taken into account in the partition function.

The fine-structure components differ in the value of the total angular momentum of the atom (with given orbital angular momentum L and spin S). Let these levels, measured from the lowest of them, be denoted by ε_J. Each level with a given J is $(2J+1)$-fold degenerate with respect to orientations of the total angular momentum.[‡] The partition function therefore becomes

$$Z = \sum_J (2J+1)e^{-\varepsilon_J/T}; \tag{46.2}$$

[†] We may write out for reference the formula for the chemical potential of a monatomic ideal gas with statistical weight (degree of degeneracy) of the ground state g:

$$\mu = T \log\left[\frac{P}{gT^{5/2}}\left(\frac{2\pi\hbar^2}{m}\right)^{3/2}\right] = T \log\left[\frac{N}{gV}\left(\frac{2\pi\hbar^2}{mT}\right)^{3/2}\right]. \tag{46.1a}$$

This applies also to a Boltzmann gas of elementary particles; for instance, in an electron gas $g = 2$.

[‡] We assume that Russell–Saunders coupling is valid in the atom; see *Quantum Mechanics*, § 72.

the summation is taken over all possible values of J for the given L and S. We obtain for the free energy

$$F = -NT \log \left[\frac{eV}{N} \left(\frac{mT}{2\pi\hbar^2} \right)^{3/2} \sum_J (2J+1)e^{-\varepsilon_J/T} \right]. \qquad (46.3)$$

This expression becomes considerably simpler in two limiting cases. Let us assume that the temperature is so high that T is large in comparison with all the fine-structure intervals: $T \gg \varepsilon_J$. Then we can put $e^{-\varepsilon_J/T} \cong 1$, and Z becomes simply the total number of fine-structure components $(2S+1)(2L+1)$. The expression for the free energy involves the constant specific heat $c_v = \frac{3}{2}$ as before, and the quantity

$$\zeta_{SL} = \log\left[(2S+1)(2L+1)\right] \qquad (46.4)$$

is added to the chemical constant (45.4).

Similar expressions for the thermodynamic quantities (with a different ζ) are obtained in the opposite limiting case where T is small compared with the fine-structure intervals.[†] In this case all terms may be neglected in the sum (46.2) except the one with $\varepsilon_J = 0$ (the lowest component of the fine structure, i.e. the ground state of the atom). In consequence the quantity added to the chemical constant (45.4) is

$$\zeta_J = \log(2J+1), \qquad (46.5)$$

where J is the total angular momentum of the atom in the ground state.

Thus, when the ground term of the atom has fine structure, the specific heat of the gas at sufficiently low and sufficiently high temperatures has the same constant value, but in the intermediate range it depends on the temperature and passes through a maximum. It must be borne in mind, however, that for gases concerned in practice (heavy-metal vapours, atomic oxygen, etc.) only the range of high temperatures, where the specific heat becomes constant, is of importance.

So far we have ignored the possibility that the atom has a non-zero nuclear spin i. The existence of such a spin causes the *hyperfine splitting* of atomic levels. The intervals in this structure are, however, so small that they may be neglected in comparison with T at all temperatures where the gas remains a gas.[‡] In calculating the partition function, the energy differences between the

---

[†] As examples, the quantities ε_J/k for the components of the triplet ground state of the oxygen atom are 230° and 320°; for those of the quintet ground state of the iron atom they are between 600° and 1400°; and for the doublet ground state of the chlorine atom 1300°.

[‡] The temperatures corresponding to the hyperfine structure intervals of various atoms range from 0.1° to 1.5°.

hyperfine multiplet components may be entirely neglected, and the splitting need be taken into account only as increasing the degree of degeneracy of each level (and therefore the sum Z) by a factor $2i+1$. Accordingly, the free energy contains an additional "nuclear" term

$$F_{\text{nuc}} = -NT \log{(2i+1)}. \tag{46.6}$$

This term does not affect the specific heat of the gas (the corresponding energy $E_{\text{nuc}} = 0$) and simply changes the entropy by $S_{\text{nuc}} = N \log{(2i+1)}$, and the chemical constant by $\zeta_{\text{nuc}} = \log{(2i+1)}$.

Because the interaction between the nuclear spin and the electron shells is extremely weak, the "nuclear" part of the thermodynamic quantities usually plays no part in the various thermal processes and does not appear in the equations. We shall therefore omit these terms, as is usually done; in other words, we shall measure the entropy not from zero but from the value S_{nuc} due to the nuclear spins.

§ 47. Diatomic gases with molecules of unlike atoms. Rotation of molecules

Turning now to the calculation of the thermodynamic quantities for a diatomic gas, we may point out first of all that, just as monatomic gases can reasonably be considered only for temperatures T which are small compared with the ionisation energy, a diatomic gas can be regarded as such only if T is small compared with the dissociation energy of the molecule.[†] This in turn means that only the lowest electronic state of the molecule need be retained in the partition function.

Let us begin with the most important case, where the gas molecule in the lowest electronic state has neither spin nor orbital angular momentum about the axis ($S = 0$, $\Lambda = 0$); such an electronic term has, of course, no fine structure. We must also distinguish molecules composed of unlike atoms (including different isotopes of the same element) from those composed of like atoms, since the latter case has certain specific properties. In the present section we shall assume that the molecule consists of unlike atoms.

The energy level of a diatomic molecule is, to a certain approximation, the sum of three independent parts: the electron energy (which includes also the energy of the Coulomb interaction of the nuclei in their equilibrium position and will be measured from the sum of the energies of the separated atoms), the rotational energy, and the vibrational energy of the nuclei within the molecule. For a singlet electronic state, these levels may be written (see

---

[†] As examples, the temperatures I_{diss}/k for some diatomic molecules are H_2 52,000°, N_2 113,000°, O_2 59,000°, Cl_2 29,000°, NO 61,000°, CO 98,000°.

Quantum Mechanics, § 82)

$$\varepsilon_{vK} = \varepsilon_0 + \hbar\omega(v + \tfrac{1}{2}) + \hbar^2 K(K+1)/2I, \tag{47.1}$$

where ε_0 is the electron energy, $\hbar\omega$ the vibrational quantum, v the vibrational quantum number, K the rotational quantum number (angular momentum of the molecule), $I = m'r_0^2$ the moment of inertia of the molecule ($m' = m_1 m_2/(m_1 + m_2)$ is the reduced mass of the two atoms and r_0 the equilibrium value of the distance between the nuclei).

When the expression (47.1) is substituted in the partition function, the latter is resolved into three independent factors:

$$Z = e^{-\varepsilon_0/T} Z_{rot} Z_{vib}, \tag{47.2}$$

where the "rotational" and "vibrational" sums are defined by

$$Z_{rot} = \sum_{K=0}^{\infty} (2K+1)e^{-\hbar^2 K(K+1)/2TI}, \tag{47.3}$$

$$Z_{vib} = \sum_{v=0}^{\infty} e^{-\hbar\omega(v+\frac{1}{2})/T}, \tag{47.4}$$

the factor $2K+1$ in Z_{rot} taking account of the degeneracy of the rotational levels with respect to the orientations of the angular momentum $\mathbf{K}$. Accordingly, the free energy is the sum of three parts:

$$F = -NT \log \left[\frac{eV}{N} \left(\frac{mT}{2\pi\hbar^2} \right)^{3/2} \right] + F_{rot} + F_{vib} + N\varepsilon_0, \tag{47.5}$$

where $m = m_1 + m_2$ is the mass of the molecule. The first term may be called the *translational part* F_{tr} (since it arises from the degrees of freedom of the translational motion of the molecules), and

$$F_{rot} = -NT \log Z_{rot}, \quad F_{vib} = -NT \log Z_{vib} \tag{47.6}$$

the *rotational* and *vibrational* parts. The translational part is always given by a formula of the type (43.1) with a constant specific heat $c_{tr} = \tfrac{3}{2}$ and chemical constant

$$\zeta_{tr} = \frac{3}{2} \log \frac{m}{2\pi\hbar^2}. \tag{47.7}$$

The total specific heat of the gas is the sum of several terms:

$$c_v = c_{tr} + c_{rot} + c_{vib},$$
$$c_p = c_{tr} + c_{rot} + c_{vib} + 1, \tag{47.8}$$

which arise respectively from the thermal excitation of the translational motion of the molecule, its rotation and the vibrations of atoms within the molecule.

Let us next calculate the rotational free energy. If the temperature is so high that $T \gg \hbar^2/2I$ (i.e. the "rotational quantum" $\hbar^2/2I$ is small compared with T),[†] then the terms with large K are the most important in the sum (47.3). For large values of K, the rotation of the molecule is quasi-classical. In this case, therefore, the partition function Z_{rot} can be replaced by the corresponding classical integral:

$$Z_{rot} = \int e^{-\varepsilon(\mathbf{M})/T} \, d\tau_{rot}, \tag{47.9}$$

where $\varepsilon(\mathbf{M})$ is the classical expression for the kinetic energy of rotation as a function of the angular momentum $\mathbf{M}$. Using a system of coordinates ξ, η, ζ rotating with the molecule, with the ζ-axis along the axis of the molecule, and bearing in mind that a diatomic molecule has two rotational degrees of freedom and the rotational angular momentum of a linear mechanical system is perpendicular to its axis, we can write

$$\varepsilon(\mathbf{M}) = (M_\xi^2 + M_\eta^2)/2I.$$

The element $d\tau_{rot}$ is the product of the differentials dM_ξ, dM_η and the differentials $d\phi_\xi$, $d\phi_\eta$ of the "generalised coordinates" corresponding to M_ξ, M_η (i.e. the infinitesimal angles of rotation about the ξ and η axes), divided by $(2\pi\hbar)^2$.[‡] The product of two infinitesimal angles of rotation about the ξ and η axes is just the element of solid angle do_ζ for the direction of the third axis ζ, and integration over the solid angle gives 4π. Thus[§]

$$Z_{rot} = \frac{4\pi}{(2\pi\hbar)^2} \int_{-\infty}^{\infty} \int_{-\infty}^{\infty} \exp\left[-\frac{1}{2TI}(M_\xi^2 + M_\eta^2)\right] dM_\xi \, dM_\eta = 2IT/\hbar^2.$$

Hence the free energy is

$$F_{rot} = -NT \log T - NT \log (2I/\hbar^2). \tag{47.10}$$

[†] In practice this condition is always satisfied for all gases except the two isotopes of hydrogen. As examples, the values of $\hbar^2/2kI$ are: H_2 85.4°, D_2 43°, HD 64°, N_2 2.9°, O_2 2.1°, Cl_2 0.36°, NO 2.4°, HCl 15.2°.

[‡] It must be remembered that this notation is to some extent arbitrary, since $d\phi_\xi$ and $d\phi_\eta$ are not total differentials of any function of the position of the axes.

[§] This value of Z_{rot} can also be derived in another way: assuming that the numbers K in the sum (47.3) are large and replacing the summation by integration with respect to K, we have

$$Z_{rot} \cong \int_0^{\infty} 2Ke^{-K^2\hbar^2/2IT} \, dK = 2TI/\hbar^2.$$

Thus, at the relatively high temperatures under consideration, the rotational part of the specific heat is a constant, $c_{rot} = 1$, in accordance with the general results of the classical treatment in § 44 ($\frac{1}{2}$ for each rotational degree of freedom). The rotational part of the chemical constant is $\zeta_{rot} = \log(2I/\hbar^2)$. We shall see below that there is a considerable range of temperatures over which the condition $T \gg \hbar^2/2I$ holds and at the same time the vibrational part of the free energy, and therefore the vibrational part of the specific heat, are zero. Over this range the specific heat of a diatomic gas $c_v = c_{tr} + c_{rot}$, i.e.

$$c_v = \tfrac{5}{2}, \quad c_p = \tfrac{7}{2}, \tag{47.11}$$

and the chemical constant $\zeta = \zeta_{tr} + \zeta_{rot}$:

$$\zeta = \log[(2I/\hbar^5)(m/2\pi)^{3/2}]. \tag{47.12}$$

In the opposite limiting case of low temperatures, $T \ll \hbar^2/2I$, it is sufficient to retain the first two terms of the sum:

$$Z_{rot} = 1 + 3e^{-\hbar^2/IT},$$

and for the free energy we have in the same approximation

$$F_{rot} = -3NTe^{-\hbar^2/IT}. \tag{47.13}$$

Hence the entropy is

$$S_{rot} = \frac{3N\hbar^2}{IT} e^{-\hbar^2/IT}(1 + IT/\hbar^2) \tag{47.14}$$

and the specific heat is

$$C_{rot} = 3N(\hbar^2/IT)^2 e^{-\hbar^2/IT}. \tag{47.15}$$

Thus the rotational entropy and specific heat of the gas tend to zero essentially exponentially as $T \to 0$. At low temperatures, therefore, a diatomic gas behaves like a monatomic one; both the specific heat and the chemical constant have the same values as in a monatomic gas of particles of mass m.

In the general case of arbitrary temperatures the sum Z_{rot} must be calculated numerically. Fig. 4 shows c_{rot} as a function of $2TI/\hbar^2$. The rotational specific heat has a maximum of 1.1 at $T = 0.81(\hbar^2/2I)$, and then tends asymptotically to the classical value 1.[†]

[†] An asymptotic expansion of the thermodynamic quantities for large values of $2TI/\hbar^2$ may be obtained. The first two terms of the expansion for the specific heat are

$$c_{rot} = 1 + \frac{1}{45}\left(\frac{\hbar^2}{2TI}\right)^2.$$

It must be remembered, however, that this expansion gives only a poor approximation to the function $c_{rot}(T)$.

FIG. 4

§ 48. Diatomic gases with molecules of like atoms. Rotation of molecules

Diatomic molecules consisting of like atoms have certain specific properties which necessitate changes in some of the formulae derived in § 47.

First of all, let us consider the limiting case of high temperatures, where
a classical treatment is possible. Since the two nuclei are identical, two opposite positions of the axis of the molecule (differing only in that the two nuclei
are interchanged) now correspond to the same physical state of the molecule.
The classical partition function (47.9) must therefore be halved, and so the
chemical constant becomes

$$\zeta_{\text{rot}} = \log{(I/\hbar^2)}; \tag{48.1}$$

accordingly the factor 2 disappears from the argument of the logarithm in the
sum $\zeta_{\text{tr}} + \zeta_{\text{rot}}$ (47.12).

More important changes are needed at temperatures where the quantum
treatment has to be used. Since in practice the entire problem is of interest
only in its application to the two isotopes of hydrogen (H_2 and D_2), we shall
consider these gases in what follows. The requirement of quantum-mechanical symmetry in the nuclei (see *Quantum Mechanics*, § 86) results in the
electronic state $^1\Sigma_g^+$ (the ground term of the hydrogen molecule) having
rotational levels of different nuclear spin degeneracy for even and odd values
of K: levels with even and odd K respectively occur only for even and odd
total spin of the two nuclei, and have relative degrees of degeneracy $g_g =
i/(2i+1)$, $g_u = (i+1)/(2i+1)$ for a half-integral spin i of the nuclei, and $g_g =
(i+1)/(2i+1)$, $g_u = i/(2i+1)$ for integral i. For hydrogen there is an accepted
terminology whereby the molecules in states of greater nuclear statistical
weight are called *orthohydrogen* molecules, and those in states of smaller

statistical weight are called *parahydrogen* molecules. Thus for the H_2 and D_2 molecules the statistical weights are

$$H_2 \ (i = \tfrac{1}{2}) \begin{cases} \text{ortho} & g_u = \tfrac{3}{4}, \\ \text{para} & g_g = \tfrac{1}{4}, \end{cases} \qquad D_2 \ (i = 1) \begin{cases} \text{ortho} & g_g = \tfrac{2}{3}, \\ \text{para} & g_u = \tfrac{1}{3}. \end{cases}$$

The suffix g denotes that the molecule has an even total nuclear spin (0 for H_2, 0 or 2 for D_2) and even rotational angular momenta K; the suffix u signifies odd total nuclear spins (1 for H_2 and D_2) and odd values of K.

Whereas, in molecules with unlike nuclei, the nuclear degrees of degeneracy of all the rotational levels are the same, and so the allowance for this degeneracy simply gives an unimportant change in the chemical constant, here it causes a change in the form of the partition function, which must now be written[†]

$$Z_{rot} = g_g Z_g + g_u Z_u, \tag{48.2}$$

where

$$Z_g = \sum_{K=0,2,\ldots} (2K+1)e^{-\hbar^2 K(K+1)/2IT}, \tag{48.3}$$
$$Z_u = \sum_{K=1,3,\ldots} (2K+1)e^{-\hbar^2 K(K+1)/2IT}.$$

Similarly the free energy becomes

$$F_{rot} = -NT \log (g_g Z_g + g_u Z_u), \tag{48.4}$$

and the remaining thermodynamic quantities are likewise changed. At high temperatures,

$$Z_g \cong Z_u \cong \tfrac{1}{2} Z_{rot} = TI/\hbar^2,$$

so that the previous classical expression is obtained for the free energy, as it should be.

As $T \to 0$ the sum Z_g tends to unity and Z_u tends exponentially to zero; at low temperatures, therefore, the gas behaves as if monatomic (the specific heat $c_{rot} = 0$) and the chemical constant simply contains a nuclear part $\zeta_{nuc} = \log g_g$.

The above formulae relate, of course, to a gas in complete thermal equilibrium. In such a gas the ratio of the numbers of molecules of parahydrogen and orthohydrogen is a definite function of temperature, which from the Boltzmann distribution is

$$x_{H_2} = N_{ortho-H_2}/N_{para-H_2} = g_u Z_u/g_g Z_g = 3Z_u/Z_g,$$
$$1/x_{D_2} = N_{ortho-D_2}/N_{para-D_2} = g_g Z_g/g_u Z_u = 2Z_g/Z_u. \tag{48.5}$$

[†] The normalisation of the nuclear statistical weights which we use (such that $g_g + g_u = 1$) signifies that the entropy is measured from $\log (2i+1)^2$, in accordance with the condition stated at the end of § 46.

As the temperature varies from 0 to ∞, the ratio x_{H_2}, varies from 0 to 3, and x_{D_2} from 0 to $\frac{1}{2}$ (at $T = 0$ all the molecules are, of course, in the state with the lowest value of K, namely $K = 0$, corresponding to pure para-H_2 and ortho-D_2).

It must be borne in mind, however, that the probability of a change in the total nuclear spin in a collision between molecules is very small. The molecules of orthohydrogen and parahydrogen consequently behave practically as different modifications of hydrogen and are not[†] converted into each other. In practice, therefore, we are concerned not with a gas in equilibrium but with a non-equilibrium mixture of the ortho and para modifications, the relative amounts of which have given constant values.[‡] The free energy of such a mixture is equal to the sum of the free energies of the two components.

In particular, for $x = \infty$ (pure ortho-H_2 or para-D_2) we have

$$F_{rot} = -NT \log (g_u Z_u).$$

At low temperatures ($\hbar^2/2IT \gg 1$) only the first term in the sum need be retained in Z_u, so that $Z_u = 3e^{-\hbar^2/IT}$, and the free energy is

$$F_{rot} = N\hbar^2/I - NT \log (3g_u).$$

This means that the gas will behave as if monatomic ($c_{rot} = 0$), the chemical constant including an additional term $\log (3g_u)$, and the energy a constant term $N\hbar^2/I$, corresponding to the rotational energy of all the molecules, with $K = 1$.

§ 49. Diatomic gases. Vibrations of atoms

The vibrational part of the thermodynamic quantities for a gas becomes important at considerably higher temperatures than the rotational part, because the intervals in the vibrational structure of the terms are large compared with those in the rotational structure.[§]

We shall suppose, however, that the temperature is not large enough to excite the very high vibrational levels. Then the vibrations are small, and therefore harmonic, and the energy levels are given by the usual expression $\hbar\omega(v + \frac{1}{2})$ as in (47.4).

The calculation of the vibrational partition function Z_{vib} (47.4) is elementary. Owing to the very rapid convergence of the series, the summation

---

[†] In the absence of suitable catalysts.

[‡] For an ordinary gas which has been at room temperature for a considerable time the ratios are $x_{H_2} = 3$, $x_{D_2} = \frac{1}{2}$.

[§] As examples, the values of $\hbar\omega/k$ for some diatomic gases are H_2 6100°, N_2 3340°, O_2 2230°, NO 2690°, HCl 4140°.

may be formally extended to $v = \infty$. We shall measure the energy of the molecule from the lowest vibrational level ($v = 0$), i.e. include $\frac{1}{2}\hbar\omega$ in the constant ε_0 in (47.1). Then

$$Z_{vib} = \sum_{v=0}^{\infty} e^{-\hbar\omega v/T} = 1/(1-e^{-\hbar\omega/T}),$$

and hence the free energy is

$$F_{vib} = NT \log (1-e^{-\hbar\omega/T}), \tag{49.1}$$

the entropy

$$S_{vib} = -N \log (1-e^{-\hbar\omega/T}) + N\hbar\omega/T(e^{\hbar\omega/T}-1), \tag{49.2}$$

the energy

$$E_{vib} = N\hbar\omega/(e^{\hbar\omega/T}-1), \tag{49.3}$$

and the specific heat

$$c_{vib} = \left(\frac{\hbar\omega}{T}\right)^2 \frac{e^{\hbar\omega/T}}{(e^{\hbar\omega/T}-1)^2}. \tag{49.4}$$

Figure 5 shows c_{vib} as a function of $T/\hbar\omega$.

FIG. 5

At low temperatures ($\hbar\omega \gg T$) all these quantities tend exponentially to zero:

$$F_{vib} = -NTe^{-\hbar\omega/T},$$
$$c_{vib} = (\hbar\omega/T)^2 e^{-\hbar\omega/T}. \tag{49.5}$$

At high temperatures ($\hbar\omega \ll T$) we have

$$F_{vib} = -NT \log T + NT \log (\hbar\omega) - N\cdot\frac{1}{2}\hbar\omega, \tag{49.6}$$

corresponding to a constant specific heat $c_{vib} = 1$[†] and a chemical constant

[†] Again in accordance with the classical results of § 44.

$\zeta_{vib} = -\log(\hbar\omega)$. Adding these to the values (47.11), (47.12), we find that at temperatures $T \gg \hbar\omega$ the total specific heat of a diatomic gas is[†]

$$c_v = \tfrac{7}{2}, \quad c_p = \tfrac{9}{2}, \tag{49.7}$$

and the chemical constant is

$$\zeta = \log\left[\frac{(2)I}{\omega\hbar^6}\left(\frac{m}{2\pi}\right)^{3/2}\right]; \tag{49.8}$$

the factor (2) must be omitted for molecules consisting of like atoms. The first two terms in the expansion of E_{vib} are

$$E_{vib} = NT - \tfrac{1}{2}N\hbar\omega. \tag{49.9}$$

The constant term $-\tfrac{1}{2}N\hbar\omega$ appears here because the energy is measured from the lowest quantum level (i.e. from the energy of the "zero-point" vibrations), whereas the classical energy would have to be measured from the minimum of the potential energy.

The expression (49.6) for the free energy can also be derived classically, of course, since for $T \gg \hbar\omega$ the important quantum numbers v are the large ones, where the motion is quasi-classical. The classical energy of small oscillations of frequency ω is

$$\varepsilon_{vib}(p, q) = \frac{p^2}{2m'} + \frac{1}{2}m'\omega^2q^2,$$

where m' is the reduced mass. The integration with this expression for ε gives for the partition function

$$Z_{vib} = \frac{1}{2\pi\hbar}\int_{-\infty}^{\infty}\int_{-\infty}^{\infty} e^{-\varepsilon_{vib}/T}\, dp\, dq = T/\hbar\omega, \tag{49.10}$$

which corresponds to (49.6);[‡] owing to the rapid convergence of the integral the integration with respect to q may be taken from $-\infty$ to ∞.

At sufficiently high temperatures, when vibrations with large v are excited, the anharmonicity of the vibrations and their interaction with the rotation of the molecule may become important. These effects are in principle of the same order of magnitude. Since v is large, the corresponding correction to the thermodynamic quantities may be determined classically.

[†] As Fig. 5 shows, c_{vib} actually approaches its limiting value of 1 when $T \cong \hbar\omega$; for $T/\hbar\omega = 1$, $c_{vib} = 0.93$. As a practical condition for the applicability of the classical expressions we may write $T \gg \hbar\omega/3$.

[‡] The same result is obtained on replacing the summation over v by an integration.

Ideal Gases

Let us consider a molecule as a mechanical system of two particles interacting in accordance with the law $U(r)$ in a coordinate system in which their
centre of mass is at rest. The energy (Hamiltonian) which gives a precise
classical description of the rotation and vibrations of the system is the sum
of the kinetic energy (the energy of a particle with the reduced mass m') and
the potential energy $U(r)$. The partition function, after integration over the
momenta, reduces to an integral over the coordinates: $\int e^{-U(r)/T}\,dV$, and
after integration over the angles (in spherical polar coordinates) there
remains the integral

$$\int_0^\infty e^{-U(r)/T} r^2 \, dr.$$

The approximation corresponding to independent harmonic vibrations and
rotation of the molecule is obtained by putting $U(r) = U_0 + \frac{1}{2}m'\omega^2(r-r_0)^2$
and, in the integration, replacing the slowly varying factor r^2 by r_0^2, where r_0
is the equilibrium distance between the particles: $U_0 = U(r_0)$. In order to
take into account the anharmonicity of the vibrations and their interaction
with the rotation we now write

$$U(r) = U_0 + \tfrac{1}{2}m'\omega^2 r_0^2(\xi^2 - \alpha\xi^3 + \beta\xi^4), \qquad (49.11)$$

where $\xi = r/r_0 - 1$, and α and β are constants,[†] and then expand the whole
integrand in powers of ξ, separating the factor $\exp\left\{-\left(U_0 + \frac{1}{2}m'\omega^2 r_0^2\xi^2\right)/T\right\}$.
In the expansion, only those terms need be retained which after integration
give the highest and next highest powers of the temperature; the integration
over ξ is taken from $-\infty$ to ∞. The zero-order term in the expansion gives
the usual value of the partition function, and the remaining terms give the
required correction. Omitting the calculations, we shall state the final result
for the correction to the free energy:

$$F_{\text{anh}} = -NT^2 \frac{1}{2I\omega^2}\left[1 + 3\alpha - \frac{3}{2}\beta + \frac{15}{8}\alpha^2\right]. \qquad (49.12)$$

Thus the anharmonicity of the vibrations and their interaction with the rotation give a correction to the free energy which is proportional to the square of
the temperature. Accordingly the specific heat has a further term proportional to the first power of the temperature.

§ 50. Diatomic gases. The effect of the electronic angular momentum

Some types of molecule, though not many, have a non-zero orbital angular
momentum or spin in their electronic ground state.

[†] These constants can be expressed in terms of the spectroscopic constants of
the molecule; see *Quantum Mechanics*, § 82.

The presence of a non-zero orbital angular momentum Λ causes a twofold degeneracy of the electronic term, corresponding to the two possible directions of this angular momentum with respect to the axis of the molecule.[†] This affects the thermodynamic quantities: because of the doubling of the partition function, a quantity

$$\zeta_\Lambda = \log 2 \tag{50.1}$$

is added to the chemical constant.

The presence of a non-zero spin S causes a splitting into $2S+1$ levels, but the intervals in this fine structure are so small (when $\Lambda = 0$) that they can always be neglected in calculating the thermodynamic quantities. The presence of the spin simply increases the degree of degeneracy of each level by a factor $2S+1$, and so the chemical constant is increased by

$$\zeta_S = \log(2S+1). \tag{50.2}$$

The fine structure which occurs when $S \neq 0$, $\Lambda \neq 0$ requires special consideration. Here the fine-structure intervals may reach values which have to be taken into account in calculating the thermodynamic quantities. We shall derive the formulae for the case of a doublet electron term.[‡] Each component of the electron doublet has its vibrational and rotational structure, the parameters of which may be regarded as the same for each component. The partition function (47.2) therefore contains a further factor

$$Z_{el} = g_0 + g_1 e^{-\Delta/T},$$

where g_0, g_1 are the degrees of degeneracy of the components of the doublet, and Δ their separation. The free energy must accordingly contain an "electronic part"

$$F_{el} = -NT \log(g_0 + g_1 e^{-\Delta/T}). \tag{50.3}$$

[†] Strictly speaking, the term is split into two levels (Λ-*doubling*), but the separation between these is so small that it may be entirely neglected here.

[‡] This case occurs for NO; the electronic ground state of the NO molecule is the doublet $\Pi_{1/2,\,3/2}$ with width $\Delta = 178°$. Each component of the doublet is doubly degenerate.

An unusual case occurs for oxygen. The electronic ground state of the O_2 molecule is a very narrow triplet $^3\Sigma$, the width of which may be neglected, but it happens by chance that the next (excited) state $^1\Delta$ (doubly degenerate) is relatively near, at $\Delta = 11{,}300°$, and at high temperatures it may be excited, with a consequent effect on the thermodynamic quantities.

We may also give the "electronic" specific heat which must be added to the other parts of the specific heat:

$$c_{el} = \frac{(\Delta/T)^2}{[1+(g_0/g_1)e^{\Delta/T}]\,[1+(g_1/g_0)e^{-\Delta/T}]}.$$ (50.4)

In the limits $T \to 0$ and $T \to \infty$, c_{el} is of course zero, and it has a maximum at some temperature $T \sim \Delta$.

PROBLEM

Determine the correction to the free energy for oxygen due to the first excited electronic state of the O_2 molecule (see the last footnote). The temperature is large compared with the vibrational quantum, but small compared with the distance Δ between the ground state $^3\Sigma$ and the excited state $^1\Delta$.

SOLUTION. The partition function is

$$Z = 3\,\frac{T}{\hbar\omega}\,\frac{TI}{\hbar^2} + 2e^{-\Delta/T}\,\frac{T}{\hbar\omega'}\,\frac{TI'}{\hbar^2},$$

where the two terms on the right are the partition functions for the ground and excited states, each of which is the product of electronic, vibrational and rotational factors. The required correction to the free energy is therefore.

$$F_{1\Delta} = -NT\log\left(1+\frac{2\omega r_0'^2}{3\omega' r_0^2}e^{-\Delta/T}\right) \cong -NT\cdot\frac{2\omega r_0'^2}{3\omega' r_0^2}e^{-\Delta/T},$$

where $\omega, r_0, \omega', r_0'$ are the frequencies and equilibrium distances between the nuclei in the ground and excited electronic states.

§ 51. Polyatomic gases

The free energy of a polyatomic gas, like that of a diatomic gas, can be written as the sum of translational, rotational and vibrational parts. The translational part, as before, is characterised by values of the specific heat and chemical constant

$$c_{tr} = \tfrac{3}{2}, \qquad \zeta_{tr} = \tfrac{3}{2}\log(m/2\pi\hbar^2).$$ (51.1)

Owing to the large moments of inertia of polyatomic molecules (and the corresponding smallness of their rotational quanta) their rotation may always be treated classically.[†] The polyatomic molecule has three rotational degrees

---

[†] Rotation quantisation effects would be observable only in methane CH_4, where they should occur at temperatures of about 50°K; see the Problem at the end of this section.

of freedom and three principal moments of inertia I_1, I_2, I_3, which are in general different; its kinetic energy of rotation is therefore

$$\varepsilon_{\text{rot}} = \frac{M_\xi^2}{2I_1} + \frac{M_\eta^2}{2I_2} + \frac{M_\zeta^2}{2I_3}, \tag{51.2}$$

where ξ, η, ζ are coordinates in a rotating system whose axes coincide with the principal axes of inertia of the molecule; for the present we disregard the special case of molecules consisting of collinear atoms. This expression is to be substituted in the partition function

$$Z_{\text{rot}} = \int' e^{-\varepsilon_{\text{rot}}/T}\, d\tau_{\text{rot}}, \tag{51.3}$$

where

$$d\tau_{\text{rot}} = \frac{1}{(2\pi\hbar)^3}\, dM_\xi\, dM_\eta\, dM_\zeta\, d\phi_\xi\, d\phi_\eta\, d\phi_\zeta,$$

and the prime denotes, as usual, that the integration is to be taken only over the physically different orientations of the molecule.

If the molecule has axes of symmetry, rotations about these axes leave the molecule unchanged, and amount to an interchange of identical atoms. It is clear that the number of physically indistinguishable orientations of the molecule is equal to the number of possible different rotations about the axes of symmetry, including a rotation through 360° (the identical transformation). Denoting this number[†] by σ, we can take the integration in (51.3) simply over all orientations and divide by σ.

In the product $d\phi_\xi\, d\phi_\eta\, d\phi_\zeta$ of three infinitesimal angles of rotation, $d\phi_\xi\, d\phi_\eta$ may be regarded as an element do_ζ of solid angle for directions of the ζ-axis. The integration over o_ζ is independent of that over rotations $d\phi_\zeta$ about the ζ-axis, and gives 4π. The integration over ϕ_ζ gives a further 2π. Integrating also over M_ξ, M_η, M_ζ from $-\infty$ to ∞, we finally have

$$Z_{\text{rot}} = \frac{8\pi^2}{\sigma(2\pi\hbar)^3}(2\pi T)^{3/2}\,(I_1 I_2 I_3)^{1/2} = (2T)^{3/2}\,(\pi I_1 I_2 I_3)^{1/2}/\sigma\hbar^3.$$

Hence the free energy is

$$F = -\frac{3}{2}NT \log T - NT \log \frac{(8\pi I_1 I_2 I_3)^{1/2}}{\sigma\hbar^3}. \tag{51.4}$$

Thus we have for the rotational specific heat, in accordance with § 44,

$$c_{\text{rot}} = \tfrac{3}{2}, \tag{51.5}$$

[†] For instance, in H_2O (an isosceles triangle) $\sigma = 2$, in NH_3 (an equilateral triangular pyramid) $\sigma = 3$, in CH_4 (a tetrahedron) $\sigma = 12$, and in C_6H_6 (a regular hexagon) $\sigma = 12$.

and the chemical constant is

$$\zeta_{rot} = \log \frac{(8\pi I_1 I_2 I_3)^{1/2}}{\sigma \hbar^3}. \tag{51.6}$$

For a *linear molecule*, i.e. one where all the atoms are collinear, there are, as in the diatomic molecule, only two rotational degrees of freedom and one moment of inertia I. The rotational specific heat and the chemical constant are, as in a diatomic gas,

$$c_{rot} = 1, \quad \zeta_{rot} = \log(2I/\sigma\hbar^2), \tag{51.7}$$

where $\sigma = 1$ for an asymmetric molecule (such as NNO) and $\sigma = 2$ for a molecule symmetrical about its midpoint (such as OCO).

The vibrational part of the free energy of a polyatomic gas is calculated in a similar way to that for a diatomic gas, given above. The only difference is that a polyatomic molecule has not one but several vibrational degrees of freedom: a non-linear molecule of n atoms clearly has $r_{vib} = 3n-6$ vibrational degrees of freedom, while for a linear molecule of n atoms $r_{vib} = 3n-5$ (see § 44). The number of vibrational degrees of freedom determines the number of *normal modes of vibration* of the molecule, to each of which there corresponds a frequency ω_α (the suffix α numbering the normal modes). It must be remembered that some of the frequencies ω_α may be equal, in which case the frequency concerned is said to be *degenerate*.

In the harmonic approximation, where the vibrations are assumed small (only temperatures for which this is so will be considered), all the normal modes are independent, and the vibrational energy is the sum of the energies of the individual modes. The vibrational partition function therefore falls into a product of partition functions of the individual modes, and the free energy F_{vib} is a sum of expressions of the type (49.1):

$$F_{vib} = NT \sum_\alpha \log(1 - e^{-\hbar\omega_\alpha/T}). \tag{51.8}$$

Each frequency appears in this sum a number of times equal to its degeneracy. Similar sums are obtained for the vibrational parts of the other thermodynamic quantities.

Each of the normal modes gives, in its own classical limit ($T \gg \hbar\omega_\alpha$), a contribution $c_{vib}^{(\alpha)} = 1$ to the specific heat; for T greater than the greatest $\hbar\omega_\alpha$ we should obtain

$$c_{vib} = r_{vib}. \tag{51.9}$$

In practice, however, this limit is not reached, since polyatomic molecules usually decompose at considerably lower temperatures.

The various frequencies ω_α for a polyatomic molecule generally range over a very wide interval. As the temperature increases, the various normal modes successively contribute to the specific heat. In consequence the specific heat of polyatomic gases may often be regarded as approximately constant over fairly wide intervals of temperature.

We may mention the possibility of a curious change from vibration to rotation, an instance of which is afforded by the ethane molecule C_2H_6. This molecule consists of two CH_3 groups at a certain distance apart and oriented in a certain way to each other. One of the normal vibrations of the molecule is a "torsional" vibration, in which one of the CH_3 groups is twisted relative to the other. As the energy of the vibrations increases, their amplitude increases and ultimately, at sufficiently high temperatures, the vibration becomes a free rotation. The contribution of this degree of freedom to the specific heat, which is approximately 1 when the vibrations are fully excited, therefore begins to decrease as the temperature increases further, approaching asymptotically the value $\frac{1}{2}$ typical of a rotation.

Finally, it may be mentioned that, if the molecule has a non-zero spin S (for example, the molecules NO_2 and ClO_2), the chemical constant includes a term

$$\zeta_S = \log(2S+1). \tag{51.10}$$

PROBLEM

Determine the rotational partition function for methane at low temperatures.

SOLUTION. As already mentioned in the first footnote to this section, a quantum calculation of Z_{rot} for methane is required at sufficiently low temperatures.

The CH_4 molecule is a tetrahedron of the spherical-top type, and so its rotational levels are $\hbar^2 J(J+1)/2I$, where I is the common value of the three principal moments of inertia, and J the rotational quantum number. Since the spin i of the H nucleus is $\frac{1}{2}$, and that of the C^{12} nucleus is zero, the total nuclear spin of the CH_4 molecule may be 0, 1 or 2, the corresponding nuclear statistical weights being 1, 3 or 5; see *Quantum Mechanics*, § 105, Problem 5. For any given value of J there are definite numbers of states corresponding to values of the total nuclear spin. The following table gives these numbers for the first five values of J.

Nuclear spin	0	1	2
$J = 0$	—	—	1
1	—	1	—
2	2	1	—
3	—	2	1
4	2	2	1

The value of the sum Z_{rot} which is obtained by taking into account the total degree of degeneracy with respect to orientations of the rotational angular momentum and nuclear spin must be divided by 16 if the entropy is to be measured from the value $\log(2i+1)^4 = \log 16$ (cf. the first footnote to § 48). The result is

$$Z_{rot} = \frac{5}{16} + \frac{9}{16}e^{-\hbar^2/IT} + \frac{25}{16}e^{-3\hbar^2/IT} + \frac{77}{16}e^{-6\hbar^2/IT} + \frac{117}{16}e^{-10\hbar^2/IT} + \cdots .$$

§ 52. Magnetism of gases

A body in an external magnetic field **H** is characterised by a further macroscopic quantity, the magnetic moment $\mathfrak{M}$ which it acquires in the field. For an ideal gas, this is $\mathfrak{M} = N\overline{\mathfrak{m}}$ (where $\overline{\mathfrak{m}}$ is the mean magnetic moment of an individual atom or molecule), and so its calculation calls for a consideration only of the behaviour of these individual particles of the gas in the magnetic field. Note also that, since the magnetisation of a rarefied medium such as a gas is small in proportion to its density, we can neglect the influence of the medium on the field, i.e. assume that the field acting on each particle is just the external field **H**.

The change in the Hamiltonian of the gas corresponding to a small change $\delta\mathbf{H}$ in the external field is $\delta\hat{H} = -\hat{\mathfrak{M}}\cdot\delta\mathbf{H}$, where $\hat{\mathfrak{M}}$ is the operator of the magnetic moment of the gas.[†] According to (15.11) (cf. also (11.4)), in which the external parameter λ is now to be taken as the field **H**, we therefore have

$$\mathfrak{M} = -(\partial F/\partial \mathbf{H})_{T,V,N}. \tag{52.1}$$

In calculating the free energy of the gas in the magnetic field, we must first determine the corrections, due to the field, to the energy levels of the gas particles. Let us first assume that the gas is monatomic. The Hamiltonian of an atom in a magnetic field is

$$\hat{H} = \hat{H}_0 - \hat{\mathfrak{m}}\cdot\mathbf{H} + \frac{e^2}{8mc^2}\sum_a (\mathbf{H}\times\mathbf{r}_a)^2, \tag{52.2}$$

where $\hat{H}_0$ is the Hamiltonian of the atom in the absence of the field, e and m the electron charge and mass, $\mathbf{r}_a$ the electron coordinates (the summation being over all the electrons), $\hat{\mathfrak{m}} = -\beta(2\hat{\mathbf{S}}+\hat{\mathbf{L}})$ is the operator of the "intrinsic" magnetic moment of the atom ($\hat{\mathbf{S}}$ and $\hat{\mathbf{L}}$ being its spin and orbital angular momentum operators), and $\beta = |e|\hbar/2mc$ is the Bohr magneton (see *Quantum Mechanics*, § 113). Regarding the second and third terms in (52.2) as a small perturbation relative to $\hat{H}_0$, we can determine the correction to the energy levels as far as the terms quadratic in the field. It is

$$\Delta\varepsilon_k \equiv \varepsilon_k - \varepsilon_k^{(0)} = -A_k H - \tfrac{1}{2}B_k H^2, \tag{52.3}$$

[†] In classical mechanics, the small change in the Lagrangian of a system of particles due to a change $\delta\mathbf{H}$ in the field is $\delta L = \mathfrak{M}(q,\dot{q})\cdot\delta\mathbf{H}$, where $\mathfrak{M}(q,\dot{q})$ is the magnetic moment of the system as a function of its dynamical variables, the coordinates and velocities: see *Fields*, (45.3). The change in the Hamiltonian for given coordinates q and momenta p differs from δL only in sign (see *Mechanics*, (40.7)): $\delta H = -\mathfrak{M}(q,p)\cdot\delta\mathbf{H}$. In quantum mechanics there is a corresponding expression for the change in the Hamiltonian, $\hat{\mathfrak{M}}$ being the magnetic moment operator expressed in terms of the coordinates and momentum operators (and spins) of the particles.

where

$$A_k = (\mathfrak{m}_z)_{kk}, \tag{52.4}$$

$$B_k = 2 \sum_{k'}{}' \frac{|(\mathfrak{m}_z)_{kk'}|^2}{\varepsilon_{k'}^{(0)} - \varepsilon_k^{(0)}} - \frac{e^2}{4mc^2} \sum_a (x_a^2 + y_a^2)_{kk}, \tag{52.5}$$

and the z-axis is along $\mathbf{H}$; the first term in (52.5) results in the second order of perturbation theory from the term linear in $\mathbf{H}$ in (52.2), and the second term in the first order, from the quadratic term in the Hamiltonian.

In calculating the free energy, we shall assume that the gas temperature is not too low, so that $\Delta\varepsilon_k \ll T$. Then the partition function can be expanded in powers of H, and as far as the quadratic terms we have

$$Z \equiv \sum_k e^{-\varepsilon_k/T} = \sum_k e^{-\varepsilon_k^{(0)}/T}\left[1 + \frac{A_k H}{T} + \frac{A_k^2 H^2}{2T^2} + \frac{B_k H^2}{2T}\right].$$

The summation over k includes in particular an averaging over directions of the intrinsic magnetic moment $\mathfrak{m}$ of the atom (on which the unperturbed levels do not depend); from symmetry it is evident that the mean value $\bar{A}$ is zero, leaving

$$Z = \left[1 + \frac{H^2}{2T}\left(\frac{\overline{A^2}}{T} + \bar{B}\right)\right]\sum_k e^{-\varepsilon_k^{(0)}/T},$$

where the bar denotes averaging over the Boltzmann distribution unperturbed by the field. Substituting this expression in (41.4) and then differentiating the free energy with respect to $\mathbf{H}$, we obtain the magnetic moment in the form $\mathfrak{M} = N\chi\mathbf{H}$, where

$$\chi = \frac{1}{T}\overline{A^2} + \bar{B} \tag{52.6}$$

is the molecular magnetic susceptibility of the gas (J. H. Van Vleck, 1927). Let us consider some particular cases.

We shall suppose that the temperature T is small in comparison with the interval between the ground state and even the nearest excited level (including the fine-structure components of the ground state). Then we can say that only the ground state of the atom ($k = 0$) contributes to the mean values $\overline{A^2}$ and $\bar{B}$.

In the simplest case, where the atom (in the ground state) has neither spin nor orbital angular momentum (for example, the atoms of the inert gases), all matrix elements of the intrinsic magnetic moment of the atom are also zero. Then $A_0 = 0$, and only the second term in B_0 is non-zero. Because the wave function of a state with $L = S = 0$ is spherically symmetrical, the diagonal matrix elements (i.e. the mean values with respect to the state of the atom) are $(x_a^2)_{00} = (y_a^2)_{00} = \frac{1}{3}(r_a^2)_{00}$. The result is

$$\chi = -\frac{e^2}{6mc^2} \sum_a (r_a^2)_{00},$$ (52.7)

i.e. the gas is diamagnetic and its susceptibility is independent of the temperature (P. Langevin, 1905).[†]

However, if the intrinsic magnetic moment of the atom is not zero, then $A_0 \neq 0$ and (with the assumption made above about the temperature) the first term in (52.6) is large compared with the second term. A calculation according to the definition (52.4) gives

$$A_0 = -\beta g M_J, \quad g = 1 + \frac{J(J+1)-L(L+1)+S(S+1)}{2J(J+1)},$$

where g is the Landé factor and M_J the projection of the total angular momentum J of the atom (see *Quantum Mechanics*, § 113). The averaging in (52.6) is equivalent to averaging over the values of M_J. Since

$$\overline{M_J^2} = \frac{1}{2J+1} \sum_{M_J=-J}^{J} M_J^2 = \frac{1}{3} J(J+1),$$

we find

$$\chi = \frac{\beta^2 g^2}{3T} J(J+1).$$ (52.8)

Thus the gas is paramagnetic and its susceptibility follows Curie's law, being inversely proportional to the temperature (P. Langevin, 1905).[‡]

[†] Note that this diamagnetism, already mentioned in *Quantum Mechanics*, § 113, is a quantum effect. Although the quantum constant does not appear explicitly in (52.7), it in fact determines the "size" of the atom. In classical statistics, there are no macroscopic magnetic properties of matter, since in classical mechanics the Hamiltonian function of a system in a magnetic field differs from that in the absence of the field only in that the particle momenta $\mathbf{p}$ are replaced by $\mathbf{P} - e\mathbf{A}(\mathbf{r})/c$, where $\mathbf{P}$ are the generalised momenta and $\mathbf{A}(\mathbf{r})$ is the vector potential of the field. In the partition function, the integration is over all momenta $\mathbf{P}$ (and coordinates $\mathbf{r}$). After a change of variables, with integration over $\mathbf{p} = \mathbf{P} - e\mathbf{A}/c$ instead of over $\mathbf{P}$, we find that the partition function does not involve the magnetic field, and this therefore does not appear in any of the thermodynamic quantities.

[‡] Formula (52.8) can be applied not only to a gas, but also to a solid or a liquid where the magnetic moments of the atoms can for any reason be regarded as "free". This is true, for instance, of the magnetism of rare-earth elements in solid salts and in solutions. The paramagnetism of these ions is due to the incomplete $4f$ shell. These relatively deep electrons are screened from the influence of neighbouring atoms by electrons further out, and the ions can therefore behave magnetically like atoms of a rarefied gas.

If the orbital angular momentum and the spin of the atom are not zero, but equal in magnitude ($L = S \neq 0$), and combine to give a total angular momentum $J = 0$, then the diagonal matrix elements of the intrinsic magnetic moment are zero, whereas the non-diagonal ones (for transitions $L, S, J \to L, S, J \pm 1$ within a multiplet) are not zero. Then $A_0 = 0$, and the second (diamagnetic) term in B_0 (52.5) is small compared with the first term, whose denominators contain the relatively small spacings of the ground-state fine structure. Here $B_0 > 0$: for the ground state in each term of the sum over k', both the numerator and the denominator are positive. Thus in this case the gas is paramagnetic and its susceptibility is independent of the temperature: $\chi = B_0$ (J. H. Van Vleck, 1928).[†]

The magnetic susceptibility of molecular gases may be calculated similarly. At ordinary temperatures the rotation of the molecules is classical. The matrix elements of the magnetic moment may therefore be calculated first for fixed nuclei, and the averaging over orientations of the molecule then carried out as if it were a rigid classical magnetic dipole (see the following Problems).[‡]

PROBLEMS

PROBLEM 1. Determine the magnetic susceptibility of a monatomic gas when the fine-structure spacing of the ground state of the atom is small in comparison with T.

SOLUTION. In this case the averaging in (52.6) must be taken over all the components of the ground state multiplet of the atom, and the Boltzmann factors $\exp(-\varepsilon_k^{(0)}/T)$ may be taken as equal for all components. Then

$$\overline{A^2} = \overline{|\langle JM_J|\mathfrak{m}_z| JM_J\rangle|^2},$$

where the averaging is over all values of J and M_J (for given values of S and L). The result of such an averaging, however, does not depend on whether it is carried out before or after the angular momenta $\mathbf{S}$ and $\mathbf{L}$ are added to give $\mathbf{J}$. It may therefore be calculated as

$$\overline{A^2} = \overline{|\langle M_L M_S|\mathfrak{m}_z| M_L M_S\rangle|^2}$$

with independent averaging over M_L and M_S. Since

$$\overline{M_S M_L} = \overline{M_S}\,\overline{M_L} = 0, \quad \overline{M_S^2} = \tfrac{1}{3}S(S+1), \quad \overline{M_L^2} = \tfrac{1}{3}L(L+1),$$

we find

$$\overline{A^2} = \beta^2[4S(S+1)+L(L+1)].$$

[†] This case occurs for the Eu^{+++} ions in europium salts (see the previous footnote).

[‡] The magnetic moment arising from the motion of the nuclei is very small in comparison with the electronic magnetic moment, and may therefore always be neglected.

In the expression (52.5) for $\bar{B}$, the second term may be neglected. The first term (which might be large because its denominators, the multiplet spacings, are small) gives zero on averaging over the components of the multiplet: in the sum

$$\sum{}' \frac{|\langle JM_J | \mathfrak{m}_z | J'M_J'\rangle|^2}{\varepsilon_J^{(0)} - \varepsilon_{J'}^{(0)}},$$

which is now taken over all numbers J, J', M_J, M_J', the terms which differ by the interchange of J and J' cancel. The susceptibility is therefore

$$\chi = \frac{\beta^2}{3T}[4S(S+1) + L(L+1)].$$

PROBLEM 2. Determine the magnetic susceptibility of a diatomic gas when the fine-structure spacings of the electron ground state of the molecule are large in comparison with T.[†]

SOLUTION. In this case it is sufficient to consider only the ground level of the molecule, i.e. the lowest component of the ground-state multiplet. The mean value of the magnetic moment of the molecule in a state where the projections of the orbital angular momentum and the spin on the axis of the molecule are Λ and Σ is

$$\langle \Lambda\Sigma | \mathfrak{m} | \Lambda\Sigma \rangle = -\beta\mathbf{n}(\Lambda + 2\Sigma),$$

where $\mathbf{n}$ is the unit vector along the axis of the molecule. For classical rotation, $\overline{n_z^2} = \frac{1}{3}$, and the magnetic susceptibility is

$$\chi = \frac{\beta^2}{3T}(\Lambda + 2\Sigma)^2.$$

PROBLEM 3. The same as Problem 2, but with the fine-structure spacings small in comparison with T (molecular term case b).

SOLUTION. In this case the averaging must be taken over all components of the multiplet. The diagonal matrix elements of the z-component of the magnetic moment for given values of Λ and the spin z-component M_S are

$$\langle \Lambda M_S | \mathfrak{m}_z | \Lambda M_S \rangle = -\beta(n_z\Lambda + 2M_S).$$

Averaging the square of this over the values of M_S and the directions of $\mathbf{n}$, we find the susceptibility

$$\chi = \frac{\beta^2}{3T}[\Lambda^2 + 4S(S+1)].$$

PROBLEM 4. Determine the magnetic susceptibility of the gas NO. The electron ground term of the molecule is $^2\Pi$ (i.e. $\Lambda = 1$, $S = \frac{1}{2}$), and the spacing Δ between

---

[†] At ordinary temperatures the multiplet spacings are certainly large in comparison with the rotational structure of the levels, so that the molecular term belongs to case *a* coupling (see *Quantum Mechanics*, § 83).

the doublet components is comparable[†] to the temperature T (J. H. Van Vleck, 1928).

SOLUTION. Here, in the averaging in (52.6), we have to take account of both components of the doublet level with different Boltzmann factors. The diagonal matrix elements of the magnetic moment for the two states $|A\Sigma\rangle$ are

$$\langle 1, -\tfrac{1}{2}|L+2S| 1, -\tfrac{1}{2}\rangle = 1n-2\cdot\tfrac{1}{2}n = 0,$$
$$\langle 1, \tfrac{1}{2}|L+2S| 1, \tfrac{1}{2}\rangle = 2n.$$

Hence

$$\overline{A^2} = \frac{4\beta^2}{3}\,\frac{e^{-\Delta/T}}{1+e^{-\Delta/T}}.$$

The operator $\hat{L}$ has non-zero matrix elements for transitions between these two states, since Σ changes in the transition but not Λ. The non-diagonal matrix elements of the operator $2\hat{S}_z$ are

$$\langle 1, \tfrac{1}{2}|2S_z| 1, -\tfrac{1}{2}\rangle = \langle 1, -\tfrac{1}{2}|2S_z| 1, \tfrac{1}{2}\rangle = -1\cdot\sin\theta,$$

where θ is the angle between n and the z-axis.[‡] According to (52.5) (where we again neglect the second term),

$$B = \frac{2\beta^2}{\Delta}\cdot\frac{2}{3}\,\frac{1-e^{-\Delta/T}}{1+e^{-\Delta/T}},$$

the factor $\tfrac{2}{3}$ coming from the averaging of $\sin^2\theta$. The complete expression for the susceptibility becomes

$$\chi = \frac{\beta^2}{3T}\,f(\Delta/T), \qquad f(x) = \frac{4[1-e^{-x}(1-x)]}{x(1+e^{-x})}.$$

[†] It is 180°. The lower and upper doublet components correspond to axial spin components $\Sigma = -\tfrac{1}{2}$ and $+\tfrac{1}{2}$ respectively. The term is case a.

[‡] The operator $\mathbf{S} = \tfrac{1}{2}\boldsymbol{\sigma}$, where $\boldsymbol{\sigma}$ represents the Pauli matrices with quantisation along the axis of the molecule, i.e.

$$\sigma_\zeta = \begin{pmatrix} 1 & 0 \\ 0 & -1 \end{pmatrix}$$

if the coordinates are ξ, η, ζ with the ζ-axis along n.

φ↗ ⌐ ┬   φₚ↗↖↓   φₓ↗↖↗

A   BCD   E

<u>MB</u> ∥ $\psi = \phi_\alpha(A)\,\phi_\beta(B)\,\phi_\beta(c)\ldots$ Product Wave fns.
                                                              All permutations

<u>Bosons (s=1)</u>

$\psi = \sum_{permutations} \phi_\alpha(A)\phi_\beta(B)\ldots$ CHAPTER V
                                                         Symmetric

THE FERMI AND BOSE DISTRIBUTIONS

<u>Fermions (s=½)</u> $\psi = \sum_{permutations}(-1)^P \phi_\alpha(A)\phi_\beta(B)\ldots$ anti symmetric

§ 53. The Fermi distribution

2-particles A,B IF THE temperature of an ideal gas (at a given density) is sufficiently low,
3-states α, β, δ Boltzmann statistics becomes inapplicable, and a different statistics must be
Write ψ's devised, in which the mean occupation numbers of the various quantum
        states of particles are not assumed small.

NB   This statistics, however, differs according to the type of wave functions by
which the gas is described when regarded as a system of N identical particles.
These functions must be either antisymmetrical or symmetrical with respect
to interchanges of any pair of particles, the former case occurring for particles
with half-integral spin, and the latter case for those with integral spin.

For a system of particles described by antisymmetrical wave functions,
Pauli's principle applies: in each quantum state there cannot simultaneously
be more than one particle. The statistics based on this principle is called
Fermi statistics, or Fermi–Dirac statistics.[†]

As in § 37, we shall apply the Gibbs distribution to the set of all particles
in the gas which are in a given quantum state; as already mentioned in § 37,
this may be done even if there is an exchange interaction between the particles. We again denote by Ω_k the thermodynamic potential of this set of
particles; by the general formula (35.3),

$$\Omega_k = -T\log\sum_{n_k}(e^{(\mu-\varepsilon_k)/T})^{n_k},\qquad(53.1)$$

since the energy of n_k particles in the kth state is just $n_k\varepsilon_k$. According to
Pauli's principle, the occupation numbers of each state can take only the
values 0 and 1. Hence

$$\Omega_k = -T\log(1+e^{(\mu-\varepsilon_k)/T}).$$

Since the mean number of particles in the system is equal to minus the
derivative of the potential Ω with respect to the chemical potential μ, the

[†] It was proposed by E. Fermi (1926) for electrons, and its relation to quantum
mechanics was elucidated by P. A. M. Dirac (1926).

158

Distinguishable particles

MB

α(A) α(B)
   :

9 states

configuration does
not
specify the state.

	states α	β	δ	n_k	$g(n_k)$
AB		—	—	2 0 0	1
—	AB	—	0 2 0	1
—	—	AB	0 0 2	1
A	B	—	1 1 0	} 2
B	A	—	1 1 0	
A	—	B	1 0 1	} 2
B	—	A	1 0 1	
—	A	B	0 1 1	} 2
—	B	A	0 1 1	

required mean number of particles in the kth quantum state is here obtained as the derivative

$$\bar{n}_k = -\frac{\partial \Omega_k}{\partial \mu} = \frac{e^{(\mu-\varepsilon_k)T}}{1+e^{(\mu-\varepsilon_k)/T}},$$

or finally

$$\bar{n}_k = \frac{1}{e^{(\varepsilon_k-\mu)/T}+1}. \tag{53.2}$$

This is the distribution function for an ideal gas obeying Fermi statistics, or in short a *Fermi gas*. All the $\bar{n}_k \leqslant 1$, as is to be expected. When $e^{(\mu-\varepsilon_k)/T} \ll 1$, formula (53.2) tends to the Boltzmann distribution function, as it should.

The Fermi distribution is normalised by the condition

$$\sum_k \frac{1}{e^{(\varepsilon_k-\mu)/T}+1} = N, \tag{53.3}$$

where N is the total number of particles in the gas. This equation implicitly determines the chemical potential as a function of T and N.

The thermodynamic potential Ω of the gas as a whole is obtained by summation of Ω_k over all quantum states:

$$\Omega = -T \sum_k \log{(1+e^{(\mu-\varepsilon_k)/T})}. \tag{53.4}$$

§ 54. The Bose distribution

Let us now consider the statistics obeyed by an ideal gas consisting of particles described by symmetrical wave functions, namely *Bose statistics* or Bose–Einstein statistics.[†]

The occupation numbers of the quantum states when the wave functions are symmetrical are unrestricted and can take any values. The distribution function may be derived as in § 53; we put

$$\Omega_k = -T \log \sum_{n_k=0}^{\infty} (e^{(\mu-\varepsilon_k)/T})^{n_k}. \tag{54.0}$$

This geometric progression is convergent only if $e^{(\mu-\varepsilon_k)/T} < 1$. Since this condition must be satisfied for all ε_k, including $\varepsilon_k = 0$, it is clear that we must certainly have

$$\mu < 0. \tag{54.1}$$

[†] This was introduced by S. N. Bose (1924) for light quanta, and generalised by Einstein.

In this connection it may be recalled that for a Boltzmann gas the chemical potential always has large negative values; for a Fermi gas μ may be either negative or positive.

Summation of the geometric progression gives

$$\Omega_k = T \log (1 - e^{(\mu - \varepsilon_k)/T}).$$

Hence we find the mean occupation numbers $\bar{n}_k = -\partial \Omega_k / \partial \mu$:

$$\bar{n}_k = \frac{1}{e^{(\varepsilon_k - \mu)/T} - 1}. \tag{54.2}$$

This is the distribution function for an ideal gas which obeys Bose statistics (or, as it is called for brevity, a *Bose gas*). It differs from the Fermi distribution function in the sign of unity in the denominator. Like that function, it tends of course to the Boltzmann distribution function when $e^{(\mu - \varepsilon_k)/T} \ll 1$. The total number of particles in the gas is given by the formula

$$N = \sum_k \frac{1}{e^{(\varepsilon_k - \mu)/T} - 1}, \tag{54.3}$$

and the thermodynamic potential Ω of the gas as a whole is obtained by summation of Ω_k over all quantum states:

$$\Omega = T \sum_k \log (1 - e^{(\mu - \varepsilon_k)/T}). \tag{54.4}$$

§ 55. Fermi and Bose gases not in equilibrium

As in § 40, we can calculate the entropy also for Fermi and Bose gases not in equilibrium, and again derive the Fermi and Bose distribution functions from the condition that the entropy be a maximum.

In the Fermi case there can be no more than one particle in each quantum state, but the numbers N_j are not small, and are in general of the same order of magnitude as the numbers G_j. (The notation is as in § 40.)

The number of possible ways of distributing N_j identical particles among G_j states with not more than one particle in each is just the number of ways of selecting N_j of the G_j states, i.e. the number of combinations of G_j things N_j at a time. Thus

$$\Delta \Gamma_j = G_j! / N_j! (G_j - N_j)!. \tag{55.1}$$

Taking the logarithm of this expression and using for the logarithm of each factorial the formula (40.3), we find

$$S = \sum_j \{ G_j \log G_j - N_j \log N_j - (G_j - N_j) \log (G_j - N_j) \}. \tag{55.2}$$

Find z

$N = \sum_k n_k$

$\bar{E} = \sum_k \varepsilon_k n_k$ — energy of a system state.

$g(n_k) = 1$

$Z = \sum \sum e^{\frac{\mu N - \varepsilon_s}{T}}$
$N \; \varepsilon_s$

$= \sum \sum e^{\frac{\mu \sum n_k}{T}}$
$N=0 \; \{n_k\} \; -\frac{\sum \varepsilon_k n_k}{T}$

$= \sum_{N=0}^{\infty} \sum_{n_k} e^{\frac{(\mu - \varepsilon_k) n_k}{T}}$

$= \left[\sum_{n_\alpha} e^{\frac{\mu - \varepsilon_\alpha}{T} n_\alpha} \right]$

$= \prod_k \sum (e^{\frac{\mu - \varepsilon_k}{T}})^{n_k}$

$= \prod_k Z_k$

to 54.0, 53

BE

$\mu(T) \quad \varepsilon_k$ High temp

Again using the mean occupation numbers of the quantum states, $\bar{n}_j = N_j/G_j$, we finally have the following expression for the entropy of a Fermi gas not in equilibrium:

$$S = -\sum_j G_j[\bar{n}_j \log \bar{n}_j + (1-\bar{n}_j) \log (1-\bar{n}_j)]. \tag{55.3}$$

From the condition for this expression to be a maximum according to (40.8) we easily find that the equilibrium distribution is given by the formula

$$\bar{n}_j = 1/(e^{\alpha + \beta e_j} + 1),$$

which is the Fermi distribution, as it should be.

Finally, for Bose statistics, each quantum state may contain any number of particles, so that the statistical weight $\Delta\Gamma_j$ is the total number of ways of distributing N_j particles among G_j states. This number is[†]

$$\Delta\Gamma_j = (G_j + N_j - 1)!/(G_j - 1)! N_j!. \tag{55.4}$$

Taking the logarithm of this expression and neglecting unity in comparison with the very large numbers $G_j + N_j$ and G_j, we obtain

$$S = \sum_j \{(G_j + N_j) \log (G_j + N_j) - N_j \log N_j - G_j \log G_j\}. \tag{55.5}$$

In terms of the numbers $\bar{n}_j$ we can write the entropy of a Bose gas not in equilibrium as

$$S = \sum_j G_j[(1 + \bar{n}_j) \log (1 + \bar{n}_j) - \bar{n}_j \log \bar{n}_j]. \tag{55.6}$$

It is easily seen that the condition for this expression to be a maximum in fact gives the Bose distribution.

The two formulae (55.2) and (55.5) for the entropy naturally tend, in the limiting case $N_j \ll G_j$, to the Boltzmann formula (40.4), and the statistical weights (55.1) and (55.4) for Fermi and Bose statistics tend to the Boltzmann expression (40.2); to see this, we must put $G_j! \cong (G_j - N_j)! G_j^{N_j}$, $(G_j + N_j - 1)!$

[†] The problem is to find the number of ways of distributing N_j identical balls among G_j urns. Let us imagine the balls as a line of N_j points, and number the urns; let us then imagine the latter to be separated by $G_j - 1$ vertical strokes placed at intervals along the line of points. For example, the diagram

$$.|...||....|..$$

represents ten balls distributed among five urns: one in the first, three in the second, none in the third, four in the fourth and two in the fifth. The total number of places (occupied by points and strokes) in the line is $G_j + N_j - 1$. The required number of distributions of the balls among the urns is the number of ways of choosing $G_j - 1$ positions for the strokes, i.e. the number of combinations of $N + G_j - 1$ things $G_j - 1$ at a time, and this gives the result (55.4).

$\simeq (G_j-1)!\,G_j^{N_j}$. It must be remembered, however, that, in going to the limit, terms of order N_j^2/G_j are neglected in the statistical weights, and these terms are not in general small; but when the logarithm is taken these terms give a correction to the entropy which is of the relatively small order N_j/G_j.

Finally, we shall give a formula for the entropy of a Bose gas in the important limiting case where the number of particles in each quantum state is large (so that $N_j \gg G_j, \bar{n}_j \gg 1$). We know from quantum mechanics that this case corresponds to the classical wave picture of the field. The statistical weight (55.4) becomes

$$\Delta\Gamma_j = N_j^{G_j-1}/(G_j-1)! \tag{55.7}$$

and the entropy is

$$S = \sum_j G_j \log\,(eN_j/G_j). \tag{55.8}$$

We shall make use of this formula in § 71.

§ 56. Fermi and Bose gases of elementary particles

Let us consider a gas consisting of elementary particles, or of particles which under certain conditions may be regarded as elementary. As has already been mentioned, the Fermi or Bose distribution need not be used for ordinary atomic or molecular gases, since these gases are in practice always described with sufficient accuracy by the Boltzmann distribution.

All the formulae derived in the present section are exactly similar in form for both Fermi and Bose statistics, differing only as regards one sign. The upper sign will always correspond to Fermi statistics and the lower sign to Bose statistics.

The energy of an elementary particle is just the kinetic energy of its translational motion, which is always quasi-classical. We therefore have

$$\varepsilon = (p_x^2+p_y^2+p_z^2)/2m, \tag{56.1}$$

and in the distribution function we make the usual change to the distribution in the phase space of the particle. Here it must be borne in mind that, for a given value of the momentum, the state of the particle still depends on the orientation of its spin. Hence the number of particles in a volume element $dp_x\,dp_y\,dp_z\,dV$ in phase space is found by multiplying the distribution (53.2) or (54.2) by

$$g\,d\tau = g\,dp_x\,dp_y\,dp_z\,dV/(2\pi\hbar)^3,$$

where $g = 2s+1$ (s being the spin of the particle), giving

$$dN = \frac{g\,d\tau}{e^{(\varepsilon-\mu)/T}\pm 1}. \qquad = \frac{g\,V4\pi p^2\,dp}{(2\pi\hbar)^3\left(e^{\frac{\varepsilon-\mu}{T}}\pm 1\right)} \tag{56.2}$$

Integrating over V (which simply involves replacing dV by the total volume V of the gas) we find the distribution for the components of the particle momentum; using spherical polar coordinates in momentum space and integrating over angles, we find the distribution for the absolute magnitude of the momentum:

$$dN_p = \frac{gVp^2\,dp}{2\pi^2\hbar^3(e^{(\varepsilon-\mu)/T}\pm 1)}, \tag{56.3}$$

where $\varepsilon = p^2/2m$, or the energy distribution

$$dN_\varepsilon = \frac{gVm^{3/2}}{2^{1/2}\pi^2\hbar^3}\frac{\sqrt{\varepsilon}\,d\varepsilon}{e^{(\varepsilon-\mu)/T}\pm 1}. \tag{56.4}$$

These formulae take the place of the classical Maxwellian distribution.

Integrating (56.4) with respect to ε, we obtain the total number of particles in the gas:

$$N = \frac{gVm^{3/2}}{2^{1/2}\pi^2\hbar^3}\int_0^\infty \frac{\sqrt{\varepsilon}\,d\varepsilon}{e^{(\varepsilon-\mu)/T}\pm 1}.$$

In terms of a new variable of integration $z = \varepsilon/T$, this equation can be written

$$\frac{N}{V} = \frac{g(mT)^{3/2}}{2^{1/2}\pi^2\hbar^3}\int_0^\infty \frac{\sqrt{z}\,dz}{e^{z-\mu/T}\pm 1}. \tag{56.5}$$

This formula implicitly determines the chemical potential μ of the gas as a function of its temperature T and density N/V.

With the same change from summation to integration in formulae (53.4), (54.4), we find for the potential Ω the expression

$$\Omega = \mp\frac{VgTm^{3/2}}{2^{1/2}\pi^2\hbar^3}\int_0^\infty \sqrt{\varepsilon}\log\left(1\pm e^{(\mu-\varepsilon)/T}\right)d\varepsilon.$$

Integration by parts gives

$$\Omega = -\frac{2}{3}\frac{gVm^{3/2}}{2^{1/2}\pi^2\hbar^3}\int_0^\infty \frac{\varepsilon^{3/2}\,d\varepsilon}{e^{(\varepsilon-\mu)/T}\pm 1}. \tag{56.6}$$

This expression is the same, apart from the factor $-\frac{2}{3}$, as the total energy of the gas,

$$E = \int_0^\infty \varepsilon\,dN_\varepsilon = \frac{gVm^{3/2}}{2^{1/2}\pi^2\hbar^3}\int_0^\infty \frac{\varepsilon^{3/2}\,d\varepsilon}{e^{(\varepsilon-\mu)/T}\pm 1}. \tag{56.7}$$

Since $\Omega = -PV$, we have therefore

$$PV = \tfrac{2}{3}E. \tag{56.8}$$

This result is exact, and so must hold good in the limiting case of a Boltzmann gas also; and in fact, on substituting the Boltzmann value $E = 3NT/2$, we obtain Clapeyron's equation.

From formula (56.6), substituting $\varepsilon/T = z$, we obtain

$$\Omega = -PV = VT^{5/2}f(\mu/T), \tag{56.9}$$

where f is a function of a single variable, i.e. Ω/V is a homogeneous function of order $\tfrac{5}{2}$ in μ and T.[†] Hence

$$\frac{S}{V} = -\frac{1}{V}\left(\frac{\partial \Omega}{\partial T}\right)_{V,\mu} \quad \text{and} \quad \frac{N}{V} = -\frac{1}{V}\left(\frac{\partial \Omega}{\partial \mu}\right)_{T,V}$$

are homogeneous functions of order $\tfrac{3}{2}$ in μ and T, and their ratio S/N is a homogeneous function of order zero, i.e. $S/N = \phi(\mu/T)$. Hence we see that in an adiabatic process ($S =$ constant) the ratio μ/T remains constant, and since $N/VT^{3/2}$ is also a function of μ/T only we have

$$VT^{3/2} = \text{constant}. \tag{56.10}$$

Then (56.9) shows that

$$PV^{5/3} = \text{constant}, \tag{56.11}$$

and also $T^{5/2}/P =$ constant. These equations are the same as that of the Poisson adiabatic (43.9) for an ordinary monatomic gas, but it must be emphasised that the exponents in (56.10), (56.11) are here unrelated to the ratio of specific heats, since the relations $c_p/c_v = \tfrac{5}{3}$ and $c_p - c_v = 1$ are not valid.

Formula (56.6), in the form

$$P = \frac{g \cdot 2^{1/2}m^{3/2}T^{5/2}}{3\pi^2\hbar^3} \int_0^\infty \frac{z^{3/2}\,dz}{e^{z-\mu/T}\pm 1}, \tag{56.12}$$

together with (56.5) determines the equation of state of the gas (in parametric form, with parameter μ), i.e. the relation between P, V and T. In the limiting case of a Boltzmann gas ($e^{\mu/T} \ll 1$) these formulae give Clapeyron's equa-

---

[†] If the energy is calculated from (56.9) as

$$E = N\mu + TS - PV = -\mu\,\partial\Omega/\partial\mu - T\,\partial\Omega/\partial T + \Omega,$$

we again obtain (56.8).

tion, as they should. We shall show this by means of a calculation which also gives the first correction term in the expansion in the equation of state.

For $e^{\mu/T} \ll 1$ we expand the integrand in (56.12) as a series of powers of $e^{\mu/T-z}$ and, retaining only the first two terms, obtain

$$\int_0^\infty \frac{z^{3/2}\,dz}{e^{z-\mu/T} \pm 1} \cong \int_0^\infty z^{3/2} e^{\mu/T-z}(1 \mp e^{\mu/T-z})\,dz$$

$$= \frac{3}{4}\sqrt{\pi} e^{\mu/T}\left(1 \mp \frac{1}{2^{5/2}} e^{\mu/T}\right).$$

Substitution in (56.12) gives

$$\Omega = -PV = -\frac{gVm^{3/2}T^{5/2}}{(2\pi)^{3/2}\hbar^3} e^{\mu/T}\left(1 \mp \frac{1}{2^{5/2}} e^{\mu/T}\right).$$

If only the first term of the expansion is retained, we obtain precisely the Boltzmann value of the chemical potential of a monatomic gas (formula (46.1a)). The next term gives the required correction, so that we can put

$$\Omega = \Omega_{\text{Bol}} \pm \frac{gVm^{3/2}T^{5/2}}{16\pi^{3/2}\hbar^3} e^{2\mu/T}. \tag{56.13}$$

But the small additions to all the thermodynamic potentials (expressed in terms of the appropriate variables; see (24.16)) are the same. Hence, expressing the correction term in Ω in terms of T and V (which can be done to the same accuracy by means of the Boltzmann expressions), we obtain the correction to the free energy:

$$F = F_{\text{Bol}} \pm \frac{\pi^{3/2}}{2g} \cdot \frac{N^2\hbar^3}{VT^{1/2}m^{3/2}}. \tag{56.14}$$

Finally, differentiating with respect to volume, we obtain the required equation of state:

$$PV = NT\left[1 \pm \frac{\pi^{3/2}}{2g} \frac{N\hbar^3}{V(mT)^{3/2}}\right]. \tag{56.15}$$

The condition for the correction term in this formula to be small is naturally the same as the condition (45.6) for Boltzmann statistics to be applicable. Thus we see that the deviations of an ideal gas from classical properties, occurring when the temperature is lowered at constant density (the gas then being said to become *degenerate*), cause in Fermi statistics an increase in pressure as compared with its value in an ordinary gas; we may say that in this case the quantum exchange effects lead to the occurrence of an additional effective repulsion between the particles.

$$PV = NT\left[1 \overset{FD}{\underset{BE}{\pm}} \frac{\pi^{3/2}}{2g} \frac{V_Q}{V}\right]$$

In Bose statistics, on the other hand, the value of the gas pressure changes in the opposite direction, becoming less than the classical value; we may say that here there is an effective attraction between the particles.

§ 57. A degenerate electron gas

The study of the properties of a Fermi gas at sufficiently low temperatures is of fundamental significance. As we shall see below, the temperatures concerned may in practice be very high in other respects.

In what follows we shall discuss an electron gas, with a view to the most important applications of Fermi statistics, and accordingly put $g = 2$ (the spin $s = \frac{1}{2}$).

Let us first consider an electron gas at a temperature of absolute zero (a *completely degenerate* Fermi gas). In such a gas, the electrons will be distributed among the the various quantum states so that the total energy of the gas has its least possible value. Since no more than one electron can be in each quantum state, the electrons occupy all states with energies from the least value (zero) to some greatest value which depends on the number of electrons in the gas.

With the twofold ($g = 2$) spin degeneracy of the levels, the number of quantum states of an electron moving in a volume V with absolute magnitude of momentum in the interval from p to $p + dp$ is

$$2 \cdot \frac{4\pi p^2 \, dp \cdot V}{(2\pi\hbar)^3} = V \frac{p^2 \, dp}{\pi^2\hbar^3}. \tag{57.1}$$

The electrons occupy all states with momenta from zero to a limit $p = p_F$, called the radius of the *Fermi sphere* in momentum space. The total number of electrons in these states is

$$N = \frac{V}{\pi^2\hbar^3} \int_0^{p_F} p^2 \, dp = \frac{V p_F^3}{3\pi^2\hbar^3},$$

whence the limiting momentum is given by

$$p_F = (3\pi^2)^{1/3} (N/V)^{1/3} \hbar, \tag{57.2}$$

and the limiting energy by

$$\varepsilon_F = p_F^2/2m$$
$$= (3\pi^2)^{2/3} (\hbar^2/2m) (N/V)^{2/3}. \tag{57.3}$$

This energy has a simple thermodynamic significance. In accordance with the foregoing discussion, the Fermi distribution function over quantum states (with fixed values of the momentum **p** and the spin component) is

$$\bar{n}_p = \frac{1}{e^{(\varepsilon - \mu)/T} + 1},$$ (57.4)

and in the limit $T \to 0$ it becomes a step function, equal to unity for $\varepsilon < \mu$ and zero for $\varepsilon > \mu$ (the continuous line in Fig. 6). Hence we see that the

FIG. 6

chemical potential of the gas for $T = 0$ is the same as the limiting energy of the electrons:

$$\mu = \varepsilon_F.$$ (57.5)

The total energy of the gas is obtained by multiplying the number of states (57.1) by $p^2/2m$ and integrating over all momenta:

$$E = \frac{V}{2m\pi^2\hbar^3} \int_0^{p_F} p^4 \, dp = \frac{V p_F^5}{10m\pi^2\hbar^3},$$

$$\varepsilon_F = (3\pi^2)^{2/3} \frac{\hbar^2}{2m} \left(\frac{N}{V}\right)^{2/3}$$

or, substituting (57.2),

$$E = \frac{3(3\pi^2)^{2/3}}{10} \frac{\hbar^2}{m} \left(\frac{N}{V}\right)^{2/3} N. \qquad = \frac{3}{5} N \varepsilon_F$$ (57.6)

Finally, from the general relation (56.8), we find the equation of state of the gas:

$$P = \frac{(3\pi^2)^{2/3}}{5} \frac{\hbar^2}{m} \left(\frac{N}{V}\right)^{5/3}.$$ (57.7)

Thus the pressure of a Fermi gas at absolute zero is proportional to the $\frac{5}{3}$ power of its density.

Formulae (57.6), (57.7) are approximately valid also at temperatures which are sufficiently close to absolute zero (for a given gas density). The condition for them to be applicable (for the gas to be "strongly degenerate") is clearly that T should be small in comparison with the limiting energy ε_F:

$$T \ll (\hbar^2/m)(N/V)^{2/3}.$$ (57.8)

This condition is, as we should expect, the opposite of the condition (45.6) for Boltzmann statistics to be valid. The temperature defined by the relation $T_F \cong \varepsilon_F$ is called the *degeneracy temperature*.

A degenerate electron gas has the peculiar property that it increasingly approaches the ideal gas state as its density increases. This is easily seen as follows.

Let us consider a *plasma*, i.e. a gas consisting of electrons and a corresponding number of positively charged nuclei which balance the charge on the electrons; a gas composed of electrons alone would obviously be entirely unstable, but we have not mentioned the nuclei hitherto, because the assumption of ideal-gas properties means that the presence of the nuclei does not affect the thermodynamic quantities for the electron gas. The energy (per electron) of the Coulomb interaction between the electrons and the nuclei is of the order of Ze^2/a, where Ze is the nuclear charge and $a \sim (ZV/N)^{1/3}$ is the mean distance between the electrons and the nuclei. The condition for an ideal gas is that this energy should be small compared with the mean kinetic energy of the electrons, which in order of magnitude is equal to the limiting energy ε_F. The inequality $Ze^2/a \ll \varepsilon_F$, after the substitution of $a \sim (ZV/N)^{1/3}$ and the expression (57.3) for ε_F, gives the condition

$$N/V \gg (e^2 m/\hbar^2)^3 Z^2. \tag{57.9}$$

We see that this condition is more nearly met as the density N/V of the gas increases.[†]

PROBLEM

Determine the number of collisions with a wall in an electron gas at absolute zero.

SOLUTION. The number of electrons per unit volume with momenta in the interval dp at an angle θ to the normal to the wall in the interval $d\theta$ is $2 \cdot 2\pi \sin \theta \, d\theta \, p^2 \, dp/(2\pi\hbar)^3$. The required number of collisions ν (per unit area of wall) is obtained by multiplying by $v \cos \theta$ ($v = p/m$) and integrating with respect to θ from 0 to $\frac{1}{2}\pi$ and with respect to p from 0 to p_F. The result is

$$\nu = \frac{3(3\pi^2)^{1/3}}{16} \frac{\hbar}{m} \left(\frac{N}{V}\right)^{4/3}.$$

§ 58. The specific heat of a degenerate electron gas

At temperatures which are low compared with the degeneracy temperature T_F, the distribution function (57.4) has the form shown by the broken line in Fig. 6: it is appreciably different from unity or zero only in a narrow range of

---

[†] The degeneracy temperature corresponding to the electron gas density $(e^2 m/\hbar^2)^3 Z^2$ is $40Z^{4/3}$ eV $\cong 0.5 \times 10^6 Z^{4/3}$ degrees.

values of the energy ε close to the limiting energy ε_F. The width of this "transition zone" of the Fermi distribution is of the order of T.

The expressions (57.6), (57.7) are the first terms in the expansions of the corresponding quantities in powers of the small ratio T/T_F. Let us now determine the next terms in the expansions.

Formula (56.6) involves an integral of the form

$$I = \int_0^\infty \frac{f(\varepsilon)\,d\varepsilon}{e^{(\varepsilon-\mu)/T}+1},$$

where $f(\varepsilon)$ is a function such that the integral converges; in (56.6), $f(\varepsilon) = \varepsilon^{3/2}$. We transform this integral by the substitution $\varepsilon - \mu = Tz$:

$$I = \int_{-\mu/T}^\infty \frac{f(\mu+Tz)}{e^z+1} T\,dz$$

$$= T\int_0^{\mu/T} \frac{f(\mu-Tz)}{e^{-z}+1}\,dz + T\int_0^\infty \frac{f(\mu+Tz)}{e^z+1}\,dz.$$

In the first integral we put $1/(e^z+1) = 1-1/(e^z+1)$, obtaining

$$I = \int_0^\mu f(\varepsilon)\,d\varepsilon - T\int_0^{\mu/T} \frac{f(\mu-Tz)}{e^z+1}\,dz + T\int_0^\infty \frac{f(\mu+Tz)}{e^z+1}\,dz.$$

In the second of these integrals we replace the upper limit by infinity, since $\mu/T \gg 1$ and the integral is rapidly convergent.[†] This gives

$$I = \int_0^\mu f(\varepsilon)\,d\varepsilon + T\int_0^\infty \frac{f(\mu+Tz)-f(\mu-Tz)}{e^z+1}\,dz.$$

We now expand the numerator of the second integrand as a Taylor series of powers of z and integrate term by term:

$$I = \int_0^\mu f(\varepsilon)\,d\varepsilon + 2T^2 f'(\mu)\int_0^\infty \frac{z\,dz}{e^z+1}$$

$$+ \frac{1}{3}T^4 f'''(\mu)\int_0^\infty \frac{z^3\,dz}{e^z+1} + \cdots .$$

[†] This amounts to neglecting exponentially small terms. It must be remembered that the expansion (58.1) derived below is an asymptotic, not a convergent, series.

$$E(T) = E(0) + \eta\varepsilon \qquad\qquad \eta = N\frac{I}{\varepsilon_F}$$

$$= E(0) + \frac{NT^2}{\varepsilon_F} \qquad\qquad \varepsilon = T$$

Substituting the values[†] of the integrals, we have finally

$$I = \int_0^\mu f(\varepsilon)\, d\varepsilon + \frac{\pi^2}{6} T^2 f'(\mu) + \frac{7\pi^4}{360} T^4 f'''(\mu) + \cdots. \qquad (58.1)$$

[†] Integrals of this type are calculated as follows:

$$\int_0^\infty \frac{z^{x-1}\, dz}{e^z+1} = \int_0^\infty z^{x-1} e^{-z} \sum_{n=0}^\infty (-)^n e^{-nz}\, dz = \Gamma(x) \sum_{n=1}^\infty (-)^{n+1} \frac{1}{n^x}$$

$$= (1-2^{1-x})\Gamma(x) \sum_{n=1}^\infty \frac{1}{n^x},$$

or

$$\int_0^\infty \frac{z^{x-1}\, dz}{e^z+1} = (1-2^{1-x})\,\Gamma(x)\,\zeta(x) \qquad (x > 0),$$

where $\zeta(x) = \sum_{n=1}^\infty 1/n^x$ is the Riemann zeta function.

For $x = 1$, the expression becomes indeterminate; the value of the integral is

$$\int_0^\infty \frac{dz}{e^z+1} = \log 2.$$

For x an even integer $(= 2n)$ the zeta function can be expressed in terms of the Bernoulli numbers B_n:

$$\int_0^\infty \frac{z^{2n-1}\, dz}{e^z+1} = \frac{2^{2n-1}-1}{2n}\,\pi^{2n}B_n.$$

The following integrals are calculated similarly:

$$\int_0^\infty \frac{z^{x-1}\, dz}{e^z-1} = \Gamma(x)\,\zeta(x) \qquad (x > 1).$$

For x an even integer $(= 2n)$,

$$\int_0^\infty \frac{z^{2n-1}\, dz}{e^z-1} = \frac{(2\pi)^{2n} B_n}{4n}.$$

For reference we shall give the values of the first few Bernoulli numbers and of some zeta and gamma functions:

$$B_1 = \tfrac{1}{6}, \qquad B_2 = \tfrac{1}{30}, \qquad B_3 = \tfrac{1}{42}, \qquad B_4 = \tfrac{1}{30};$$
$$\zeta(\tfrac{3}{2}) = 2.612, \qquad \zeta(\tfrac{5}{2}) = 1.341, \qquad \zeta(3) = 1.202,$$
$$\zeta(5) = 1.037, \qquad \Gamma(\tfrac{3}{2}) = \tfrac{1}{2}\sqrt{\pi}, \qquad \Gamma(\tfrac{5}{2}) = \tfrac{3}{4}\sqrt{\pi}.$$

The third term in the expansion is given for reference; it will not needed here.

Putting $f = \varepsilon^{3/2}$ in formula (58.1) and substituting in (56.6), we obtain the required next term in the expansion of the potential Ω at low temperatures:

$$\Omega = \Omega_0 - VT^2 \frac{\sqrt{(2\mu)}m^{3/2}}{6\hbar^3}, \qquad (58.2)$$

where Ω_0 denotes the value of Ω at absolute zero.

Regarding the second term as a small correction to Ω_0 and expressing μ in it in terms of T and V by means of the "zero-order approximation" (57.5), we can immediately write down an expression for the free energy (according to the theorem of small increments (24.16)):

$$F = F_0 - \tfrac{1}{2}\beta NT^2(V/N)^{2/3}, \qquad (58.3)$$

using for brevity the notation

$$\beta = \left(\frac{\pi}{3}\right)^{2/3} \frac{m}{\hbar^2}. \qquad (58.4)$$

Hence we find the entropy

$$S = \beta NT(V/N)^{2/3}, \qquad (58.5)$$

specific heat[†]

$$C = \beta NT(V/N)^{2/3}, \qquad (58.6)$$

and energy of the gas:

$$E = E_0 + \tfrac{1}{2}\beta NT^2(V/N)^{2/3}$$
$$= E_0[1 + 0.18(mT/\hbar^2)^2 (V/N)^{4/3}]. \qquad (58.7)$$

Thus the specific heat of a degenerate Fermi gas at low temperatures is proportional to the temperature.

§ 59. Magnetism of an electron gas. Weak fields

The magnetisation of an electron gas in weak magnetic fields is made up of two independent parts: a paramagnetic part due to the intrinsic (spin) magnetic moment of the electrons (the *Pauli paramagnetism*: W. Pauli, 1927) and a diamagnetic part due to the quantisation of the orbital motion of the electrons in the magnetic field (the *Landau diamagnetism*: L. D.

[†] The suffix v or p to the specific heat is omitted, since C_v and C_p are the same in this approximation. We have seen in § 23 that, if S tends to zero as T^n when $T \to 0$, the difference $C_p - C_v$ tends to zero as T^{2n+1}, and so in this case $C_p - C_v \propto T^3$.

Landau, 1930). Let us calculate the corresponding magnetic susceptibilities, assuming the gas to be degenerate (the temperature $T \ll \varepsilon_F$). The condition for the magnetic field to be weak is (see below) $\beta H \ll T$, where $\beta = |e|\hbar/2mc$ is the Bohr magneton.[†]

For a degenerate gas, the thermodynamic calculations are more conveniently performed with T, V and μ as independent variables (instead of T, V and N). Accordingly, in place of (52.1) for the magnetic moment of a Boltzmann gas we shall use the derivative

$$\mathfrak{M} = -(\partial\Omega/\partial\mathbf{H})_{T, V, \mu} \tag{59.1}$$

of the thermodynamic potential Ω.

Let us first find the paramagnetic part of the susceptibility. The additional (spin) energy of the electron in the magnetic field is $\pm\beta H$, the two signs corresponding to the two values $\pm\frac{1}{2}$ of the spin component along the field. The statistical distribution of electrons in the magnetic field consequently differs from that in the absence of the field in that the energy $\varepsilon = p^2/2m$ is replaced by $\varepsilon = p^2/2m\pm\beta H$. But, since ε appears in the distribution together with the chemical potential in the combination $\varepsilon-\mu$, this change is equivalent to replacing μ by $\mu\mp\beta H$. The potential Ω of an electron gas in a magnetic field may therefore be written

$$\Omega(\mu) = \tfrac{1}{2}\Omega_0(\mu+\beta H)+\tfrac{1}{2}\Omega_0(\mu-\beta H), \tag{59.2}$$

where $\Omega_0(\mu)$ is the potential in the absence of the field (the arguments T and V are omitted, for brevity); the two terms in (59.2) correspond to the sets of electrons with different spin components, and the factors $\frac{1}{2}$ take account of the halving of the number of quantum states of the electron when the value of its spin component is fixed.

Expanding (59.2) in powers of βH, we have

$$\Omega(\mu) \cong \Omega_0(\mu)+\tfrac{1}{2}\beta^2 H^2\, \partial^2\Omega_0(\mu)/\partial\mu^2, \tag{59.3}$$

and the magnetic moment is therefore $\mathfrak{M} = -H\beta^2\, \partial^2\Omega_0/\partial\mu^2$. The derivative $\partial\Omega_0/\partial\mu = -N$, and the paramagnetic susceptibility, which in this section is taken relative to unit volume of the gas, is

$$\chi_{\text{para}} = -\frac{\beta^2}{V}\frac{\partial^2\Omega_0}{\partial\mu^2} = \frac{\beta^2}{V}\left(\frac{\partial N}{\partial\mu}\right)_{T, V}. \tag{59.4}$$

[†] In the opposite case of high temperatures ($T \gg \varepsilon_F$), the electrons form a Boltzmann gas, and the paramagnetic part of the susceptibility per unit volume is $\chi_{\text{para}} = N\beta^2/VT$ (formula (52.8) with $g = 2$, $J = \frac{1}{2}$).

Neglecting the temperature effect, which is small when $T \ll \mu$, i.e. assuming the gas to be completely degenerate, we have from (57.3)

$$N = V \frac{(2m\mu)^{3/2}}{3\pi^2 \hbar^3},$$

and differentiation gives

$$\chi_{\text{para}} = \frac{\beta^2 (2m)^{3/2} \sqrt{\mu}}{2\pi^2 \hbar^3} = \frac{\beta^2 p_F m}{\pi^2 \hbar^3}. \tag{59.5}$$

Let us now calculate the diamagnetic susceptibility. The energy levels of the orbital motion of an electron in a magnetic field are

$$\varepsilon = \frac{p_z^2}{2m} + (2n+1)\beta H, \tag{59.6}$$

where $n = 0, 1, 2, \ldots$, and p_z is the momentum in the direction of the field, taking a continuous range of values from $-\infty$ to ∞ (see *Quantum Mechanics*, § 112). The number of states in the interval dp_z for any given value of n is

$$2 \frac{V |e| H}{(2\pi\hbar)^2 c} \, dp_z, \tag{59.7}$$

the factor 2 taking account of the two directions of the spin. The expression (53.4) for the potential Ω is

$$\Omega = 2\beta H \sum_{n=0}^{\infty} f[\mu - (2n+1)\beta H], \tag{59.8}$$

$$f(\mu) = -\frac{TmV}{2\pi^2 \hbar^3} \int_{-\infty}^{\infty} \log \left[1 + \exp\left(\frac{\mu}{T} - \frac{p_z^2}{2mT}\right)\right] dp_z. \tag{59.9}$$

The sum (59.8) can be calculated to the necessary accuracy by means of the formula[†]

$$\sum_{n=0}^{\infty} F(n+\tfrac{1}{2}) \cong \int_0^{\infty} F(x) \, dx + \tfrac{1}{24} F'(0). \tag{59.10}$$

[†] According to the familiar Euler–Maclaurin sum formula,

$$\tfrac{1}{2} F(a) + \sum_{n=1}^{\infty} F(a+n) \cong \int_a^{\infty} F(x) \, dx - \tfrac{1}{12} F'(a). \tag{59.10a}$$

Formula (59.10) is obtained from this by putting $a = \tfrac{1}{2}$ and representing the function $F(x)$ in the range $0 \leqslant x \leqslant \tfrac{1}{2}$ by $F(x) \cong F(0) + x F'(0)$.

The condition for this to be applicable is that the relative change of F in one step $(n \to n+1)$ is small. In respect of the function (59.9), the condition is $\beta H \ll T$.[†]

Applying (59.10) to the sum (59.8), we have

$$\Omega = 2\beta H \int_0^\infty f(\mu - 2\beta Hx) \, dx + \frac{2\beta H}{24} \left[\frac{\partial f(\mu - 2n\beta H)}{\partial n} \right]_{x=0}$$

$$= \int_{-\infty}^\mu f(x) \, dx - \frac{(2\beta H)^2}{24} \frac{\partial f(\mu)}{\partial \mu}.$$

The first term does not contain H, i.e. it is the potential $\Omega_0(\mu)$ of the gas in the absence of the field. Thus

$$\Omega = \Omega_0(\mu) - \frac{1}{6} \beta^2 H^2 \frac{\partial^2 \Omega_0(\mu)}{\partial \mu^2}, \qquad (59.11)$$

and hence the susceptibility is[‡]

$$\chi_{\text{dia}} = \frac{\beta^2}{3V} \frac{\partial^2 \Omega_0}{\partial \mu^2} = -\frac{1}{3} \chi_{\text{para}}. \qquad (59.12)$$

Altogether, the gas is paramagnetic, with susceptibility $\chi = 2\chi_{\text{para}}/3$. Here we have calculated its two parts separately in order to clarify their origin. It would, of course, be possible to calculate the total susceptibility χ directly. To do so, we should have to write the electron energy levels in the form $\varepsilon = p_z^2/2m + (2n+1)\beta H \pm \beta H$, obtained by adding the spin magnetic energy $\pm \beta H$ to (59.6). This set of values of ε may also be written

$$\varepsilon = \frac{p_z^2}{2m} + 2n\beta H, \qquad n = 0, 1, 2, \ldots, \qquad (59.13)$$

each value with $n \neq 0$ occurring twice, and that with $n = 0$ once; in other words, the number density of states with $n \neq 0$ is given by the same formula

---

[†] Otherwise, the condition is violated in the "dangerous" range of values of n where $\mu - (2n+1)\beta H$ is close to zero. This range causes (see § 60) the occurrence in Ω of terms that oscillate rapidly as functions of H. These terms disappear if the series (59.8) is averaged over a range ΔH such that the change of the argument $\mu - 2\beta nH$ near the point where it is zero is much greater than the difference of its two neighbouring values:

$$\beta H \ll n\beta \, \Delta H \sim \mu \, \Delta H/H, \quad \text{or} \quad \Delta H/H \gg \beta H/\mu.$$

Formula (59.10) then becomes again valid, and the result given by it is subject only to the condition $\beta H \ll \mu$.

[‡] This relation is valid for any degree of degeneracy of the gas.

(59.7), and that with $n = 0$ by half the amount. The potential Ω is then given by the sum

$$\Omega = 2\beta H \left\{ \tfrac{1}{2} f(\mu) + \sum_{n=1}^{\infty} f(\mu - 2\beta H n) \right\},\qquad (59.14)$$

and is calculated by means of the formula[†]

$$\tfrac{1}{2} F(0) + \sum_{n=1}^{\infty} F(n) = \int_0^{\infty} F(x)\, dx - \tfrac{1}{12} F'(0).\qquad (59.15)$$

§ 60. Magnetism of an electron gas. Strong fields

Let us now consider fields for which βH is again small in comparison with μ, but need not be small in comparison with T:

$$T \lesssim \beta H \ll \mu.\qquad (60.1)$$

Under these conditions, the effects of the quantisation of orbital motion cannot be separated from the spin effects, and both must be considered together; thus we must start from (59.14) in calculating Ω.

We shall see that the magnetisation of the electron gas for $\beta H \gtrsim T$ contains a part which oscillates with a large amplitude as a function of H, and it is this oscillatory part of the magnetisation that will concern us here.

To separate the oscillatory parts of the thermodynamic quantities, it is convenient to transform the sum (59.14) by means of Poisson's formula:[‡]

$$\tfrac{1}{2} F(0) + \sum_{n=1}^{\infty} F(n) = \int_0^{\infty} F(x)\, dx + 2\,\mathrm{re} \sum_{k=1}^{\infty} \int_0^{\infty} F(x) e^{2\pi i k x}\, dx,\qquad (60.2)$$

after which it becomes

$$\Omega = \Omega_0(\mu) + \frac{TmV}{\pi^2 \hbar^3}\, \mathrm{re} \sum_{k=1}^{\infty} I_k,\qquad (60.3)$$

[†] Obtained from the Euler–Maclaurin formula by putting $a = 0$.
[‡] This formula follows from the equation

$$\sum_{n=-\infty}^{\infty} \delta(x-n) = \sum_{k=-\infty}^{\infty} e^{2\pi i k x};$$

the sum of the delta functions on the left is a periodic function of x with unit period, and the sum on the right is the expansion of this function as a Fourier series. On multiplying the equation by any function $F(x)$ and integrating over x from 0 to ∞, we get equation (60.2); the integral $\int F(x)\, \delta(x)\, dx$, which is the term with $n = 0$, extended only over the region on one side of $x = 0$, gives $\tfrac{1}{2} F(0)$.

where

$$I_k = -2\beta H \int_{-\infty}^{\infty} \int_0^{\infty} \log\left[1 + \exp\left(\frac{\mu}{T} - \frac{p_z^2}{2mT} - \frac{2x\beta H}{T}\right)\right] e^{2\pi i k x} \, dx \, dp_z, \quad (60.4)$$

and $\Omega_0(\mu)$ is the thermodynamic potential in the absence of the field.

In the integrals I_k, we replace the variable x by $\varepsilon = p_z^2/2m + 2x\beta H$. For the required oscillatory part of the integrals, denoted by $\tilde{I}_k$, we have

$$\tilde{I}_k = -\int_{-\infty}^{\infty} \int_0^{\infty} \log\left[1 + \exp\left(\frac{\mu - \varepsilon}{T}\right)\right] \exp\left(\frac{i\pi k\varepsilon}{\beta H}\right) \exp\left(-\frac{i\pi k p_z^2}{2m\beta H}\right) d\varepsilon \, dp_z.$$

In the integral over p_z, the important values are $p_z^2/2m \sim \beta H$. The oscillatory part of the integral, however, comes from the range of values of ε near μ (see below); the lower limit of integration over ε is therefore taken as zero instead of $p_z^2/2m$.

The integration over p_z is separable and is effected by means of the formula[†]

$$\int_{-\infty}^{\infty} e^{-i\alpha p^2} \, dp = e^{-i\pi/4} \sqrt{\frac{\pi}{\alpha}},$$

leaving

$$\tilde{I}_k = -e^{-i\pi/4} \sqrt{\frac{2m\beta H}{k}} \int_0^{\infty} \log\left[1 + e^{(\mu - \varepsilon)/T}\right] e^{i\pi k\varepsilon/\beta H} \, d\varepsilon.$$

In this integral, we integrate twice by parts, and in the remaining integral make the change of variable $(\varepsilon - \mu)/T = \xi$. Omitting the non-oscillatory part, we have

$$\tilde{I}_k = \frac{\sqrt{(2m)}\,(\beta H)^{5/2}}{T\pi^2 k^{5/2}} \exp\left(\frac{i\pi k\mu}{\beta H} - \frac{i\pi}{4}\right) \int_{-\infty}^{\infty} \frac{e^{\xi}}{(e^{\xi} + 1)^2} \exp\left(\frac{i\pi kT}{\beta H}\xi\right) d\xi.$$

The lower limit $-\mu/T$ of the integral over ξ is replaced by $-\infty$, since $\mu \gg T$ When $\beta H \gtrsim T$, the range $\xi \sim 1$ predominates in the integral, i.e. the range

[†] This is derived by rotating the path of integration in the complex p-plane: we put $p = e^{-i\pi/4} u$ and integrate over real u from $-\infty$ to ∞.

of ε values near μ ($\varepsilon - \mu \sim T$). The integral is calculated from the formula[†]

$$\int_{-\infty}^{\infty} \frac{e^{\xi}}{(e^{\xi}+1)^2} e^{i\alpha\xi} \, d\xi = \frac{\pi\alpha}{\sinh \pi\alpha}.$$

We finally have for the oscillatory part of Ω

$$\tilde{\Omega} = \frac{\sqrt{2}(m\beta H)^{3/2} TV}{\pi^2\hbar^3} \sum_{k=1}^{\infty} \frac{\cos\left(\pi\mu k/\beta H - \frac{1}{4}\pi\right)}{k^{3/2} \sinh\left(\pi^2 kT/\beta H\right)}. \tag{60.5}$$

In calculating the magnetic moment as the derivative of (60.5), only the most rapidly varying factors need be differentiated, namely the cosines in the numerators. This gives

$$\tilde{\mathfrak{M}} = -\frac{\sqrt{(2\beta)}m^{3/2}\mu TV}{\pi\hbar^3\sqrt{H}} \sum_{k=1}^{\infty} \frac{\sin\left(\pi\mu k/\beta H - \frac{1}{4}\pi\right)}{\sqrt{k} \sinh\left(\pi^2 kT/\beta H\right)} \tag{60.6}$$

(L. D. Landau, 1939). This function oscillates with high frequency. Its "period" in the variable $1/H$ is constant,

$$\Delta(1/H) = 2\beta/\mu, \tag{60.7}$$

independent of the temperature. Here $\Delta H/H \sim \beta H/\mu \ll 1$.[‡]

When $\beta H \sim T$, the amplitude of oscillation of the magnetic moment $\tilde{\mathfrak{M}} \sim V\mu H^{1/2}(m\beta)^{3/2}\hbar^{-3}$. The "monotonic" part $\overline{\mathfrak{M}}$ of the magnetisation, determined from the susceptibility calculated in § 59, is $\overline{\mathfrak{M}} \sim V\mu^{1/2}Hm^{3/2}\beta^2\hbar^{-3}$. Hence $\tilde{\mathfrak{M}}/\overline{\mathfrak{M}} \sim (\mu/\beta H)^{1/2}$; the amplitude of the oscillating part is large compared with the monotonic part. However, if $\beta H \ll T$, this amplitude is exponentially small, as $\exp(-\pi^2 T/\beta H)$, and becomes negligible.

[†] By the substitution $u = 1/(e^{\xi}+1)$, the integral is reduced to a beta function:

$$\int_0^1 (1-u)^{i\alpha} u^{-i\alpha} \, du = \Gamma(1+i\alpha)\Gamma(1-i\alpha)/\Gamma(2)$$

and the result given in the text follows, since

$$\Gamma(1-z)\Gamma(1+z) = \pi z/\sin\pi z.$$

[‡] The oscillations of the magnetisation were qualitatively predicted by Landau (1930). In metals, this phenomenon is called the *de Haas–van Alphen effect*.

§ 61. A relativistic degenerate electron gas

As the gas is compressed, the mean energy of the electrons increases (ε_F increases); when it becomes comparable with mc^2, relativistic effects begin to be important. Here we shall discuss in detail a completely degenerate extreme relativistic electron gas, the energy of whose particles is large compared with mc^2. In this case the relation between the energy and momentum of a particle is

$$\varepsilon = cp. \tag{61.1}$$

The previous formulae (57.1) and (57.2) give the number of quantum states and hence the limiting momentum. The limiting energy (i.e. the chemical potential of the gas) is now

$$\varepsilon_F = cp_F = (3\pi^2)^{1/3} \hbar c \left(\frac{N}{V}\right)^{1/3}. \tag{61.2}$$

The total energy of the gas is

$$E = \frac{cV}{\pi^2 \hbar^3} \int_0^{p_F} p^3 \, dp = \frac{cp_F^4}{4\pi^2 \hbar^3} V,$$

or

$$E = \frac{3}{4} (3\pi^2)^{1/3} \hbar c N \left(\frac{N}{V}\right)^{1/3}. \tag{61.3}$$

The gas pressure can be obtained by differentiating the energy with respect to the volume at constant entropy (equal to zero). This gives

$$P = \frac{E}{3V} = \frac{1}{4} (3\pi^2)^{1/3} \hbar c \left(\frac{N}{V}\right)^{4/3}. \tag{61.4}$$

The pressure of an extreme relativistic electron gas is proportional to the 4/3 power of the density.

It should be mentioned that the relation

$$PV = \tfrac{1}{3}E \tag{61.5}$$

is actually valid for an extreme relativistic gas not only at absolute zero but at all temperatures. This is easily seen by exactly the same method as that used to derive the relation (56.8), with the energy given by $\varepsilon = cp$ instead of $\varepsilon = p^2/2m$. With $\varepsilon = cp$, formula (53.4) leads to the following expression for Ω:

$$\Omega = -\frac{TV}{\pi^2 c^3 \hbar^3} \int_0^\infty \varepsilon^2 \log\left(1 + e^{(\mu-\varepsilon)/T}\right) d\varepsilon,$$

or, integrating by parts,

$$\Omega = -\frac{1}{3}\frac{V}{\pi^2 c^3 \hbar^3}\int_0^\infty \frac{\varepsilon^3\,d\varepsilon}{e^{(\varepsilon-\mu)/T}+1} = -\frac{1}{3}E. \tag{61.6}$$

Thus the limiting value that the pressure of any macroscopic body can have for a given E (see § 27) is reached for an extreme relativistic Fermi gas.

Using the variable of integration $z = \varepsilon/T$, we have

$$\Omega = -\frac{VT^4}{3\pi^2 c^3 \hbar^3}\int_0^\infty \frac{z^3\,dz}{e^{z-\mu/T}+1}.$$

This shows that

$$\Omega = VT^4 f(\mu/T). \tag{61.7}$$

Hence, as in § 56, we find that in an adiabatic process the volume, pressure and temperature of an extreme relativistic Fermi gas are related by

$$PV^{4/3} = \text{constant}, \quad VT^3 = \text{constant}, \quad T^4/P = \text{constant}. \tag{61.8}$$

These are the same as the usual equation of the Poisson adiabatic with $\gamma = \frac{4}{3}$; but it must be emphasised that γ here is not the ratio of the specific heats of the gas.

PROBLEMS

PROBLEM 1. Determine the number of collisions with a wall in an extreme relativistic completely degenerate electron gas.

SOLUTION. The calculation is as in § 57, Problem; it must be remembered that the electron velocity $v \cong c$. The result is $v = \frac{1}{4}cN/V$.

PROBLEM 2. Determine the specific heat of a degenerate extreme relativistic electron gas.

SOLUTION. Applying the formula (58.1) to the integral in (61.6), we find

$$\Omega = \Omega_0 - \frac{(\mu T)^2}{6(c\hbar)^3}V.$$

Hence the entropy

$$S = \frac{\mu^2}{3(c\hbar)^3}VT = N\frac{(3\pi^2)^{2/3}}{3c\hbar}T\left(\frac{V}{N}\right)^{1/3}$$

and the specific heat

$$C = N\frac{(3\pi^2)^{2/3}}{3c\hbar}\left(\frac{V}{N}\right)^{1/3}T.$$

PROBLEM 3. Determine the equation of state of a relativistic completely degenerate electron gas (the electron energy and momentum being related by $\varepsilon^2 = c^2 p^2 + m^2 c^4$).

SOLUTION. The previous formulae (57.1) and (57.2) give the number of states and the limiting momentum, and the total energy is

$$E = \frac{Vc}{\pi^2\hbar^3} \int_0^{p_F} p^2 \sqrt{(m^2c^2+p^2)}\, dp,$$

whence

$$E = \frac{cV}{8\pi^2\hbar^3} \{p_F(2p_F^2+m^2c^2)\sqrt{(p_F^2+m^2c^2)}-(mc)^4 \sinh^{-1}(p_F/mc)\}.$$

The pressure $P = -(\partial E/\partial V)_{S=0}$ is

$$P = \frac{c}{8\pi^2\hbar^3} \left\{p_F\left(\frac{2}{3}p_F^2-m^2c^2\right)\sqrt{(p_F^2+m^2c^2)}+(mc)^4 \sinh^{-1}(p_F/mc)\right\}.$$

These formulae are conveniently put in parametric form, using as parameter the quantity $\xi = 4\sinh^{-1}(p_F/mc)$. Then

$$N/V = (mc/\hbar)^3 \cdot (1/3\pi^2) \sinh^3 \tfrac{1}{4}\xi,$$
$$P = (m^4c^5/32\pi^2\hbar^3)(\tfrac{1}{3}\sinh\xi-\tfrac{8}{3}\sinh\tfrac{1}{2}\xi+\xi),$$
$$E/V = (m^4c^5/32\pi^2\hbar^3)(\sinh\xi-\xi).$$

The chemical potential μ of the gas (including the rest energy of the particle) is equal to the limiting energy $\varepsilon_F = \varepsilon(p_F)$. It is related to the density by

$$\frac{N}{V} = \frac{1}{3\pi^2\hbar^3}\left(\frac{\mu^2}{c^2}-m^2c^2\right)^{3/2}.$$

§ 62. A degenerate Bose gas

At low temperatures the properties of a Bose gas bear no resemblance to those of a Fermi gas. This is evident from the fact that for a Bose gas the state of lowest energy, occupied by the gas at $T = 0$, must be that with $E = 0$ (all the particles being in the quantum state with $\varepsilon = 0$), whereas a Fermi gas has a non-zero energy at absolute zero.

If the temperature of the gas is lowered at constant density N/V, the chemical potential μ given by equation (56.5) (with the lower sign) will increase, i.e. its absolute magnitude will decrease (since μ is negative). It reaches the value $\mu = 0$ at a temperature determined by the equation

$$\frac{N}{V} = \frac{g(mT)^{3/2}}{2^{1/2}\pi^2\hbar^3} \int_0^\infty \frac{\sqrt{z}\, dz}{e^z-1}. \tag{62.1}$$

The integral in (62.1) can be expressed in terms of the zeta function; see the second footnote to § 58. Denoting the required temperature by T_0, we obtain

$$T_0 = \frac{3.31}{g^{2/3}} \frac{\hbar^2}{m} \left(\frac{N}{V}\right)^{2/3}. \tag{62.2}$$

For $T < T_0$, equation (56.5) has no negative solutions, whereas in Bose statistics the chemical potential must be negative at all temperatures.

This apparent contradiction arises because under the conditions in question it is not legitimate to go from the summation in formula (54.3) to the integration in (56.5): in this process the first term in the sum (with $\varepsilon_k = 0$) is multiplied by $\sqrt{\varepsilon} = 0$ and so disappears from the sum; but, as the temperature decreases, more and more particles must occupy that state of lowest energy, until at $T = 0$ they are all in it. The mathematical effect of this is that, when the limit $\mu \to 0$ is taken in the sum (54.3), the sum of all the terms in the series except the first tends to a finite limit given by the integral (56.5), but the first term (with $\varepsilon_k = 0$) tends to infinity. Consequently, by letting μ tend not to zero but to some small finite value, we can make this first term in the sum take the desired finite value.

In reality, therefore, the situation for $T < T_0$ is as follows. Particles with energy $\varepsilon > 0$ are distributed according to formula (56.4) with $\mu = 0$:

$$dN_\varepsilon = \frac{gm^{3/2}V}{2^{1/2}\pi^2\hbar^3} \frac{\sqrt{\varepsilon}\, d\varepsilon}{e^{\varepsilon/T}-1}. \tag{62.3}$$

The total number of particles with energies $\varepsilon > 0$ will thus be

$$N_{\varepsilon>0} = \int dN_\varepsilon = \frac{gV(mT)^{3/2}}{2^{1/2}\pi^2\hbar^3} \int_0^\infty \frac{\sqrt{z}\, dz}{e^z-1} = N(T/T_0)^{3/2}.$$

The remaining

$$N_{\varepsilon=0} = N[1-(T/T_0)^{3/2}] \tag{62.4}$$

particles are in the lowest state, i.e. have energy $\varepsilon = 0$.[†] The energy of the gas for $T < T_0$ is, of course, determined only by the particles with $\varepsilon > 0$; putting $\mu = 0$ in (56.7), we have

$$E = \frac{gV(mT)^{3/2}T}{2^{1/2}\pi^2\hbar^3} \int_0^\infty \frac{z^{3/2}\, dz}{e^z-1}.$$

[†] The steady increase of particles in the state with $\varepsilon = 0$ is called *Bose–Einstein condensation*. It should be emphasised that this refers only to "condensation" in momentum space; no condensation actually occurs in the gas, of course.

This integral reduces to $\zeta\left(\frac{5}{2}\right)$ (see the second footnote to § 58), and we obtain

$$E = 0.770NT\,(T/T_0)^{3/2}$$
$$= 0.128g(m^{3/2}T^{5/2}/\hbar^3)V. \tag{62.5}$$

The specific heat is therefore

$$C_v = 5E/2T, \tag{62.6}$$

i.e. is proportional to $T^{3/2}$. Integration of the specific heat gives the entropy:

$$S = 5E/3T, \tag{62.7}$$

and the free energy is

$$F = E - TS = -\tfrac{2}{3}E. \tag{62.8}$$

This is obvious, since for $\mu = 0$

$$F = \Phi - PV = N\mu + \Omega = \Omega.$$

The pressure is

$$P = -(\partial F/\partial V)_T = 0.0851gm^{3/2}T^{5/2}/\hbar^3. \tag{62.9}$$

We see that for $T < T_0$ the pressure is proportional to $T^{5/2}$ and is independent of the volume. This is the natural consequence of the fact that particles in a state with $\varepsilon = 0$ have no momentum and make no contribution to the pressure.

At the point $T = T_0$ itself, all the above-mentioned thermodynamic quantities are continuous, but it may be shown that the derivative of the specific heat with respect to temperature is discontinuous there (see the Problem). The curve of the specific heat itself as a function of temperature has a change in slope at $T = T_0$, and the specific heat has its maximum value there (equal to $1.28 \times 3N/2$).[†]

PROBLEM

Determine the discontinuity of the derivative $(\partial C_v/\partial T)_V$ at $T = T_0$.

SOLUTION. To solve this problem we must determine the energy of the gas for small positive $T - T_0$. The equation (56.5) is identical with

$$N = N_0(T) + \frac{gVm^{3/2}}{2^{1/2}\pi^2\hbar^3} \int\limits_0^\infty \left[\frac{1}{e^{(\varepsilon - \mu)/T} - 1} - \frac{1}{e^{\varepsilon/T} - 1}\right]\sqrt{\varepsilon}\,d\varepsilon,$$

[†] Note, however, that this behaviour of the specific heat is the result of entirely neglecting the interaction of the gas particles; even a weak interaction changes the situation.

$$C_v = \frac{\partial E}{\partial T}$$

$$C_v = \int C_v\,dT$$

$$S = \int \frac{dQ}{T} = \int_0 \frac{C_v\,dT}{T}$$

where $N_0(T)$ is given by (62.1). Expanding the integrand and using the fact that μ is small near the point $T = T_0$, and therefore the important part of the integral arises from the region where ε is small, we find that the integral is equal to

$$T\mu \int_0^\infty \frac{d\varepsilon}{\sqrt{\varepsilon(\varepsilon+|\mu|)}} = -\pi T \sqrt{|\mu|}. \tag{1}$$

Substituting this value and then expressing μ in terms of $N-N_0$, we have

$$-\mu = \frac{2\pi^2\hbar^6}{g^2m^3} \left(\frac{N_0-N}{TV}\right)^2.$$

To the same accuracy we can write

$$\frac{\partial E}{\partial \mu} = -\frac{3}{2}\frac{\partial \Omega}{\partial \mu} = \frac{3}{2}N \approx \frac{3}{2}N_0,$$

whence

$$E = E_0 + \frac{3}{2}N_0\mu = E_0 - \frac{3\pi^2\hbar^6}{g^2m^3}N_0\left(\frac{N_0-N}{TV}\right)^2,$$

where $E_0 = E_0(T)$ denotes the energy for $\mu = 0$, i.e. the function (62.5). The second derivative of the second term with respect to temperature will clearly give the required discontinuity. The result of the calculation is

$$\Delta\left(\frac{\partial C_v}{\partial T}\right)_V = -\frac{6\pi^2\hbar^6}{g^2m^3V^2}\left[N_0\left(\frac{1}{T}\frac{\partial N_0}{\partial T}\right)^2\right]_{T=T_0} = -3.66N/T_0. \tag{2}$$

The value of the derivative $(\partial C_v/\partial T)_V$ for $T = T_0-0$ is, from (62.5), $+2.89N/T_0$, and for $T = T_0+0$ it is therefore $-0.77N/T_0$.

§ 63. Black-body radiation

The most important application of Bose statistics relates to electromagnetic radiation which is in thermal equilibrium—called *black-body radiation*. Such radiation may be regarded as a gas consisting of photons. The linearity of the equations of electrodynamics expresses the fact that photons do not interact with one another (the principle of superposition for the electromagnetic field), so that the photon gas is an ideal gas. Because the angular momentum of the photons is integral, this gas obeys Bose statistics.

If the radiation is not in a vacuum but in a material medium, the condition for an ideal photon gas requires also that the interaction between radiation and matter should be small. This condition is satisfied in gases throughout the radiation spectrum except for frequencies in the neighbourhood of absorption lines of the material, but at high densities of matter it may be violated except at very high temperatures.

It should be remembered that at least a small amount of matter must be present if thermal equilibrium is to be reached in the radiation, since the interaction between the photons themselves may be regarded as completely absent.[†] The mechanism by which equilibrium can be established consists in the absorption and emission of photons by matter. This results in an important specific property of the photon gas: the number of photons N in it is variable, and not a given constant as in an ordinary gas. Thus N itself must be determined from the conditions of thermal equilibrium. From the condition that the free energy of the gas should be a minimum (for given T and V), we obtain as one of the necessary conditions $\partial F/\partial N = 0$. Since $(\partial F/\partial N)_{T, V} = \mu$, this gives

$$\mu = 0, \tag{63.1}$$

i.e. the chemical potential of the photon gas is zero.

The distribution of photons among the various quantum states with definite values of the momentum $\hbar\mathbf{k}$ and energies $\varepsilon = \hbar\omega = \hbar ck$ (and definite polarisations) is therefore given by formula (54.2) with $\mu = 0$:

$$\overline{n_{\mathbf{k}}} = 1/(e^{\hbar\omega/T} - 1). \tag{63.2}$$

This is called *Planck's distribution.*

Assuming that the volume is sufficiently large, we can make the usual change (see *Fields,*§ 52) from the discrete to the continuous distribution of eigenfrequencies of the radiation. The number of modes of oscillation for which the components of the wave vector $\mathbf{k}$ lie in the intervals $d^3k = dk_x\, dk_y\, dk_z$ is $V\, d^3k/(2\pi)^3$, and the number of modes for which the absolute magnitude of the wave vector lies in the range dk is correspondingly $V \cdot 4\pi k^2\, dk/(2\pi)^3$. Using the frequency $\omega = ck$ and multiplying by 2 (for the two independent directions of polarisation of the oscillations), we obtain the number of quantum states of photons with frequencies between ω and $\omega + d\omega$:

$$V\omega^2\, d\omega/\pi^2 c^3. \tag{63.3}$$

Multiplying the distribution (63.2) by this quantity, we find the number of photons in this frequency interval:

$$dN_\omega = \frac{V}{\pi^2 c^3}\,\frac{\omega^2\, d\omega}{e^{\hbar\omega/T} - 1}, \tag{63.4}$$

and a further multiplication by $\hbar\omega$ gives the radiation energy in this segment of the spectrum:

$$dE_\omega = \frac{V\hbar}{\pi^2 c^3}\,\frac{\omega^3\, d\omega}{e^{\hbar\omega/T} - 1}. \tag{63.5}$$

[†] Apart from the entirely negligible interaction (the scattering of light by light) which is due to the possible production of virtual electron–positron pairs.

This formula for the spectral energy distribution of black-body radiation is called *Planck's formula.*[†] In terms of the wavelength $\lambda = 2\pi c/\omega$, it becomes

$$dE_\lambda = \frac{16\pi^2 c\hbar V}{\lambda^5} \frac{d\lambda}{e^{2\pi\hbar c/T\lambda} - 1}. \tag{63.6}$$

At low frequencies ($\hbar\omega \ll T$), formula (63.5) gives

$$dE_\omega = V(T/\pi^2 c^3)\omega^2 \, d\omega. \tag{63.7}$$

This is the *Rayleigh–Jeans formula.* It should be noticed that formula (63.7) does not contain the quantum constant $\hbar$, and can be derived by multiplying by T the number of modes (63.3); in this sense it corresponds to classical statistics, in which an energy T must correspond to each "vibrational degree of freedom"—the law of equipartition (§ 44).

In the opposite limiting case of high frequencies ($\hbar\omega \gg T$), formula (63.5) becomes

$$dE_\omega = V(\hbar/\pi^2 c^3)\omega^3 e^{-\hbar\omega/T} \, d\omega. \tag{63.8}$$

This is *Wien's formula.*

Figure 7 shows a graph of the function $x^3/(e^x - 1)$, corresponding to the distribution (63.5).

FIG. 7

The density of the spectral frequency distribution of the energy of black-body radiation, $dE_\omega/d\omega$, has a maximum at a frequency ω_m given by

$$\hbar\omega_m/T = 2.822. \tag{63.9}$$

[†] The discovery of this law by M. Planck (1900) was the foundation of the quantum theory.

Thus, when the temperature rises, the position of the maximum of the distribution is displaced towards higher frequencies in proportion to T (the *displacement law*).[†]

Let us calculate the thermodynamic quantities for black-body radiation. For $\mu = 0$, the free energy is the same as Ω (since $F = \Phi - PV = N\mu + \Omega$). According to formula (54.4), in which we put $\mu = 0$ and change in the usual way (by means of (63.3)) from summation to integration, we obtain

$$F = T \frac{V}{\pi^2 c^3} \int_0^\infty \omega^2 \log\left(1 - e^{-\hbar\omega/T}\right) d\omega. \tag{63.10}$$

With the new variable of integration $x = \hbar\omega/T$, integration by parts gives

$$F = -V \frac{T^4}{3\pi^2 \hbar^3 c^3} \int_0^\infty \frac{x^3\, dx}{e^x - 1}.$$

The integral is equal to $\pi^4/15$ (see the second footnote to § 58). Thus

$$F = -V \cdot \pi^2 T^4/45(\hbar c)^3$$
$$= -4\sigma V T^4/3c. \tag{63.11}$$

If T is measured in degrees, the coefficient σ (called the *Stefan–Boltzmann constant*) is

$$\sigma = \pi^2 k^4/60\hbar^3 c^2$$
$$= 5.67 \times 10^{-5}\ \text{g/sec}^3\ \text{deg}^4. \tag{63.12}$$

The entropy is

$$S = -\partial F/\partial T = 16\sigma V T^3/3c, \tag{63.13}$$

and is proportional to the cube of the temperature. The total radiation energy $E = F + TS$ is

$$E = 4\sigma V T^4/c = -3F. \tag{63.14}$$

This expression could, of course, be derived also by direct integration of the distribution (63.5). Thus the total energy of black-body radiation is proportional to the fourth power of the temperature. This is *Boltzmann's law*.

For the specific heat of the radiation C_v we have

$$C_v = (\partial E/\partial T)_V = 16\sigma T^3 V/c. \tag{63.15}$$

[†] The wavelength distribution density $dE_\lambda/d\lambda$ also has a maximum, but at a different value of the corresponding ratio: $2\pi\hbar c/T\lambda_m = 4.965$. Thus the maximum λ_m of the wavelength distribution is displaced in inverse proportion to the temperature.

Finally, the pressure is

$$P = -(\partial F/\partial V)_T = 4\sigma T^4/3c, \qquad (63.16)$$

$$PV = \tfrac{1}{3}E. \qquad (63.17)$$

Thus for a photon gas the same limiting value of the pressure is obtained as for an extreme relativistic electron gas (§ 61); this is as it should be, since the relation (63.17) is a direct consequence of the linear relation ($\varepsilon = cp$) between the energy and momentum of a particle.

The total number of photons in black-body radiation is

$$N = \frac{V}{\pi^2 c^3} \int\limits_0^\infty \frac{\omega^2 \, d\omega}{e^{\hbar\omega/T} - 1} = \frac{VT^3}{\pi^2 c^3 \hbar^3} \int\limits_0^\infty \frac{x^2 \, dx}{e^x - 1}.$$

The integral can be expressed in terms of $\zeta(3)$; see the second footnote to § 58. Thus

$$N = \frac{2\zeta(3)}{\pi^2} \left(\frac{T}{\hbar c}\right)^3 V = 0.244 \left(\frac{T}{\hbar c}\right)^3 V. \qquad (63.18)$$

In an adiabatic expansion (or compression) of the photon gas, the volume and temperature are related by $VT^3 = $ constant. From (63.16), the pressure and volume are then related by $PV^{4/3} = $ constant. A comparison with (61.8) shows that the equation of the adiabatic for a photon gas coincides (as we should expect) with that for an extreme relativistic gas.

Let us consider a body in thermal equilibrium with black-body radiation around it. The body continually reflects and absorbs photons incident on it, and at the same time emits new ones, and in equilibrium all these processes balance in such a way that the distribution of photons in frequency and direction remains unchanged on the average.

Owing to the complete isotropy of the black-body radiation, each volume element emits a flux of energy uniformly in all directions. We use the notation

$$e_0(\omega) = \frac{1}{4\pi V} \frac{dE_\omega}{d\omega} = \frac{\hbar\omega^3}{4\pi^3 c^3 (e^{\hbar\omega/T} - 1)} \qquad (63.19)$$

for the spectral density of black-body radiation per unit volume and unit solid angle. Then the energy flux density with frequencies in the interval $d\omega$ leaving each point and entering the solid angle element do is $ce_0(\omega) \, do \, d\omega$. The radiation energy (with frequencies in $d\omega$) incident in unit time on unit area of the surface of the body at an angle θ to the normal is therefore $ce_0(\omega) \cos \theta \, do \, d\omega$, $do = 2\pi \sin \theta \, d\theta$.

Let $A(\omega, \theta)$ denote the absorbing power of the body as a function of the frequency and direction of incidence of the radiation; this quantity is defined as the fraction of the radiation energy incident on the surface of the body, in the given frequency interval, which is absorbed by the body, not including the radiation (if any) which passes through the body. Then the quantity of radiation absorbed per unit time and surface area will be

$$ce_0(\omega) \, A(\omega, \theta) \cos \theta \, do \, d\omega. \tag{63.20}$$

Let us assume that the body does not scatter radiation and is not fluorescent, i.e. that the reflection occurs without change in the angle θ or in the frequency. We shall also suppose that the radiation does not pass through the body; in other words, all radiation not reflected is completely absorbed. Then the quantity of radiation (63.20) must be balanced by the radiation emitted by the body itself in the same directions at the same frequencies. Denoting by $J(\omega, \theta) \, d\omega \, do$ the intensity of emission from unit area of the surface and equating it to the absorbed energy, we obtain

$$J(\omega, \theta) = ce_0(\omega) \, A(\omega, \theta) \cos \theta. \tag{63.21}$$

The functions $J(\omega, \theta)$ and $A(\omega, \theta)$ are, of course, different for different bodies, but we see that their ratio is independent of the properties of the body and is a universal function of frequency and direction:

$$J(\omega, \theta)/A(\omega, \theta) = ce_0(\omega) \cos \theta,$$

which is determined by the energy distribution in the black-body radiation spectrum (at a temperature equal to that of the body). This is *Kirchhoff's law*.

If the body scatters radiation, Kirchhoff's law can be formulated only in a more restricted way. Since in this case reflection occurs with a change in the angle θ, we can derive from the condition of equilibrium only the requirement that the radiation (of a given frequency) absorbed from all directions should be equal to the total emission from the body in all directions:

$$\int J(\omega, \theta) \, do = ce_0(\omega) \int A(\omega, \theta) \cos \theta \, do. \tag{63.22}$$

The angle θ also changes, in general, when radiation can pass through the body (because of refraction on entering and leaving the body). In this case the relation (63.22) must be integrated over the entire surface of the body; the functions $A(\omega, \theta)$ and $J(\omega, \theta)$ now depend not only on the material of the body but also on its shape and on the point considered on its surface.

Finally, when there is scattering with change of frequency (fluorescence), Kirchhoff's law applies only to the integrals over both direction and frequency of the radiation:

$$\iint J(\omega, \theta) \, do \, d\omega = c \iint e_0(\omega) \, A(\omega, \theta) \cos \theta \, do \, d\omega. \tag{63.23}$$

A body which completely absorbs all radiation incident on it is called a *black body*.[†] For such a body, $A(\omega, \theta) = 1$ by definition, and its emissive

[†] Such a body can be realised in the form of a cavity with highly absorbing internal walls and a small aperture. Any ray entering through the aperture can return to it and leave the cavity only after repeated reflection from the walls of the cavity. When the aperture is sufficiently small, therefore, the cavity will absorb practically all the radiation incident on the aperture, and so the surface of the aperture will be a black body.

power is entirely determined by the function

$$J_0(\omega, \theta) = ce_0(\omega) \cos \theta, \tag{63.24}$$

which is the same for all black bodies. It may be noted that the intensity of emission from a black body is a very simple function of direction, being proportional to the cosine of the angle to the normal to the surface of the body. The total intensity of emission from a black body, J_0, is obtained by integrating (63.24) over all frequencies and over a hemisphere:

$$J_0 = c \int_0^\infty e_0(\omega) \, d\omega \int_0^{\pi/2} 2\pi \cos \theta \sin \theta \, d\theta = cE/4V,$$

where E is given by (63.14). Thus

$$J_0 = \sigma T^4, \tag{63.25}$$

i.e. the total intensity of emission from a black body is proportional to the fourth power of its temperature.

Finally, let us consider radiation not in thermal equilibrium, having a non-equilibrium spectral or directional distribution. Let $e(\omega, \mathbf{n}) \, d\omega \, do$ be the volume density of this radiation in the frequency interval $d\omega$ and with the direction $\mathbf{n}$ of the wave vector lying in the solid-angle element do. We can use the concept of the temperature of the radiation in each small interval of frequency and direction, defined as the temperature for which the density $e(\omega, \mathbf{n})$ is equal to that given by Planck's formula, i.e. $e(\omega, \mathbf{n}) = e_0(\omega)$. Denoting this temperature by $T_{\omega, \mathbf{n}}$, we have

$$T_{\omega, \mathbf{n}} = \frac{\hbar\omega}{\log\left\{1 + \dfrac{\hbar\omega^3}{4\pi^3 c^3} \cdot \dfrac{1}{e(\omega, \mathbf{n})}\right\}}. \tag{63.26}$$

Let us imagine a black body emitting into a surrounding vacuum. The radiation is propagated freely along straight lines and will not be in thermal equilibrium outside the body; it is by no means isotropic, as equilibrium radiation must be. Since the photons are propagated in a vacuum and do not interact with one another, we are in a position to apply Liouville's theorem rigorously to the photon distribution function in the corresponding phase space of coordinates and wave-vector components.[†] According to this theorem, the distribution function remains constant along the phase trajectories. But the distribution function is, apart from a factor dependent on frequency, the same as the volume density of radiation of a given frequency and

---

[†] When considering the limiting case of geometrical optics, we can speak of coordinates of a photon.

direction, $e(\omega, \mathbf{n}, \mathbf{r})$. Since the radiation frequency is also constant during propagation, we have the following important result: in every solid-angle element where radiation is propagated (from a given point in space) the radiation density $e(\omega, \mathbf{n}, \mathbf{r})$ is equal to the density within the emitting black body, i.e. to the black-body radiation density $e_0(\omega)$. Whereas, however, for equilibrium radiation the density exists for all directions, here it exists only for a certain interval of directions.

Defining the temperature of non-equilibrium radiation by (63.26), we can express the result differently by saying that the temperature $T_{\omega, \mathbf{n}}$ is equal to the temperature T of the emitting black body for all directions in which radiation is being propagated (at any given point in space). If the radiation temperature is defined from the density averaged over all directions, however, it is of course less than the temperature of the black body.

All these consequences of Liouville's theorem remain fully valid when reflecting mirrors and refracting lenses are present, provided, of course, that the conditions for geometrical optics to be applicable are still satisfied. By means of lenses or mirrors the radiation can be focused, i.e. the range of directions from which rays reach a given point in space can be enlarged. This may increase the mean radiation temperature at the point considered, but the foregoing discussion shows that there is no means of raising it above the temperature of the black body which emitted the radiation.

CHAPTER VI

SOLIDS

§ 64. Solids at low temperatures

SOLIDS form another suitable topic for the application of statistical methods of calculating the thermodynamic quantities. A characteristic property of solids is that the atoms in them execute only small vibrations about certain equilibrium positions, the crystal lattice sites. The configuration of the lattice sites which corresponds to thermal equilibrium of the body is preferred, i.e. distinguished from all other possible distributions, and must therefore be regular. In other words, a solid in thermal equilibrium must be *crystalline*.

According to classical mechanics, all the atoms are at rest at absolute zero, and the potential energy of their interaction must be a minimum in equilibrium. At sufficiently low temperatures, therefore, the atoms must always execute only small vibrations, i.e. all bodies must be solid. In reality, however, quantum effects may bring about exceptions to this rule. One such is liquid helium, the only substance which remains liquid at absolute zero (at pressures that are not too high); all other substances solidify well before quantum effects become important.[†]

For a body to be solid its temperature must certainly be small in comparison with the energy of interaction of the atoms (in practice, all solids melt or decompose at higher temperatures). From this it results that the vibrations of atoms in a solid about their equilibrium positions are always small.

As well as crystals, there exist in Nature also *amorphous* solids, in which the atoms vibrate about randomly situated points. Such bodies are thermodynamically metastable, and must ultimately become crystalline. In practice, however, the relaxation times are so long that amorphous bodies behave as if stable for an almost unlimited time. All the following calculations apply equally to both crystalline and amoprhous substances. The only difference is that, since amorphous bodies are not in equilibrium, Nernst's theorem does not apply to them, and as $T \rightarrow 0$ their entropy tends to a non-zero value.

[†] Quantum effects become important when the de Broglie wavelength corresponding to the thermal motion of the atoms becomes comparable with the distances between atoms. In liquid helium this occurs at 2–3°K.

Consequently, for amorphous bodies the formula (64.7) derived below for the entropy has to be augmented by some constant S_0 (and the free energy by a corresponding term $-TS_0$); we shall omit this unimportant constant, which, in particular, does not affect the specific heats of a body.

The residual entropy, which does not vanish as $T \to 0$, may also be observed in crystalline solids, because of what is called *ordering* of crystals. If the number of crystal lattice sites at which atoms of a given kind can be situated is equal to the number of such atoms, there will be one atom near each site; that is, the probability of finding an atom (of the kind in question) in the neighbourhood of each site is equal to unity. Such crystals are said to be *completely ordered*. There are also, however, crystals in which the atoms may be not only at their "own" positions (i.e. those which they occupy in complete ordering) but also at certain "other" positions. In that case the number of sites that may be occupied by an atom of the given kind is greater than the number of such atoms, and the probability of finding atoms of this kind at either the old or the new sites will not be unity.

For example, solid carbon monoxide is a molecular crystal, in which the CO molecule can have two opposite orientations differing by interchange of the two atoms; the number of sites that may be occupied by carbon (or oxygen) atoms is here equal to twice the number of these atoms.

In a state of complete thermodynamic equilibrium at absolute zero, any crystal must be completely ordered, and the atoms of each kind must occupy entirely definite positions.[†] However, because the processes of lattice rearrangement are slow, especially at low temperatures, a crystal which is incompletely ordered at a high temperature may in practice remain so even at very low temperatures. This "freezing" of the disorder leads to the existence of a constant residual term in the entropy of the crystal. For instance, in the example of the CO crystal mentioned above, if the CO molecules have the two orientations with equal probability, the residual entropy will be $S_0 = \log 2$.

Let N be the number of unit cells in the lattice, and ν the number of atoms in each cell. Then the number of atoms is $N\nu$. Of the total number of degrees of freedom $3N\nu$, three correspond to translational and three to rota-

---

[†] Strictly speaking, this statement also is valid only if quantum effects are neglected. The latter may become important (at $T = 0$) if the amplitude of the zero-point vibrations of the atoms in the lattice is comparable with the interatomic distances. In a "quantum crystal" of this kind, a situation is in principle possible where the number of sites in the ground state (the state at $T = 0$) exceeds the number of atoms. The "zero-point" defects (free vacancies) then present in the lattice are, however, not localised at any particular sites as they would be in a "classical" crystal, and represent a collective property of the lattice without disturbing its strict periodicity; see A. F. Andreev and I. M. Lifshits, *Soviet Physics JETP* **29**, 1107, 1969.

tional motion of the body as a whole. The number of vibrational degrees of freedom is therefore $3N\nu - 6$, but since $3N\nu$ is extremely large we can, of course, neglect 6 and assume that the number of vibrational degrees of freedom is just $3N\nu$.

It should be emphasised that in discussing solids we shall entirely ignore the "internal" (electronic) degrees of freedom of the atoms. Hence, if these degrees of freedom are important (as they may be, for example, in metals), the following formulae will relate only to the *lattice part* of the thermodynamic quantities for the solid, which is due to the vibrations of the atoms. In order to obtain the total values of these quantities, the electronic part must be added to the lattice part.

In mechanical terms, a system with $3N\nu$ vibrational degrees of freedom may be regarded as an assembly of $3N\nu$ independent oscillators, each corresponding to one normal mode of vibration. The thermodynamic quantities relating to one vibrational degree of freedom have already been calcuated in § 49. From the formulae there we can immediately write down the free energy of the solid as[†]

$$F = N\varepsilon_0 + T \sum_{\alpha} \log (1 - e^{-\hbar\omega_\alpha/T}). \qquad (64.1)$$

The summation is over all $3N\nu$ normal vibrations, which are labelled by the suffix α.[‡] We have added to the sum over vibrations a term $N\varepsilon_0$ which represents the energy of all the atoms in the body in their equilibrium positions (more precisely, when executing their "zero-point" vibrations); this term depends on the density, but not on the temperature: $\varepsilon_0 = \varepsilon_0(N/V)$.

Let us now consider the limiting case of low temperatures. For small T, only the terms with low frequencies ($\hbar\omega_\alpha \sim T$) are of importance in the sum over α. But vibrations with low frequencies are just ordinary *sound waves*, whose wavelength is related to the frequency by $\lambda \sim u/\omega$, where u is the velocity of sound. In sound waves the wavelength is large in comparison with the lattice constant ($\lambda \gg a$), and so $\omega \ll u/a$. In other words, the vibrations can be regarded as sound waves at temperatures such that

$$T \ll \hbar u/a. \qquad (64.2)$$

Let us assume that the body is isotropic (an amorphous solid). In an isotropic solid, longitudinal sound waves can be propagated (with velocity u_l) and so can transverse waves with two independent directions of polarisation and equal velocities of propagation u_t; see *Elasticity*, § 22. The frequency

---

[†] Quantised vibrations were first used by Einstein (1907) to calculate the thermodynamic quantities for a solid.

[‡] See (71.7) for an integral form of this.

of these waves is linearly related to the absolute magnitude of the wave vector $\mathbf{k}$ by $\omega = u_l k$ or $\omega = u_t k$.

The number of vibrational modes in the spectrum of sound waves with absolute magnitude of the wave vector lying in the interval dk and with a given polarisation is $V \cdot 4\pi k^2 \, dk/(2\pi)^3$, where V is the volume of the body. Putting for one of the three independent polarisations $k = \omega/u_l$ and for the other two $k = \omega/u_t$, we find that the interval $d\omega$ contains altogether

$$V \frac{\omega^2 \, d\omega}{2\pi^2} \left(\frac{1}{u_l^3} + \frac{2}{u_t^3} \right) \tag{64.3}$$

vibrations.

A mean velocity of sound $\bar{u}$ can be defined according to the formula

$$\frac{3}{\bar{u}^3} = \frac{2}{u_t^3} + \frac{1}{u_l^3}.$$

Then the expression (64.3) becomes

$$V \cdot 3\omega^2 \, d\omega / 2\pi^2 \bar{u}^3. \tag{64.4}$$

In this form it is applicable not only to isotropic bodies but also to crystals, where $\bar{u} = \bar{u}(V/N)$ must be understood as the velocity of propagation of sound in the crystal, averaged in a certain way. The determination of the averaging procedure requires the solution of the problem (which belongs to the theory of elasticity) of the propagation of sound in a crystal of given symmetry.[†]

By means of (64.4) we can change from the summation in (64.1) to integration, obtaining

$$F = N\varepsilon_0 + T \frac{3V}{2\pi^2 \bar{u}^3} \int_0^\infty \log (1 - e^{-\hbar\omega/T}) \omega^2 \, d\omega; \tag{64.5}$$

because of the rapid convergence of the integral when T is small, the integration can be taken from 0 to ∞. This expression (apart from the term $N\varepsilon_0$) differs from the formula (63.10) for the free energy of black-body radiation only in that the velocity of light c is replaced by the velocity of sound $\bar{u}$ and a factor $\frac{3}{2}$ appears. This resemblance is not surprising, since the frequency of sound vibrations is related to their wave number by the same type of linear formula as is valid for photons. The integers v_α in the energy levels $\Sigma v_\alpha \hbar\omega_\alpha$ of a system of sound oscillators may be regarded as occupation numbers of the

---

[†] In an anisotropic medium, there are in general three different branches of the sound-wave spectrum, in each of which the velocity of propagation is dependent on direction (see *Elasticity*, § 23).

various quantum states with energies $\varepsilon_\alpha = \hbar\omega_\alpha$, the values of these numbers being arbitrary (as in Bose statistics). The appearance of the extra factor $\frac{3}{2}$ in (64.5) is due to the fact that sound vibrations have three possible directions of polarisation instead of two as for photons.

Thus, without having to repeat the calculations, we can use the expression (63.11) derived in § 63 for the free energy of black-body radiation, if c is replaced by $\bar{u}$ and a factor $\frac{3}{2}$ included. The free energy of a solid is therefore

$$F = N\varepsilon_0 - V \cdot \pi^2 T^4 / 30(\hbar\bar{u})^3 ; \qquad (64.6)$$

the entropy is

$$S = V \cdot 2\pi^2 T^3 / 15(\hbar\bar{u})^3 , \qquad (64.7)$$

the energy

$$E = N\varepsilon_0 + V \cdot \pi^2 T^4 / 10(\hbar\bar{u})^3 , \qquad (64.8)$$

and the specific heat

$$C = 2\pi^2 T^3 V / 5(\hbar\bar{u})^3 . \qquad (64.9)$$

Thus the specific heat of a solid at low temperatures is proportional to the cube of the temperature[†] (P. Debye, 1912). We write the specific heat as C simply (not distinguishing C_v and C_p), since at low temperatures the difference $C_p - C_v$ is a quantity of a higher order of smallness than the specific heat itself (see § 23; here $S \propto T^3$ and so $C_p - C_v \propto T^7$).

For solids having a simple crystal lattice (elements and simple compounds) the T^3 law for the specific heat does in fact begin to hold at temperatures of the order of tens of degrees, but for bodies with a complex lattice this law may be expected to be satisfactorily obeyed only at much lower temperatures.

§ 65. Solids at high temperatures

Let us now turn to the opposite limiting case of high temperatures (of order $T \gg \hbar u/a$, where a is the lattice constant). In this case we can put $1 - e^{-\hbar\omega_\alpha/T} \cong \hbar\omega_\alpha/T$, and formula (64.1) becomes

$$F = N\varepsilon_0 + T \sum_\alpha \log\,(\hbar\omega_\alpha/T). \qquad (65.1)$$

The sum over α contains altogether $3N\nu$ terms. We define the "geometric mean" frequency $\bar{\omega}$ by

$$\log \bar{\omega} = \frac{1}{3N\nu} \sum_\alpha \log \omega_\alpha . \qquad (65.2)$$

[†] It may be recalled that when "electronic degrees of freedom" are present these formulae give only the lattice part of the thermodynamic quantities. However, even when there is an electronic part (as in metals) it begins to affect the specific heat, for example, only at temperatures of a few degrees.

Then the free energy of the solid is given by

$$F = N\varepsilon_0 - 3N\nu T \log T + 3N\nu T \log \hbar\bar{\omega}. \tag{65.3}$$

The mean frequency $\bar{\omega}$, like $\bar{u}$, is a function of the density, $\bar{\omega}(V/N)$.

From (65.3) we find the energy of the body, $E = F - T\,\partial F/\partial T$:

$$E = N\varepsilon_0 + 3N\nu T. \tag{65.4}$$

The case of high temperatures corresponds to the classical treatment of the vibrations of the atoms; it is therefore clear why formula (65.4) accords exactly with the law of equipartition (§ 44): apart from the constant $N\varepsilon_0$, an energy T corresponds to each of the $3N\nu$ vibrational degrees of freedom.

For the specific heat we have

$$C = Nc = 3N\nu, \tag{65.5}$$

where $c = 3\nu$ is the specific heat per cell. We again write the specific heat as C simply, since in solids the difference between C_p and C_v is always negligible (see the end of § 67).

Thus at sufficiently high temperatures the specific heat of a solid is constant and depends only on the number of atoms in the body. In particular, the specific heat per atom must be the same for different elements with a simple crystal lattice ($\nu = 1$) and equal to 3; this is *Dulong and Petit's law*. At ordinary temperatures this law is well satisfied for many elements. Formula (65.5) is valid at high temperatures for a number of simple compounds also, but for more complex compounds it gives a limiting value of the specific heat which in general is not reached before the substance melts or decomposes.

Substituting (65.5) in (65.3) and (65.4), we can write the free energy and energy of a solid as

$$F = N\varepsilon_0 - NcT \log T + NcT \log \hbar\bar{\omega}, \tag{65.6}$$

$$E = N\varepsilon_0 + NcT. \tag{65.7}$$

The entropy $S = -\partial F/\partial T$ is

$$S = Nc \log T - Nc \log (\hbar\bar{\omega}/e). \tag{65.8}$$

Formula (65.1) can also, of course, be derived directly from classical statistics, using the general formula (31.5)

$$F = -T \log \int' e^{-E(p,\,q)/T}\,d\Gamma. \tag{65.9}$$

For a solid, the integration over the coordinates in this integral is carried out as follows. Each atom is regarded as being situated near a particular lattice site, and the integration over its coordinates is taken only over a small neighbourhood of that site. It is clear that all the points in the region of integration thus defined will correspond to physically different microstates, and no additional factor is needed in the integral.[†]

We substitute in (65.9) the energy expressed in terms of the coordinates and momenta of the normal modes:

$$E(p, q) = \tfrac{1}{2} \sum_{\alpha} (p_\alpha^2 + \omega_\alpha^2 q_\alpha^2), \qquad (65.10)$$

and write dI' in the form

$$dI' = \frac{1}{(2\pi\hbar)^{3N\nu}} \prod_\alpha dp_\alpha \, dq_\alpha .$$

Then the integral becomes a product of $3N\nu$ integrals, all of the form

$$\int_{-\infty}^{\infty} \int_{-\infty}^{\infty} \exp \{-(p_\alpha^2 + \omega_\alpha^2 q_\alpha^2)/2T\} \, dp_\alpha \, dq_\alpha = 2\pi T/\omega_\alpha ,$$

leading to formula (65.1); because of the rapid convergence of the integral, the integration over q_α may be extended from $-\infty$ to ∞.

At sufficiently high temperatures (provided that the solid does not melt or decompose) the effects of anharmonic vibrations of the atoms may become appreciable. The nature of these effects as regards the thermodynamic quantities for the body may be investigated as follows; cf. the similar calculations for gases in § 49. Taking into account the terms following the quadratic terms in the expansion of the potential energy of the vibrations in powers of q_α, we have

$$E(p, q) = f_2(p, q) + f_3(q) + f_4(q) + \cdots,$$

where $f_2(p, q)$ denotes the harmonic expression (65.10) (a quadratic form in q_α and p_α), and $f_3(q), f_4(q), \ldots$ are forms homogeneous in all the coordinates q_α, of degree three, four, etc. Substituting in the partition function in (65.9) $q_\alpha = q_\alpha' \sqrt{T}, p_\alpha = p_\alpha' \sqrt{T}$, we obtain

$$Z = \int' e^{-E(p, q)/T} \, dI$$
$$= T^{3N\nu} \int' \exp \{-f_2(p', q') - \sqrt{T} f_3(q') - T f_4(q') - \ldots\} \, dI'.$$

We see that, when the integrand is expanded in powers of the temperature, all odd powers of $\sqrt{T}$ are multiplied by odd functions of the coordinates, which give zero on integration over the coordinates. Hence Z is a series $Z = Z_0 + T Z_1 + T^2 Z_2 + \ldots$ which contains only integral powers of the

[†] Whereas it was for a gas, where the integration over the coordinates of each particle was taken over the whole volume (cf. the end of § 31).

temperature. On substitution in (65.9), the first correction term to the free energy will accordingly be of the form

$$F_{\text{anh}} = AT^2, \qquad (65.11)$$

i.e. proportional to the square of the temperature. In the specific heat it gives a correction[†] proportional to the temperature itself. It should be emphasised that the expansion under discussion here is essentially one in powers of the ratio T/ε_0, which is always small, and not, of course, in powers of the ratio $T/\hbar\bar{\omega}$, which in the present case is large.

PROBLEMS

PROBLEM 1. Determine the maximum work which can be obtained from two identical solid bodies at temperatures T_1 and T_2 when their temperatures are made equal.

SOLUTION. The solution is similar to that in § 43, Problem 12, and gives

$$|R|_{\max} = Nc(\sqrt{T_1} - \sqrt{T_2})^2.$$

PROBLEM 2. Determine the maximum work which can be obtained from a solid when it is cooled from a temperature T to the temperature T_0 of the medium (at constant volume).

SOLUTION. From formula (20.3) we have

$$|R|_{\max} = Nc(T - T_0) + NcT_0 \log (T_0/T).$$

§ 66. Debye's interpolation formula

Thus in both the limiting cases of low and high temperatures it is possible to make a sufficiently complete calculation of the thermodynamic quantities for a solid. In the intermediate temperature range, such a calculation in a general form is impossible, since the sum over frequencies in (64.1) depends considerably on the actual frequency distribution over the whole spectrum of vibrations of the body concerned.

It is therefore of interest to construct a single interpolation formula giving the correct values of the thermodynamic quantities in the two limiting cases. More than one such formula can be found, of course, but we should expect that a reasonable interpolation formula will give at least a qualitatively correct description of the behaviour of the body throughout the intermediate range.

The form of the thermodynamic quantities for a solid at low temperatures is given by the distribution (64.4) of the frequencies in the vibration spectrum.

[†] This correction is usually negative (corresponding to positive A in (65.11)).

At high temperatures it is important that all the $3N\nu$ vibrations are excited. To construct the required interpolation formula, therefore, it is reasonable to start from a model in which the law (64.4) (which in reality is valid only at low frequencies) governs the frequency distribution over the whole vibration spectrum, the spectrum beginning at $\omega = 0$ and terminating at some finite frequency ω_m determined by the condition that the total number of vibrations is equal to the correct value $3N\nu$:

$$\frac{3V}{2\pi^2 \bar{u}^3} \int_0^{\omega_m} \omega^2 \, d\omega = \frac{V\omega_m^3}{2\pi^2 \bar{u}^3} = 3N\nu,$$

whence

$$\omega_m = \bar{u}(6\pi^2 N\nu/V)^{1/3}. \tag{66.1}$$

Thus the frequency distribution in this model is given by the formula

$$9N\nu\omega^2 \, d\omega/\omega_m^3 \quad (\omega < \omega_m) \tag{66.2}$$

for the number of vibrations with frequencies in the interval $d\omega$ (here $\bar{u}$ has been expressed in terms of ω_m).

Changing from the sum in (64.1) to an integral, we now have

$$F = N\varepsilon_0 + T \cdot \frac{9N\nu}{\omega_m^3} \int_0^{\omega_m} \omega^2 \log\left(1 - e^{-\hbar\omega/T}\right) d\omega.$$

The *Debye temperature* or *characteristic temperature* Θ of the body is defined by

$$\Theta = \hbar\omega_m \tag{66.3}$$

(and is, of course, dependent on the density of the body). Then

$$F = N\varepsilon_0 + 9N\nu T(T/\Theta)^3 \int_0^{\Theta/T} z^2 \log\left(1 - e^{-z}\right) dz. \tag{66.4}$$

Integrating by parts and using the *Debye function*

$$D(x) = \frac{3}{x^3} \int_0^x \frac{z^3 \, dz}{e^z - 1}, \tag{66.5}$$

we can rewrite this formula as

$$F = N\varepsilon_0 + N\nu T[3 \log\left(1 - e^{-\Theta/T}\right) - D(\Theta/T)]. \tag{66.6}$$

Solids

Hence the energy $E = F - T\,\partial F/\partial T$ is

$$E = N\varepsilon_0 + 3N\nu T D(\Theta/T) \tag{66.7}$$

and the specific heat is

$$C = 3N\nu\{D(\Theta/T) - (\Theta/T)\,D'(\Theta/T)\}. \tag{66.8}$$

Figure 8 shows a graph of $C/3N\nu$ as a function of T/Θ.

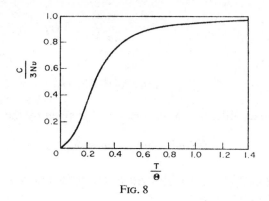

FIG. 8

Formulae (66.6)–(66.8) are the required interpolation formulae for the thermodynamic quantities for a solid (P. Debye, 1912).

It is easy to see that in both limiting cases these formulae in fact give the correct results. For $T \ll \Theta$ (low temperatures) the argument Θ/T of the Debye function is large. In a first approximation we can replace x by ∞ in the upper limit of the integral in the definition (66.5) of $D(x)$; the resulting definite integral is $\pi^4/15$, and so[†]

$$D(x) \cong \pi^4/5x^3 \qquad (x \gg 1).$$

Substituting this in (66.8), we obtain

$$C = (12N\nu\pi^4/5)\,(T/\Theta)^3, \tag{66.9}$$

which is the same as (64.9). At high temperatures $(T \gg \Theta)$ the argument of the Debye function is small; for $x \ll 1$ we have $D(x) \cong 1$ to a first approx-

---

[†] Replacing $\displaystyle\int_0^x$ by $\displaystyle\int_0^\infty - \int_x^\infty$, expanding $(e^z - 1)^{-1}$ in the second integrand in powers of e^{-z}, and integrating term by term, we find that, for $x \gg 1$,

$$D(x) = \frac{\pi^4}{5x^3} - 3e^{-x}\{1 + O(1/x)\}.$$

The value given in the text is therefore correct to within exponentially small terms.

imation,[†] and (66.8) gives $C = 3N\nu$, again in full agreement with the previous result (65.5).[‡]

It is useful to point out that the actual form of the function $D(x)$ is such that the criterion of applicability of the limiting expressions for the specific heat is the relative magnitude of T and $\frac{1}{4}\Theta$: the specific heat may be regarded as constant for $T \gg \frac{1}{4}\Theta$ and proportional to T^3 for $T \ll \frac{1}{4}\Theta$.[§]

According to Debye's formula, the specific heat is some universal function of the ratio Θ/T. In other words, according to this formula, the specific heats of bodies must be the same if the bodies are in *corresponding states*, i.e. have the same value of Θ/T.

Debye's formula gives a good description of the variation of specific heat with temperature (as far as can be expected from an interpolation formula) only for certain substances with simple crystal lattices: most of the elements, and some simple compounds such as the halides. It is inapplicable in practice to substances of more complex structure; this is quite reasonable, since in such substances the vibration spectrum is extremely complicated.

§ 67. Thermal expansion of solids

The term proportional to T^4 in the free energy at low temperatures (64.6) can be regarded as a small correction to $F_0 = N\varepsilon_0(V/N)$. The small correction to the free energy (for given V and T) is equal to the small correction to the thermodynamic potential Φ (for given P and T; see (15.12)). We can therefore write immediately

$$\Phi = \Phi_0(P) - \pi^2 T^4 V_0(P)/30(\hbar\bar{u})^3. \tag{67.1}$$

Here $\Phi_0(P)$ is the temperature-independent part of the thermodynamic potential, $V_0(P)$ the volume expressed as a function of pressure by means of the relations $P = -\partial F_0/\partial V = -N\,d\varepsilon_0/dV$, and $\bar{u} = \bar{u}(P)$ is the mean velocity of sound, expressed in terms of the pressure by means of the same relations. The

---

[†] For $x \ll 1$ a direct expansion of the integrand in powers of x and integration term by term gives

$$D(x) = 1 - \tfrac{3}{8}x + \tfrac{1}{20}x^2 - \cdots .$$

[‡] The specific heat at high temperatures accurate to the next term in the expansion is

$$C = 3N\nu\{1 - \tfrac{1}{20}(\Theta/T)^2\}.$$

[§] As examples, the values of Θ for a number of substances, derived from their specific-heat values, are Pb 90°, Ag 210°, Al 400°, KBr 180°, NaCl 280°. For diamond, Θ is particularly large, $\sim 2000°$.

dependence of the volume of the body on the temperature is given by $V = \partial\Phi/\partial P$:

$$V = V_0(P) - \frac{\pi^2 T^4}{30\hbar^3} \frac{d}{dP}\left(\frac{V_0}{u^3}\right). \tag{67.2}$$

The thermal expansion coefficient is

$$\alpha = (1/V)(\partial V/\partial T)_P = -\frac{2\pi^2 T^3}{15\hbar^3 V_0} \frac{d}{dP}\left(\frac{V_0}{u^3}\right). \tag{67.3}$$

We see that at low temperatures α is proportional to the cube of the temperature. This result is already obvious from Nernst's theorem (§ 23) together with the T^3 law for the specific heat.

Similarly, at high temperatures we can consider the second and third terms in (65.6) as small corrections to the first term (we must always have $T \ll \varepsilon_0$ if the body is solid), and obtain

$$\Phi = \Phi_0(P) - NcT \log T + NcT \log \hbar\bar{\omega}(P), \tag{67.4}$$

whence

$$V = V_0(P) + (NcT/\bar{\omega})\, d\bar{\omega}/dP. \tag{67.5}$$

The thermal expansion coefficient is

$$\alpha = (Nc/V_0\bar{\omega})\, d\bar{\omega}/dP, \tag{67.6}$$

and is independent of the temperature.

When the pressure increases, the atoms in a solid come closer together, and the amplitude of their vibrations (for a given energy) decreases, i.e. the frequency increases. Thus $d\bar{\omega}/dP > 0$, so that $\alpha > 0$, and solids expand when the temperature rises. Similar considerations show that the coefficient α given by formula (67.3) is also positive.

Finally, we can make use of the law of corresponding states given at the end of § 66. The statement that the specific heat is a function only of the ratio T/Θ is equivalent to saying that the thermodynamic potential, for example, is of the form

$$\Phi = \Phi_0(P) + \Theta f(T/\Theta). \tag{67.7}$$

The volume is

$$V = V_0(P) + (d\Theta/dP)\,[f - (T/\Theta)f'],$$

and the thermal expansion coefficient is

$$\alpha = -(T/V_0\Theta^2)\,(d\Theta/dP)f''.$$

Similarly, we find the heat function $W = \Phi - T\,\partial\Phi/\partial T$ and the specific heat $C = \partial W/\partial T$:

$$C = -(T/\Theta)f''.$$

Taking the ratio of the two expressions for α and C, we obtain

$$\frac{\alpha}{C} = \frac{1}{\Theta V_0(P)} \frac{d\Theta}{dP}. \tag{67.8}$$

Thus, within the limits of validity of the law of corresponding states, the ratio of the thermal expansion coefficient to the specific heat of a solid is independent of temperature (*Grüneisen's law*).

It has already been mentioned that in solids the difference between the specific heats C_p and C_v is very slight. At low temperatures this is a general consequence of Nernst's theorem, which applies to all bodies. At high temperatures we have, using the thermodynamic relation (16.9),

$$C_p - C_v = -T \frac{\alpha^2 V_0^2}{dV_0/dP},$$

where $\alpha = \alpha(P)$ is the thermal expansion coefficient (67.6). We see that the difference $C_p - C_v$ is proportional to T; essentially this means that its expansion in powers of T/ε_0 begins with a first-order term, whereas that of the specific heat itself begins with a zero-order (constant) term. Hence it follows that in solids $C_p - C_v \ll C$ at high temperatures also.

§ 68. Highly anisotropic crystals

It has been noted at the end of § 66 that Debye's formula is inapplicable in practice to crystals of complex structure. These include, in particular, highly anisotropic crystal structures of the "layer" and "chain" types. The former may be described as consisting of parallel layers of atoms, the interaction energy of the atoms within each layer being large in comparison with the coupling energy of adjacent layers. Similarly, the chain structures consist of parallel chains of atoms that are relatively weakly coupled together. The acoustic vibration spectrum of such crystals has not one but several Debye temperatures with different orders of magnitude. The T^3 law for the specific heat then applies only at temperatures that are low compared with the lowest Debye temperature; in the intermediate regions, different limiting laws occur (I. M. Lifshitz, 1952).

Let us first consider layer structures. Such a lattice is most rigid with regard to vibrations of atoms in the plane of the layers, which we take as the xy-plane; the rigidities for vibrations of one layer as a whole relative to the others are relatively very small. These properties bring about a dependence of the frequency on the wave vector (the dispersion relation) in three branches of the acoustic wave spectrum, as shown by the following formulae (which are written here for the case of a crystal with hexagonal symmetry):

$$\omega_1^2 = U_1^2\varkappa^2 + u_3^2 k_z^2, \qquad \omega_2^2 = U_2^2\varkappa^2 + u_3^2 k_z^2,$$
$$\omega_3^2 = u_3^2\varkappa^2 + u_4^2 k_z^2, \qquad (\varkappa^2 = k_x^2 + k_y^2), \tag{68.1}$$

with U_1, $U_2 \gg u_3$, u_4. Here the velocities of propagation U_1, U_2 refer to vibrations of atoms in the plane of the layers, u_3 (in the branches ω_1 and ω_2) to shear vibrations of layers with respect to one another, and u_4 to vibrations of the relative distance between the layers.[†]

The expressions (68.1) are, however, insufficient for studying the thermal properties of the crystal. They are in fact only the first terms in the expansion of the functions $\omega^2(\mathbf{k})$ in powers of the wave vector. In view of the "anomalous" smallness of some of the coefficients in the quadratic terms in such expansions, terms of the next (fourth) order begin to be important.[‡] To ascertain their form, we note that, when the coupling between layers is entirely neglected, the dispersion relations of the waves become

$$\omega_1^2 = U_1^2\varkappa^2, \qquad \omega_2^2 = U_2^2\varkappa^2, \qquad \omega_3^2 = \gamma^2\varkappa^4. \tag{68.2}$$

The frequencies ω_1 and ω_2 correspond to longitudinal vibrations in the plane of the layers, and ω_3 to transverse vibrations, which in this case are bending waves of the layers, regarded as free elastic thin plates (cf. *Elasticity*, § 25). Hence, neglecting the small fourth-order terms that depend on the coupling between the layers, we can finally write the wave dispersion relation as

$$\omega_{1,2}^2 = U_{1,2}^2\varkappa^2 + u_3^2 k_z^2, \qquad \omega_3^2 = u_3^2\varkappa^2 + u_4^2 k_z^2 + \gamma^2\varkappa^4. \tag{68.3}$$

We shall assume that $U_1 \sim U_2$, $u_3 \sim u_4$, and use the notation $\eta \sim u/U$ for the small ratio which represents the coupling energy between the layers relative to that between the atoms in one layer. We shall also use the Debye temperature, or more precisely the highest of the Debye temperatures,

---

[†] The assumption of hexagonal symmetry of the crystal has no fundamental significance, and is made only in order to give formulae (68.1) a more definite form. The velocities $U_1, \ldots, u_4$ are expressed in terms of the elastic moduli λ_{iklm} of such a crystal by

$$U_1^2 = \lambda_{xyxy}/\varrho, \qquad U_2^2 = \lambda_{xxxx}/\varrho, \qquad u_3^2 = \lambda_{xzxz}/\varrho, \qquad u_4^2 = \lambda_{zzzz}/\varrho,$$

where ϱ is the density; these can be derived from the expressions in *Elasticity*, § 23, Problem, by expanding in powers of the moduli λ_{xzxz} and λ_{zzzz}, which for a layer crystal are small in comparison with λ_{xxxx} and λ_{xyxy}. The type of vibration mentioned in the text is evident from the sense of the various components λ_{iklm}.

[‡] The equation giving the dispersion relation of waves is an algebraic equation for ω^2 (see § 69). It is therefore the function $\omega^2(\mathbf{k})$ that has a regular expansion in powers of k_x, k_y, k_z. Since this function is even (see § 69), the expansion contains only even powers.

$\Theta = \hbar\omega_m$, where $\omega_m \sim U/a$ is the limiting frequency of the "hard" vibrations (a being the lattice constant); the limiting frequency of the "soft" vibrations is small in comparison with ω_m, in the ratio η. Lastly, it is reasonable to suppose that the limiting frequency of the bending waves is of the same order as or less than ω_m; let it be $\sim \omega_m$.[†] Under these conditions, let us find the nature of the temperature dependence of the specific heat of the crystal when $T \ll \Theta$.[‡]

With inclusion of the contribution from acoustic vibrations, the free energy of the body is given by

$$F = N\varepsilon_0 + T \sum_{\alpha=1}^{3} \int \log\left(1 - e^{-\hbar\omega_\alpha/T}\right) \frac{V\, dk_x\, dk_y\, dk_z}{(2\pi)^3}, \qquad (68.4)$$

where the summation is taken over the three branches of the spectrum, and the integration over the whole range of variation of the wave vector.[§]

If $T \gg \eta\Theta$, we can neglect the coupling between layers, and therefore use the spectrum (68.2). The main contribution to the free energy comes from the "bending" branch ω_3. Because of the rapid convergence when $T \ll \Theta$, the integration over k_x and k_y can be extended from $-\infty$ to ∞. Replacing it by integration over $2\pi\varkappa\, d\varkappa$, we find by an obvious substitution

$$\int_0^\infty \log\left(1 - e^{-\hbar\gamma\varkappa^2/T}\right) 2\pi\varkappa\, d\varkappa \doteq \frac{\pi T}{\hbar\gamma} \int_0^\infty \log\left(1 - e^{-x}\right) dx.$$

The integration over k_z, with range $|k_z| \leqslant k_{z,\,\text{max}} \sim 1/a$, gives a temperature-independent factor $\sim 1/a$. The result is that the temperature-dependent part of the free energy is proportional to T^2 and correspondingly, for the specific heat,

$$C \propto T \quad \text{for} \quad \eta\Theta \ll T \ll \Theta. \qquad (68.5)$$

If $T \ll \eta\Theta$, we must use the full expressions (68.3) for $\omega_\alpha(\mathbf{k})$ in the integrals (68.4), and the integration over each component of $\mathbf{k}$ can be extended from $-\infty$ to ∞. The resulting temperature dependence of the free energy is fairly complicated, but two further limiting cases may be distinguished. If $T \gg \eta^2\Theta$, the main contribution again comes from the branch ω_3, and the term in $\varkappa^2$ may be omitted, leaving

$$\omega_3^2 = u_4^2 k_z^2 + \gamma^2\varkappa^4.$$

[†] That is, we take $\gamma \sim \omega_m a^2 \sim Ua$. It must be emphasised that the coefficient γ, which is related to the "transverse rigidity" of the layers, is not expressible in terms of the elastic moduli λ_{iklm} alone.

[‡] High temperatures $T \gg \Theta$ form the classical range, in which the specific heat C is constant.

[§] That is, over one reciprocal lattice cell; see (71.7).

This follows since the most important range in the integral over $\varkappa\, d\varkappa$ is that where $\hbar\gamma\varkappa^2 \sim T$, for which $\hbar u \varkappa \sim \hbar u(T/\hbar\gamma)^{1/2} \sim T(\eta^2\Theta/T)^{1/2} \ll T$. We then have

$$\int_{-\infty}^{\infty}\int_{0}^{\infty} \log\left[1-\exp\left\{-\frac{\hbar}{T}\sqrt{(u_4^2 k_z^2 + \gamma^2\varkappa^4)}\right\}\right]\cdot 2\pi\varkappa\, d\varkappa\, dk_z$$

$$= \text{constant}\times T^2/u_4\gamma,$$

and so for the specific heat

$$C \propto T^2 \quad \text{for} \quad \eta^2\Theta \ll T \ll \eta\Theta. \tag{68.6}$$

Lastly, when $T \ll \eta^2\Theta$ we can show by the same method that the term in $\varkappa^4$ can be omitted from (68.3), and we then return to the acoustic spectrum (68.1) with ω linearly dependent on k and the specific heat given by Debye's law

$$C \propto T^3 \quad \text{for} \quad T \ll \eta^2\Theta. \tag{68.7}$$

Crystals with a chain structure can be treated similarly; we shall take the direction of the chains as the z-axis. In this case, the dispersion relations in the three branches of the acoustic-wave spectrum are

$$\omega_{1,2}^2 = u_{1,2}^2\varkappa^2 + u_3^2 k_z^2 + \gamma^2 k_z^4, \quad \omega_3^2 = u_3^2\varkappa^2 + U_4^2 k_z^2, \tag{68.8}$$

where now $u_1, u_2, u_3 \ll U_4$.[†] If the interaction between the chains is neglected, the relations (68.8) become

$$\omega_{1,2}^2 = \gamma^2 k_z^4, \quad \omega_3^2 = U_4^2 k_z^2;$$

the branch ω_3 corresponds to longitudinal vibrations of the atoms in the chains, and the branches ω_1 and ω_2 to bending waves in the chains, regarded as elastic strings. If we assume $u_1 \sim u_2 \sim u_3$ and again use the small parameter $\eta \sim u/U$ and the Debye temperature $\Theta \sim \hbar U/a$, we can derive the following limiting forms for the temperature dependence of the specific heat:

$$\left.\begin{array}{lll} C \propto T^{1/2} & \text{for} & \eta\Theta \ll T \ll \Theta, \\ C \propto T^{5/2} & \text{for} & \eta^2\Theta \ll T \ll \eta\Theta, \\ C \propto T^3 & \text{for} & T \ll \eta^2\Theta. \end{array}\right\} \tag{68.9}$$

[†] Here we again take the particular case of hexagonal symmetry, in this instance about the direction of the chains The velocities $u_1, \ldots, U_4$ are expressed in terms of the elastic moduli by the same formulae as in the first footnote to this section, but λ_{xxxx}, λ_{xyxy} and λ_{xzxz} are now small in comparison with λ_{zzzz}.

§ 69. Crystal lattice vibrations

In the preceding sections, we have considered the thermal motion of atoms in a solid as a set of normal modes of small vibrations of the crystal lattice. Let us now consider more closely the mechanical properties of these vibrations.

Each unit cell of the crystal generally contains several atoms. It is therefore necessary to specify each atom by stating which unit cell contains it and giving the number of the atom in that cell. The position of the unit cell can be defined by the radius vector $\mathbf{r}_n$ of any particular vertex of it; this takes values given by

$$\mathbf{r}_n = n_1\mathbf{a}_1 + n_2\mathbf{a}_2 + n_3\mathbf{a}_3, \tag{69.1}$$

where n_1, n_2, n_3 are integers and $\mathbf{a}_1$, $\mathbf{a}_2$, $\mathbf{a}_3$ the basis lattice vectors corresponding to the edge lengths of the unit cell.

We shall denote the displacements of the atoms in their vibrations by $\mathbf{u}_s$, where the suffix s is the number of the atom in the cell ($s = 1, 2, \ldots, \nu$, with ν the total number of atoms per cell). The Lagrangian of a crystal lattice as a mechanical system of particles executing small vibrations about their equilibrium positions (lattice sites) is

$$L = \tfrac{1}{2} \sum_{n, s} m_s \dot{u}_s^2(\mathbf{n}) - \tfrac{1}{2} \sum_{\substack{n, n' \\ s, s'}} \Lambda_{ik}^{ss'}(\mathbf{n} - \mathbf{n}') u_{si}(\mathbf{n}) u_{s'k}(\mathbf{n}'). \tag{69.2}$$

where the "vector" $\mathbf{n} = (n_1, n_2, n_3)$, m_s are the masses of the atoms, and i, k are vector suffixes taking the values x, y, z; summation over repeated suffixes is, as usual, implied. The coefficients Λ depend only on the differences $\mathbf{n} - \mathbf{n}'$, since the interaction forces between atoms can depend only on the relative position of the lattice cells, not on their absolute position in space. These coefficients are symmetrical:

$$\Lambda_{ik}^{ss'}(\mathbf{n}) = \Lambda_{ki}^{s's}(-\mathbf{n}), \tag{69.3}$$

as is evident from the form of the function (69.2).

The equations of motion

$$m_s \ddot{u}_{si} = -\sum_{n', s'} \Lambda_{ik}^{ss'}(\mathbf{n} - \mathbf{n}') u_{s'k}(\mathbf{n}') \tag{69.4}$$

follow from the Lagrangian (69.2). Note that the coefficients Λ satisfy certain relations which express the fact that a parallel displacement or a rotation of the lattice as a whole gives rise to no forces on the atoms. For a parallel displacement, all the $\mathbf{u}_s(\mathbf{n})$ are constant, and therefore

$$\sum_{n, s'} \Lambda_{ik}^{ss'}(\mathbf{n}) = 0. \tag{69.5}$$

SP - H

We shall not pause to write out the relations that follow from the invariance under rotations.

We shall seek solutions of the equations (69.4) in the form of a monochromatic plane wave

$$u_s(n) = e_s(k) \exp [i(k \cdot r_n - \omega t)]. \qquad (69.6)$$

The (complex) amplitude e_s depends only on the suffix s, i.e. differs only for different atoms in the same cell, but not for equivalent atoms in different cells. The vectors e_s determine both the vibration amplitude and the direction of *polarisation* of the vibrations.

Substitution of (69.6) in (69.4) gives

$$\omega^2 m_s e_{si} \exp (ik \cdot r_n) = \sum_{n', s'} A_{ik}^{ss'}(n - n') e_{s'k} \exp (ik \cdot r_{n'}).$$

On dividing each side by $\exp (ik \cdot r_n)$ and replacing the summation over n' by one over $n' - n$, we find

$$\sum_{s'} A_{ik}^{ss'}(k) e_{s'k} - \omega^2 m_s e_{si} = 0, \qquad (69.7)$$

where

$$A_{ik}^{ss'}(k) = \sum_n A_{ik}^{ss'}(n) \exp (-ik \cdot r_n). \qquad (69.8)$$

The set of linear homogeneous algebraic equations (69.7) for the amplitudes has non-zero solutions if the consistency equation

$$\det |A_{ik}^{ss'}(k) - \omega^2 m_s \delta_{ik} \delta_{ss'}| = 0 \qquad (69.9)$$

is satisfied. Since the suffixes i and k take 3 values each, and s and s' v values each, the order of the determinant is $3v$, so that (69.9) is an algebraic equation of degree $3v$ in ω^2.

Each of the $3v$ solutions of this equation determines the frequency ω as a function of the wave vector k; this is called the *dispersion relation* of the waves, and equation (69.9), which determines it, is called the *dispersion equation*. Thus for any given value of the wave vector the frequency can in general take $3v$ different values. We can say that the frequency is a many-valued function of the wave vector, with $3v$ branches: $\omega = \omega_\alpha(k)$, where the suffix α labels the branches of the function.

From the definition (69.8) and the equations (69.3), it follows that

$$A_{ik}^{ss'}(k) = A_{ki}^{s's}(-k)$$
$$= [A_{ki}^{s's}(k)]^*. \qquad (69.10)$$

Thus the quantities $A_{ik}^{ss'}(k)$ form an Hermitian matrix, and the solution of equations (69.7) is mathematically one of determining the eigenvalues and corresponding "eigenvectors" of such a matrix. According to the known

properties of Hermitian matrices, the eigenvectors corresponding to different eigenvalues are orthogonal. In the present case, this means that

$$\sum_{s=1}^{v} m_s \mathbf{u}_s^{(\alpha)} \cdot \mathbf{u}_s^{(\alpha')*} = 0 \quad \text{for} \quad \alpha \neq \alpha', \tag{69.11}$$

where the superscript (α) to the displacement vector indicates the branch of
the vibration spectrum to which it belongs.[†] The equations (69.11) express
the orthogonality of the polarisations in different branches of the spectrum.

Owing to the symmetry of the mechanical equations of motion under
time reversal, if the propagation of a wave (69.6) is possible, then so is that
of a similar wave in the opposite direction. This change of direction is equiv-
alent to a change in the sign of $\mathbf{k}$. The function $\omega(\mathbf{k})$ must therefore be even:

$$\omega(-\mathbf{k}) = \omega(\mathbf{k}). \tag{69.12}$$

The wave vector of the lattice vibrations has the following important
property. The vector $\mathbf{k}$ appears in the expression (69.6) only through the
exponential factor $\exp(i\mathbf{k}\cdot\mathbf{r_n})$. But this factor is unchanged when $\mathbf{k}$ is replaced
according to

$$\mathbf{k} \rightarrow \mathbf{k}+\mathbf{b}, \quad \mathbf{b} = p_1\mathbf{b}_1+p_2\mathbf{b}_2+p_3\mathbf{b}_3, \tag{69.13}$$

where $\mathbf{b}$ is any vector of the reciprocal lattice, $\mathbf{b}_1$, $\mathbf{b}_2$, $\mathbf{b}_3$ are the basic vectors
of that lattice, and p_1, p_2, p_3 are integers.[‡] This means that the wave vector
of the lattice vibrations is physically indeterminate: values of $\mathbf{k}$ differing by
$\mathbf{b}$ are physically equivalent. The function $\omega(\mathbf{k})$ is periodic in the reciprocal
lattice:

$$\omega(\mathbf{k}+\mathbf{b}) = \omega(\mathbf{k}),$$

and in each branch it is therefore sufficient to consider values of the vector $\mathbf{k}$
that lie in a certain finite range, a single cell of the reciprocal lattice. If the
coordinate axes (in the general case, oblique) are along the three basic vectors
of the reciprocal lattice, we can take, for example, just the range

$$-\tfrac{1}{2}b_i < k_i \leqslant \tfrac{1}{2}b_i. \tag{69.14}$$

When $\mathbf{k}$ takes values in this range, the frequency $\omega(\mathbf{k})$ in each branch of
the spectrum takes values that occupy a band or *zone* of finite width. Different zones may, of course, partly overlap.

In geometrical terms, the functional relation $\omega = \omega(\mathbf{k})$ is represented by a
four-dimensional hypersurface, different sheets of which correspond to
the different branches of the function. These sheets may be not completely

---

[†] The occurrence of the "weighting" factor m_s in the relations (69.11) is due to
the fact that the ω_α^2 are eigenvalues not of the matrix $\Lambda_{ik}^{ss'}(\mathbf{k})$ itself, but of
$\Lambda_{ik}^{ss'}/\sqrt{(m_s m_{s'})}$, and the corresponding eigenvectors are $\sqrt{m_s}\mathbf{u}_s^{(\alpha)}$.
[‡] The concepts used here are considered in detail in § 133.

separate, i.e. they may intersect. The possible types of such intersections depend considerably on the specific symmetry of the crystal lattice. The study of this topic involves the application of group theory (see § 136).

Among the 3ν branches of the vibration spectrum, there must be some which for wavelengths large compared with the lattice constant correspond to ordinary elastic (i.e. sound) waves in the crystal. It is known (see *Elasticity*, § 23) that waves of three types with different dispersion relations can be propagated in a crystal regarded as a continuous medium, and for all three types ω is a homogeneous function of the first order in the components of the vector $\mathbf{k}$, and vanishes when $\mathbf{k} = 0$. Thus the 3ν branches of the function $\omega(\mathbf{k})$ must include three for which the dispersion relation when $\mathbf{k}$ is small has the form

$$\omega = kf(\mathbf{k}/k). \qquad (69.15)$$

These three types are called *acoustic waves*; they are characterised by the fact that (when $\mathbf{k}$ is small) the lattice vibrates as a whole, as a continuous medium. In the limit as $\mathbf{k} \to 0$, these vibrations become a simple parallel displacement of the entire lattice.

In complex lattices containing more than one atom per cell, there are a further $3(\nu-1)$ types of wave. In these branches of the spectrum, the frequency does not vanish when $\mathbf{k} = 0$, but tends to a constant limit as $\mathbf{k} \to 0$. These vibrations of the lattice are called *optical vibrations*. In this case the atoms in each unit cell are in relative motion, and in the limit $\mathbf{k} = 0$ the centre of mass of the cell remains fixed.[†]

The $3\nu-3$ *limiting frequencies* (at $\mathbf{k} = 0$) of the optical vibrations need not be all different. When the crystal has certain symmetry properties, the limiting frequencies of some of the optical branches of the spectrum may coincide or be *degenerate* (see § 136).

The function $\omega(\mathbf{k})$ with a non-degenerate limiting frequency can be expanded (near $\mathbf{k} = 0$) in powers of the components of the vector $\mathbf{k}$. Since the function $\omega(\mathbf{k})$ is even, this expansion can contain only even powers of k_i, and its leading terms are therefore

$$\omega = \omega_0 + \tfrac{1}{2}\gamma_{ik}k_ik_k, \qquad (69.16)$$

where ω_0 is the limiting frequency and the γ_{ik} are constants.

[†] This last fact can be formally deduced directly from the equations of motion (69.7), (69.8). For $\mathbf{k} = 0$, these become

$$\sum_{\mathbf{n}, s'} A_{ik}^{ss'}(\mathbf{n})e_{s'k} = m_s\omega^2 e_{si}.$$

Summing both sides over s, we have zero on the left, by (69.5), and so we must also have $\Sigma m_s e_s = 0$ if the equations are consistent when $\mathbf{k} = 0$.

If, however, the limiting frequencies of several branches coincide, the functions $\omega(\mathbf{k})$ for these branches cannot be expanded in powers of $\mathbf{k}$, since the point $\mathbf{k} = 0$ is a singular one (a branch point) for them. We can say only that, near $\mathbf{k} = 0$, the difference $\omega - \omega_0$ will be a homogeneous function of the components of $\mathbf{k}$, of either the first or the second order (depending on the symmetry of the crystal).

Concerning the whole of the foregoing discussion, it must be emphasised once more that this has related only to the *harmonic* approximation, in which only those terms are taken into account, in the potential energy of the vibrating particles, which are quadratic in the displacements of the atoms. It is only in this approximation that the various monochromatic waves (69.6) do not interact but are freely propagated through the lattice. When the subsequent *anharmonic* terms are taken into account, various processes of decay and scattering of these waves by one another appear. The interaction may also lead to the formation of "bound states" of waves (phonons; see below), as new branches of the spectrum that do not exist in the harmonic approximation.

Moreover, it is assumed that the lattice is perfectly periodic, but it must be borne in mind that the perfect periodicity is to some extent perturbed, even without allowing for possible impurities and other lattice defects, if the crystal contains randomly distributed atoms of different isotopes. This perturbation, however, is comparatively small if the relative difference of atomic weights of the isotopes is small or if the abundance of one isotope greatly exceeds those of the others. In such cases the above description remains valid in a first approximation, and in subsequent approximations there occur various processes of scattering of waves by inhomogeneities in the lattice.[†]

§ 70. Number density of vibrations

The number of vibrations corresponding to a range $d^3k \equiv dk_x\,dk_y\,dk_z$ of the values of components of the wave vector is $d^3k/(2\pi)^3$ per unit volume of the crystal. The frequency distribution function $g(\omega)$ of the vibrations, which gives the number $g(\omega)\,d\omega$ of vibrations whose frequencies lie in a given range between ω and $\omega + d\omega$, is a characteristic of the vibration spectrum of a particular lattice. This number is, of course, different for different branches

---

[†] The presence of lattice defects also brings about certain changes in its vibration spectrum, with new frequencies that correspond to "local" vibrations near the defects. For a study of these effects see I.M. Lifshitz and A.M. Kosevich, *Reports on Progress in Physics* **29**, 217, 1966.

of the spectrum, but to simplify the notation we shall omit in this section the appropriate suffix α to the functions $\omega(\mathbf{k})$ and $g(\omega)$.

The number $g(\omega)\,d\omega$ is the volume (divided by $8\pi^3$) in $\mathbf{k}$-space that lies between two infinitesimally close surfaces of constant frequency $\omega(\mathbf{k}) =$ constant. At each point in $\mathbf{k}$-space, the gradient of the function $\omega(\mathbf{k})$ is along the normal to the constant-frequency surface through the point. It is therefore clear from the expression $d\omega = d\mathbf{k}\cdot\nabla_{\mathbf{k}}\omega(\mathbf{k})$ that the distance between two such infinitesimally close surfaces (measured along the segment of the normal between them) is $d\omega/|\nabla_{\mathbf{k}}\omega|$. Multiplying this by the area $df_{\mathbf{k}}$ of the constant-frequency surface element and integrating over the whole surface within one cell of the reciprocal lattice, we find the required part of the volume of $\mathbf{k}$-space, and division by $(2\pi)^3$ gives the frequency distribution density:

$$g(\omega) = \frac{1}{(2\pi)^3} \int \frac{df_{\mathbf{k}}}{|\nabla_{\mathbf{k}}\omega(\mathbf{k})|}. \qquad (70.1)$$

In each zone (range of values traversed by some branch $\omega(\mathbf{k})$ in one cell of the reciprocal lattice of $\mathbf{k}$) the function $\omega(\mathbf{k})$ must have at least one minimum and one maximum. Hence it follows that this function must also have saddle-points.[†] The existence of these various stationary points leads to certain properties of the frequency distribution function $g(\omega)$ (L. Van Hove, 1953).

Near an extremum at some $\mathbf{k} = \mathbf{k}_0$, the difference $\omega(\mathbf{k}) - \omega_0$, where $\omega_0 = \omega(\mathbf{k}_0)$, has the form

$$\omega - \omega_0 = \tfrac{1}{2}\gamma_{ik}(k_i - k_{0i})(k_k - k_{0k}).$$

Taking the axes in $\mathbf{k}$-space along the principal axes of this quadratic form, we can write it as

$$\omega - \omega_0 = \tfrac{1}{2}[\gamma_1(k_x - k_{0x})^2 + \gamma_2(k_y - k_{0y})^2 + \gamma_3(k_z - k_{0z})^2], \qquad (70.2)$$

where γ_1, γ_2, γ_3 are the principal values of the symmetrical tensor γ_{ik}.

Let us first consider a minimum or maximum of $\omega(\mathbf{k})$. Then γ_1, γ_2, γ_3 have the same sign. Using instead of k_x, k_y, k_z the new variables $\varkappa_x$, $\varkappa_y$, $\varkappa_z$ according to $\varkappa_x = \sqrt{|\gamma_1|}\,(k_x - k_{0x})$, etc., we have

$$\omega - \omega_0 = \pm\tfrac{1}{2}(\varkappa_x^2 + \varkappa_y^2 + \varkappa_z^2) = \pm\tfrac{1}{2}\varkappa^2. \qquad (70.3)$$

In this case the constant-frequency surfaces in $\varkappa$-space are spheres. Changing to integration in $\varkappa$-space in (70.1), we have

$$g(\omega) = \frac{1}{(2\pi)^3\sqrt{\gamma}} \int \frac{df_{\mathbf{k}}}{|\nabla_{\varkappa}\omega(\varkappa)|}, \qquad \gamma = |\gamma_1\gamma_2\gamma_3|. \qquad (70.4)$$

[†] It can be shown (but we shall not pause to do so) that there must exist at least six saddle-points, three of each of two types, corresponding to the two signs in (70.8).

The surface element of the sphere is $df_\varkappa = \varkappa^2 \, do_\varkappa$, where $do_\varkappa$ is the element of solid angle. The gradient of the function (70.3) is $\nabla_\varkappa \omega(\varkappa) = \pm \varkappa$. The integral in (70.4) is therefore equal to $4\pi\varkappa$; expressing $\varkappa$ in terms of $\omega - \omega_0$ by (70.3), we have finally

$$g(\omega) = \frac{1}{\pi^2 \sqrt{(2\gamma)}} \sqrt{|\omega - \omega_0|} \; . \tag{70.5}$$

Thus the vibration number density has a square-root singularity; the derivative $dg/d\omega$ becomes infinite as $\omega \to \omega_0$.

It must be borne in mind, however, that in the general case (if $\omega = \omega_0$ lies within and not on the edge of the frequency band) the constant-frequency surfaces for values of ω close to ω_0 may include other sheets in other parts of the k-space cell, as well as the ellipsoids round $\mathbf{k} = \mathbf{k}_0$. In the general case, therefore, the expression (70.5) gives only the "singular" part of the vibration number density, and it would be more correct to write

$$g(\omega) = g(\omega_0) + \frac{\sqrt{|\omega - \omega_0|}}{\pi^2 \sqrt{(2\gamma)}} \tag{70.6}$$

on one side of $\omega = \omega_0$ ($\omega < \omega_0$ for a maximum, or $\omega > \omega_0$ for a minimum), and $g(\omega) = g(\omega_0)$ on the other side.

Note also that formula (70.5) does not, of course, relate to the neighbourhood of the lower edge ($\omega = 0$) of the acoustic vibration zone, where the dispersion relation has the form (69.15). It is easily seen that in this case

$$g(\omega) = \text{constant} \times \omega^2 . \tag{70.7}$$

Let us now consider the neighbourhood of a saddle-point. In this case two of the quantities γ_1, γ_2, γ_3 in (70.2) are positive and one negative, or vice versa. Instead of (70.3) we now have

$$\omega - \omega_0 = \pm \tfrac{1}{2}(\varkappa_x^2 + \varkappa_y^2 - \varkappa_z^2). \tag{70.8}$$

Let us take the specific case of the upper sign. Then the constant-frequency surfaces for $\omega < \omega_0$ are hyperboloids of two sheets, and those for $\omega > \omega_0$ are hyperboloids of one sheet; the limiting surface $\omega = \omega_0$ is a double cone (Fig. 9, p. 214).

The integration in (70.4) is now conveniently performed in cylindrical polar coordinates in $\varkappa$-space; $\varkappa_\perp$, $\varkappa_z$, ϕ, where $\varkappa_\perp = \sqrt{(\varkappa_x^2 + \varkappa_y^2)}$ and ϕ is the polar angle in the $\varkappa_x \varkappa_y$-plane. The magnitude of the gradient is $|\nabla_\varkappa \omega| = \varkappa$. For $\omega < \omega_0$ the integral is taken over the two sheets of the hyperboloid:

$$df_\varkappa = \frac{2\pi\varkappa_\perp \varkappa}{|\varkappa_z|} \, d\varkappa_\perp, \quad g(\omega) = \frac{1}{(2\pi)^3 \sqrt{\gamma}} \, 2 \int_0^K \frac{2\pi\varkappa_\perp \, d\varkappa_\perp}{\sqrt{[\varkappa_\perp^2 + 2(\omega_0 - \omega)]}} \; ;$$

FIG. 9

the upper limit K (whose value does not affect the form of the resulting singularity) may be taken as any value of $\varkappa$ large compared with $\sqrt{(\omega_0-\omega)}$, but also so small that the expression (70.8) still gives the form of the constant-frequency surface. The result is

$$g(\omega) = \frac{1}{2\pi^2\sqrt{\gamma}}\,[K - \sqrt{\{2(\omega_0-\omega)\}}].$$

For $\omega > \omega_0$, we similarly find

$$g(\omega) = \frac{2}{(2\pi)^3\sqrt{\gamma}}\int_{\varkappa_{\perp\min}}^{K}\frac{2\pi\varkappa_\perp\,d\varkappa_\perp}{\sqrt{[\varkappa_\perp^2 - 2(\omega-\omega_0)]}} = \frac{K}{2\pi^2\sqrt{\gamma}},$$

where $\varkappa_{\perp\min}^2 = 2(\omega-\omega_0)$. Thus the vibration number density near the saddle-point is

$$g(\omega) = \begin{cases} g(\omega_0) - \sqrt{|\omega_0-\omega|}/\pi^2\sqrt{(2\gamma)} & \text{for} \quad \omega < \omega_0, \\ g(\omega_0) & \text{for} \quad \omega > \omega_0. \end{cases} \tag{70.9}$$

Here again $g(\omega)$ has a square-root singularity.

For a saddle-point with the lower sign in (70.8), the result is similar with the regions $\omega < \omega_0$ and $\omega > \omega_0$ interchanged (square-root singularity for $\omega > \omega_0$).

§ 71. Phonons

Let us now consider the quantum treatment of lattice vibrations.

Instead of the waves (69.6), in which the atoms have definite displacements at any instant, quantum theory uses the concept of *phonons*, which are "quasi-particles" propagated through the lattice, with definite energies and directions of motion. Since the energy of an oscillator in quantum mechanics is an integral multiple of $\hbar\omega$ (where ω is the frequency of the classical wave), the phonon energy ε is related to the frequency ω by

$$\varepsilon = \hbar\omega, \tag{71.1}$$

in the same way as for light quanta or photons. The wave vector **k** determines the *quasi-momentum* **p** of the phonon:

$$\mathbf{p} = \hbar\mathbf{k}. \tag{71.2}$$

This is a quantity in many ways analogous to the ordinary momentum, but there is an important difference between them due to the fact that the quasi-momentum is defined only to within an arbitrary additive constant vector of the form $\hbar\mathbf{b}$; values of **p** differing by such a quantity are physically equivalent.

The velocity of a phonon is given by the group velocity of the corresponding classical waves, $\mathbf{v} = \partial\omega/\partial\mathbf{k}$. This formula may also be written

$$\mathbf{v} = \partial\varepsilon(\mathbf{p})/\partial\mathbf{p}, \tag{71.3}$$

which is exactly analogous to the usual relation between the energy, momentum and velocity of particles.

The whole of the discussion in §§ 69 and 70 of the properties of the classical vibration spectrum of a crystal lattice is entirely applicable (with the appropriate change of terminology) to the phonon energy spectrum, i.e. the dependence of the phonon energy on the quasi-momentum. In particular, the phonon energy spectrum $\varepsilon(\mathbf{p})$ has 3ν branches, including three acoustic branches. The vibration number density considered in § 70 here becomes the phonon quantum state number density.

In the quantum picture, free movement of non-interacting phonons takes the place of free wave propagation in the harmonic approximation. In subsequent approximations, various elastic and inelastic phonon collision processes occur. These collisions provide the mechanism for the establishment of thermal equilibrium in the phonon gas, i.e. of an equilibrium thermal motion in the lattice.

The laws of conservation of energy and quasi-momentum must be satisfied in all such processes. The latter law, however, demands the conservation of the total quasi-momentum of the phonons only to within any additive vector of the form $\hbar \mathbf{b}$, because of the non-uniqueness of the quasi-momentum itself. Thus the initial and final quasi-momenta $\mathbf{p}$ and $\mathbf{p}'$ in any phonon collision process must be related by[†]

$$\Sigma \mathbf{p} = \Sigma \mathbf{p}' + \hbar \mathbf{b}. \tag{71.4}$$

Any number of identical phonons may be created simultaneously in the lattice. That is, any number of phonons may be in each of the phonon quantum states (in the classical picture, this corresponds to arbitrary wave intensity). This means that the phonon gas obeys Bose statistics. Since, furthermore, the total number of particles in this gas is not given and is itself determined by the equilibrium conditions, its chemical potential is zero (see § 63). The mean number of phonons in a given quantum state (with quasi-momentum $\mathbf{p}$ and energy ε) is therefore determined in thermal equilibrium by Planck's distribution function:

$$\bar{n}_\mathbf{p} = 1/(e^{\varepsilon(\mathbf{p})/T} - 1). \tag{71.5}$$

It may be noted that at high temperatures ($T \gg \varepsilon$) this expression becomes

$$\bar{n}_\mathbf{p} = T/\varepsilon(\mathbf{p}), \tag{71.6}$$

i.e. the number of phonons in a given state is proportional to the temperature.

The concept of phonons is a particular case of a more general one that is fundamental in the theory of quantum energy spectra of all macroscopic bodies. Any low excited state of a macroscopic body may be regarded in quantum mechanics as an assembly of separate *elementary excitations*, which behave as quasi-particles moving in the volume occupied by the body. So long as the number of elementary excitations is sufficiently small, they do not "interact" with one another (i.e. their energies are additive), and so the assembly of them may be regarded as an ideal gas of quasi-particles. It should again be emphasised that the concept of elementary excitations arises as a means of quantum description of the collective motion of the atoms in the body, and they cannot in any way be identified with the individual atoms or molecules.

For phonons, their interaction corresponds (in the classical picture) to anharmonic vibrations of the atoms in the lattice. But, as already mentioned in § 64, these vibrations in solids are in practice always small, and are therefore "almost harmonic". The interaction of phonons in solids is therefore always weak in practice.

[†] Processes in which the total quasi-momentum does not remain constant but changes by $\hbar \mathbf{b}$ are called *Umklapp processes*.

In conclusion, we shall give the formulae for the thermodynamic quantities of a solid in terms of the phonon spectrum in it.

The free energy of a solid in thermodynamic equilibrium is given by (64.1). Changing from summation to integration over a continuum of phonon states, we have

$$F = N\varepsilon_0 + T \sum_{\alpha=1}^{3\nu} \int \log \left[1 - \exp\left(-\frac{\hbar\omega_\alpha(\mathbf{k})}{T}\right)\right] \frac{V \, d^3k}{(2\pi)^3}. \qquad (71.7)$$

where the summation is over all branches of the spectrum and the integration over the values of $\mathbf{k}$ in one reciprocal lattice cell.[†] In terms of the number density of states $g_\alpha(\omega)$ in each branch of the spectrum, with a change to integration over frequencies, we can also write this formula as

$$F = N\varepsilon_0 + TV \sum_{\alpha=1}^{3\nu} \int \log\left(1 - e^{-\hbar\omega/T}\right) g_\alpha(\omega) \, d\omega. \qquad (71.8)$$

The non-equilibrium macroscopic state of a solid is described by some non-equilibrium distribution of phonons among their quantum states, as for an ideal gas. The entropy of a body in such a state can be calculated by means of the formulae derived in § 55 (for a Bose gas). In particular, when there are many phonons in each state, the entropy is

$$S = \sum_j G_j \log\left(eN_j/G_j\right),$$

where N_j is the number of phonons in a group of G_j neighbouring states (see (55.8)). This case corresponds to high temperatures $(T \gg \Theta)$.

We can rewrite this formula in an integral form corresponding to the classical picture of thermal vibrations. The number of phonon states (of each branch of the spectrum) which correspond to the interval d^3k of values of the wave vector and the space volume element dV is

$$d\tau = \frac{d^3k \, dV}{(2\pi)^3}.$$

Let $U_\alpha(\mathbf{r}, \mathbf{k}) \, d\tau$ be the energy of thermal vibrations in the same phase-space element $d\tau$. The corresponding number of phonons is $U_\alpha(\mathbf{r}, \mathbf{k}) \, d\tau/\hbar\omega_\alpha(\mathbf{k})$. Substituting these expressions for G_j and N_j and changing to integration, we have the following formula for the entropy of a solid with a given non-equilibrium distribution of energy in the spectrum of thermal vibrations:

$$S = \sum_{\alpha=1}^{3\nu} \int \log\left[eU_\alpha(\mathbf{r}, \mathbf{k})/\hbar\omega_\alpha(\mathbf{k})\right] d\tau. \qquad (71.9)$$

[†] This formula has already been used in § 68 for the free energy contribution from the acoustic branches of the spectrum.

§ 72. Phonon creation and annihilation operators

We shall now show how the ideas defined in § 71 occur in the consistent quantisation of the lattice vibrations. The formulae obtained also have independent significance, since they form the basis for the mathematical formalism in the analysis of phonon interaction events.

Any vibrational motion of the crystal lattice can be represented as a super-position of travelling plane waves.[†] If the volume of the lattice is taken to be large but finite, the wave vector $\mathbf{k}$ will take a series of close but discrete values. The atomic displacements $\mathbf{u}_s(t, \mathbf{n})$ are then given by the discrete sum

$$\mathbf{u}_s(t, \mathbf{n}) = \frac{1}{\sqrt{N}} \sum_{\alpha=1}^{3\nu} \sum_{\mathbf{k}} \left[a_{\mathbf{k}\alpha} \mathbf{e}_s^{(\alpha)}(\mathbf{k}) e^{i\mathbf{k}\cdot\mathbf{r}_\mathbf{n}} + a_{\mathbf{k}\alpha}^* \mathbf{e}_s^{(\alpha)*}(\mathbf{k}) e^{-i\mathbf{k}\cdot\mathbf{r}_\mathbf{n}} \right], \qquad (72.1)$$

where N is the number of unit cells in the lattice. The summation is taken over all (non-equivalent) values of $\mathbf{k}$ and over all branches of the vibration spectrum; the remaining notation is as follows.

The vectors $\mathbf{e}_s^{(\alpha)}$ in (72.1) are the vibration polarisation vectors, i.e. the amplitudes which not only satisfy equations (69.7), but are now assumed normalised in a particular way. The normalisation condition is written (together with the orthogonality relations (69.11)) as

$$\sum_{s=1}^{\nu} \frac{m_s}{m} \mathbf{e}_s^{(\alpha)}(\mathbf{k}) \cdot \mathbf{e}_s^{(\alpha')*}(\mathbf{k}) = \delta_{\alpha\alpha'}, \qquad (72.2)$$

where $m = \Sigma m_s$ is the total mass of the atoms in one cell. The conditions (72.2) still leave arbitrary the common phase factor (independent of s) in the vectors $\mathbf{e}_s^{(\alpha)}$. This arbitrariness enables us to impose on the vectors the further conditions

$$\mathbf{e}_s^{(\alpha)}(-\mathbf{k}) = \mathbf{e}_s^{(\alpha)*}(\mathbf{k}); \qquad (72.3)$$

the possibility of such a choice is evident, since by (69.10) the vectors on each side of equation (72.3) satisfy the same equations.

The coefficients $a_{\mathbf{k}\alpha}$ in (72.1) are functions of time, satisfying the equations

$$\ddot{a}_{\mathbf{k}\alpha} + \omega_\alpha^2(\mathbf{k}) a_{\mathbf{k}\alpha} = 0, \qquad (72.4)$$

which are obtained by substituting (72.1) in equations (69.4). We put

$$a_{\mathbf{k}\alpha} \propto \exp\left[-i\omega_\alpha(\mathbf{k})t\right]; \qquad (72.5)$$

then each term in the sum depends only on the difference $\mathbf{k}\cdot\mathbf{r}_\mathbf{n} - \omega_\alpha t$, i.e. is a wave travelling in the direction of $\mathbf{k}$.

[†] Exactly as for a free electromagnetic field (cf. *Fields*, § 52).

The vibrational energy of the lattice is expressed in terms of the displacements and velocities of the atoms by

$$E = \tfrac{1}{2} \sum_{\mathbf{n}, s} m_s \dot{u}_s^2(\mathbf{n}) + \tfrac{1}{2} \sum_{\substack{\mathbf{n}, \mathbf{n}' \\ s, s'}} A_{ik}^{ss'}(\mathbf{n} - \mathbf{n}') \, u_{si}(\mathbf{n}) \, u_{s'k}(\mathbf{n}'). \tag{72.6}$$

Here we substitute the expansion (72.1). All terms in the resulting sums that contain factors $\exp[\pm i(\mathbf{k} \pm \mathbf{k}') \cdot \mathbf{r}_{\mathbf{n}}]$ with $\mathbf{k} \pm \mathbf{k}' \neq 0$ vanish on summation over $\mathbf{n}$, since

$$\sum_{\mathbf{n}} e^{i\mathbf{q} \cdot \mathbf{r}_{\mathbf{n}}} = N \quad \text{for} \quad \mathbf{q} = 0,$$
$$= 0 \quad \text{for} \quad \mathbf{q} \neq 0,$$

where $\mathbf{q}$ takes all non-equivalent values (see § 133). Using also the conditions (72.2) and (72.3), we can transform the kinetic energy into

$$\sum_{\alpha, \mathbf{k}} m\omega_\alpha^2 \{ a_{\mathbf{k}\alpha} a_{\mathbf{k}\alpha}^* + \tfrac{1}{2}(a_{\mathbf{k}\alpha} a_{-\mathbf{k}\alpha} + a_{\mathbf{k}\alpha}^* a_{-\mathbf{k}\alpha}^*) \}.$$

The potential energy in (72.6) can be rewritten by means of the equations of motion (69.4), as

$$-\tfrac{1}{2} \sum_{\mathbf{n}, s} m_s \ddot{u}_s(\mathbf{n}) \cdot u_s(\mathbf{n})$$

and then transformed similarly; it then takes a form that differs from the kinetic energy only in the sign of the second term in the braces. Addition of the two parts of the energy gives

$$E = \sum_{\alpha, \mathbf{k}} 2m\omega_\alpha^2(\mathbf{k}) \, |a_{\mathbf{k}\alpha}|^2. \tag{72.7}$$

Thus the total energy of the lattice vibrations is expressed as the sum of the energies associated with each wave separately.

Let us now carry out a transformation of the equations of motion of the lattice into the canonical equations of mechanics. To do so, we use real "canonical variables" $Q_{\mathbf{k}\alpha}$ and $P_{\mathbf{k}\alpha}$ defined by

$$Q_{\mathbf{k}\alpha} = \sqrt{m}(a_{\mathbf{k}\alpha} + a_{\mathbf{k}\alpha}^*),$$
$$P_{\mathbf{k}\alpha} = -i\omega_\alpha(\mathbf{k})\sqrt{m}(a_{\mathbf{k}\alpha} - a_{\mathbf{k}\alpha}^*) = \dot{Q}_{\mathbf{k}\alpha}. \tag{72.8}$$

Expressing $a_{\mathbf{k}\alpha}$ and $a_{\mathbf{k}\alpha}^*$ in terms of these and substituting in (72.7), we obtain the Hamiltonian of the lattice:

$$H = \tfrac{1}{2} \sum_{\alpha, \mathbf{k}} [P_{\mathbf{k}\alpha}^2 + \omega_\alpha^2(\mathbf{k}) Q_{\mathbf{k}\alpha}^2]. \tag{72.9}$$

The Hamilton's equations $\partial H/\partial P_{\mathbf{k}\alpha} = \dot{Q}_{\mathbf{k}\alpha}$ are the same as $P_{\mathbf{k}\alpha} = \dot{Q}_{\mathbf{k}\alpha}$, and from $\partial H/\partial Q_{\mathbf{k}\alpha} = -\dot{P}_{\mathbf{k}\alpha}$ we obtain

$$\ddot{Q}_{\mathbf{k}\alpha} + \omega_\alpha^2(\mathbf{k}) Q_{\mathbf{k}\alpha} = 0,$$

which are the equations of motion of the lattice.

Thus the Hamiltonian is represented as a sum of independent terms, each in the form of the Hamiltonian of a one-dimensional harmonic oscillator. This way of describing a classical vibrational motion makes evident the means of the transition to the quantum theory.[†] We must now consider the canonical variables (the generalised coordinates $Q_{k\alpha}$ and generalised momenta $P_{k\alpha}$) as operators with the commutation rule

$$\hat{P}_{k\alpha}\hat{Q}_{k\alpha} - \hat{Q}_{k\alpha}\hat{P}_{k\alpha} = -i\hbar. \tag{72.10}$$

The Hamiltonian (72.9) is replaced by a similar operator, whose eigenvalues are known from quantum mechanics:

$$E = \sum_{\alpha, \mathbf{k}} \hbar\omega_\alpha(\mathbf{k})\,(n_{k\alpha} + \tfrac{1}{2}), \qquad n_{k\alpha} = 0, 1, 2 \dots . \tag{72.11}$$

This formula enables us to define the concept of phonons as in § 71: the excited state of the lattice may be regarded as a set of elementary excitations (quasi-particles), each with an energy $\hbar\omega_\alpha(\mathbf{k})$ that is a definite function of the parameter (quasi-momentum) $\mathbf{k}$. The quantum numbers $n_{k\alpha}$ then become the occupation numbers of the various quasi-particle states.[‡]

In accordance with the known properties of the harmonic oscillator in quantum mechanics, the quantities $\omega_\alpha(\mathbf{k})Q_{k\alpha} \pm iP_{k\alpha}$ have non-zero matrix elements only for transitions in which the numbers $n_{k\alpha}$ change by unity (see *Quantum Mechanics*, § 23). If we use the operators

$$\hat{c}_{k\alpha} = \frac{1}{\sqrt{[2\hbar\omega_\alpha(\mathbf{k})]}}\,[\omega_\alpha(\mathbf{k})\hat{Q}_{k\alpha} + i\hat{P}_{k\alpha}],$$

$$\hat{c}_{k\alpha}^+ = \frac{1}{\sqrt{[2\hbar\omega_\alpha(\mathbf{k})]}}\,[\omega_\alpha(\mathbf{k})\hat{Q}_{k\alpha} - i\hat{P}_{k\alpha}], \tag{72.12}$$

the non-zero matrix elements are

$$\langle n_{k\alpha} - 1 \,|\, \hat{c}_{k\alpha} \,|\, n_{k\alpha} \rangle = \langle n_{k\alpha} \,|\, \hat{c}_{k\alpha}^+ \,|\, n_{k\alpha} - 1 \rangle$$
$$= \sqrt{n_{k\alpha}}. \tag{72.13}$$

The commutation rules for these operators are obtained from the definition (72.12) and the rule (72.10):

$$\hat{c}_{k\alpha}\hat{c}_{k\alpha}^+ - \hat{c}_{k\alpha}^+\hat{c}_{k\alpha} = 1. \tag{72.14}$$

[†] As in the transition from the classical description of a free electromagnetic field to the quantum picture of photons (see *RQT*, § 2).

[‡] The "zero-point" energy $\sum \tfrac{1}{2}\hbar\omega_\alpha$ which remains in (72.11) when all the $n_{k\alpha} = 0$ is to be included in the ground-state energy of the body. It is finite (because the sum contains only a finite number of terms) and its existence does not here lead to any difficulties of principle (in contrast to quantum electrodynamics, where the sum $\sum \hbar\omega$ diverges).

It is seen from (72.13) that, when acting on functions of the occupation numbers, the operators $\hat{c}_{k\alpha}$ and $\hat{c}_{k\alpha}^+$ are phonon annihilation and creation operators. The rule (72.14) corresponds to Bose statistics, as it should.

Together with the quantities $c_{k\alpha}$, the displacement vectors[†] also become operators (in second quantisation):

$$\hat{u}_s(n) = \sqrt{\frac{\hbar}{2mN}} \sum_{\alpha,k} \frac{1}{\sqrt{[\omega_\alpha(k)]}} [\hat{c}_{k\alpha} e_s^{(\alpha)}(k) e^{i k \cdot r_n} + \hat{c}_{k\alpha}^+ e_s^{(\alpha)*}(k) e^{-i k \cdot r_n}]. \quad (72.15)$$

From this, the anharmonic terms in the Hamiltonian (terms of the third and higher powers in the displacements) can be expressed in terms of products of various numbers of phonon creation and annihilation operators. These terms represent the perturbation that leads to various phonon-scattering processes, with changes of the phonon occupation numbers.

§ 73. Negative temperatures

Let us now consider some peculiar effects related to the properties of paramagnetic dielectrics. In such substances the atoms have angular momenta, and therefore magnetic moments, which are more or less freely oriented. The interaction of these moments (magnetic or exchange interaction, depending on their distance apart) brings about a new "magnetic" spectrum which is superposed on the ordinary dielectric spectrum.

This new spectrum lies entirely within a finite interval of energy, of the order of magnitude of the energy of interaction of the magnetic moments of all the atoms of the body, lying at fixed distances apart at the crystal lattice sites; he amount of this energy per atom may be from tenths of a degree to hundreds of degrees. In this respect the magnetic energy spectrum is completely different from the ordinary spectra, which, owing to the presence of the kinetic energy of the particles, extend to arbitrarily high energy values.[‡]

Because of this property we can consider an interval of temperatures large compared with the maximum possible interval of energy values per atom. The free energy F_{mag} pertaining to the magnetic part of the spectrum is calculated in exactly the same way as in § 32.

[†] It is easily seen from the definitions (72.8) and (72.12) that the quantities $c_{k\alpha}$ differ from the $a_{k\alpha}$ only by a factor.

[‡] The electronic (including the magnetic) spectra of various categories of solids will be studied in Part 2. In this section we are considering only the purely thermodynamic consequences of the above-mentioned general property of the magnetic spectrum.

Let E_n be the energy levels of the system of interacting moments. Then we have for the required partition function

$$Z_{\text{mag}} = \sum_n e^{-E_n/T}$$

$$\cong \sum_n \left(1 - \frac{E_n}{T} + \frac{E_n^2}{2T^2}\right).$$

Here, as in § 32, a formal expansion in powers of the quantity E_n/T, which is not in general small, will give (after taking logarithms) an expansion in terms of a small quantity $\sim E_n/NT$, where N is the number of atoms. The total number of levels in the spectrum under consideration is finite and equal to the number of all possible combinations of orientations of the atomic moments; for example, if all the moments are equal, this number is g^N, where g is the number of possible orientations of an individual moment relative to the lattice. Taking the average here as the ordinary arithmetic mean, we can write Z_{mag} as

$$Z_{\text{mag}} = g^N \left(1 - \frac{1}{T}\overline{E_n} + \frac{1}{2T^2}\overline{E_n^2}\right).$$

Finally, taking logarithms and again expanding in series with the same accuracy, we obtain for the free energy the expression

$$F_{\text{mag}} = -T \log Z_{\text{mag}}$$

$$= -NT \log g + \overline{E_n} - \frac{1}{2T}\left\langle (E_n - \overline{E_n})^2 \right\rangle. \tag{73.1}$$

Hence the entropy is

$$S_{\text{mag}} = N \log g - \frac{1}{2T^2}\left\langle (E_n - \overline{E_n})^2 \right\rangle, \tag{73.2}$$

the energy

$$E_{\text{mag}} = \overline{E_n} - \frac{1}{T}\left\langle (E_n - \overline{E_n})^2 \right\rangle, \tag{73.3}$$

and the specific heat

$$C_{\text{mag}} = \frac{1}{T^2}\left\langle (E_n - \overline{E_n})^2 \right\rangle. \tag{73.4}$$

We shall regard the atomic magnetic moments fixed at the lattice sites and interacting with one another as a single isolated system, ignoring its interaction with the lattice vibrations, which is usually very weak. Formulae (73.1)–(73.4) determine the thermodynamic quantities for this system at high temperatures.

The proof given in § 10 that the temperature is positive was based on the condition for the system to be stable with respect to the occurrence of internal macroscopic motions within it. But the system of moments here considered is by its nature incapable of macroscopic motion, and so the previous arguments do not apply to it; nor does the proof based on the normalisation condition for the Gibbs distribution (§ 36), since in the present case the system has only a finite number of energy levels, themselves finite, and so the normalisation sum converges for any value of T.

Thus we have the interesting result that the system of interacting moments may have either a positive or a negative temperature. Let us examine the properties of the system at various temperatures.

At $T = 0$, the system is in its lowest quantum state, and its entropy is zero. As the temperature increases, the energy and entropy of the system increase monotonically. At $T = \infty$, the energy is $\overline{E}_n$ and the entropy reaches its maximum value $N \log g$; these values correspond to a distribution with equal probability over all quantum states of the system, which is the limit of the Gibbs distribution as $T \to \infty$.

The temperature $T = -\infty$ is physically identical with $T = \infty$; the two values give the same distribution and the same values of the thermodynamic quantities for the system. A further increase in the energy of the system corresponds to an increase in the temperature from $T = -\infty$, with decreasing absolute magnitude since the temperature is negative. The entropy decreases monotonically (Fig. 10).[†] Finally, at $T = 0-$ the energy reaches its greatest

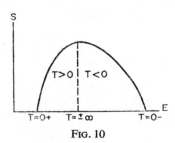

FIG. 10

value and the entropy returns to zero, the system then being in its highest quantum state.

Thus the region of negative temperatures lies not "below absolute zero" but "above infinity". In this sense we can say that negative temperatures are "higher" than positive ones. This is in accordance with the fact that, when a system at a negative temperature interacts with one at a positive temperature (i.e. the lattice vibrations), energy must pass from the former to the latter system; this is easily seen by the same method as that used in § 9 to discuss the exchange of energy between bodies at different temperatures.

[†] The curve $S = S(E)$ is symmetrical near its maximum, but in general need not be symmetrical far from the maximum.

States with negative temperature can be attained in practice in a paramagnetic system of nuclear moments in a crystal where the relaxation time t_2 for the interaction between nuclear spins is very small compared with the relaxation time t_1 for the spin–lattice interaction (E. M. Purcell and R. V. Pound, 1951). Let the crystal be magnetised in a strong magnetic field, and let the direction of the field then be reversed so quickly that the spins "cannot follow it". The system is thus in a non-equilibrium state, with an energy which is obviously higher than $\overline{E_n}$. During a time of the order of t_2, the system reaches an equilibrium state with the same energy. If the field is then adiabatically removed, the system remains in the equilibrium state, which will clearly have a negative temperature. The subsequent exchange of energy between the spin system and the lattice, whereby their temperatures are equalised, takes place in a time of the order of t_1.

CHAPTER VII

NON-IDEAL GASES

§ 74. Deviations of gases from the ideal state

THE equation of state of an ideal gas can often be applied to actual gases with sufficient accuracy. This approximation may, however, be inadequate, and it is then necessary to take account of the deviations of an actual gas from the ideal state which result from the interaction between its component molecules.

Here we shall do this on the assumption that the gas is still so rarefied that triple, quadruple, etc., collisions between molecules may be neglected, and their interaction may be assumed to occur only through binary collisions.

To simplify the formulae, let us first consider a monatomic actual gas. The motion of its particles may be treated classically, so that its energy has the form

$$E(p, q) = \sum_{a=1}^{N} \frac{p_a^2}{2m} + U, \qquad (74.1)$$

where the first term is the kinetic energy of the N atoms of the gas, and U is the energy of their mutual interaction. In a monatomic gas, U is a function only of the distances between the atoms. The partition function $\int e^{-E(p, q)/T}\, d\Gamma$ becomes the product of the integral over the momenta of the atoms and the integral over their coordinates. The latter integral is

$$\int \ldots \int e^{-U/T}\, dV_1 \ldots dV_N,$$

where the integration over each $dV_a = dx_a\, dy_a\, dz_a$ is taken over the whole volume V occupied by the gas. For an ideal gas, $U = 0$, and this integral would be simply V^N. It is therefore clear that, on calculating the free energy from the general formula (31.5), we obtain

$$F = F_{id} - T \log \frac{1}{V^N} \int \ldots \int e^{-U/T}\, dV_1 \ldots dV_N, \qquad (74.2)$$

where F_{id} is the free energy of an ideal gas.

Non-ideal Gases

Adding and subtracting unity in the integrand, we can rewrite formula (74.2) as

$$F = F_{id} - T \log \left\{ \frac{1}{V^N} \int \ldots \int (e^{-U/T} - 1) \, dV_1 \ldots dV_N + 1 \right\}. \quad (74.3)$$

For the subsequent calculations we make use of the following formal device. Let us suppose that the gas is not only sufficiently rarefied but also so small in quantity that not more than one pair of atoms may be assumed to be colliding in the gas at any one time. This assumption does not affect the generality of the resulting formulae, since we know from the additivity of the free energy that it must have the form $F = Nf(T, V/N)$ (see § 24), and therefore the formulae deduced for a small quantity of gas are necessarily valid for any quantity.

The interaction between atoms is very small except when the two atoms concerned are very close together, i.e. are almost colliding. The integrand in (74.3) is therefore appreciably different from zero only when some pair of atoms are very close together. According to the above assumption, not more than one pair of atoms can satisfy this condition at any one time, and this pair can be selected from N atoms in $\frac{1}{2}N(N-1)$ ways. Consequently, the integral in (74.3) may be written

$$\tfrac{1}{2}N(N-1) \int \ldots \int (e^{-U_{12}/T} - 1) \, dV_1 \ldots dV_N,$$

where U_{12} is the energy of interaction of the two atoms (it does not matter which two, as they are all identical); U_{12} depends only on the coordinates of two atoms, and we can therefore integrate over the remaining coordinates, obtaining V^{N-2}. We can also, of course, write N^2 instead of $N(N-1)$, since N is very large; substituting the resulting expression (74.3) in place of the integral, and using the fact that $\log(1+x) \cong x$ for $x \ll 1$, we have[†]

$$F = F_{id} - \frac{TN^2}{2V^2} \int \int (e^{-U_{12}/T} - 1) \, dV_1 \, dV_2,$$

where $dV_1 \, dV_2$ is the product of differentials of the coordinates of the two atoms.

But U_{12} is a function only of the distance between the two atoms, i.e. of the differences of their coordinates. Thus, if the coordinates of the two atoms are expressed in terms of the coordinates of their centre of mass and their relative coordinates, U_{12} will depend only on the latter (the product of

---

[†] We shall see later that the first term in the argument of the logarithm in (74.3) is proportional to N^2/V. The expansion in question therefore depends on precisely the assumption made above, that not only the density N/V but also the quantity of the gas is small.

whose differentials will be denoted by dV). We can therefore integrate with respect to the coordinates of the centre of mass, again obtaining the volume V. The final result is

$$F = F_{id} + N^2 T B(T)/V,\tag{74.4}$$

where

$$B(T) = \tfrac{1}{2}\int(1-e^{-U_{12}/T})\,dV.\tag{74.5}$$

From this we find the pressure $P = -\partial F/\partial V$:

$$P = \frac{NT}{V}\left(1+\frac{NB(T)}{V}\right),\tag{74.6}$$

since $P_{id} = NT/V$. Equation (74.6) is the equation of state of the gas in the approximation considered.

As we know from the theorem of small increments (§ 15), the changes in the free energy and the thermodynamic potential resulting from small changes in the external conditions or properties of a body are equal, one being taken at constant volume and the other at constant pressure.

If we regard the deviation of a gas from the ideal state as such a change, we can change directly to Φ from (74.4). To do so, we need only express the volume in terms of the pressure in the correction term in (74.4) by means of the equation of state for an ideal gas, obtaining

$$\Phi = \Phi_{id} + NBP.\tag{74.7}$$

The volume may hence be expressed as a function of the pressure:

$$V = \frac{NT}{P} + NB.\tag{74.8}$$

The whole of the foregoing discussion applies to monatomic gases. The same formulae remain valid, however, for polyatomic gases also. In this case the potential energy of interaction of the molecules depends not only on their distance apart but also on their relative orientation. If (as almost always happens) the rotation of the molecules may be treated classically, we can say that U_{12} is a function of the coordinates of the centres of mass of the molecules and of rotational coordinates (angles) which define the spatial orientation of the molecules. It is easy to see that the only difference from the case of a monatomic gas amounts to the fact that dV_a must be taken as the product of the differentials of all these coordinates of the molecule. But the rotational coordinates can always be so chosen that the integral $\int dV_a$ is again equal to the volume V of the gas. For the integration over the coordinates of the centre of mass gives this volume V, while the integration over angles gives a constant, and the angles can always be normalised so that this constant is unity. Thus all the formulae derived in this section have the same form for polyatomic gases, the only difference being that in (74.5) dV is now the product of the differentials of coordinates defining both the distance between two molecules and their relative orientation.[†]

All the above formulae are meaningful, of course, only if the integral (74.5) converges. For this to be so it is certainly necessary that the forces of interaction between the molecules should decrease sufficiently rapidly with increasing distance: U_{12} must decrease at large distances more rapidly than $1/r^3$.[‡]

If this condition is not satisfied, a gas consisting of identical particles cannot exist as a homogeneous body. In this case every region of matter will be subject to very large forces exerted by distant parts of the gas. The regions near to and far from the boundary of the volume occupied by the gas will therefore be in quite different conditions, and so the gas is no longer homogeneous.

FIG. 11

For monatomic gases the function $U_{12}(r)$ has the form shown in Fig. 11; the abscissa is the distance r between the atoms. At small distances, U_{12} increases with decreasing distance, corresponding to repulsive forces between the atoms; beginning approximately at the place where the curve crosses the abscissa axis, it rises steeply, so that U_{12} rapidly becomes very large, corresponding to the mutual "impenetrability" of the atoms (for which reason the distance r_0 is sometimes called the radius of the atom). At large distances, U_{12} increases slowly, approaching zero asymptotically. The increase of U_{12} with distance corresponds to a mutual attraction of the atoms. The minimum

---

[†] If the particles in the gas have spin, the form of the function U_{12} depends in general on the orientation of the spins. In that case a summation over spin orientations must be added to the integration with respect to dV.

[‡] This condition is always satisfied for atomic and molecular gases: the forces of interaction between electrically neutral atoms or molecules (including dipoles), when averaged over the relative orientations of the particles, decrease at large distances as $U_{12} \propto 1/r^6$; see *Quantum Mechanics*, § 89.

of U_{12} corresponds to a stable equilibrium. The absolute value U_0 of the energy at this point is usually small, being of the order of the critical temperature of the substance.

For a polyatomic gas, the interaction energy has a similar form, but it cannot, of course, be represented by the curve in Fig. 11, since it is a function of a larger number of variables.

This information as to the nature of the function U_{12} is sufficient to determine the sign of $B(T)$ in the limiting cases of high and low temperatures. At high temperatures $(T \gg U_0)$ we have $|U_{12}|/T \ll 1$ throughout the region $r > 2r_0$, and the integrand in $B(T)$ (74.5) is almost zero. Thus the value of the integral is mainly determined by the region $r < 2r_0$, where U_{12}/T is large and positive; in this region, therefore, the integrand is positive, and so the integral itself is positive. Thus $B(T)$ is positive at high temperatures.

At low temperatures $(T \ll U_0)$, on the other hand, the region $r > 2r_0$ is the important one in the integral, and in this region U_{12}/T is large and negative. At sufficiently low temperatures, therefore, $B(T)$ must be negative, and its temperature dependence is essentially given by the exponential factor $-e^{U_0/T}$.

Being positive at high temperatures and negative at low temperatures, $B(T)$ must pass through zero at some temperature.[†]

Finally, let us consider a Joule–Thomson process involving a non-ideal gas. The variation of temperature during the process is given by the derivative

$$\left(\frac{\partial T}{\partial P}\right)_W = \frac{1}{C_p}\left[T\left(\frac{\partial V}{\partial T}\right)_P - V\right]; \tag{74.9}$$

see (18.2). For an ideal gas this derivative is of course zero, but for a gas with the equation of state (74.8) we have

$$\left(\frac{\partial T}{\partial P}\right)_W = \frac{N}{C_p}\left(T\frac{dB}{dT} - B\right) = \frac{N}{2C_p}\int\left[e^{-U_{12}/T}\left(1 - \frac{U_{12}}{T}\right) - 1\right] dV. \tag{74.10}$$

As in the discussion of $B(T)$, it is easy to see that at high temperatures $(\partial T/\partial P)_W < 0$, i.e. when the gas goes from a higher to a lower pressure in a Joule–Thomson process its temperature rises; at low temperatures, $(\partial T/\partial P)_W > 0$, i.e. the gas temperature falls with the pressure. At a definite temperature for each gas, called the *inversion point*, the Joule–Thomson effect must therefore change sign.[‡]

[†] The temperature T_B for which $B(T_B) = 0$ is called the *Boyle point*. If PV/T is plotted against P for various given T, the isotherm $T = T_B$ has a horizontal tangent as $P \to 0$, and separates isotherms with positive and negative initial slopes; all the isotherms start from the point $PV/T = 1$, $P = 0$.

[‡] It will be recalled that we are considering a gas which is only slightly non-ideal, so that the pressure is relatively low. The result that the inversion point is independent of pressure is valid only in this approximation; cf. § 76, Problem 4.

PROBLEMS

PROBLEM 1. Determine $B(T)$ for a gas whose particles repel one another according to $U_{12} = \alpha/r^n$ $(n > 3)$.

SOLUTION. In (74.5) we put $dV = 4\pi r^2\, dr$ and integrate by parts with respect
to r from 0 to ∞; the substitution $\alpha/Tr^n = x$ then reduces the integral to a gamma
function:

$$B(T) = \frac{2\pi}{3}\left(\frac{\alpha}{T}\right)^{3/n}\Gamma\left(1-\frac{3}{n}\right).$$

PROBLEM 2. The *fugacity* of a gas is the pressure P^* which it would have for given
values of the temperature and chemical potential if so rarefied that it could be
regarded as an ideal gas. Determine the fugacity of a gas with the thermodynamic
potential (74.7).

SOLUTION. The chemical potential of the gas is (with μ_{id} given by (42.6))

$$\mu = \mu_{\mathrm{id}} + BP = T\log P + \chi(T) + BP.$$

Equating this to $T\log P^* + \chi(T)$ by the definition of the fugacity, we have to the
same accuracy as that of (74.7)

$$P^* = P\left(1+\frac{BP}{T}\right) = \frac{NT}{V}\left(1+\frac{2NB}{V}\right).$$

§ 75. Expansion in powers of the density

The equation of state (74.6) derived in § 74 consists essentially of the first
two terms in an expansion of the pressure in powers of $1/V$:

$$P = \frac{NT}{V}\left(1+\frac{NB(T)}{V}+\frac{N^2C(T)}{V^2}+\ \cdots\right). \tag{75.1}$$

The first term in the expansion corresponds to an ideal gas. The second term
is obtaining by taking into account the interaction between pairs of molecules, and the subsequent terms must involve the interactions between
groups of three, four, etc., molecules.[†]

The coefficients B, C, ... in the expansion (75.1) are called the second,
third, etc., *virial coefficients*. To determine these quantities, it is convenient to
begin by calculating the potential Ω, not the free energy. Let us again consider a monatomic gas, and start from the general formula (35.5), which for a
gas consisting of identical particles becomes

$$e^{-\Omega/T} = \sum_{N=0}^{\infty}\frac{1}{N!}\, e^{\mu N/T}\int e^{-E_N(p,\,q)/T}\, d\Gamma_N. \tag{75.2}$$

[†] The dimensionless small parameter with respect to which the expansion is
made is actually the ratio Nv_0/V of the "volume" v_0 of one molecule to the gas
volume per molecule V/N.

The factor $1/N!$ is included and the integration is then taken simply over the whole phase space of the system of N particles; cf. (31.7).

In the successive terms of the sum over N, the energy $E_N(p, q)$ has the following forms. For $N = 0$, of course, $E_0(p, q) \equiv 0$. For $N = 1$, it is simply the kinetic energy of one atom:

$$E_1(p, q) = p^2/2m.$$

For $N = 2$ it consists of the kinetic energy of the two atoms and the energy of their interaction:

$$E_2(p, q) = \sum_{a=1}^{2} \frac{p_a^2}{2m} + U_{12}.$$

Similarly,

$$E_3(p, q) = \sum_{a=1}^{3} \frac{p_a^2}{2m} + U_{123},$$

where U_{123} is the interaction energy of three atoms (which in general is not equal to the sum $U_{12}+U_{13}+U_{23}$), and so on.

We substitute these expressions in (75.2) and use the notation

$$\xi = \frac{e^{\mu/T}}{(2\pi\hbar)^3} \int e^{-p^2/2mT} \, d^3p = \left(\frac{mT}{2\pi\hbar^2}\right)^{3/2} e^{\mu/T}. \tag{75.3}$$

We shall see below that this is simply equal to P_{id}/T, where P_{id} is the pressure of an ideal gas for given T and V. We obtain

$$\Omega = -T \log \left\{ 1 + \xi V + \frac{\xi^2}{2!} \int\!\!\int e^{-U_{12}/T} \, dV_1 \, dV_2 \right.$$

$$\left. + \frac{\xi^3}{3!} \int\!\!\int\!\!\int e^{-U_{123}/T} \, dV_1 \, dV_2 \, dV_3 + \cdots \right\}.$$

Each of the U_{12}, U_{123}, ... is a function only of the distances between the atoms; hence, by using the relative coordinates of the atoms (relative to the first atom, say), we reduce the multiplicity of the integrals by one, with a further factor V entering:

$$\Omega = -PV = -T \log \left\{ 1 + \xi V + \frac{\xi^2 V}{2!} \int e^{-U_{12}/T} \, dV_2 \right.$$

$$\left. + \frac{\xi^3 V}{3!} \int\!\!\int e^{-U_{123}/T} \, dV_2 \, dV_3 + \cdots \right\}.$$

Finally, we expand this expression in powers of ξ; the resulting series can be written as

$$P = T \sum_{n=1}^{\infty} \frac{J_n}{n!} \xi^n, \tag{75.4}$$

where

$$J_1 = 1, \quad J_2 = \int (e^{-U_{12}/T} - 1) \, dV_2,$$
$$J_3 = \int\int (e^{-U_{123}/T} - e^{-U_{12}/T} - e^{-U_{13}/T} - e^{-U_{23}/T} + 2) \, dV_2 \, dV_3, \tag{75.5}$$

etc. The structure of the integrals J_n is evident; the integrand in J_n is appreciably different from zero only if n atoms are close together, i.e. in a collision of n atoms.

Differentiating (75.4) with respect to μ, we obtain the number of particles in the gas, since $N = -(\partial\Omega/\partial\mu)_{T, V} = V(\partial P/\partial\mu)_{T, V}$. Bearing in mind that by definition (75.3) $\partial\xi/\partial\mu = \xi/T$, we have

$$N = V \sum_{n=1}^{\infty} \frac{J_n}{(n-1)!} \xi^n. \tag{75.6}$$

The two equations (75.4) and (75.6) give in parametric form (with parameter ξ) the relation between P, V and T, i.e. the equation of state of the gas. Eliminating ξ, we can obtain the equation of state in the form of the series (75.1) with any desired number of terms.[†]

§ 76. Van der Waals' formula

In gases the interaction between molecules is very weak. As this interaction increases, the properties of the gas differ more and more from those of ideal gases, and finally the gas condenses into a liquid. In the latter, the interaction between molecules is strong, and the properties of this interaction (and therefore those of the liquid) depend considerably on the particular liquid concerned. For this reason it is, as already mentioned, impossible to derive any general formulae giving a quantitative description of the properties of a liquid.

We can, however, find an interpolation formula which gives a qualitative description of the transition between the liquid and gaseous states. This formula must give the correct results in the two limiting cases. For rarefied gases it must become the formulae valid for ideal gases. When the density increases and the gas approaches the liquid state, it must take account of the finite compressibility of the substance. Such a formula will then give

---

[†] In the first approximation, $P = T\xi, N = V\xi$, whence $P = NT/V = P_{\text{id}}$. In the second approximation, $P = T\xi(1 + \frac{1}{2}J_2\xi)$, $N = V\xi(1 + J_2\xi)$; eliminating ξ from these equations (with the same accuracy), we have

$$P = \frac{NT}{V} - \frac{N^2 T}{2V^2} J_2,$$

in agreement with (74.6).

a qualitative description of the behaviour of the gas in the intermediate range also.

To derive such a formula, let us examine in more detail the deviations from the ideal state at high temperatures. As in the preceding sections, we shall first consider a monatomic gas; by the same arguments as used previously, all the resulting formulae will be equally applicable to polyatomic gases.

The type of interaction between gas atoms described in § 74 (Fig. 11) enables us to determine the form of the leading terms in the expansion of $B(T)$ in inverse powers of the temperature; here we shall assume that the ratio U_0/T is small:

$$U_0/T \ll 1. \tag{76.1}$$

Since U_{12} depends only on the distance r between the atoms, we write in the integral (74.5) $dV = 4\pi r^2 \, dr$. Dividing the range of integration with respect to r into two parts, we write

$$B(T) = 2\pi \int_0^{2r_0} (1 - e^{-U_{12}/T}) r^2 \, dr + 2\pi \int_{2r_0}^{\infty} (1 - e^{-U_{12}/T}) r^2 \, dr.$$

For values of r between 0 and $2r_0$, the potential energy U_{12} is in general very large. In the first integral we can therefore neglect the term $e^{-U_{12}/T}$ in comparison with unity. Then this integral is equal to the positive quantity $b = 16\pi r_0^3/3$; if r_0 for a monatomic gas is regarded as the radius of the atom, then b is four times its volume. In the second integral, we have everywhere $|U_{12}|/T \leqslant U_0/T \ll 1$. The integrand can therefore be expanded in powers of U_{12}/T as far as the first non-vanishing term. The second integral then becomes

$$-\frac{2\pi}{T} \int_{2r_0}^{\infty} |U_{12}| \, r^2 \, dr = -\frac{2a}{T},$$

where a is a positive constant.

Thus we find

$$B(T) = b - a/T, \tag{76.2}$$

and substitution of this in (74.4) and (74.7) gives for the free energy of the gas

$$F = F_{id} + N^2(Tb - a)/V, \tag{76.3}$$

and for the thermodynamic potential

$$\Phi = \Phi_{id} + NP(b - a/T). \tag{76.4}$$

The desired interpolation formula can be obtained from (76.3), which itself does not satisfy the necessary conditions, since it does not take account of the finite compressibility of the substance. In (76.3) we substitute the expression for F_{id} from (42.4); this gives

$$F = Nf(T) - NT \log (e/N) - NT(\log V - Nb/V) - N^2 a/V. \qquad (76.5)$$

In deriving formula (74.4) for the free energy of a gas we assumed that the gas, though not sufficiently rarefied to be regarded as an ideal gas, nevertheless occupies such a large volume that we can neglect triple and higher-order collisions between molecules, so that the distances between molecules are in general considerably larger than their dimensions. We may say that the gas volume V is always considerably greater than Nb; hence

$$\log (V - Nb) = \log V + \log (1 - Nb/V)$$
$$\cong \log V - Nb/V.$$

Hence (76.5) may be written

$$F = Nf(T) - NT \log [e(V - Nb)/N] - N^2 a/V$$
$$= F_{id} - NT \log (1 - Nb/V) - N^2 a/V. \qquad (76.6)$$

In this form the conditions stated above are satisfied, since when V is large the formula for the free energy of an ideal gas is obtained, and when V is small the formula shows that the gas cannot be indefinitely compressed, since the argument of the logarithm becomes negative when $V < Nb$.

If the free energy is known, we can determine the gas pressure:

$$P = -\partial F/\partial V = \frac{NT}{V - Nb} - \frac{N^2 a}{V^2}$$

or

$$\left(P + \frac{N^2 a}{V^2}\right)(V - Nb) = NT. \qquad (76.7)$$

This is the required interpolation formula for the equation of state of an actual gas. It is called *van der Waals' equation*.

Van der Waals' formula is, of course, only one of an infinity of possible interpolation formulae which satisfy the requirements stated, and there is no physical reason to select any one of them; van der Waals' formula is merely the simplest and most convenient.[†]

From (76.6) we can find the entropy of the gas:

$$S = S_{id} + N \log (1 - Nb/V), \qquad (76.8)$$

and thence the energy $E = F + TS$:

$$E = E_{id} - N^2 a/V. \qquad (76.9)$$

[†] In actual applications of this formula, the values of the constants a and b must be chosen so as to give the best agreement with experiment. The constant b cannot then be regarded as four times the molecular volume, even for a monatomic gas.

Thus the specific heat $C_v = (\partial E/\partial T)_V$ of a van der Waals gas is equal to that of an ideal gas; it depends only on the temperature and, in particular, may be constant. The specific heat C_p is easily seen (cf. Problem 1) to depend not only on the temperature but also on the volume, and so cannot be constant.

The second term in (76.9) corresponds to the energy of interaction of the gas molecules; it is, of course, negative, since on average the forces between molecules are attractive.

PROBLEMS

PROBLEM 1. Find $C_p - C_v$ for a non-ideal gas described by van der Waals' formula.

SOLUTION. Using formula (16.10) and van der Waals' equation, we find

$$C_p - C_v = \frac{N}{1 - 2Na(V-Nb)^2/TV^3} \cdot$$

PROBLEM 2. Find the equation of an adiabatic process for a van der Waals gas of constant specific heat C_v.

SOLUTION. Substituting in (76.8) $S_{id} = N \log V + Nc_v \log T$ (omitting unimportant constants) and putting $S = $ constant, we obtain the relation $(V-Nb)T^{c_v} = $ constant. This differs from the corresponding equation for an ideal gas in that V is replaced by $V-Nb$.

PROBLEM 3. For a gas of the same kind as in Problem 2, find the change in temperature on expansion into a vacuum from volume V_1 to V_2.

SOLUTION. In an expansion into a vacuum, the energy of the gas remains constant. Thus formula (76.9), with $E_{id} = NC_vT$, gives

$$T_2 - T_1 = \frac{Na}{C_v}\left(\frac{1}{V_2} - \frac{1}{V_1}\right).$$

PROBLEM 4. For a van der Waals gas find the temperature dependence of the inversion point for the Joule–Thomson effect.

SOLUTION. The inversion point is determined by the equation $(\partial T/\partial V)_p = T/V$ (see (74.9)). Substitution of T from (76.7) leads to an equation which has to be solved simultaneously with (76.7). Algebraic calculation gives the following dependence of the inversion point on pressure:

$$T_{inv} = \frac{2a}{9b}(2 \pm \sqrt{(1 - 3b^2P/a)})^2.$$

For any given pressure $P < a/3b^2$ there are two inversion points, between which the derivative $(\partial T/\partial P)_W$ is positive, while outside this temperature interval it is negative. When $P > a/3b^2$ there are no inversion points and $(\partial T/\partial P)_W < 0$ everywhere.[†]

† The upper inversion point as $P \to 0$ ($T_{inv} = 2a/b$) corresponds to the case considered at the end of § 74. The lower inversion point for small P may not occur in a gas owing to its condensation into a liquid.

§ 77. Relationship of the virial coefficient and the scattering amplitude

In calculating the virial coefficients in §§ 74–76 we have used classical statistics, as is practically always justifiable. There is, however, methodological interest in the problem of calculating these coefficients in the quantum case; and such a case may actually occur for helium at sufficiently low temperatures. We shall show how the second virial coefficient may be calculated with allowance for the quantisation of the binary interaction of the gas particles (E. Beth and G. E. Uhlenbeck, 1937). We shall consider a monatomic gas whose atoms have no electronic angular momentum; and having in mind the case of helium, we shall also suppose for definiteness that the nuclei of the atoms have no spin and that the atoms obey Bose statistics.

In the approximation concerned, it is sufficient to retain only the first three terms in the sum over N in formula (35.3), which determines the potential Ω:

$$\Omega = -T \log \left\{ 1 + \sum_n e^{(\mu - E_{1n})/T} + \sum_n e^{(2\mu - E_{2n})/T} \right\}. \tag{77.1}$$

Here the E_{1n} are the energy levels of a single atom, and the E_{2n} are those of a system of two interacting atoms. Our object is to calculate only those correction terms in the thermodynamic quantities which are due to the direct interaction of the atoms; the corrections due to quantum exchange effects exist even in an ideal gas and are given by formula (56.15), according to which the exchange part of the second virial coefficient is (in the case of Bose statistics)

$$B_{\text{exch}} = -\tfrac{1}{2}(\pi \hbar^2 / mT)^{3/2}. \tag{77.2}$$

Thus the problem reduces to the calculation of

$$Z^{(2)} = \sum_n e^{(2\mu - E_{2n})/T},$$

and from this must be subtracted the expression which would be obtained for two non-interacting atoms.

The energy levels E_{2n} consist of the kinetic energy of the motion of the centre of mass of the two atoms ($p^2/4m$, where $\mathbf{p}$ is the momentum of that motion, and m the mass of an atom) and the energy of their relative motion. Let the latter be denoted by ε; this component is given by the energy levels of a particle of mass $\tfrac{1}{2}m$ (the reduced mass of the two atoms) moving in a central field $U_{12}(\mathbf{r})$, where U_{12} is the potential energy of the interaction of the atoms. The motion of the centre of mass is always quasi-classical; integrating over its coordinates and momenta in the usual manner (cf. § 42), we obtain

$$Z^{(2)} = V e^{2\mu/T} (mT/\pi \hbar^2)^{3/2} \Sigma e^{-\varepsilon/T}.$$

If we denote by Z_{int} the part of $Z^{(2)}$ which depends on the interaction of the particles, we can write Ω in the form

$$\Omega = \Omega_{id} - TVe^{2\mu/T}(mT/\pi\hbar^2)^{3/2} Z_{int}.$$

Regarding the second term as a small correction to the first term, and expressing it in terms of T, V and N by means of formula (45.5) for the chemical potential of an ideal gas, we obtain for the free energy the expression

$$F = F_{id} - T\frac{8N^2}{V}\left(\frac{\pi\hbar^2}{mT}\right)^{3/2} Z_{int}.$$

Differentiation with respect to V gives the pressure, and the required part of the virial coefficient that is due to the interaction of the atoms is

$$B_{int}(T) = -8(\pi\hbar^2/mT)^{3/2} Z_{int}. \tag{77.3}$$

The spectrum of energy levels ε consists of a discrete spectrum of negative values (corresponding to a finite relative motion of the atoms) and a continuous spectrum of positive values (infinite motion). We denote the former by ε_n; the latter may be written in the form p^2/m, where $\mathbf{p}$ is the momentum of the relative motion of the atoms when the distance between them has become very large. The whole of the sum $\Sigma e^{|\varepsilon_n|/T}$ over the discrete spectrum appears in Z_{int}; from the integral over the continuous spectrum we must separate the part corresponding to the free motion of non-interacting particles. To do this we proceed as follows.

At large distances r, the wave function of a stationary state with orbital angular momentum l and positive energy p^2/m has the asymptotic form (see *Quantum Mechanics*,§ 33)

$$\psi = \frac{\text{constant}}{r} \times \sin\left(\frac{p}{\hbar}r - \frac{1}{2}l\pi + \delta_l\right),$$

where the phase shifts $\delta_l = \delta_l(p)$ depend on the specific form of the field $U_{12}(r)$. Let us formally suppose that the range of variation of the distance r is bounded by some very large but finite value R. Then the momentum p can take only a discrete series of values given by the boundary condition that $\psi = 0$ for $r = R$:

$$\frac{p}{\hbar}R - \frac{1}{2}l\pi + \delta_l = s\pi,$$

where s is an integer. For large R these values are very close together and the summation in

$$\sum_p e^{-p^2/mT}$$

may be replaced by an integration. To do so, we multiply the summand (for a given l) by

$$ds = \frac{1}{\pi} \left(\frac{R}{\hbar} + \frac{d\delta_l}{dp} \right) dp$$

and integrate over p; the result must then be multiplied by $2l+1$ (the degree of degeneracy with respect to orientations of the orbital angular momentum) and summed over l:

$$\sum_p e^{-p^2/mT} = \frac{1}{\pi} \sum_l (2l+1) \int_0^\infty \left(\frac{R}{\hbar} + \frac{d\delta_l}{dp} \right) e^{-p^2/mT} \, dp.$$

For particles obeying Bose statistics and having no spin, the coordinate wave functions must be symmetric; this means that only even values of l are possible, and so the summation over l is over all even integers.

In free motion, all the phase shifts $\delta_l = 0$. The expression remaining when $\delta_l = 0$ is therefore the part of the sum which is unrelated to the interaction of the atoms and is to be omitted. Thus we obtain the following expression for Z_{int}:

$$Z_{\text{int}} = \sum_n e^{|\varepsilon_n|/T} + \frac{1}{\pi} \sum_l \int_0^\infty (2l+1) \frac{d\delta_l}{dp} e^{-p^2/mT} \, dp, \qquad (77.4)$$

and the virial coefficient $B = B_{\text{exch}} + B_{\text{int}}$ is

$$B(T) = -\tfrac{1}{2} (\pi\hbar^2/mT)^{3/2} (1 + 16 Z_{\text{int}}). \qquad (77.5)$$

The phase shifts δ_l determine the scattering amplitude for particles moving in the field $U_{12}(r)$ by means of the formula[†]

$$f(\theta) = \frac{\hbar}{2ip} \sum_l (2l+1)(e^{2i\delta_l} - 1) P_l(\cos\theta),$$

where the P_l are Legendre polynomials, and θ the angle between the directions of incidence and scattering; in the present case the summation is over all even values of l. It is therefore possible to express the integral in (77.4) in terms of the scattering amplitude. By direct substitution of the expression for $f(\theta)$ the following relation may easily be verified:

$$\sum_l (2l+1) \frac{d\delta_l}{dp} = \frac{1}{2\hbar} \frac{d}{dp} \{ p[f(0) + f^*(0)] \} + \frac{i}{4\pi\hbar^2} \int p^2 \left(f \frac{\partial f^*}{\partial p} - f^* \frac{\partial f}{\partial p} \right) do.$$

[†] See *Quantum Mechanics*, § 123. The cross-section for scattering into the solid-angle element do is $|f(\theta)|^2 \, do$.

The sum on the left-hand side appears in the integrand in (77.4), and on sub-stituting it and integrating by parts in one of the terms we find

$$Z_{int} = \sum_n e^{|\varepsilon_n|/T} + \frac{1}{\pi \hbar m T} \int_0^\infty p^2 e^{-p^2/mT} [f(0) + f^*(0)] \, dp$$

$$+ \frac{i}{(2\pi\hbar)^2} \iint p^2 e^{-p^2/mT} \left(f \frac{\partial f^*}{\partial p} - f^* \frac{\partial f}{\partial p} \right) dp \, do. \qquad (77.6)$$

If there are discrete levels in the field $U_{12}(r)$, then at sufficiently low temper-atures the dependence of $B(T)$ on temperature will be mainly governed by the sum over the discrete levels, which increases exponentially with decreas-ing T. There may, however, be no discrete levels; then the virial coefficient will vary as a power of the temperature (if we bear in mind that the scattering amplitude tends to a constant limit as $p \to 0$, we easily find that at sufficiently low temperatures B will be determined mainly by the term B_{exch}).

It may be noted that in the case of a weak interaction, when particle colli-sions can be described by the Born approximation, the scattering amplitude is small and the third term in (77.6), which is quadratic in the amplitude, may be omitted. For weak interaction there are no bound states, and so the first term in (77.6) is also absent. Using the familiar expression for the scattering amplitude $f(0)$ in the Born approximation, it is easy to see that the expression for F agrees exactly with (32.3) (without the quadratic term), as it should in this case.

PROBLEM

Determine the quantum correction (of the order of $\hbar^2$) in the quasi-classical case in the virial coefficient $B(T)$ for a monatomic gas.

SOLUTION. The correction to the classical free energy is given by formula (33.15). Bearing in mind that in the present case only binary interaction of atoms occurs, and that U_{12} depends only on the distance between atoms, we find

$$B_{qu} = \frac{\pi \hbar^2}{6mT^3} \int_0^\infty \left(\frac{dU_{12}}{dr} \right)^2 e^{-U_{12}/T} r^2 \, dr.$$

This expression is the correction to the classical value given by (74.5). It may be noted that $B_{qu} > 0$.

§ 78. Thermodynamic quantities for a classical plasma

The method given in § 75 for calculating the thermodynamic quantities for a non-ideal gas is certainly inapplicable to a gas consisting of charged particles with Coulomb interaction, since the integrals which appear in the formu-lae then diverge. Such gases must therefore be treated separately.

Let us consider a completely ionised gas or *plasma*. The charges on its particles will be denoted by $z_a e$, where the suffix a refers to the different kinds of ion; e is the unit charge and the z_a are positive or negative integers. Also, let n_{a0} be the number of ions of the ath kind per unit volume of the gas. The gas as a whole is, of course, electrically neutral, so that

$$\sum_a z_a n_{a0} = 0. \tag{78.1}$$

We shall suppose that the gas does not deviate greatly from the ideal state. For this to be so it is certainly necessary that the mean energy of the Coulomb interaction between two ions ($\sim (ze)^2/r$, where $r \sim n^{-1/3}$ is the mean distance between ions) should be small compared with the mean kinetic energy of the ions ($\sim T$). Thus we must have $(ze)^2 n^{1/3} \ll T$ or

$$n \ll (T/z^2 e^2)^3. \tag{78.2}$$

Since the plasma is electrically neutral, the mean value of the Coulomb interaction energy of its particles would be zero if they were all uniformly and independently distributed in space. The first correction terms in the thermodynamic quantities for a plasma (in comparison with the values for an ideal gas) therefore arise only when the correlation between the positions of the various particles is taken into account. As a reminder of this, we shall call these the *correlation corrections*.

Let us first determine the correction E_{corr} to the plasma energy. As we know from electrostatics, the electrical interaction energy of a system of charged particles can be written as half the sum of the products of each charge and the potential of the field at the position of that charge due to all the other charges. In the present case

$$E_{\text{corr}} = V \cdot \tfrac{1}{2} \sum_a e z_a n_{a0} \phi_a, \tag{78.3}$$

where ϕ_a is the potential of the field acting on an ion of the ath kind due to the other charges. To calculate these potentials we proceed as follows.[†]

Each ion creates around itself a non-uniformly charged *ion cloud*, which on average is spherically symmetrical. In other words, if we select any particular ion in the gas and consider the density of distribution of the other ions relative to that ion, this density will depend only on the distance r from the centre. Let the density of distribution of ions (of the ath kind) in this ion cloud be denoted by n_a. The potential energy of each ion of the ath kind in the electric field around the ion considered is $z_a e \phi$, where ϕ is the potential of this field.

[†] This method was used by P. Debye and E. Hückel to calculate the thermodynamic quantities for strong electrolytes (1923).

Hence Boltzmann's formula (38.6) gives

$$n_a = n_{a0}e^{-z_a e\phi/T}.$$ (78.4)

The constant coefficient is put equal to n_{a0}, since at a large distance from the centre (where $\phi \to 0$) the density of the ion cloud must become equal to the mean ion density in the gas.

The potential ϕ of the field in the ion cloud is related to the charge density in it (equal to $\Sigma e z_a n_a$) by the electrostatic Poisson's equation:

$$\triangle\phi = -4\pi e \sum_a z_a n_a.$$ (78.5)

Formulae (78.4) and (78.5) together give the equations of the self-consistent electric field of the electrons and ions.

With the above assumption that the interaction of the ions is relatively weak, the energy $e z_a \phi$ is small in comparison with T, and formula (78.4) may be written in the approximate form

$$n_a = n_{a0} - \frac{n_{a0} e z_a}{T}\phi.$$ (78.6)

Substituting this in equation (78.5) and using the condition (78.1) for the gas to be neutral as a whole, we obtain

$$\triangle\phi - \varkappa^2\phi = 0,$$ (78.7)

where

$$\varkappa^2 = \frac{4\pi e^2}{T}\sum_a n_{a0} z_a^2.$$ (78.8)

The quantity $\varkappa$ has the dimensions of reciprocal length.

The spherically symmetric solution of equation (78.7) is $\phi = \text{constant}\times e^{-\varkappa r}/r$. In the immediate neighbourhood of the centre, the field must become the purely Coulomb field of the charge considered; this charge will be denoted by $z_b e$. In other words, for sufficiently small r we must have $\phi \cong e z_b/r$. This shows that the constant must be taken as $z_b e$, and so the required potential distribution is given by

$$\phi = e z_b e^{-\varkappa r}/r.$$ (78.9)

Hence we see, incidentally, that the field becomes very small at distances large compared with $1/\varkappa$. The length $1/\varkappa$ can therefore be regarded as determining the dimensions of the ion cloud due to a given ion; it is also called the *Debye length*. All the calculations given here assume, of course, that this length is large in comparison with the mean distances between ions. This condition is clearly identical with (78.2).

Expanding the potential (78.9) in series for small $\varkappa r$, we have

$$\phi = \frac{ez_b}{r} - ez_b\varkappa + \ \ldots \ .$$

The terms omitted vanish when $r = 0$. The first term is the Coulomb field of the ion itself; the second term is clearly the potential produced by all the other ions in the cloud at the point occupied by the ion considered, and is the quantity to be substituted in formula (78.3): $\phi_a = -ez_a\varkappa$.

Thus we have the following expression for the correlation part of the plasma energy:

$$E_{\text{corr}} = -\frac{1}{2} V\varkappa e^2 \sum_a n_{a0}z_a^2 = -Ve^3 \sqrt{\frac{\pi}{T}} \left(\sum_a n_{a0}z_a^2\right)^{3/2}, \qquad (78.10)$$

or, in terms of the total numbers of different ions in the gas $N_a = n_{a0}V$,

$$E_{\text{corr}} = -e^3 \sqrt{\frac{\pi}{TV}} \left(\sum_a N_az_a^2\right)^{3/2}. \qquad (78.11)$$

This energy is inversely proportional to the square root of the temperature and to that of the volume of the gas.

Integrating the thermodynamic relation $E/T^2 = -(\partial/\partial T)(F/T)$, we can deduce from E_{corr} the corresponding change in the free energy:

$$F = F_{\text{id}} - \frac{2e^3}{3} \sqrt{\frac{\pi}{TV}} \left(\sum_a N_az_a^2\right)^{3/2}; \qquad (78.12)$$

the constant of integration must be taken as zero, since when $T \to \infty$ we must have $F = F_{\text{id}}$. Hence the pressure is

$$P = \frac{NT}{V} - \frac{e^3}{3V^{3/2}} \sqrt{\frac{\pi}{T}} \left(\sum_a N_az_a^2\right)^{3/2}, \qquad (78.13)$$

where $N = \Sigma N_a$. The thermodynamic potential Φ can be obtained from F by means of the theorem of small increments (as in § 74), i.e. by regarding the second term in (78.12) as a small increment to F_{id} and expressing it with the appropriate accuracy in terms of P and T:[†]

$$\Phi = \Phi_{\text{id}} - \frac{2e^3}{3T} \left(\frac{\pi P}{N}\right)^{1/2} \left(\sum_a N_az_a^2\right)^{3/2}. \qquad (78.14)$$

[†] This method was not available for going from (78.11) to (78.12), since the energy (78.11) was not expressed in terms of the requisite variables S and V.

§ 79. The method of correlation functions

The advantage of the Debye–Hückel method described in § 78 lies in its simplicity and physical clarity. Its basic drawback, however, is that it cannot be generalised to calculate further approximations with respect to the concentration. We shall therefore also give a brief description of another method (proposed by N. N. Bogolyubov, 1946), which, though more complicated, allows in principle the calculation of further terms in the expansion of the thermodynamic quantities.

This method is based on a consideration of what are called *correlation functions* between the simultaneous positions of several particles at given points in space. The simplest and most important of these is the binary correlation function w_{ab}, which is proportional to the probability of finding two particles (ions) simultaneously at given points r_a and r_b; the ions a and b may be of either the same or different kinds. Because the gas is isotropic and homogeneous, this function of course depends only on $r = |r_a - r_b|$. We choose the normalisation coefficient in the function w_{ab} such that this function tends to unity as $r \to \infty$.

If the function w_{ab} is known, the required energy E_{corr} can be obtained by integration, using the obvious formula[†]

$$E_{corr} = \frac{1}{2V^2} \sum_{a,b} N_a N_b \iint u_{ab} w_{ab} \, dV_a \, dV_b, \qquad (79.1)$$

where the summation is over all the kinds of ions, and u_{ab} is the Coulomb interaction energy of a pair of ions at distance r.

According to the Gibbs distribution formula, the function w_{ab} is given by

$$w_{ab} = \frac{1}{V^{N-2}} \int \exp \left\{ \frac{F - F_{id} - U}{T} \right\} dV_1 \, dV_2 \ldots dV_{N-2}, \qquad (79.2)$$

where U is the Coulomb interaction energy of all the ions, and the integration is over the coordinates of all the ions except the two considered. For an approximate calculation of this integral we proceed as follows.

We differentiate equation (79.2) with respect to the coordinates of ion b:

$$\frac{\partial w_{ab}}{\partial r_b} = -\frac{w_{ab}}{T} \frac{\partial u_{ab}}{\partial r_b} - \frac{1}{TV} \sum_c N_c \int \frac{\partial u_{bc}}{\partial r_b} w_{abc} \, dV_c, \qquad (79.3)$$

[†] This formula itself is, of course, independent of the Coulomb nature of the interaction between the particles, and assumes only a binary interaction.

where the summation in the last term is over all the kinds of ions, and w_{abc} is the ternary correlation function, defined by

$$w_{abc} = \frac{1}{V^{N-3}} \int \exp\left\{\frac{F-F_{id}-U}{T}\right\} dV_1\, dV_2 \ldots dV_{N-3}$$

analogously to (79.2).

Assuming the gas sufficiently rarefied and considering only the first-order terms, we can express the ternary correlation function in terms of binary correlations: neglecting the possibility that all three ions are close together, we have $w_{abc} = w_{ab} w_{bc} w_{ac}$. In the same approximation we can suppose that even the pairs of particles are not so close together that the w_{ab} are appreciably different from unity. We define the small quantities

$$\omega_{ab} = w_{ab} - 1 \tag{79.4}$$

and write

$$w_{abc} = \omega_{ab} + \omega_{bc} + \omega_{ac} + 1, \tag{79.5}$$

neglecting the higher powers of the ω_{ab}.

When this expression is substituted in the integral on the right-hand side of (79.3), only the term in ω_{ac} remains; the others give zero identically, because of the isotropy of the gas. In the first term on the right of (79.3) it is sufficient to put $w_{ab} = 1$. Thus

$$\frac{\partial \omega_{ab}}{\partial \mathbf{r}_b} = -\frac{1}{T}\frac{\partial u_{ab}}{\partial \mathbf{r}_b} - \frac{1}{TV}\sum_c N_c \int \omega_{ac} \frac{\partial u_{bc}}{\partial \mathbf{r}_b}\, dV_c.$$

We now take the divergence of both sides of this equation, using the facts that $u_{ab} = z_a z_b e^2/r$, $\mathbf{r} = \mathbf{r}_b - \mathbf{r}_a$, and the well-known formula $\triangle(1/r) = -4\pi\,\delta(\mathbf{r})$. The integration is then trivial because of the presence of the delta function, and we have

$$\triangle\omega_{ab}(\mathbf{r}) = \frac{4\pi z_a z_b e^2}{T}\,\delta(\mathbf{r}) + \frac{4\pi e^2 z_b}{TV}\sum_c N_c z_c \omega_{ac}(\mathbf{r}). \tag{79.6}$$

The solution of this system of equations can be sought in the form

$$\omega_{ab}(\mathbf{r}) = z_a z_b \omega(\mathbf{r}), \tag{79.7}$$

whereby they are reduced to a single equation

$$\triangle\omega(\mathbf{r}) - \varkappa^2\omega(\mathbf{r}) = \frac{4\pi e^2}{T}\,\delta(\mathbf{r}). \tag{79.8}$$

This final equation has the same form as equation (78.7) in Debye and Hückel's method; the term containing the delta function in (79.8) corresponds to the boundary condition as $r \to 0$ imposed on the function $\phi(r)$ in (78.7). The solution of equation (79.8) is

$$\omega(r) = -\frac{e^2}{T}\frac{e^{-\varkappa r}}{r},$$
(79.9)

and this determines the binary correlation functions in the plasma.

To calculate the energy, it is now sufficient to substitute w_{ab} from (79.4), (79.7) and (79.9) in (79.1). Changing to integration with respect to the relative coordinates of the two particles, we find

$$E_{corr} = -\frac{1}{2}V\sum_{a,b}n_a n_b\int_0^\infty \frac{z_a z_b e^2}{r}\frac{z_a z_b e^2}{Tr}e^{-\varkappa r}\cdot 4\pi r^2\,dr;$$

the term 1 in w_{ab} makes no contribution to the energy because of the electrical neutrality of the plasma. On carrying out the integration, we return to the previous result (78.11).

In the next approximation the calculations become more laborious. In particular, the assumption (79.5) is now insufficient, and ternary correlations which do not reduce to binary ones must be introduced. For these we obtain an equation analogous to (79.3) but involving quaternary correlations; in this (the second) approximation the latter reduce to ternary ones.[†]

§ 80. Thermodynamic quantities for a degenerate plasma

In the theory given in § 78, it was assumed that the plasma is far from degeneracy, i.e. obeys Boltzmann statistics. Let us now consider a situation where the plasma temperature is so low that its electron component is degenerate:

$$T \lesssim \frac{\hbar^2}{m}n^{2/3},$$
(80.1)

where m is the electron mass (cf. (57.8)); the ion component may still be far from degeneracy, because of the large mass of the ions. It will be recalled that the condition for the plasma to be only slightly non-ideal is

$$mz^{2/3}e^2/\hbar^2 n^{1/3} \ll 1$$
(80.2)

(see (57.9)); this is better satisfied at high plasma densities.

For a degenerate gas, convenient variables are (together with the temperature T and the volume V) its chemical potentials μ_a instead of the particle

---

[†] The terms of next higher order in the thermodynamic quantities for a plasma have in fact been calculated (using a different method) by A. A. Vedenov and A. I. Larkin, *Soviet Physics JETP* (36) **9**, 806, 1959.

numbers N_a.[†] Accordingly, we shall calculate Ω, the thermodynamic potential with respect to these variables. Note that the chemical potentials are not all independent variables here: they are connected by a relation that follows from the electrical neutrality of the plasma,

$$\sum_a z_a N_a = \sum_a z_a \partial\Omega/\partial\mu_a = 0. \tag{80.3}$$

The formula

$$\left(\frac{\partial\Omega}{\partial\lambda}\right)_{T, V, \mu_a} = \left\langle\frac{\partial\hat{H}}{\partial\lambda}\right\rangle$$

expresses the derivative of Ω with respect to a parameter λ in terms of the mean value of the corresponding derivative of the Hamiltonian of the system; cf. the analogous formulae (11.4), (15.11). In the present case, we take as the parameter λ the squared charge e^2. The plasma Hamiltonian contains e^2 as a common coefficient in the operator $\hat{U}$ of the Coulomb interaction of the particles. Hence

$$\left(\frac{\partial\Omega}{\partial e^2}\right)_{T, V, \mu_a} = \left\langle\frac{\partial\hat{H}}{\partial e^2}\right\rangle = \frac{1}{e^2}\langle\hat{U}\rangle, \tag{80.4}$$

so that the calculation of Ω is equivalent to that of the mean value $\langle\hat{U}\rangle$.

We shall see that, in a degenerate slightly non-ideal plasma, the main contribution to the corrections to the thermodynamic quantities for an ideal gas comes from the exchange part of the electrical interaction of the electrons (which in the classical case is unimportant and was ignored in § 78). We therefore retain in the operator $\hat{U}$ only the terms that describe the Coulomb interaction of the electrons.

The calculation of $\langle\hat{U}\rangle$ is most simply carried out by the second quantisation method. According to this (see *Quantum Mechanics*, §§ 64, 65), we define a set of normalised wave functions $\psi_{p\sigma}$ which describe the states of free electrons moving in a volume V with momenta $\mathbf{p}$ and spin components $\sigma\left(=\pm\tfrac{1}{2}\right)$. The momentum $\mathbf{p}$ takes an infinity of discrete values, the intervals between which tend to zero as $V \to \infty$. We also define the operators $\hat{a}_{p\sigma}$ and $\hat{a}_{p\sigma}^+$ of annihilation and creation of electrons in the state $\psi_{p\sigma}$, and use them to construct the ψ-operators

$$\hat{\psi} = \sum_{p, \sigma} \psi_{p\sigma}\hat{a}_{p\sigma}, \qquad \hat{\psi}^+ = \sum_{p, \sigma} \psi_{p\sigma}^*\hat{a}_{p\sigma}^+. \tag{80.5}$$

The Coulomb interaction of the particles is a binary interaction; the operator of such an interaction in the second quantisation method is written as the

---

[†] See § 85 for the definition of the chemical potentials of components of a mixture.

integral

$$\hat{U} = \frac{1}{2} \int\int \hat{\psi}^+(\mathbf{r}_1)\, \hat{\psi}^+(\mathbf{r}_2)\, \frac{e}{|\mathbf{r}_1 - \mathbf{r}_2|}\, \hat{\psi}(\mathbf{r}_2)\, \hat{\psi}(\mathbf{r}_1)\, dV_1\, dV_2. \tag{80.6}$$

The required averaging of this operator is done in two stages: first an averaging for a specified quantum state of the system and then an averaging over the equilibrium statistical distribution among the various quantum states. In a slightly non-ideal plasma, $\hat{U}$ acts as a small perturbation. We shall calculate the mean value of this quantity in the first approximation of perturbation theory, i.e. with respect to the states of a system of non-interacting particles (an ideal gas).

The quantum averaging amounts to taking the appropriate diagonal matrix element. After substitution of the ψ-operators (80.5), the operator (80.6) is a sum of terms containing various products of four creation and annihilation operators:

$$\hat{U} = \frac{1}{2} \Sigma \langle \mathbf{p}_1' \mathbf{p}_2' | U_{12} | \mathbf{p}_1 \mathbf{p}_2 \rangle \hat{a}^+_{\mathbf{p}_1' \sigma_1} \hat{a}^+_{\mathbf{p}_2' \sigma_2} \hat{a}_{\mathbf{p}_2 \sigma_2} \hat{a}_{\mathbf{p}_1 \sigma_1}, \tag{80.7}$$

where the summation is taken over all momenta and spin components, and the $\langle \mathbf{p}_1' \mathbf{p}_2' | U_{12} | \mathbf{p}_1 \mathbf{p}_2 \rangle$ are the matrix elements of the interaction energy of two electrons, $U_{12} = e^2/|\mathbf{r}_1 - \mathbf{r}_2|$; since the Coulomb interaction is independent of the spins, these matrix elements are taken for transitions without change of the electron spin component, i.e. can be calculated from the purely orbital functions

$$\psi_\mathbf{p} = \frac{1}{\sqrt{V}} e^{i\mathbf{p} \cdot \mathbf{r}/\hbar}.$$

Of all the terms in the sum (80.7), the only ones that have non-zero diagonal matrix elements are those which contain two pairs of operators $\hat{a}_{\mathbf{p}\sigma}$, $\hat{a}^+_{\mathbf{p}\sigma}$ with the same suffixes, and the product $\hat{a}^+_{\mathbf{p}\sigma}\hat{a}_{\mathbf{p}\sigma}$ is replaced simply by the occupation number of the relevant quantum state of the electrons.[†] Putting $\mathbf{p}_1 = \mathbf{p}_1'$, $\mathbf{p}_2 = \mathbf{p}_2'$, we have the terms

$$\frac{e^2}{2V^2} \sum_{\mathbf{p}_1 \neq \mathbf{p}_2} \sum_{\sigma_1, \sigma_2} n_{\mathbf{p}_1 \sigma_1} n_{\mathbf{p}_2 \sigma_2} \int \frac{dV_1\, dV_2}{|\mathbf{r}_1 - \mathbf{r}_2|}, \tag{80.8}$$

and putting $\mathbf{p}_1' = \mathbf{p}_2$, $\mathbf{p}_2' = \mathbf{p}_1$, $\sigma_1 = \sigma_2$, the terms

$$-\frac{e^2}{2V^2} \sum_{\mathbf{p}_1 \neq \mathbf{p}_2} \sum_\sigma n_{\mathbf{p}_1 \sigma} n_{\mathbf{p}_2 \sigma} \int e^{i(\mathbf{p}_1 - \mathbf{p}_2) \cdot (\mathbf{r}_1 - \mathbf{r}_2)/\hbar} \frac{dV_1\, dV_2}{|\mathbf{r}_1 - \mathbf{r}_2|}; \tag{80.9}$$

[†] The number of terms containing products of four operators with the same suffixes is infinitesimal compared with the number having two pairs of like suffixes, and they may therefore be omitted (the contribution of such terms to Ω would contain an extra factor $1/V$).

the minus sign arises from the interchange of the operators $\hat{a}^+_{\mathbf{p}_1\sigma}$ and $\hat{a}_{\mathbf{p}_2\sigma}$ that is needed to bring the product $\hat{a}^+_{\mathbf{p}_2\sigma}\hat{a}^+_{\mathbf{p}_1\sigma}\hat{a}_{\mathbf{p}_2\sigma}\hat{a}_{\mathbf{p}_1\sigma}$ to the form $\hat{a}^+_{\mathbf{p}_2\sigma}\hat{a}_{\mathbf{p}_2\sigma}\hat{a}^+_{\mathbf{p}_1\sigma}\hat{a}_{\mathbf{p}_1\sigma}$ (for fermions, these operators anticommute).

The terms (80.8) give just the energy of the direct Coulomb interaction of electrons uniformly distributed in space. As already noted in § 78, because of the electrical neutrality of the plasma these terms in fact cancel exactly the corresponding terms giving the energy of interaction of the other particles (ions) with one another and with the electrons (and for this reason the divergence of the integral in (80.8) is unimportant). The terms (80.9), which contain non-diagonal matrix elements of the Coulomb potential, express the required exchange effect.[†]

Bearing in mind that with a macroscopic volume V the electron momenta have an almost continuous sequence of values, we can change from summation over $\mathbf{p}_1$ and $\mathbf{p}_2$ to integration over $V^2\,d^3p_1\,d^3p_2/(2\pi\hbar)^6$; the limitation $\mathbf{p}_1 \neq \mathbf{p}_2$ then becomes unimportant. The integral in (80.9) is[‡]

$$V \int e^{i(\mathbf{p}_1-\mathbf{p}_2)\cdot\mathbf{r}/\hbar}\,dV/r = V\cdot 4\pi\hbar^2/(\mathbf{p}_1-\mathbf{p}_2)^2.$$

The expression (80.9) then becomes

$$-2\pi e^2 V \sum_\sigma \int\int \frac{n_{\mathbf{p}_1\sigma}n_{\mathbf{p}_2\sigma}}{(\mathbf{p}_1-\mathbf{p}_2)^2}\,\frac{d^3p_1\,d^3p_2}{(2\pi)^6\,\hbar^4}.$$

The statistical averaging of this expression is taken (in the approximation considered) over the equilibrium distribution for an ideal gas. Because the particles of an ideal gas in different quantum states are statistically independent, we then have

$$\langle n_{\mathbf{p}_1\sigma}n_{\mathbf{p}_2\sigma}\rangle = \overline{n_{\mathbf{p}_1\sigma}}\,\overline{n_{\mathbf{p}_2\sigma}};$$

the mean values on the right are given by the Fermi distribution

$$\overline{n_{\mathbf{p}\sigma}} = [e^{(\varepsilon-\mu_e)/T}+1]^{-1},$$

where μ_e is the chemical potential of the electrons. Lastly, since the resulting expression is simply proportional to e^2, according to (80.4) it gives

---

† For a better understanding of the structure of the terms (80.8) and (80.9), it may be noted that in the former the pairs of operators $\hat{a}_{\mathbf{p}\sigma}$ and $\hat{a}^+_{\mathbf{p}\sigma}$ with the same indices come from ψ-operators taken at the same point in space ($\mathbf{r}_1$ or $\mathbf{r}_2$); in (80.9) they come from ψ-operators taken at different points.

‡ Here we make use of the familiar expression for the Fourier component of the Coulomb potential:

$$\int e^{i\mathbf{k}\cdot\mathbf{r}}\,dV/r = 4\pi/k^2;$$

see the footnote to § 117, Problem 3.

immediately the required correction to the thermodynamic potential of the plasma:

$$\Omega_{\text{exch}} = -\frac{4\pi e^2}{\hbar^4} V \iint \frac{\overline{n_{p_1}}\,\overline{n_{p_2}}}{(\mathbf{p}_1-\mathbf{p}_2)^2} \frac{d^3p_1\,d^3p_2}{(2\pi)^6} \tag{80.10}$$

(E. Wigner and F. Seitz, 1934).

In the limit of a highly degenerate electron gas ($T \ll \hbar^2 n^{2/3}/m$), the distribution $\overline{n_p}$ becomes a step function ($\overline{n_p} = 1$ for $p < p_F$, $= 0$ for $p > p_F$). The calculation of the integral then gives[†]

$$\Omega_{\text{exch}} = -V e^2 p_F^4/4\pi^3\hbar^4 = -V e^2 m^2 \mu_e^2/\pi^3\hbar^4. \tag{80.11}$$

This quantity, if the chemical potential in it is expressed in terms of the number density of electrons $n_e = N_e/V$ according to (57.3), gives the correction to the free energy:

$$F_{\text{exch}} = -N_e \frac{3^{4/3}}{4\pi^{1/3}} e^2 n_e^{1/3}. \tag{80.12}$$

In the opposite limiting case of a Boltzmann gas ($\mu_e < 0$, $|\mu_e| \gg T$), a calculation from formula (80.10) gives[‡]

$$\Omega_{\text{exch}} = -V \frac{e^2 m^2 T^2}{4\pi^2\hbar^4} e^{2\mu_e/T} \tag{80.13}$$

or, expressing μ_e in terms of n_e according to (46.1a),

$$F_{\text{exch}} = -V \frac{\pi e^2 h^2 n_e^2}{2mT}. \tag{80.14}$$

† The integral

$$I = \iint \frac{d^3p_1\,d^3p_2}{(\mathbf{p}_1-\mathbf{p}_2)^2}, \qquad p_1,\,p_2 < p_F,$$

is converted by the substitutions $\mathbf{p}_1-\mathbf{p}_2 = \mathbf{q}$, $\frac{1}{2}(\mathbf{p}_1+\mathbf{p}_2) = \mathbf{s}$ into $I = \iint q^{-2}\,d^3q\,d^3s$, taken over the range $|\mathbf{s}\pm\frac{1}{2}\mathbf{q}| < p_F$. The integral $\int d^3s$ (for a given $\mathbf{q}$) is the volume between two spheres of radius p_F whose centres are at a distance q apart:

$$\int d^3s = \frac{4\pi}{3} h^2(3p_F-h), \qquad h = p_F - \frac{1}{2}q.$$

Integrating over q from 0 to $2p_F$, we get $I = 4\pi^2 p_F^4$.

‡ In this case

$$\overline{n_{p_1}}\,\overline{n_{p_2}} = \exp\left\{\frac{2\mu_e}{T} - \frac{p_1^2+p_2^2}{2mT}\right\}$$

$$= \exp\left\{\frac{2\mu_e}{T} - \frac{4s^2+q^2}{4mT}\right\}$$

and the integration over $d^3s\,d^3q$ extends to the whole of $\mathbf{q}$-space and $\mathbf{s}$-space.

When $T \sim \mu_e$, the exchange correction $F_{\text{exch}} \sim Ve^2n^{4/3}$, whereas the correlation correction found in § 78 is $F_{\text{corr}} \sim Ve^3n^{3/2}/T^{1/2}$; from the condition for slight non-ideality,

$$F_{\text{corr}}/F_{\text{exch}} \sim (e^2n^{1/3}/T)^{1/2} \ll 1,$$

i.e. the electron exchange correction is in fact the principal one. As the temperature increases, however, F_{exch} decreases more rapidly than F_{corr} (for $T \gg \mu_e$, $F_{\text{exch}} \propto T^{-1}$, but $F_{\text{corr}} \propto T^{-1/2}$). There is consequently a region in which the two contributions are of the same order of magnitude, but in this region the degeneracy of the plasma is already slight and the classical formulae (78.11)–(78.14) are therefore applicable for the correlation correction.[†]

In the foregoing discussion we have assumed that the ion component of the plasma is not degenerate and also that it is nearly ideal, i.e. that the ion interaction energy is small in comparison with the ion thermal energy: $n^{1/3}e^2 \ll T$.[‡] If the plasma density is not too large,

$$me^2/\hbar^2 \ll n^{1/3} \ll e^2M/\hbar^2, \tag{80.15}$$

where M is the ion mass, the temperature $T \sim n^{1/3}e^2$ exceeds the ion degeneracy temperature:

$$T \sim e^2n^{1/3} \gg \hbar^2n^{2/3}/M \tag{80.16}$$

(and $T \ll e^4M/\hbar^2$). Under these conditions, the ion component forms a non-degenerate but far from ideal system. Then an ordered configuration of the nuclei corresponds to the minimum interaction energy of the ions with one another and with the electrons, i.e. the nuclei form a crystal lattice (A. A. Abrikosov, 1960). This leads to the result that the energies of the direct Coulomb interaction of various particles do not completely balance. In each lattice cell, the ion field is balanced by the electrons in the cell, but the interaction energy of the particles within one cell (dimensions $\sim n^{-1/3}$) is not zero. A rough estimate of this energy is $\sim e^2n^{1/3}$, and for the whole lattice (with $N \sim Vn$ cells) the binding energy is

$$|E_{\text{latt}}| \sim Ne^2n^{1/3} \sim Ve^2n^{4/3}. \tag{80.17}$$

This agrees in order of magnitude with the exchange energy of the degenerate electron component of the plasma. For a stable lattice the binding energy is, of course, negative.[§]

[†] The question of calculating the correlation correction for any degree of degeneracy of the electrons has some methodological interest, nevertheless. This will be discussed in Part 2.

[‡] In this and the following estimates we put for simplicity $z = 1$ (a hydrogen plasma).

[§] For a quantitative calculation of the lattice binding energy see A. A. Abrikosov, *Soviet Physics JETP* **12**, 1254, 1961.

CHAPTER VIII

PHASE EQUILIBRIUM

§ 81. Conditions of phase equilibrium

THE (equilibrium) state of a homogeneous body is determined by specifying any two thermodynamic quantities, for example the volume V and the energy E. There is, however, no reason to suppose that for every given pair of values of V and E the state of the body corresponding to thermal equilibrium will be homogeneous. It may be that for a given volume and energy in thermal equilibrium the body is not homogeneous, but separates into two homogeneous parts in contact which are in different states.

Such states of matter which can exist simultaneously in equilibrium with one another and in contact are described as different *phases*.

Let us write down the conditions for equilibrium between two phases. First of all, as for any bodies in equilibrium, the temperatures T_1 and T_2 of the two phases must be equal:

$$T_1 = T_2.$$

The pressures in the two phases must also be equal:

$$P_1 = P_2,$$

since the forces exerted by the two phases on each other at their surface of contact must be equal and opposite. Finally, the chemical potentials of the two phases must be equal:

$$\mu_1 = \mu_2;$$

this condition is derived for the two phases in exactly the same way as in § 25 for any two adjoining regions of a body. If the potentials are expressed as functions of pressure and temperature, and the common temperature and pressure are denoted by T and P, we have

$$\mu_1(P, T) = \mu_2(P, T), \tag{81.1}$$

whence the pressure and temperature of phases in equilibrium can be expressed as functions of each other. Thus two phases can not be in equilibrium

251

with each other at all pressures and temperatures; when one of these is given, the other is completely determined.

If the pressure and temperature are plotted as coordinates, the points at which phase equilibrium is possible will lie on a curve (the phase equilibrium curve), and the points lying on either side of the curve will represent homogeneous states of the body. When the state of the body varies along a line which intersects the equilibrium curve, the phases separate at the point of intersection and the body then changes to the other phase. It may be noted that, when the state of the body changes slowly, it may sometimes remain homogeneous even when the phases should separate in complete equilibrium. Examples are supercooled vapours and superheated liquids, but such states are only metastable.

Fig. 12

If the equilibrium of phases is plotted in a diagram with temperature and volume (of a fixed quantity of matter) as coordinates, then the states in which two phases exist simultaneously will occupy a whole region of the plane, and not simply a curve. This difference from the (P, T) diagram arises because the volume V, unlike the pressure, is not the same for the two phases. The resulting diagram is of the kind shown in Fig. 12. Points in the regions I and II on either side of the hatched area correspond to homogeneous first and second phases. The hatched area represents states in which the two phases are in equilibrium: at any point a the phases I and II are in equilibrium, with specific volumes given by the abscissae of the points 1 and 2 which lie on a horizontal line through a. It is easily deduced directly from the mass balance that the quantities of phases I and II are inversely proportional to the lengths of the segments $a1$ and $a2$; this is called the *lever rule*.

In a similar way to the conditions for equilibrium of two phases, the equilibrium of three phases of the same substance is governed by the equations

$$P_1 = P_2 = P_3, \quad T_1 = T_2 = T_3, \quad \mu_1 = \mu_2 = \mu_3. \tag{81.2}$$

If the common values of the pressure and temperature of the three phases are
again denoted by P and T, we have the conditions

$$\mu_1(P, T) = \mu_2(P, T) = \mu_3(P, T). \tag{81.3}$$

[handwritten: ⟹ P,T specified only one point]

These give two equations in the two unknowns P and T, and their solutions
are specific pairs of values of P and T. The states in which three phases are
simultaneously present (called *triple points*) in the (P, T) diagram are represented by isolated points which are the points of intersection of the equilibrium curves of each pair of phases (Fig. 13, where regions I, II, III are those
of the three homogeneous phases). The equilibrium of more than three phases of the same substance is clearly impossible.

FIG. 13   FIG. 14

[handwritten margin notes: equill. does not imply homogenity; All P's = ; All T's = ; heat evolved δω; L≷D]

In the (T, V) diagram the neighbourhood of the triple point has the appearance shown in Fig. 14, where the hatched areas are those of equilibrium of
two phases; the specific volumes of the three phases in equilibrium at the
triple point (at the temperature T_{tr}) are given by the abscissae of the points
1, 2, 3.

The change from one phase to another is accompanied by the evolution or
absorption of a certain quantity of heat, called the *latent heat of transition*
or simply the *heat of transition*. According to the conditions of equilibrium
such a transition occurs at constant pressure and temperature. But in a process occurring at constant pressure the quantity of heat absorbed by the body
is equal to the change in its heat function. The heat of transition q per molecule is therefore

$$q = w_2 - w_1, \tag{81.4}$$

where w_1 and w_2 are the heat functions per molecule of the two phases. The
quantity q is positive if heat is absorbed by the body in changing from the
first to the second phase, and negative if heat is evolved.

Since, for bodies consisting of a single substance, μ is the thermodynamic
potential per molecule, we can write $\mu = \varepsilon - Ts + Pv$ (where ε, s, v are the

[handwritten: $\mu_1 = \mu_2$; $\frac{\Phi_1}{N_1} = \frac{\Phi_2}{N_2} \Rightarrow \frac{E_1 - TS_1 + PV_1}{N_1} = \frac{E_2 - TS_2 + PV_2}{N_2}$; next page]

molecular energy, entropy and volume). The condition $\mu_1 = \mu_2$ therefore gives

$$(\varepsilon_2 - \varepsilon_1) - T(s_2 - s_1) + P(v_2 - v_1) = (w_2 - w_1) - T(s_2 - s_1) = 0,$$

where T and P are the temperature and pressure of both phases; hence

$$q = T(s_2 - s_1) \tag{81.5}$$

We may note that this formula also follows directly from $q = \int T\, ds$ with T constant; the latter formula is applicable here, since the transition is reversible: the two phases remain in equilibrium during the transition process.

FIG. 15

Let the two curves in Fig. 15 represent the chemical potentials of the two phases as functions of temperature (at a given pressure). The point of intersection of the two curves gives the temperature T_0 for which (at the given pressure) the two phases can exist in equilibrium. At any other temperature only one or the other phase can exist. It is easy to see that at temperatures below T_0 the first phase exists, i.e. is stable, and at temperatures above T_0 the second phase. This follows because the stable state is the one where μ is smaller, since the thermodynamic potential tends to a minimum for given P and T. At the point of intersection of the two curves, the derivative $\partial \mu_1/\partial T$ is greater than $\partial \mu_2/\partial T$, i.e. the entropy of the first phase, $s_1 = -\partial \mu_1/\partial T$, is less than that of the second phase, $s_2 = -\partial \mu_2/\partial T$. The heat of transition $q = T(s_2 - s_1)$ is therefore positive. Thus we conclude that, if the body goes from one phase to another when the temperature is raised, heat is absorbed in the process. This result could also be derived from Le Chatelier's principle.

PROBLEMS

PROBLEM 1. Determine the temperature dependence of the saturated vapour pressure above a solid. The vapour is regarded as an ideal gas, and both the gas and the solid have constant specific heats.

SOLUTION. The chemical potential of the vapour is given by formula (43.3) and that of the solid by (65.6); since the saturated vapour pressure is relatively small, the quantity PV may be neglected for the solid, and Φ taken as equal to F. Equating the two expressions, we find

$$P = \text{constant} \times T^{(c_{p2}-c_1)} e^{(\varepsilon_{01}-\varepsilon_{02})/T}$$

where the suffix 1 refers to the solid and 2 to the vapour.

In the same approximation, the heat function of the solid may be taken as equal to its energy; the heat of transition (heat of sublimation) $q = w_2 - w_1$ is

$$q = (c_{p2}-c_1)T + (\varepsilon_{02}-\varepsilon_{01}).$$

In particular, for $T = 0$ the heat of transition is $q_0 = \varepsilon_{02}-\varepsilon_{01}$, so that we can write

$$P = \text{constant} \times T^{(c_{p2}-c_1)} e^{-q_0/T}.$$

PROBLEM 2. Determine the rate of evaporation from a condensed state into a vacuum.

SOLUTION. The rate of evaporation into a vacuum is determined by the number of particles which leave unit surface area of the body per unit time. Let us consider a body in equilibrium with its saturated vapour. Then the number of particles leaving the surface is equal to the number which strike and "adhere to" this surface in the same time, i.e. $P_0(1-R)/\sqrt{(2\pi mT)}$, where $P_0 = P_0(T)$ is the saturated vapour pressure, and R a mean reflection coefficient for gas particles colliding with the surface (see (39.2)). If P_0 is not too large, the number of particles leaving the surface of the body is independent of whether there is vapour in the surrounding space, so that the above expression gives the required rate of evaporation into a vacuum.

§ 82. The Clapeyron–Clausius formula

Let us differentiate both sides of the equilibrium condition $\mu_1(P, T) = \mu_2(P, T)$ with respect to temperature, bearing in mind, of course, that the pressure P is not an independent variable but a function of temperature determined by this same equation. We therefore write

$$\frac{\partial \mu_1}{\partial T} + \frac{\partial \mu_1}{\partial P}\frac{dP}{dT} = \frac{\partial \mu_2}{\partial T} + \frac{\partial \mu_2}{\partial P}\frac{dP}{dT};$$

since $(\partial \mu/\partial T)_P = -s$, $(\partial \mu/\partial P)_T = v$ (see (24.12)), this gives

$$\frac{dP}{dT} = \frac{s_1 - s_2}{v_1 - v_2}, \tag{82.1}$$

where s_1, v_1, s_2, v_2 are the molecular entropies and volumes of the two phases.

In this formula the difference $s_1 - s_2$ may conveniently be expressed in terms of the heat of transition from one phase to the other. Substituting

$q = T(s_2 - s_1)$, we obtain the *Clapeyron–Clausius formula*:

$$\frac{(T)}{(T)}\left(\frac{s_1 - s_2}{N_i - N_2}\right) = \frac{dP}{dT} = \frac{q}{T(v_2 - v_1)}. \qquad (82.2)$$

but $q = q(T)$

This gives the change in the pressure of phases in equilibrium when the temperature changes or, in other words, the change in pressure with temperature along the phase equilibrium curve. The same formula written as

$$\frac{dT}{dP} = \frac{T(v_2 - v_1)}{q}$$

gives the change in the temperature of the transition between phases (e.g. freezing point or boiling point) when the pressure changes. Since the molecular volume of the gas is always greater than that of the liquid, and heat is absorbed in the passage from liquid to vapour, it follows that the boiling point always rises when the pressure increases ($dT/dP > 0$). The freezing point may rise or fall with increasing pressure, according as the volume increases or decreases on melting.[†]

All these consequences of formula (82.2) are in full agreement with Le Chatelier's principle. Let us consider, for example, a liquid in equilibrium with its saturated vapour. If the pressure is increased, the boiling point must rise, and so some of the vapour will become liquid; this in turn will cause a decrease in pressure, so that the system acts as if to oppose the interaction which disturbs its equilibrium.

Let us consider the particular case of formula (82.2) which relates to equilibrium between a solid or liquid and its vapour. Then formula (82.2) determines the change in the saturated vapour pressure with temperature.

The volume of a gas is usually much greater than that of a liquid or solid containing the same number of particles. We can therefore neglect the volume v_1 in (82.2) in comparison with v_2 (the second phase being taken to be a gas), i.e. write $dP/dT = q/Tv_2$. Regarding the vapour as an ideal gas, we can express its volume in terms of the pressure and temperature by $v_2 = T/P$; then $dP/dT = qP/T^2$, or

$$d\log P/dT = q/T^2. \qquad (82.3)$$

We may note that, in temperature intervals over which the heat of transition may be regarded as constant, the saturated vapour pressure varies exponentially with the temperature ($\propto e^{-q/T}$).

[†] The liquid helium isotope He^3 is an exception. Its heat of fusion is negative over a certain temperature range.

(margin handwritten notes)

$\mu_1 = \mu_2$

$d\mu_1 = d\mu_2$

$\frac{\partial\mu_1}{\partial T}dT + \frac{\partial\mu_1}{\partial P}dP = \frac{\partial\mu_2}{\partial T}dT + \frac{\partial\mu_2}{\partial P}dP$

$-s_1\,dT + N_1\,dP = -s_2\,dT + N_2\,dP$

water

$N_L < N_S$

other stuff

$N_S > N_L$

PROBLEMS

PROBLEM 1. Determine the specific heat of a vapour along the equilibrium curve of the liquid and its saturated vapour (i.e. the specific heat for a process in which the liquid is always in equilibrium with its saturated vapour). The vapour is regarded as an ideal gas.

SOLUTION. The required specific heat $h = T\,ds/dT$, where ds/dT is the derivative along the equilibrium curve:

$$h = T\frac{ds}{dT} = T\left(\frac{\partial s}{\partial T}\right)_P + T\left(\frac{\partial s}{\partial P}\right)_T \frac{dP}{dT} = c_p - T\left(\frac{\partial v}{\partial T}\right)_P \frac{dP}{dT}.$$

Substituting the expression given by (82.3) for dP/dT, and $v = T/P$, we find

$$h = c_p - q/T.$$

At low temperatures, h is negative, i.e. if heat is removed in such a way that the vapour is always in equilibrium with the liquid, its temperature can increase.

PROBLEM 2. Determine the change in the volume of a vapour with temperature in a process where the vapour is always in equilibrium with the liquid (i.e. along the equilibrium curve of the liquid and its vapour).

SOLUTION. We have to determine the derivative dv/dT along the equilibrium curve:

$$\frac{dv}{dT} = \left(\frac{\partial v}{\partial T}\right)_P + \left(\frac{\partial v}{\partial P}\right)_T \frac{dP}{dT}.$$

Substituting from (82.3), and $v = T/P$, we find

$$\frac{dv}{dT} = \frac{1}{P}\left(1 - \frac{q}{T}\right).$$

At low temperatures $dv/dT < 0$, i.e. the vapour volume decreases with increasing temperature in the process considered.

§ 83. The critical point

The phase equilibrium curve (in the PT-plane) may terminate at a certain point (Fig. 16), called the *critical point*; the corresponding temperature and pressure are the *critical temperature* and the *critical pressure*. At temperatures above T_c and pressures higher than P_c, no difference of phases exists, the substance is always homogeneous, and we can say that at the critical point the two phases become identical. The concept of the critical point was first used by D. I. Mendeleev (1860).

In the coordinates T, V, when there is a critical point, the equilibrium diagram appears as in Fig. 17. As the temperature approaches its critical value, the specific volumes of the phases in equilibrium become closer, and at the critical point (K in Fig. 17) they coincide. The diagram in the coordinates P, V has a similar form.

$$\left(V_{II} - V_{I}\right) \xrightarrow{V} 0$$

FIG. 16

FIG. 17

When there is a critical point, a continuous transition can be effected between any two states of the substance without its ever separating into two phases. To achieve this, the state must be varied along a curve which passes round the critical point and nowhere intersects the equilibrium curve. In this sense, when there is a critical point, the concept of different phases is itself arbitrary, and it is not possible to say in every case which states have one phase and which have the other. Strictly speaking, there can be said to be two phases only when they exist simultaneously and in contact—that is, at points lying on the equilibrium curve.

It is clear that the critical point can exist only for phases such that the difference between them is purely quantitative, for example a liquid and a gas differing only in the degree of interaction between the molecules.

On the other hand, such phases as a liquid and a solid (crystal), or different crystal modifications of a substance, are qualitatively different, since they have different internal symmetry. It is clear that we can say only that a particular symmetry property (symmetry element) exists or does not exist; it can appear or disappear only as a whole, not gradually. In each state the body will have one symmetry or the other, and so we can always say to which of the two phases it belongs. The critical point therefore cannot exist for such phases, and the equilibrium curve must either go to infinity or terminate by intersecting the equilibrium curves of other phases.

An ordinary phase transition point is not a mathematical singularity of the thermodynamic quantities of the substance. For each of the phases can exist (though in a metastable state) beyond the transition point; the thermodynamic inequalities are not violated at that point. At the transition point the chemical potentials of the two phases are equal: $\mu_1(P, T) = \mu_2(P, T)$; but this point has no special property with respect to either one of the functions $\mu_1(P, T)$ and $\mu_2(P, T)$.[†]

Let us plot in the *PV*-plane an *isotherm* of the liquid and gas, i.e. the curve of *P* as a function of *V* in an isothermal expansion of a homogeneous body (*abc* and *def* in Fig. 18). According to the thermodynamic inequality $(\partial P/\partial V)_T < 0$,

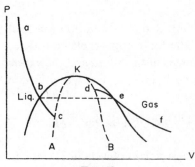

FIG. 18

P is a decreasing function of *V*. This slope of the isotherms must continue for some distance beyond their intersections (*b* and *e*) with the liquid–gas equilibrium curve; the segments *bc* and *ed* of the isotherms correspond to metastable superheated liquid and supercooled vapour, in which the thermodynamic inequalities are still satisfied. (A complete-equilibrium isothermal change of state between the points *b* and *e* corresponds, of course, to the horizontal segment *be*, on which separation into two phases occurs.) If we use the fact that the points *b* and *e* have the same ordinate *P*, it is clear that the two parts of the isotherm cannot pass continuously into each other; there must be a discontinuity between them. The isotherms terminate at points (*c* and *d*) where the thermodynamic inequality ceases to hold, i.e.

$$(\partial P/\partial V)_T = 0. \tag{83.1}$$

By constructing the locus of the points of termination of the isotherms of the liquid and gas, we obtain a curve *AKB* on which the thermodynamic inequalities are violated (for a homogeneous body), and which is the boundary of a region in which the body can never exist in a homogeneous state. The regions

---

[†] It must be noted, however, that there is some degree of arbitrariness in these statements, due to an indeterminateness of $\mu(P, T)$ in the region of metastability. The metastable state is one of partial equilibrium, having a certain relaxation time, in this case for the process of formation of nuclei of a new phase (see § 162). The thermodynamic functions in such a state can therefore be defined only without taking account of these processes, and they cannot be regarded as the analytic continuation of the functions from the region of stability corresponding to the complete equilibrium states of the substance.

between this curve and the phase equilibrium curve correspond to super-heated liquid and supercooled vapour.[†] It is evident that at the critical point the two curves must touch.

Of the points lying on the curve *AKB* itself, only the critical point *K* corresponds to an actually existing state of the homogeneous body; this is the only point where the curve reaches the region of stable homogeneous states.

It is worth mentioning that the condition (83.1) at the critical point can also be derived from the following simple considerations. Near the critical point, the specific volumes of the liquid and the vapour are almost the same; denoting them by V and $V+\delta V$, we can write the condition for equal pressures of the two phases as

$$P(V, T) = P(V+\delta V, T).\tag{83.2}$$

Expanding the right-hand side in powers of δV and dividing by the small but finite quantity δV, we have

$$\left(\frac{\partial P}{\partial V}\right)_T + \frac{1}{2}\,\delta V\left(\frac{\partial^2 P}{\partial V^2}\right)_T + \ldots = 0.\tag{83.3}$$

Hence we see that, when δV tends to zero, i.e. at the critical point, $(\partial P/\partial V)_T$ must tend to zero.

In contrast to the ordinary phase-equilibrium points, the critical point is a mathematical singularity of the thermodynamic functions of the substance (and the same applies to the whole curve *AKB* which bounds the region where homogeneous states of the body exist). The nature of this singularity and the behaviour of matter near the critical point will be discussed in § 153.

§ 84. The law of corresponding states

Van der Waals' interpolation formula for the equation of state,

$$P = \frac{NT}{V-Nb} - \frac{N^2 a}{V^2},\tag{84.1}$$

[†] The segment of the isotherm that corresponds to superheated liquid (*bc* in Fig. 18) may lie partly below the abscissa axis. Thus a superheated liquid may have a negative pressure. Such a liquid exerts an inward force on its boundary surface. Thus the pressure is not necessarily positive, and there can exist in Nature states (though only metastable ones) of a body with negative pressures, as already mentioned in § 12.

is in qualitative agreement with the properties of the liquid–vapour transition which have been described in the preceding sections. The isotherms determined by this equation are shown in Fig. 19. Curves passing above the critical point K represent monotonically decreasing functions $P(V)$ for $T > T_c$. The isotherm that passes through the critical point has an inflexion there. At temperatures $T < T_c$, each isotherm has a minimum and a maximum, and between them is a section with $(\partial P/\partial V)_T > 0$; these sections (shown by the broken lines in Fig. 19) do not correspond to any homogeneous states of matter that actually exist in Nature.

FIG. 19

As has been explained in § 83, a straight horizontal segment intersecting the isotherm corresponds to the equilibrium liquid–gas transition. The level at which this segment must be drawn is determined by the phase equilibrium condition $\mu_1 = \mu_2$, which we write as

$$\int_1^2 d\mu = 0,$$

the integral being taken along the path of the transition from a state of one phase to a state of the other phase. Integrating along the isotherm, we have $d\mu = v \, dP$, and hence

$$\int_1^2 V \, dP = 0. \tag{84.2}$$

Geometrically, this condition signifies that the areas shown hatched in Fig. 19 for one isotherm are equal (*Maxwell's rule*).

The critical temperature, critical pressure and critical volume can be expressed in terms of the parameters a and b which appear in van der Waals' equation. To do this, we differentiate (84.1) and put

$$\left(\frac{\partial P}{\partial V}\right)_T = 0,$$

$$\left(\frac{\partial^2 P}{\partial V^2}\right)_T = 0,$$

which determine the point of inflexion on the isotherm. These equations with (84.1) give

$$T_c = \frac{8}{27}\frac{a}{b}, \quad V_c = 3Nb, \quad P_c = \frac{1}{27}\frac{a}{b^2}. \tag{84.3}$$

We now use the *reduced* temperature, pressure and volume,

$$T' = T/T_c, \quad P' = P/P_c, \quad V' = V/V_c. \tag{84.4}$$

In terms of these quantities, van der Waals' equation becomes

$$\left(P' + \frac{3}{V'^2}\right)(3V' - 1) = 8T'. \tag{84.5}$$

It contains only V', P' and T', and not quantities pertaining to a given substance. Equation (84.5) is therefore the equation of state for all bodies to which van der Waals' equation is applicable. The states of two bodies for which their values of T', P', V' are equal are called *corresponding states* (clearly the critical states of all bodies are corresponding states). It follows from (84.5) that, if two bodies have equal values of two of the three quantities T', P', V', then the values of the third quantity are also equal, i.e. they are in corresponding states (the *law of corresponding states*).

The "reduced" isotherms $P' = P'(V')$ given by equation (84.5) are the same for all substances. The positions of the straight segments which give the liquid–gas transition points are therefore also the same. We can therefore conclude that for equal reduced temperatures the following quantities must be the same for all substances: (1) the reduced saturated vapour pressure, (2) the reduced specific volume of the saturated vapour, (3) the reduced specific volume of the liquid in equilibrium with the saturated vapour.

The law of corresponding states can also be applied to the heat of transition from the liquid to the gaseous state. Here the "reduced heat of evaporation" must be represented by a dimensionless quantity, q/T_c. We can therefore write[†]

$$q/T_c = f(T/T_c). \tag{84.6}$$

In conclusion, we may note that the law of corresponding states does not apply only to van der Waals' equation. The parameters of a specific substance disappear when reduced quantities are used in any equation of state containing only two such parameters. The law of corresponding states, taken as a general theorem not pertaining to any specific form of the equation of state, is somewhat more accurate than van der Waals' equation, but its applicability is in general very restricted.

[†] At temperatures considerably below the critical temperature, the ratio q/T_c is approximately 10 (where q is the molecular heat of evaporation).

CHAPTER IX

SOLUTIONS

§ 85. Systems containing different particles

HITHERTO we have considered only bodies consisting of identical particles. Let us now go on to discuss systems which contain different particles. These include all kinds of mixtures of more than one substance; if the mixture contains much more of one substance than of the others, it is called a *solution* of the other substances in the predominant substance (the *solvent*).

The number of *independent components* of the system customarily signifies the number of substances whose quantities in a state of complete equilibrium can be specified arbitrarily. All the thermodynamic quantities for a system in complete equilibrium are entirely determined, for example, by the temperature, the pressure and the numbers of particles of the independent components. The number of independent components may not be the same as the total number of different substances in the system if a chemical reaction can occur between the latter; if such a system is in partial equilibrium only, the determination of its thermodynamic quantities requires, in general, a knowledge of the amounts of all the substances present in it.

It is easy to generalise the results of § 24 to bodies consisting of different substances. Firstly, all the thermodynamic quantities must be homogeneous functions of the first order in all the additive variables—the numbers of the different particles and the volume.

Next, instead of the concept of a single chemical potential of the body as the derivative of some one of its thermodynamic potentials with respect to the number of particles (§ 24), there are chemical potentials μ_i for each component of the mixture, which are the derivatives of the thermodynamic potential with respect to the numbers N_i of particles of the components. Accordingly, the sum $\sum_i \mu_i \, dN_i$ must now replace $\mu \, dN$ in all the formulae (24.5) and (24.7)–(24.9).

For example, the expression for the differential $d\Phi$ becomes

$$d\Phi = -S \, dT + V \, dP + \sum_i \mu_i \, dN_i$$

and the chemical potential is

$$\mu_i = (\partial\Phi/\partial N_i)_{P,\,T}. \tag{85.1}$$

The chemical potentials are then expressed as functions of the pressure, the temperature and the *concentrations*, i.e. the ratios of the numbers of particles of the different substances. These numbers of particles can appear in μ_i only as ratios, since Φ is a homogeneous function of the first order in the N_i, and the chemical potentials must therefore be homogeneous functions of zero order in these variables.

From the fact that Φ is a homogeneous function of the first order in the N_i, we have, using Euler's theorem,

$$\Phi = \sum_i N_i \, \partial\Phi/\partial N_i = \sum_i \mu_i N_i, \tag{85.2}$$

which is a generalisation of the formula $\Phi = N\mu$.

For the potential Ω we now have

$$\Omega = F - \Sigma \mu_i N_i$$

and hence again $\Omega = -PV$. The last formula ceases to be valid only for bodies in an external field, when the pressure in different parts of the bodies is different.

The results of § 25 can also be generalised immediately: the conditions of equilibrium for a system in an external field require the temperature and also the chemical potential of each component to be constant throughout the system:

$$\mu_i = \text{constant}. \tag{85.3}$$

Finally, the Gibbs distribution for systems consisting of different particles becomes

$$w_{nN_1N_2\ldots} = \exp\left\{\frac{\Omega + \Sigma\mu_i N_i - E_{nN_1N_2\ldots}}{T}\right\}, \tag{85.4}$$

an obvious generalisation of formula (35.2).

§ 86. The phase rule

Let us now consider a system consisting of different substances and comprising r phases in contact (each phase containing, in general, all the substances).

Let the number of independent components in the system be n. Then each phase is described by its pressure, temperature and n chemical potentials. We have seen in § 81 that the condition for equilibrium of phases consisting of identical particles is that temperature, pressure and chemical potential should be equal. It is evident that, in the general case of more than one component, the phase equilibrium condition will be that the temperature, pressure and each chemical potential are equal. Let T and P be the common temperature and pressure of the phases. In order to distinguish the chemical potentials belonging to different phases and components we shall write them with a roman index for the phase and an arabic suffix for the component. Then the phase equilibrium conditions may be written

$$
\left.
\begin{aligned}
\mu_1^{\mathrm{I}} &= \mu_1^{\mathrm{II}} = \ldots = \mu_1^{r}, \\
\mu_2^{\mathrm{I}} &= \mu_2^{\mathrm{II}} = \ldots = \mu_2^{r}, \\
&\cdots \quad \cdots \quad \cdots \quad \cdots \\
\mu_n^{\mathrm{I}} &= \mu_n^{\mathrm{II}} = \ldots = \mu_n^{r}.
\end{aligned}
\right\}
\tag{86.1}
$$

Each of these potentials is a function of $n+1$ independent variables: P, T, and $n-1$ concentrations of different components in the phase concerned (each phase contains n independent numbers of particles of different kinds, giving $n-1$ independent ratios).

The conditions (86.1) form a set of $n(r-1)$ equations. The number of unknowns is $2+r(n-1)$. If these equations have solutions, the number of equations must certainly not be greater than the number of unknowns, i.e. $n(r-1) \leqslant 2+r(n-1)$, or

$$
r \leqslant n+2. \tag{86.2}
$$

In other words, in a system consisting of n independent components, not more than $n+2$ phases can be in equilibrium simultaneously. This is called *Gibbs' phase rule*. We have seen a particular case of it in § 81: when there is one component, the number of phases that exist in contact at one time cannot exceed three.

If the number r of coexisting phases is less than $n+2$, $n+2-r$ of the variables in equations (86.1) can obviously take arbitrary values. That is, we can arbitrarily vary any $n+2-r$ variables without destroying the equilibrium; the other variables must, of course, be varied in a definite manner. The number of variables which can be arbitrarily varied without destroying the equilibrium is called the number of *thermodynamic degrees of freedom* of the system. If this is denoted by f, the phase rule may be written

$$
f = n+2-r, \tag{86.3}
$$

where f cannot, of course, be less than zero. If the number of phases has its maximum possible value $n+2$, then $f = 0$, i.e. all the variables in equations (86.1) have definite values, and none of them can be varied without destroying the equilibrium and causing one of the phases to disappear.

§ 87. Weak solutions

We shall now consider (in §§ 87–91) the thermodynamic properties of weak solutions, i.e. those in which the number of molecules of the dissolved substances (the *solutes*) is much less than the number of solvent molecules. Let us first take the case of a solution with only one solute; the generalisation to a solution with more than one solute is immediate.

Let N be the number of solvent molecules in the solution, and n the number of solute molecules. The ratio $c = n/N$ is the *concentration* of the solution, and from the above hypothesis $c \ll 1$.

Let us derive an expression for the thermodynamic potential of the solution. Let $\Phi_0(P, T, N)$ be the thermodynamic potential of the pure solvent (containing no solute). According to the formula $\Phi = N\mu$ (which is valid for pure substances) it can be written $\Phi_0 = N\mu_0(P, T)$, where $\mu_0(P, T)$ is the chemical potential of the pure solvent. Let $\alpha = \alpha(P, T, N)$ denote the small change which would occur in the thermodynamic potential if one molecule of solute were added to the solvent. Since the solution is assumed weak, the solute molecules in it are comparatively far apart, and their interaction is therefore weak. Neglecting this interaction, we can then say that the change in the thermodynamic potential when n molecules are added to the solvent is $n\alpha$. But the expression $\Phi_0 + n\alpha$ thus obtained fails to take account of the fact that all the molecules of the solute are identical. This is the expression which would be obtained from formula (31.5) if all the solute particles were regarded as different in calculating the partition function. As we know (cf. 31.7)), the partition function thus calculated must in fact be divided by $n!$.[†]

This leads to an additional term $T \log n!$ in the free energy, and therefore in the potential Φ. Thus

$$\Phi = N\mu_0(P, T) + n\alpha(P, T, N) + T \log n!.$$

Next, since n is itself a very large number, though small in comparison with N, we can write $\log n! = n \log (n/e)$ in the last term. Then

$$\Phi = N\mu_0 + n[\alpha + T \log (n/e)]$$
$$= N\mu_0 + nT \log [(n/e)e^{\alpha/T}].$$

We now take into consideration the fact that Φ must be a homogeneous function of the first order in n and N. For this to be so it is clearly necessary that $e^{\alpha/T}$ in the argument of the logarithm should be of the form $f(P, T)/N$. Thus

$$\Phi = N\mu_0 + nT \log [(n/eN) f(P, T)].$$

Defining a new function $\psi(P, T) = T \log f(P, T)$, we finally have for the thermodynamic potential of the solution the expression

$$\Phi = N\mu_0(P, T) + nT \log (n/eN) + n\psi(P, T). \qquad (87.1)$$

[†] Here we neglect quantum effects, which is always permissible for a weak solution, as it is for a sufficiently rarefied gas.

The assumption, made at the beginning of this section, about the addition of a term of the form $n\alpha$ to the potential of the pure solvent amounts essentially to an expansion in powers of n, retaining only the leading terms. The term of the next order in n is proportional to n^2, and by the homogeneity in the variables n and N it must be $n^2\beta(P, T)/2N$, where β is a function of P and T only. Thus the thermodynamic potential of a weak solution as far as the second-order terms is

$$\Phi = N\mu_0(P, T) + nT \log (n/eN) + n\psi(P, T) + (n^2/2N)\beta(P, T). \quad (87.2)$$

The generalisation of this to the case of a solution of more than one substance will obviously be

$$\Phi = N\mu_0 + \sum_i n_i T \log (n_i/eN) + \sum_i n_i\psi_i + \sum_{i,k} (n_i n_k/2N)\beta_{ik}, \quad (87.3)$$

where the n_i are the numbers of molecules of the various solutes.

From (87.1) we can easily find the chemical potentials for the solvent (μ) and the solute (μ'): the former is

$$\mu = \partial\Phi/\partial N = \mu_0 - Tn/N = \mu_0 - Tc, \quad (87.4)$$

and the latter is

$$\mu' = \partial\Phi/\partial n = T \log (n/N) + \psi = T \log c + \psi. \quad (87.5)$$

§ 88. Osmotic pressure

In this and the following sections we shall discuss some properties of solutions, again assuming them weak and therefore using the results of § 87.

Let us suppose that two solutions of the same substance in the same solvent but with different concentrations c_1 and c_2 are separated by a partition through which solvent molecules can pass but solute molecules cannot (a semi-permeable membrane). The pressures on the two sides of the membrane will then be different; the argument in § 12 to prove the equality of pressures is invalid here, because of the presence of the semi-permeable membrane. The difference between the pressures is called the *osmotic pressure*.

The condition of equilibrium between the two solutions is (apart from the equality of their temperatures) that the chemical potentials of the solvent in them should be equal. The chemical potentials of the solute need not be the same, since the semi-permeability of the membrane means that there is equilibrium only with respect to the solvent.

Denoting the pressures in the two solutions by P_1 and P_2, and using the expression (87.4), we obtain the equilibrium condition in the form

$$\mu_0(P_1, T) - c_1 T = \mu_0(P_2, T) - c_2 T. \quad (88.1)$$

The pressure difference $P_2-P_1 = \Delta P$ (i.e. the osmotic pressure) is relatively small for weak solutions, and so we can expand $\mu_0(P_2, T)$ in powers of ΔP, retaining only the first two terms:

$$\mu_0(P_2, T) = \mu_0(P_1, T) + \Delta P \cdot \partial \mu_0 / \partial P.$$

Substitution in (88.1) gives

$$\Delta P \cdot \partial \mu_0 / \partial P = (c_2 - c_1)T.$$

But $\partial \mu_0 / \partial P$ is just the molecular volume v of the pure solvent. Thus

$$\Delta P = (c_2 - c_1)T/v. \tag{88.2}$$

In particular, if there is pure solvent on one side of the membrane ($c_1 = 0$, $c_2 = c$), the osmotic pressure is

$$\Delta P = cT/v = nT/V, \tag{88.3}$$

where n is the number of solute molecules in a volume V of solvent; since the solution is weak, V is almost exactly equal to the total volume of the solution. Formula (88.3) is called *van't Hoff's formula*. It should be pointed out that this formula is applicable to weak solutions independently of the particular solvent and solute concerned, and that it resembles the equation of state of an ideal gas. The gas pressure is replaced by the osmotic pressure, the gas volume by the solution volume, and the number of particles in the gas by the number of molecules of solute.

§ 89. Solvent phases in contact

Let us consider the equilibrium of two solvent phases in contact, with a certain amount of the same substance dissolved in each. The equilibrium conditions are (apart from the equality of pressures and temperatures) the equality of the chemical potentials of the solvent and those of the solute in the two phases. Here we shall use the first condition, writing it in the form

$$\mu_0^{\mathrm{I}}(P, T) - c_{\mathrm{I}}T = \mu_0^{\mathrm{II}}(P, T) - c_{\mathrm{II}}T, \tag{89.1}$$

where c_{I}, c_{II} are the concentrations and μ_0^{I}, μ_0^{II} the chemical potentials of the two phases of the pure solvent.

It must be noted that the system considered here, consisting of two components and two phases, has two thermodynamic degrees of freedom. Of the four quantities $P, T, c_{\mathrm{I}}, c_{\mathrm{II}}$, therefore, only two may be chosen arbitrarily; if we choose P or T and one of the concentrations, for example, then the other concentration has a definite value.

If the two solvent phases contained no solute, the condition for their equilibrium would be

$$\mu_0^I(P_0, T_0) = \mu_0^{II}(P_0, T_0), \tag{89.2}$$

the temperature and pressure of both phases being denoted by T_0 and P_0.

Thus, whereas in the equilibrium of pure solvent phases the relation between pressure and temperature is given by equation (89.2), when any substance is dissolved in these phases the relation is given by equation (89.1). For weak solutions the two equations are not greatly different.

Let us now expand $\mu_0^I(P, T)$ and $\mu_0^{II}(P, T)$ in equation (89.1) in powers of $P-P_0 = \Delta P$ and $T-T_0 = \Delta T$, where P_0 and T_0 are the pressure and temperature at some point on the equilibrium curve of the pure solvent phases close to a given point P, T on the equilibrium curve of the solution phases. Retaining in the expansion only the first-order terms in ΔP and ΔT, and using (89.2), we have from (89.1)

$$\frac{\partial \mu_0^I}{\partial T} \Delta T + \frac{\partial \mu_0^I}{\partial P} \Delta P - c_I T = \frac{\partial \mu_0^{II}}{\partial T} \Delta T + \frac{\partial \mu_0^{II}}{\partial P} \Delta P - c_{II} T.$$

But $-\partial \mu_0/\partial T$ and $\partial \mu_0/\partial P$ are just the entropy s and the volume v of the pure solvent (per molecule). Adding the suffix denoting the phase, we have

$$-(s_I - s_{II}) \Delta T + (v_I - v_{II}) \Delta P = (c_I - c_{II})T. \tag{89.3}$$

According to formula (81.5), we have $(s_{II} - s_I)T = q$, where q is the latent heat of transition of the solvent from phase I to phase II. Thus (89.3) may be written

$$(q/T) \Delta T + (v_I - v_{II}) \Delta P = (c_I - c_{II})T. \tag{89.4}$$

Let us examine two particular cases of this formula. We first choose the point P_0, T_0 such that $P_0 = P$. Then ΔT will be the horizontal distance between the two curves, i.e. the change in the temperature of transition between the two phases when the solute is added, or the difference between the transition temperature T (at pressure P) when both phases are solutions and the transition temperature T_0 (at the same pressure) for the pure solvent. Since $\Delta P = 0$ here, (89.4) gives

$$\Delta T = T^2(c_I - c_{II})/q. \tag{89.5}$$

If one of the phases (I, say) is the pure solvent ($c_{II} = 0$, $c_I = c$), then

$$\Delta T = T^2 c/q. \tag{89.6}$$

This formula determines, in particular, the change in the freezing point when the solute is added, if the solute is insoluble in the solid phase; the two phases are then the liquid solution and the solid solvent, and ΔT is the difference between the temperature at which the solvent freezes out of the solution and that at which the pure solvent freezes. On freezing, heat is liberated, i.e. q is negative. Hence $\Delta T < 0$ also; i.e. if the pure solvent freezes out, the addition of solute lowers the freezing point.

The relation (89.6) also determines the change in the boiling point when the solute is added, if the solute is not volatile; the two phases are then the liquid solution and the solvent vapour, and ΔT is the difference between the temperature at which the solvent boils off from the solution and that at which the pure solvent boils. Since heat is absorbed in boiling, $q > 0$ and therefore $\Delta T > 0$, i.e. the boiling point is raised by the addition of the solute.

All these consequences of formula (89.6) are fully in accordance with Le Chatelier's principle. For example, let a liquid solution be in equilibrium with the solid solvent. If the concentration of the solution is increased, then by Le Chatelier's principle the freezing point must be lowered so that part of the solid solvent is added to the solution and the concentration is thereby lowered. The system as it were counteracts its disturbance from the equilibrium state. Similarly, if the concentration of the liquid solution in equilibrium with the solvent vapour is increased, the boiling point must be raised so that part of the vapour condenses into the solution and the concentration is lowered.

Let us now consider another particular case of formula (89.4), choosing the point P_0, T_0 so that $T = T_0$. Then ΔP is the vertical distance between the two curves, i.e. the difference between the pressure of the two solution phases in equilibrium and that of the two pure solvent phases in equilibrium (at the same temperature). Here $\Delta T = 0$, and from (89. 4) we have

$$\Delta P = T(c_{\mathrm{I}} - c_{\mathrm{II}})/(v_{\mathrm{I}} - v_{\mathrm{II}}). \tag{89.7}$$

Let us apply formula (89.7) to an equilibrium between liquid and gaseous phases. Then the volume of one phase (the liquid) may be neglected in comparison with that of the other, and (89.7) becomes

$$\Delta P = T(c_{\mathrm{I}} - c_{\mathrm{II}})/v, \tag{89.8}$$

where v is the molecular volume of the gas phase (I). Noting that $Pv = T$, and substituting to the same accuracy $P \cong P_0$ (where P_0 is the saturated vapour pressure over the pure solvent), we can write this formula as

$$\Delta P = P_0(c_{\mathrm{I}} - c_{\mathrm{II}}). \tag{89.9}$$

If the gas phase is the pure solvent vapour ($c_{\mathrm{I}} = 0$, $c_{\mathrm{II}} = c$), then (89.9) becomes

$$\Delta P/P_0 = -c, \tag{89.10}$$

where c is the concentration of the solution. This formula gives the difference between the saturated vapour pressure of the solvent over the solution (P)

and over the pure solvent (P_0). The relative decrease in the saturated vapour pressure when the solute is added is equal to the concentration of the solution (*Raoult's law*).[†]

§ 90. Equilibrium with respect to the solute

Let us now consider a system consisting of two solutions in contact, the solutions being of the same substance in different solvents (for instance, in two immiscible liquids), and their concentrations being denoted by c_1 and c_2.

The equilibrium condition for this system is that the chemical potentials of the solute in the two solutions should be equal. Using (87.5), we can write this condition in the form

$$T \log c_1 + \psi_1(p, T) = T \log c_2 + \psi_2(p, T).$$

The functions ψ_1 and ψ_2 are, of course, different for the different solvents. Hence we find

$$c_1/c_2 = e^{(\psi_2 - \psi_1)/T}. \tag{90.1}$$

The right-hand side of this equation is a function of P and T only. Thus the solute is distributed between the solvents in such a way that the ratio of concentrations is always the same (for given pressure and temperature), independently of the total quantities of the solute and solvents (the *distribution law*). The same law obviously applies to a solution of one substance in two adjacent phases of the same solvent.

Now let us consider the equilibrium between a gas (assumed ideal) and a solution of it in a liquid or solid solvent. The equilibrium condition, i.e. the equality of the chemical potentials of the pure gas and the dissolved gas, can be written (using (42.6) and (87.5)) in the form

$$T \log c + \psi(P, T) = T \log P + \chi(T),$$

whence

$$c = P e^{(\chi - \psi)/T}. \tag{90.2}$$

The function $\psi(P, T)$ describes the properties of the liquid (or solid) solution. At low pressures, the properties of a liquid depend only very slightly on the pressure. Hence the dependence of $\psi(P, T)$ on the pressure is unimportant, and we can suppose that the coefficient of P in (90.2) is a constant independent of the pressure:

$$c = P \times \text{constant}. \tag{90.3}$$

[†] It will be remembered that c denotes the molecular concentration (ratio of numbers of molecules, n/N).

Thus, when a gas dissolves, the concentration of the (weak) solution is proportional to the gas pressure (*Henry's law*).[†]

PROBLEMS

PROBLEM 1. Find the variation of concentration with height for a solution in a gravitational field.

SOLUTION. We apply the equilibrium condition (85.3) in an external field, writing it for the solute: $T \log c + \psi(P, T) + mgz = $ constant, since the potential energy of a solute molecule in the gravitational field is mgz (z being the height, and m the mass of the molecule). We differentiate this equation with respect to z, noting that the temperature is constant by one of the conditions of equilibrium:

$$\frac{T}{c}\frac{dc}{dz} + mg + \frac{\partial\psi}{\partial P}\frac{dP}{dz} = 0.$$

Since the volume of the solution is

$$\frac{\partial\Phi}{\partial P} = N\frac{\partial\mu_0}{\partial P} + n\frac{\partial\psi}{\partial P}$$

(substituting for Φ the expression (87.1)), the quantity $\partial\psi/\partial P$ may be called the volume v' per molecule of solute. Hence

$$\frac{T}{c}\frac{dc}{dz} + mg + v'\frac{dP}{dz} = 0.$$

In order to find P as a function of z, we use the equilibrium condition for the solvent:[‡]

$$v\,dP/dz + Mg = 0,$$

where $v = \partial\mu_0/\partial P$ is the molecular volume and M the mass of a solvent molecule. Substituting dP/dz in the previous condition, we find

$$\frac{T}{c}\frac{dc}{dz} + mg - Mg\frac{v'}{v} = 0.$$

If the solution may be regarded as incompressible, i.e. v and v' are constants, this gives

$$c = c_0 e^{-(gz/T)\,(m - v'M/v)},$$

where c_0 is the concentration of the solution when $z = 0$, i.e. the usual barometric formula corrected in accordance with Archimedes' principle.

[†] It is assumed that the molecules dissolve unchanged. If they dissociate (as in the dissolution of hydrogen H_2 in certain metals), the dependence of the concentration on the pressure is different; see § 102, Problem 3.

[‡] In this condition the term involving the concentration ($-T\,dc/dz$) is small and may be omitted; in the condition for the solute, it contained c in the denominator and was therefore not small.

PROBLEM 2. Find the relation between the changes in the solubilities of two sub-stances simultaneously dissolved in the same solvent.[†]

SOLUTION. The interaction between the two solutes is taken into account by the quadratic term (proportional to $n_1 n_2$) in the thermodynamic potential (87.3). The chemical potentials of the solutes are

$$\mu_1' = \partial\Phi/\partial n_1 = T\log c_1 + \psi_1 + c_1\beta_{11} + c_2\beta_{12}$$

and similarly for μ_2' (the concentrations are $c_1 = n_1/N$, $c_2 = n_2/N$). The solu-bilities c_{01} and c_{02} of each substance in the absence of the other are given by the equilibrium conditions

$$\mu_{01}' = T\log c_{01} + \psi_1 + c_{01}\beta_{11},$$
$$\mu_{02}' = T\log c_{02} + \psi_2 + c_{02}\beta_{22},$$
(1)

where μ_{01}' and μ_{02}' are the chemical potentials of the pure solutes. The joint solubilities c_{01}' and c_{02}' are given by

$$\mu_{01}' = T\log c_{01}' + \psi_1 + c_{01}'\beta_{11} + c_{02}'\beta_{12},$$
$$\mu_{02}' = T\log c_{02}' + \psi_2 + c_{02}'\beta_{22} + c_{01}'\beta_{12}.$$
(2)

Subtracting (1) from (2) term by term, and using the relative smallness of the chan-ges in the solubilities ($\delta c_{01} = c_{01}' - c_{01} \ll c_{01}$, $\delta c_{02} \ll c_{02}$), we find

$$T\,\delta c_{01}/c_{01} = -c_{02}\beta_{12}, \quad T\,\delta c_{02}/c_{02} = -c_{01}\beta_{12}.$$

Hence

$$\delta c_{01} = \delta c_{02},$$

i.e. the changes in the solubilities of the two substances are equal.

PROBLEM 3. Find the relation between the changes of saturated vapour pressure of two solutes when both are present.

SOLUTION. The saturated vapour pressures above solutions of each substance separately are given by the equilibrium conditions

$$T\log P_1 + \chi_1(T) = T\log c_1 + \psi_1 + c_1\beta_{11},$$
$$T\log P_2 + \chi_2(T) = T\log c_2 + \psi_2 + c_2\beta_{22};$$

the expressions on the left-hand side are the chemical potentials of the two sub-stances in the vapour. The pressures P_1' and P_2' above the solution of both solutes are given by

$$T\log P_1' + \chi_1 = T\log c_1 + \psi_1 + c_1\beta_{11} + c_2\beta_{12},$$
$$T\log P_2' + \chi_2 = T\log c_2 + \psi_2 + c_2\beta_{22} + c_1\beta_{12}.$$

Hence, if the changes $\delta P_1 = P_1' - P_1$ and δP_2 are small, we have

$$T\,\delta P_1/P_1 = c_2\beta_{12}, \quad T\,\delta P_2/P_2 = c_1\beta_{12}.$$

and the required relation

$$\frac{\delta P_1}{P_1} \bigg/ \frac{\delta P_2}{P_2} = \frac{c_2}{c_1}.$$

[†] The solubility is the concentration of a saturated solution. It is assumed that this concentration is still so small that the formulae of the theory of weak solu-tions are applicable.

§ 91. Evolution of heat and change of volume on dissolution

The process of dissolution is accompanied by the evolution or absorption of heat. Let us now calculate the quantity of heat involved, and first determine the maximum work which can be done as a result of the dissolution process.

Let us suppose that the dissolution occurs at constant pressure and temperature. In this case the maximum work is determined by the change in the thermodynamic potential. Let us calculate it for a process in which a small number δn of solute molecules are dissolved in a solution already of concentration c. The change $\delta\Phi$ in the total thermodynamic potential of the system is equal to the sum of the changes in the potentials of the solution and the pure solute. Since δn molecules of solute are added to the solution, the change in its thermodynamic potential is

$$\delta\Phi_{sol} = \frac{\partial\Phi_{sol}}{\partial n}\,\delta n = \mu'\,\delta n,$$

where μ' is the chemical potential of the solute in the solution. The change in the potential Φ_0' of the pure solute is

$$\delta\Phi_0' = -\frac{\partial\Phi_0'}{\partial n}\,\delta n = -\mu_0'\,\delta n,$$

since the number of molecules of it decreases by δn, μ_0' being the chemical potential of the pure solute. The total change in the thermodynamic potential in this process is therefore

$$\delta\Phi = \delta n(\mu' - \mu_0'). \tag{91.1}$$

We now substitute μ' from (87.5):

$$\delta\Phi = -T\,\delta n\log\frac{c_0(P, T)}{c}, \tag{91.2}$$

where

$$c_0(P, T) = e^{(\mu_0' - \psi)/T} \tag{91.3}$$

is the solubility, i.e. the concentration of a saturated solution (that is, one which is in equilibrium with the pure solute). This is clear from the fact that in equilibrium Φ must have a minimum, i.e. we must have $\delta\Phi = 0$. Formula (91.3) can also be derived directly from the condition for equilibrium between the solution and the pure solute, i.e. from the equality of the chemical potentials of the pure solute and that in the solution. Note, however, that c_0 may be identified with the concentration of the saturated solution only if c_0 is small, since all the formulae in the last few sections are applicable only to small concentrations.

The expression obtained gives the required quantity of work: $|\delta\Phi|$ is the maximum work which can be done by the dissolution of δn molecules, and is also the minimum work which is needed to separate δn molecules of solute from a solution of concentration c.

There is now no difficulty in calculating the heat δQ_P absorbed in dissolution at constant pressure (if $\delta Q_P < 0$, this means that heat is evolved). The quantity of heat absorbed in a process which occurs at constant pressure is equal to the change in the heat function (§ 14). Since, on the other hand,

$$W = -T^2 \left(\frac{\partial}{\partial T} \frac{\Phi}{T} \right)_P ,$$

we have[†]

$$\delta Q_P = -T^2 \left(\frac{\partial}{\partial T} \frac{\delta\Phi}{T} \right)_P . \tag{91.4}$$

Substituting the expression (91.2) in this formula, we find the required quantity of heat:

$$\delta Q_P = T^2 \, \delta n \, \partial \log c_0 / \partial T. \tag{91.5}$$

Thus the quantity of heat involved in the dissolution process is related to the temperature dependence of the solubility. We see that δQ_P is simply proportional to δn; this formula is therefore applicable also to the dissolution of any finite quantity of substance (so long as the solution remains weak, of course). The quantity of heat absorbed in the dissolution of n molecules is

$$Q_P = T^2 n \, \partial \log c_0 / \partial T. \tag{91.6}$$

We may also determine the change in volume on dissolution, i.e. the difference between the volume of the solution and the sum of the volumes of the pure solute and the solvent in which it is dissolved. Let us calculate this change δV in the dissolution of δn molecules. The volume is the derivative of the thermodynamic potential with respect to the pressure. The change in volume is therefore equal to the derivative, with respect to pressure, of the change in the thermodynamic potential:

$$\delta V = \frac{\partial}{\partial P} \, \delta\Phi. \tag{91.7}$$

[†] The corresponding formula for the quantity of heat in a process which occurs at constant volume is

$$\delta Q_V = -T^2 \left(\frac{\partial}{\partial T} \frac{\delta F}{T} \right)_V . \tag{91.4a}$$

Substituting $\delta\Phi$ from (91.2), we find

$$\delta V = -T\,\delta n\,\frac{\partial}{\partial P}\log c_0. \qquad (91.8)$$

In conclusion, it may be noted that formula (91.6) is in accordance with Le Chatelier's principle. Let us suppose, for example, that Q_p is negative, i.e. that heat is evolved on dissolution, and let us consider a saturated solution. If this is cooled, then by Le Chatelier's principle the solubility must increase so that more dissolution occurs. Heat is then evolved, i.e. the system as it were counteracts the cooling which disturbs its equilibrium. The same follows from (91.6), since in this case $\partial c_0/\partial T$ is negative. Similar arguments show that formula (91.8) is also in accordance with Le Chatelier's principle.

PROBLEMS

PROBLEM 1. Find the maximum work that can be done in the formation of a saturated solution.

SOLUTION. Before dissolution, the thermodynamic potential of the pure solvent was $N\mu_0$, and that of the pure solute $n\mu_0'$. The potential of the whole system was $\Phi_1 = N\mu_0 + n\mu_0'$. After dissolution, the thermodynamic potential $\Phi_2 = N\mu_0 + nT\log(n/eN) + n\psi$. The maximum work is

$$R_{max} = \Phi_1 - \Phi_2$$
$$= -nT\log(n/eN) + n(\mu_0' - \psi)$$
$$= nT\log(ec_0/c);$$

this may also be derived by integration of (91.2). If a saturated solution is formed, i.e. $c = c_0$ and $n = Nc = Nc_0$, then

$$R_{max} = nT = Nc_0T.$$

PROBLEM 2. Find the minimum work which must be done to raise the concentration of a solution from c_1 to c_2 by removing some of the solvent.

SOLUTION. Before the removal, the thermodynamic potential of the solution was $\Phi_1 = N\mu_0 + Nc_1T\log(c_1/e) + Nc_1\psi$ (the number of solute molecules was Nc_1, where N was the original number of solvent molecules). In order to raise the concentration of the solution to c_2, we must remove from it $N(1-c_1/c_2)$ solvent molecules. The sum of the thermodynamic potentials of the remaining solution and the solvent removed gives $\Phi_2 = N\mu_0 + Nc_1T\log(c_2/e) + Nc_1\psi$. The minimum work is

$$R_{min} = \Phi_2 - \Phi_1 = Nc_1T\log(c_2/c_1).$$

§ 92. Solutions of strong electrolytes

The method of expanding the thermodynamic quantities in powers of the concentration used in the preceding sections is completely inapplicable in the important case of solutions of *strong electrolytes*, that is, substances which dissociate almost completely into ions when dissolved. The slow decrease of the Coulomb interaction forces between ions with increasing distance leads to terms proportional to a power of the concentration lower than the second (namely, the $\frac{3}{2}$ power).

It is easy to see that the problem of determining the thermodynamic quantities of a weak solution of a strong electrolyte reduces to the problem of a completely ionised gas discussed in § 78 (P. Debye and E. Hückel, 1923). This result may be derived by starting from the fundamental statistical formula (31.5) for the free energy. The integration in the partition function will be carried out in two stages, first integrating over the coordinates and momenta of the solvent molecules. Then the partition function becomes

$$\int e^{-F(p,\,q)/T}\,\mathrm{d}\Gamma,$$

where the integration is now taken only over the phase space of the electrolyte particles, and $F(p, q)$ is the free energy of the solvent with the ions "fixed" in it, the ion coordinates and momenta being regarded as parameters. We know from electrodynamics that the free energy of a system of charges in a medium (of given volume and temperature) can be deduced from the energy of the charges in empty space by dividing the products of each pair of charges by the dielectric constant ε of the medium.[†] The second step in calculating the free energy of the solution is therefore identical with the calculations given in § 78.

Thus the required contribution of the strong electrolyte to the free energy of the solution is given, according to (78.12), by

$$-\frac{2e^3}{3\varepsilon^{3/2}}\left(\frac{\pi}{TV}\right)^{1/2}\left(\sum_a n_a z_a^2\right)^{3/2},$$

where the summation is over all the kinds of ion in the solution; in accordance with the notation used in this chapter, n_a denotes the total number of ions of the ath kind (in the whole volume of the solution). The same expression gives the contribution to the thermodynamic potential for given temperature and pressure. Putting $V \cong Nv$, where $v(P, T)$ is the molecular volume

---

[†] This assumes that the distances between ions are large compared with molecular dimensions, but we know from § 78 that in the approximation considered the main contribution to the thermodynamic quantities comes in fact from these distances.

of the solvent, we can write the thermodynamic potential of the solution in the form

$$\Phi = N\mu_0 + \sum_a n_a T \log{(n_a/eN)} + \sum_a n_a \psi_a$$

$$- \frac{2e^3}{3\varepsilon^{3/2}} \left(\frac{\pi}{vT}\right)^{1/2} \left(\frac{\sum n_a z_a^2}{N}\right)^{3/2}. \tag{92.1}$$

From this we can find, by the usual rules, any of the thermodynamic properties of the electrolyte solution. For example, to calculate the osmotic pressure we write chemical potential of the solvent as

$$\mu = \mu_0 - \frac{T}{N}\sum_a n_a + \frac{e^3}{3\varepsilon^{3/2}} \left(\frac{\pi}{vT}\right)^{1/2} \left(\frac{\sum n_a z_a^2}{N}\right)^{3/2}. \tag{92.2}$$

As in § 88, we find from this the osmotic pressure (at an interface with the pure solvent)

$$\Delta P = \frac{T}{V}\sum_a n_a - \frac{e^3}{3\varepsilon^{3/2}} \left(\frac{\pi}{T}\right)^{1/2} \left(\frac{\sum n_a z_a^2}{V}\right)^{3/2}. \tag{92.3}$$

The heat function of the solution is

$$W = -T^2\left(\frac{\partial}{\partial T}\frac{\Phi}{T}\right)_P = Nw_0 - T^2\sum_a n_a \frac{\partial}{\partial T}\frac{\psi_a}{T}$$

$$+ \frac{2e^3}{3}\left(\frac{\pi}{N}\right)^{1/2}(\sum n_a z_a^2)^{3/2} T^2 \frac{\partial}{\partial T}\left(\frac{1}{\varepsilon^{3/2}T^{3/2}v^{1/2}}\right). \tag{92.4}$$

From this we can find the *heat of solution Q* which is liberated when the solution is diluted (at constant P and T) with a very large amount of solvent (so that the concentration tends to zero). This quantity of heat is given by the change in the heat function during the process. The terms linear in the number of particles obviously give zero difference, and we find from (92.4)

$$Q = \frac{2e^3\pi^{1/2}}{3}N\left(\frac{\sum n_a z_a^2}{N}\right)^{3/2} T^2 \frac{\partial}{\partial T}\left(\frac{1}{\varepsilon^{3/2}T^{3/2}v^{1/2}}\right). \tag{92.5}$$

The only condition for the above formulae to be valid is that the concentration should be sufficiently small. For the fact that the electrolyte is strong means that the energy of attraction between ions of different kinds is always less than T. Hence it follows that the interaction energy is certainly small compared with T at distances large compared with molecular distances. But the condition $n \ll N$ for the solution to be weak means precisely that the mean distance between ions is large in comparison with molecular dimensions. Thus this condition necessarily implies that the condition of weak interaction

$$n/V \ll (\varepsilon T/z^2 e^2)^3$$

(cf. (78.2)) is satisfied, and this is the basis of the approximations used in § 78.

PROBLEM

Find the change in the solubility (assumed small) of a strong electrolyte when a certain quantity of another electrolyte is added to the solution (all the ions of the second electrolyte being different from those of the first).

SOLUTION. The solubility (i.e. the concentration of a saturated solution) of the strong electrolyte is given by the equation

$$\mu_s(P, T) = \sum_a \nu_a \mu_a = T \sum_a \nu_a \log(n_a/N) + \sum_a \nu_a \psi_a$$

$$- \frac{e^3}{\varepsilon^{3/2}} \left(\frac{\pi}{N v T}\right)^{1/2} \left(\sum_a \nu_a z_a^2\right) \left(\sum_b n_b z_b^2\right)^{1/2}. \tag{1}$$

Here μ_s is the chemical potential of the pure solid electrolyte, and ν_a the number of ions of the ath kind per molecule of the electrolyte. When other ions are added to the solution, the chemical potentials of the original ions are changed because of the change in the sum $\sum n_b z_b^2$, which must include all ions present in the solution. Having defined the solubility c_0 by $n_a/N = \nu_a c_0$, we find the change in it by varying the expression (1) for given P and T:

$$\delta c_0 = \frac{\pi^{1/2} e^3 (\sum n_b z_b^2)^{1/2}}{2 \varepsilon^{3/2} v^{1/2} T^{3/2} N^{3/2} \sum \nu_a} \, \delta(\sum n_b z_b^2).$$

The sum following δ includes only the added kinds of ion. It should be noted that the solubility is raised under the conditions assumed.

§ 93. Mixtures of ideal gases

The additivity of the thermodynamic quantities (such as energy and entropy) holds good only so long as the interaction between the various parts of a body is negligible. For a mixture of several substances, e.g. a mixture of several liquids, the thermodynamic quantities are therefore not equal to the sums of the thermodynamic quantities for the individual components of the mixture.

An exception is formed by mixtures of ideal gases, since the interaction between their molecules is by definition negligible. For example, the entropy of such a mixture is equal to the sum of the entropies which each of the gases forming the mixture would have if the other gases were absent and the volume of the one gas were equal to that of the mixture, its pressure therefore being equal to its partial pressure in the mixture. The partial pressure P_i of the ith gas is expressed in terms of the pressure P of the whole mixture by

$$P_i = N_i T/V = N_i P/N, \tag{93.1}$$

where N is the total number of molecules in the mixuture, and N_i the number of molecules of the ith gas. Hence, by (42.7), the entropy of a mixture of two gases is

$$S = N_1 \log (eV/N_1) + N_2 \log (eV/N_2) - N_1 f_1'(T) - N_2 f_2'(T), \quad (93.2)$$

or, from (42.8),

$$\begin{aligned} S &= -N_1 \log P_1 - N_2 \log P_2 - N_1 \chi_1'(T) - N_2 \chi_2'(T) \\ &= -(N_1 + N_2) \log P - N_1 \log (N_1/N) - N_2 \log (N_2/N) \\ &\quad - N_1 \chi_1'(T) - N_2 \chi_2'(T). \end{aligned} \quad (93.3)$$

The free energy of the mixture is, by (42.4),

$$F = -N_1 T \log (eV/N_1) - N_2 T \log (eV/N_2) + N_1 f_1(T) + N_2 f_2(T), \quad (93.4)$$

and similarly (42.6) gives for the potential Φ

$$\begin{aligned} \Phi &= N_1 T \log P_1 + N_2 T \log P_2 + N_1 \chi_1(T) + N_2 \chi_2(T) \\ &= N_1(T \log P + \chi_1) + N_2(T \log P + \chi_2) \\ &\quad + N_1 T \log (N_1/N) + N_2 T \log (N_2/N). \end{aligned} \quad (93.5)$$

This expression shows that the chemical potentials of the two gases in the mixture are

$$\begin{aligned} \mu_1 &= T \log P_1 + \chi_1 = T \log P + \chi_1 + T \log (N_1/N), \\ \mu_2 &= T \log P_2 + \chi_2 = T \log P + \chi_2 + T \log (N_2/N), \end{aligned} \quad (93.6)$$

i.e. each has the same form as the chemical potential of a pure gas with pressure P_1 or P_2.

It may be noted that the free energy (93.4) of a mixture of gases has the form

$$F = F_1(N_1, V, T) + F_2(N_2, V, T),$$

where F_1 and F_2 are the free energies of the two gases as functions of the number of particles, volume and temperature. No similar formula is valid for the thermodynamic potential, however: the potential Φ of the mixture has the form

$$\Phi = \Phi_1(N_1, P, T) + \Phi_2(N_2, P, T) + N_1 T \log (N_1/N) + N_2 T \log (N_2/N).$$

Let us suppose that we have two different gases with numbers of particles N_1 and N_2 in vessels of volumes V_1 and V_2 at the same temperature and pressure, the two vessels then being connected and the gases mixed. The volume of the mixture becomes $V_1 + V_2$, and the pressure and temperature obviously remain the same. The entropy, however, changes: before mixing, the entropy of the two gases is equal to the sum of their entropies,

$$S_0 = N_1 \log (eV_1/N_1) + N_2 \log (eV_2/N_2) - N_1 f_1'(T) - N_2 f_2'(T),$$

while after mixing the entropy is, by (93.2),

$$S = N_1 \log [e(V_1+V_2)/N_1] + N_2 \log [e(V_1+V_2)/N_2] - N_1 f_1' - N_2 f_2'.$$

The change in entropy is

$$\Delta S = S - S_0$$
$$= N_1 \log [(V_1+V_2)/V_1] + N_2 \log [(V_1+V_2)/V_2],$$

or, since the volume is proportional to the number of particles for given pressure and temperature,

$$\Delta S = N_1 \log (N/N_1) + N_2 \log (N/N_2). \tag{93.7}$$

This quantity is positive, i.e. the entropy increases on mixing, as it should, because the process is clearly irreversible. The quantity ΔS is called the *entropy of mixing*.

If the two gases were identical, the entropy after connecting the vessels would be

$$S = (N_1+N_2) \log [(V_1+V_2)/(N_1+N_2)] - (N_1+N_2)f',$$

and, since $(V_1+V_2)/(N_1+N_2) = V_1/N_1 = V_2/N_2$ (the pressures and temperatures being equal), the change in entropy would be zero.

Thus the change in entropy on mixing is due to the difference in the molecules of the gases that are mixed. This is in accordance with the fact that some work must be done in order to separate again the molecules of the two gases.

§ 94. Mixtures of isotopes

A mixture of different isotopes (in any aggregate state) is a kind of "solution". For simplicity and definiteness we shall speak of a mixture of two isotopes of any element, but the same results apply to a mixture of any number of isotopes and also to chemical compounds in which different molecules contain different isotopes.

In classical mechanics, the difference between isotopes is simply a difference in mass, the laws of interaction between atoms of different isotopes being identical. This enables us to express the thermodynamic quantities for the mixture very simply in terms of those for the pure isotopes. In calculating the partition function for the mixture, the essential difference is that the phase volume element should be divided not by $N!$ as for a pure substance but by the product $N_1! N_2!$ of the factorials of the numbers of atoms of the two components of the system. This gives in the free energy the further terms

$$N_1 T \log (N_1/N) + N_2 T \log (N_2/N)$$

(where $N = N_1+N_2$), which correspond to the "entropy of mixing" discussed in § 93 for the case of a mixture of gases.

Similar terms appear in the thermodynamic potential of the mixture, which may be written

$$\Phi = N_1 T \log (N_1/N) + N_2 T \log (N_2/N) + N_1\mu_{01} + N_2\mu_{02}. \qquad (94.1)$$

Here μ_{01} and μ_{02} are the chemical potentials of the pure isotopes, which differ only by a constant times the temperature:

$$\mu_{01} - \mu_{02} = -\tfrac{3}{2}T \log (m_1/m_2), \qquad (94.2)$$

where m_1 and m_2 are the atomic masses of the two isotopes. This difference arises from the integration over the atomic momenta in the partition function; for gases, (94.2) is simply the difference between the chemical constants multiplied by T.

The difference (94.2) is the same for all phases of a given substance. The equation of phase equilibrium (the condition that the chemical potentials of the phases are equal) is therefore the same for every isotope. In particular, we can say that in the classical approximation the saturated vapour pressures of the various isotopes are equal.

The situation is no longer so simple when the substance cannot be described by means of classical statistics. In quantum theory, the difference between isotopes becomes considerably more profound, because of the differences in the vibrational and rotational levels, nuclear spins, etc.

It is important to note, however, that, even when the first correction terms (of order $\hbar^2$; see § 33) in the thermodynamic quantities are taken into account, the thermodynamic potential of the mixture may be written in the form (94.1), since the terms in question form a sum, with each term containing the mass of only one atom (see formula (33.15) for the free energy). These terms may therefore be grouped so as to include them in the chemical potentials μ_{01} and μ_{02}, and hence formula (94.1) (but not, of course, (94.2)) remains valid.

It should be pointed out that the thermodynamic potential (94.1) is formally identical with that of a mixture of any two gases (§ 93). Mixtures having this property are called *ideal mixtures*. Thus mixtures of isotopes are ideal mixtures up to and including terms of order $\hbar^2$. In this sense, mixtures of isotopes form an exceptional case, since condensed (solid or liquid) mixtures of different substances which are not isotopes can be ideal mixtures only to a very rough approximation.

Within the limits of validity of formula (94.1) we can draw certain conclusions about the vapour pressure of the isotopes over the condensed mixture. The chemical potentials of the two components of the mixture are

$$\mu_1 = T \log c_1 + \mu_{01},$$
$$\mu_2 = T \log c_2 + \mu_{02}$$

(where $c_1 = N_1/N$, $c_2 = N_2/N$ are the concentrations of the isotopes). Equating these to the chemical potentials in the gas phase (which have the forms $T \log P_1 + \chi_1(T)$ and $T \log P_2 + \chi_2(T)$), we find for the partial vapour pressures

$$P_1 = P_{01}c_1, \quad P_2 = P_{02}c_2, \tag{94.3}$$

where P_{01} and P_{02} denote the vapour pressures of the two pure isotopes (at a given temperature). Thus the partial vapour pressures of the two isotopes are proportional to their concentrations in the condensed mixture.

In the classical approximation we have for the saturated vapour pressures of the pure isotopes $P_{01} = P_{02}$, as already mentioned. When quantum effects are taken into account, however, the two vapour pressures are no longer equal. The difference cannot be calculated in a general form applicable to all substances. Such a calculation can be made only for monatomic elements (the inert gases) as far as the terms of order $\hbar^2$ (K. F. Herzfeld and E. Teller, 1938).

The correction to the thermodynamic potential of a liquid phase is given by formula (33.15);[†] taking the value per atom, we find the chemical potential

$$\mu = \mu_{cl} + (\hbar^2/24mT)\overline{F^2},$$

where $\overline{F^2}$ is the mean square of the force exerted on one atom by the other atoms in the liquid. The chemical potential of the gas remains equal to its classical value, since the interaction between atoms in the gas is negligible. Equating the chemical potentials of the liquid and the gas, we find the correction to the classical value of the vapour pressure, and the required difference of vapour pressures between the the the two isotopes is

$$P_{01} - P_{02} = P_0 \frac{\hbar^2 \overline{F^2}}{24T^2} \left(\frac{1}{m_1} - \frac{1}{m_2} \right), \tag{94.4}$$

where P_0 is the common classical value of P_{01} and P_{02}. We see that the sign of the difference is determined by that of the difference of the reciprocal masses of the isotopes, the vapour pressure of the lighter isotope being the greater.

§ 95. Vapour pressure over concentrated solutions

Let us consider the equilibrium of a solution with the vapour over it, which in general also contains both substances. The solution may be either weak or strong, i.e. the quantities of the two substances in it are arbitrary. It will be remembered that the results derived in § 89 apply only to weak solutions.

[†] We again make use of the fact that small increments to the various thermodynamic potentials, when expressed in terms of the corresponding variables, are equal (§ 15).

Since the solution and the vapour are in equilibrium, the chemical potentials μ_1 and μ_2 in the solution and in the vapour are equal. If the numbers of particles of the two substances in the solution are N_{1s} and N_{2s}, we can write the expression (24.14) for the solution in the form

$$d\Omega = -N_{1s}\,d\mu_1 - N_{2s}\,d\mu_2 - S_s\,dT - P\,dV_s,\qquad(95.1)$$

where S_s and V_s are the entropy and volume of the solution; the temperature T and pressure P are the same for the solution and the vapour.

We shall assume that the vapour over the solution is so rarefied that it may be regarded as an ideal gas; its pressure is small. Then we can neglect in (95.1) the terms proportional to P, viz. $P\,dV$ and $d\Omega$. Let us first consider all derivatives to be taken at constant temperature. Then (95.1) gives

$$N_{1s}\,d\mu_1 + N_{2s}\,d\mu_2 = 0.\qquad(95.2)$$

For the gas phase we have

$$\mu_{1g} = T\log P_1 + \chi_1(T),$$
$$\mu_{2g} = T\log P_2 + \chi_2(T),$$

where P_1 and P_2 are the partial pressures of the two components of the vapour. Differentiating these expressions (with T constant), we find

$$d\mu_{1g} = T\,d\log P_1,\quad d\mu_{2g} = T\,d\log P_2.$$

Substitution in (95.2) gives

$$N_{1s}\,d\log P_1 + N_{2s}\,d\log P_2 = 0.\qquad(95.3)$$

The concentration ξ of the solution can be defined as the ratio of the number of particles of the first component to the total number of particles:

$$\xi = N_{1s}/(N_{1s} + N_{2s}),$$

and we can similarly define the concentration x of the vapour. The partial pressures P_1 and P_2 are equal to the total pressure P of the vapour multiplied by the concentrations of the corresponding components, i.e. $P_1 = xP$, $P_2 = (1-x)P$. Substituting these values in (95.3) and dividing this equation by the total number of particles in the solution, $N = N_{1s} + N_{2s}$, we find

$$\xi\,d\log Px + (1-\xi)\,d\log P(1-x) = 0,$$

whence

$$d\log P = (x-\xi)\,dx/x(1-x),$$

or

$$\xi = x - x(1-x)\,\partial\log P/\partial x.\qquad(95.4)$$

This equation relates the solution and vapour concentrations to the dependence of the vapour pressure on the vapour concentration.

One further general relation can be obtained by considering the dependence of quantities on temperature. The condition for equality of the chemical potentials of one component, say the first, in the vapour and in the solution is $\mu_{1g} = \partial \Phi_s / \partial N_{1s}$. Dividing both sides by T and using the fact that the derivative with respect to the number of particles is taken at constant temperature, we write

$$\frac{\mu_{1g}}{T} = \frac{\partial}{\partial N_{1s}} \frac{\Phi_s}{T},$$

and then take the total derivative of each side with respect to temperature. In doing so we may assume with sufficient accuracy that the thermodynamic potential of the condensed phase (the solution) is independent of pressure. Noting also that the partial derivative with respect to temperature is

$$\frac{\partial}{\partial T} \frac{\Phi}{T} = -\frac{1}{T^2} \left(\Phi - T \frac{\partial \Phi}{\partial T} \right) = -\frac{W}{T^2},$$

we obtain the relation

$$T^2 \frac{\partial \log P_1}{\partial T} = w_{1g} - \frac{\partial W_s}{\partial N_{1s}}. \tag{95.5}$$

Here w_{1g} is the molecular heat function of the first substance as a gas; the derivative $\partial W_s / \partial N_{1s}$ gives the change in the heat function of the solution when one molecule of that substance is added to it. The quantity on the right of (95.5) is therefore the heat absorbed when one particle of the first substance goes from the solution to the vapour.

For the first substance in the pure state, the relation (95.5) becomes the ordinary Clapeyron–Clausius equation,

$$T^2 \frac{\partial \log P_{10}}{\partial T} = w_{1g} - w_{1l},$$

where P_{10} is the vapour pressure of the first substance in the pure state, and w_{1l} its molecular heat function when liquid. Subtracting this equation term by term from (95.5), we have finally

$$T^2 \frac{\partial}{\partial T} \log \frac{P_1}{P_{10}} = -q_1, \tag{95.6}$$

where $q_1 = \partial W_s / \partial N_{1s} - w_{1l}$ denotes the molecular *heat of dilution*, i.e. the quantity of heat absorbed when one particle from the liquid first substance goes into the solution. A similar relation can, of course, be written down for the second substance also.

§ 96. Thermodynamic inequalities for solutions

It has been shown in § 21 that a body can exist only in states for which certain conditions called *thermodynamic inequalities* are satisfied. These conditions were derived, however, for bodies consisting of identical particles. We shall now give a corresponding analysis for solutions, taking only the case of a mixture of two substances.

In § 21 the condition of equilibrium used was not the condition of maximum entropy of a closed system as a whole but the equivalent condition which requires that the minimum work needed to bring any small part of the system from the equilibrium state to any neighbouring state should be positive.

We now use a similar procedure, considering some small part of the solution, which contains N solvent and n solute particles, say. In the equilibrium state the temperature, pressure and concentration in this small part are equal to their values in the rest of the solution (which acts as an external medium). Let us determine the minimum work needed to bring the temperature, pressure and number of solute particles in the small part considered (containing a fixed number N of solvent particles) to values which differ by small but finite amounts δT, δP and δn from their equilibrium values.

The minimum work will be done if the process occurs reversibly. The work done by an external source is then equal to the change in the energy of the system, i.e.

$$\delta R_{\min} = \delta E + \delta E_0 \, ;$$

quantities without suffix refer to the small part considered, and those with suffix zero refer to the remainder of the system. We express δE_0 in terms of the changes in the independent variables:

$$\delta R_{\min} = \delta E + T_0 \, \delta S_0 - P_0 \, \delta V_0 + \mu_0' \, \delta n_0,$$

where μ_0' is the chemical potential of the solute in the medium; the number of solvent particles is unchanged in the process considered, and so the corresponding term for the solvent may be omitted.[†] From the reversibility of the process it follows that $\delta S_0 = -\delta S$, and from the conservation of the

---

[†] The differential of the energy of the medium (at constant N) is

$$dE_0 = T_0 \, dS_0 - P_0 \, dV_0 + \mu_0' \, dn_0.$$

Since the quantities T_0, P_0, μ_0' may be regarded as constant, integration of this relation leads to a similar relation between the finite variations of the quantities E_0, S_0, V_0, n_0.

The quantity μ_0' should not be confused with the chemical potential of the pure solute.

total volume and quantity of solute in the whole solution we have $\delta V = -\delta V_0$, $\delta n = -\delta n_0$. Substituting these, we obtain the final expression for the work:

$$\delta R_{\min} = \delta E - T_0 \, \delta S + P_0 \, \delta V - \mu_0' \, \delta n. \tag{96.1}$$

Thus the condition of equilibrium can be taken to be that for any small part of the solution the inequality

$$\delta E - T_0 \, \delta S + P_0 \, \delta V - \mu_0' \, \delta n > 0 \tag{96.2}$$

holds. Henceforward, as in § 21, we shall omit the suffix zero in expressions which are coefficients of the deviations of quantities from their equilibrium values; the values of these expressions in the equilibrium state will always be meant.

We expand δE in powers of δV, δS and δn (regarding E as a function of V, S and n). As far as the second-order terms this gives

$$\delta E = \frac{\partial E}{\partial S} \, \delta S + \frac{\partial E}{\partial V} \, \delta V + \frac{\partial E}{\partial n} \, \delta n$$

$$+ \frac{1}{2} \left[\frac{\partial^2 E}{\partial S^2} (\delta S)^2 + \frac{\partial^2 E}{\partial V^2} (\delta V)^2 + \frac{\partial^2 E}{\partial n^2} (\delta n)^2 \right.$$

$$\left. + 2 \frac{\partial^2 E}{\partial S \, \partial V} \, \delta S \, \delta V + 2 \frac{\partial^2 E}{\partial S \, \partial n} \, \delta S \, \delta n + 2 \frac{\partial^2 E}{\partial V \, \partial n} \, \delta V \, \delta n \right].$$

But $\partial E / \partial V = -P$, $\partial E / \partial S = T$, $\partial E / \partial n = \mu'$. Thus the first-order terms cancel on substitution in (96.2), leaving

$$2 \, \delta R_{\min} = \frac{\partial^2 E}{\partial S^2} (\delta S)^2 + \frac{\partial^2 E}{\partial V^2} (\delta V)^2 + \frac{\partial^2 E}{\partial n^2} (\delta n)^2$$

$$+ 2 \frac{\partial^2 E}{\partial S \, \partial V} \, \delta S \, \delta V + 2 \frac{\partial^2 E}{\partial S \, \partial n} \, \delta S \, \delta n + 2 \frac{\partial^2 E}{\partial V \, \partial n} \, \delta V \, \delta n > 0. \tag{96.3}$$

It is known from the theory of quadratic forms that, for a form in three variables (here δS, δV, δn) to be everywhere positive, its coefficients must satisfy three conditions, which for the form (96.3) are

$$\begin{vmatrix} \partial^2 E / \partial V^2 & \partial^2 E / \partial V \, \partial S & \partial^2 E / \partial V \, \partial n \\ \partial^2 E / \partial S \, \partial V & \partial^2 E / \partial S^2 & \partial^2 E / \partial S \, \partial n \\ \partial^2 E / \partial n \, \partial V & \partial^2 E / \partial n \, \partial S & \partial^2 E / \partial n^2 \end{vmatrix} > 0,$$

$$\tag{96.4}$$

$$\begin{vmatrix} \partial^2 E / \partial V^2 & \partial^2 E / \partial V \, \partial S \\ \partial^2 E / \partial S \, \partial V & \partial^2 E / \partial S^2 \end{vmatrix} > 0, \quad \partial^2 E / \partial S^2 > 0.$$

Substituting the values of the derivatives of E with respect to V, S and n, we can write these conditions as

$$\begin{vmatrix} \partial P/\partial V & \partial P/\partial S & \partial P/\partial n \\ \partial T/\partial V & \partial T/\partial S & \partial T/\partial n \\ \partial \mu'/\partial V & \partial \mu'/\partial S & \partial \mu'/\partial n \end{vmatrix} < 0,$$

$$\begin{vmatrix} \partial P/\partial V & \partial P/\partial S \\ \partial T/\partial V & \partial T/\partial S \end{vmatrix} < 0, \quad \partial T/\partial S > 0.$$

These determinants are Jacobians:

$$\frac{\partial(P, T, \mu')}{\partial(V, S, n)} < 0, \quad \left(\frac{\partial(P, T)}{\partial(V, S)}\right)_n < 0, \quad \left(\frac{\partial T}{\partial S}\right)_{V, n} > 0. \qquad (96.5)$$

The second and third conditions give the already known inequalities $(\partial P/\partial V)_{T, n} < 0$ and $C_v > 0$. The first condition may be transformed as follows:

$$\frac{\partial(P, T, \mu')}{\partial(V, S, n)} = \frac{\partial(P, T, \mu')/\partial(P, T, n)}{\partial(V, S, n)/\partial(P, T, n)}$$

$$= \frac{(\partial \mu'/\partial n)_{P, T}}{(\partial(V, S)/\partial(P, T))_n} < 0.$$

Since the denominator is negative by the second condition (96.5), we must have

$$(\partial \mu'/\partial n)_{P, T} > 0. \qquad (96.6)$$

Using instead of n the concentration $c = n/N$, we find (since N is constant)

$$(\partial \mu'/\partial c)_{P, T} > 0. \qquad (96.7)$$

Thus, as well as the inequalities $(\partial P/\partial V)_{T, c} < 0$, $C_v > 0$, the inequality (96.7) must also be satisfied in solutions.

It may be noted that for weak solutions $\partial \mu'/\partial c = T/c$, so that the inequality (96.7) is always satisfied.

The case where

$$(\partial \mu'/\partial c)_{P, T} = 0 \qquad (96.8)$$

needs special consideration.

The equality (96.8) corresponds to the vanishing of the first determinant in (96.4) (the third-order determinant). In this case the quadratic form (96.3) may vanish for certain values of δS, δV and δn, and higher-order terms in its expansion would have to be examined in order to ascertain the conditions for the inequality (96.2) to be satisfied.

However, we shall see in § 97 that such a state is a critical point for the equilibrium of two liquid phases (two solutions with different concentrations), which is analogous to the critical point of a liquid and a vapour. As in the latter case, the critical point in solutions is actually a singularity of the thermodynamic functions of the substance, and a regular expansion of these becomes impossible there. We shall merely mention that the regular expansion would lead (as in § 152 for the critical point of a liquid and a vapour) to the conditions

$$(\partial^2 \mu'/\partial c^2)_{P,\,T} = 0, \quad (\partial^3 \mu'/\partial c^3)_{P,\,T} > 0, \tag{96.9}$$

which would have to be satisfied together with (96.8).

§ 97. Equilibrium curves

The state of a body consisting of identical particles is defined by the values of any two quantities, for instance P and T. To define the state of a system having two components (a *binary mixture*) it is necessary to specify three quantities, for instance P, T and the concentration. In this and subsequent sections, the concentration of the mixture will be defined as the ratio of the quantity of one of the substances to the total quantity of both, and will be denoted by x; clearly x takes values from 0 to 1. The state of a binary mixture may be represented by a point in a three-dimensional coordinate system, whose axes correspond to these three quantities (just as the state of a system of identical particles was represented by a point in the PT-plane).

According to the phase rule, a two-component system can consist of not more than four phases in contact. The number of degrees of freedom of such a system is two when there are two phases, one for three phases, and none for four phases. The states in which two phases are in equilibrium are therefore represented by points forming a surface in the three-dimensional coordinate system; states with three phases (triple points) by points forming a line (called the *line of triple points* or the *three-phase line*) and states with four phases by isolated points.

It has already been shown in § 81 that, for systems with only one component, the states in which two phases are in equilibrium are represented by a curve in the PT-plane; each point on this curve determines the pressure and temperature (which are the same in both phases, from the conditions of equilibrium). Points not lying on the curve represent homogeneous states of the system. If the temperature and volume are taken as coordinates, the phase equilibrium is represented by a curve such that points within it represent states where there is separation into two phases represented by the points of intersection of a straight line $T = $ constant with the equilibrium curve.

The situation is similar for mixtures. If we take as coordinates P, T and the chemical potential of one component (i.e. quantities which have equal values for phases in contact), equilibrium of two phases is represented by a surface, each point of which determines P, T and μ for the two phases in equilibrium. When three phases are present, the points representing their equilibrium (triple points) will lie on the curves of intersection of the equilibrium surfaces for each pair of them.

The use of the variables P, T, μ is inconvenient, however, and in what follows we shall use P, T, x as independent variables. In terms of these variables the equilibrium of two phases is represented by a surface whose points of intersection with a straight line $P = $ constant, $T = $ constant represent the states of the two phases in contact for the relevant values of P and T (i.e. determine the concentrations of the phases, which may of course be different). The points on this line between the two points of intersection represent states in which a homogeneous body is unstable and therefore separates into two phases (represented by the points of intersection).

We shall generally use two-dimensional diagrams with P and x, or T and x, as coordinates; the lines of intersection of the equilibrium surface with the planes of constant temperature or pressure can then be drawn. We shall call these lines *equilibrium curves*.

Let us consider the points on an equilibrium curve at which the concentration becomes equal in the two phases. Two cases are possible: (1) at such a point all other properties of the two phases also become equal, i.e. the phases become identical, (2) at such a point two distinct phases continue to exist. In case (1) the point is said to be a *critical point*, in case (2) it will be called a *point of equal concentration*.

Near a critical point the equilibrium curve has the form shown in Fig. 20, or a similar form with a minimum at the critical point K (the abscissa being x and the ordinate P or T; the curve is then the intersection of the equilibrium surface with a plane of constant temperature or constant pressure respectively). Points lying within this curve (in the hatched region) represent states in which there is separation into two phases; the concentrations in these phases are determined by the points of intersection of the curve with the appropriate horizontal line. At the point K the two phases coalesce. A continuous passage can be effected between any two points in the non-hatched region along any path that passes round the critical point.

FIG. 20

FIG. 21

Figure 20 shows that near the critical point there exist states in which two phases are in equilibrium which have concentrations x and $x+\delta x$ differing by an arbitrarily small amount. For such phases the equilibrium condition is $\mu(P, T, x) = \mu(P, T, x+\delta x)$, where μ is the chemical potential of one of the substances in the mixture. Hence we see (cf. § 83) that at the critical point the condition

$$(\partial\mu/\partial x)_{P, T} = 0 \tag{97.1}$$

must hold.

This condition is identical with (96.8), and hence the two definitions of the critical point (here and in § 96) are equivalent. It may be noted that μ in (97.1) signifies the chemical potential of either of the two substances in the mixture; but the two conditions obtained by taking these two chemical potentials in (97.1) are actually equivalent. This is easily seen by noting that each of the chemical potentials is the derivative of Φ with respect to the corresponding number of particles, and Φ is a first-order homogeneous function in both numbers of particles.

The critical points clearly form a line on the equilibrium surface.

Near a point of equal concentration the equilibrium curves must have the form shown in Fig. 21, or a similar form with a minimum at the point K. The two curves touch at the maximum (or minimum). The region between the two curves is that where separation into phases occurs. At the point K the concentrations of the two phases in equilibrium become equal, but the different phases continue to exist, since any path between the points which coincide at K must pass through the region of separation into two phases. Like critical points, points of equal concentration lie on a curve on the equilibrium surface.

Let us now consider the properties of the equilibrium curves at low concentrations (i.e. when one of the substances is present in the mixture in a considerably smaller quantity than the other; x is close to zero or to unity).

It has been shown in § 89 that at low concentrations (weak solutions) the difference between the phase equilibrium temperatures of solutions and of the pure substance (at a given pressure) is proportional to the difference of concentrations of the two phases. The same applies to the pressure difference at a given temperature. Moreover, it has been shown in § 90 (also for low concentrations) that the ratio of concentrations in the two phases depends only on P and T, and so it may be regarded as constant in the neighbourhood of $x = 0$.

From the above it follows that at low concentrations the equilibrium curves have the form shown in Fig. 22, i.e. consist of two straight lines intersecting on the ordinate axis (or a similar form with the straight lines ascending). The region between the two lines is the region of separation into phases. The regions above and below the lines are the regions of the two different phases.

At the beginning of this section it has already been mentioned that a system with two components may consist of three phases in contact. Near a triple point the equilibrium curves appear as shown in Fig. 23. All three phases have equal pressure and temperature in equilibrium. The points A, B, C which determine their concentrations therefore lie on a straight line parallel to the axis of abscissae. The point A, which gives the concentration of the first phase at the triple point, is the point of intersection of the equilibrium curves 12 and 13 between the first and second, and first and third, phases. Similarly, the points B and C are the intersections of the equilibrium curves 12 and 23 between the first and second, and second and third,

FIG. 22

FIG. 23

phases (B), and of the equilibrium curves 23 and 13 between the second and third, and first and third, phases (C). The points A, B, C are of course, the points of intersection of the plane $P = $ constant or $T = $ constant with three lines on the equilibrium surface; we shall call the line corresponding to the point B a *line of triple points* or a *three-phase line*. The regions I, II, III represent states of the separate phases, first, second and third. The region between the two curves 13 below the line ABC is the region of separation into the first and third phases, and those between the two curves 12 and the two curves 23 (above the line ABC) are respectively the regions of separation into the first and second, and second and third, phases. Region II must obviously lie entirely above ABC (or entirely below ABC). At the points A, B and C the curves 12, 13 and 23 intersect, in general, at certain angles, and do not join smoothly. The directions of the curves 12, 13, 23 need not necessarily be as shown in Fig. 23, of course. The only essential feature is that the curves 12 and 23 and the curves 13 must lie on opposite sides of the straight line ABC.

If any of these singular lines on the equilibrium surface is projected on the PT-plane, the projection divides this plane into two parts. For a critical line, the points projected on one part are those corresponding to the two different phases and those corresponding to separation into these phases. The other part of the *PT*-plane contains the projections of points which represent homogeneous states, at none of which does separation into two phases occur. In Fig. 24 the dotted line represents the projection of a critical line on the *PT*-plane. The letters *a* and *b* denote the two phases. The symbol *a-b* signifies that this part of the plane contains the projections of the two phases and those of states where these two phases are present in equilibrium. The symbol *ab* denotes the single phase into which the phases *a* and *b* merge above the critical points.

FIG. 24

FIG. 25

FIG. 26

FIG. 27

The projection of a three-phase line similarly divides the *PT*-plane into two parts. Figure 25 shows which points are projected on the two parts. The symbol *a-b-c* signifies that this region contains the projections of points which represent states of the phases *a, b, c* and states in which there is separation into phases *a* and *b* or *b* and *c*.

Figure 26 shows a similar projection for a line of points of equal concentration, and Fig. 27 for a line of phase equilibrium for a pure substance (i.e. $x = 0$ or $x = 1$); the latter, of course, lies in the *PT*-plane. The letter *b* in Fig. 27 signifies that this part of the plane contains the projections of points corresponding to states of phase *b* only. In the sequence of letters in the symbols *a-b, a-b-c* the letter *b* will be understood to denote a phase whose concentration is higher than that of *a*, and *c* a phase whose concentration is higher than that of *b*.[†]

It may be noted that the four types of singular point on the equilibrium curves (triple point, point of equal concentration, critical point, and pure-

---

[†] To avoid misunderstanding, we should emphasise that the notation *a-b-c* for a line of equal concentration (unlike a three-phase line) is to some extent arbitrary: the letters *a* and *c* here denote states which are not two essentially different phases, since they never exist simultaneously in contact.

substance point) correspond to the four possible types of maximum (or minimum) on these curves.

If any phase has the same fixed composition everywhere (i.e. independently of the values of P and T), the equilibrium curves become somewhat simpler near the points here considered. Such phases are a chemical compound of the two components or pure-substance phases, which always have concentration $x = 0$ (or $x = 1$).

Let us consider the form of the equilibrium curves when there are phases of constant composition, near points where the lines corresponding to these phases terminate. It is evident that such points must be maxima or minima of the equilibrium curves, and thus are among the types of point considered in this section.

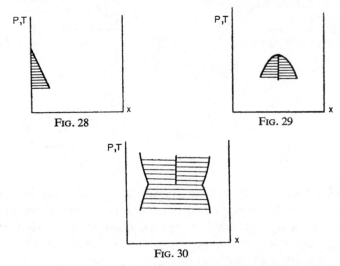

FIG. 28   FIG. 29

FIG. 30

If the phase of constant composition is a pure-substance phase with concentration $x = 0$, the corresponding line coincides with the P-axis or the T-axis and can terminate at a point of the kind shown in Fig. 28. This diagram gives the form of the equilibrium curve near such a point; one of the lines in Fig. 22 coincides with the axis of ordinates.

If one phase is a chemical compound of fixed composition, then near a point of equal concentration the equilibrium curve has the form shown in Fig. 29, i.e. the inner region in Fig. 21 becomes a vertical line. The hatched region on either side of this line is the region of separation into two phases, one a chemical compound whose composition is given by the vertical line. The curve has no break at the maximum, as in Fig. 21.

Similarly, near a triple point the equilibrium curves have the form shown in Fig. 30. The phase which is a chemical compound is represented by a vertical line, to which region II (Fig. 23) here reduces.

§ 98. Examples of phase diagrams

In this section we shall enumerate the principal types of equilibrium curve; in contrast to § 97, their form will be considered in general, and not only near the singular points. These curves (also called *phase diagrams*) can have many forms, but in most cases they belong to one of the types given below, or are a combination of more than one of these. The hatched regions in all these diagrams are the regions of separation into phases, and the remaining regions are those of homogeneous states. The points of intersection of horizontal lines with the curves bounding the regions of separation into phases determine the composition of the phases into which separation occurs (for given P and T). The relative amounts of the two phases are determined by the "lever rule" already mentioned in § 81.

Fig. 31

In what follows we shall for definiteness discuss Tx-diagrams; similar types are possible in the coordinates P nd x. The concentration x is taken as abscissa, and varies from 0 to 1.

1. There are two phases; each can have any concentration (i.e. the two components mix in any proportion in both phases). In the simplest case, where the curves have no maxima or minima (apart from the pure-substance point), the phase diagram has the "cigar" form shown in Fig. 31.

Let one of the phases be a liquid (the region below the cigar), and the other a vapour (the region above the cigar). Then the upper curve of the cigar is called the *condensation point curve*, and the lower curve the *boiling point curve*.[†]

If a liquid mixture of given composition is heated, then the liquid will begin to boil at a temperature determined by the intersection B of the vertical line AD (corresponding to the given concentration) with the lower curve of the cigar. Vapour boils off, whose composition is given by the point C, i.e.

[†] The laws of boiling and condensation of liquid mixtures were established by D. P. Konovalov (1884).

has a lower concentration than the liquid. The concentration of the remaining liquid is obviously increased, and its boiling point accordingly rises. On further heating, the point which represents the state of the liquid phase will move upwards along the lower curve, and the point which represents the vapour leaving the liquid will move upwards along the upper curve. Boiling ceases at various temperatures, depending on the way in which the process takes place. If boiling occurs in a closed vessel, so that all the vapour generated remains permanently in contact with the liquid, the liquid will obviously boil away completely at a temperature where the concentration of the vapour is equal to the original concentration of the liquid (the point D). In this case, therefore, boiling begins and ends at temperatures given by the intersection of the vertical line AD with the lower and upper curves of

FIG. 32

the cigar. If the vapour boiling off is steadily removed (boiling in an open vessel), then only the vapour just evolved will be in equilibrium with the liquid at any given time. In this case, it is evident that boiling will cease at the boiling point G of the pure substance, where the liquid and vapour compositions are the same. The condensation of vapour into liquid occurs in a similar manner.

The situation is exactly analogous when the two phases are a liquid (above the cigar) and a solid (below the cigar).

2. The two components mix in any proportion in both phases (as in case 1), but there is a point of equal concentration. The phase diagram then has the form shown in Fig. 32 (or a similar form with a minimum). At the point of equal concentration, the two curves touch, and both have a maximum or a minimum.

The transition from one phase to another occurs in the same way as described for case 1, except that the process can terminate (if one phase is steadily removed, as for example by boiling a liquid in an open vessel) not only at the pure-substance point but also at the point of equal concentration. At the composition corresponding to this point, the entire process occurs at a single temperature.[†]

[†] A mixture corresponding to the point of equal concentration is also said to be *azeotropic*.

3. There are two phases, liquid and gas, in which the two components mix in any proportion, and there is a critical point. The phase diagram is as shown in Fig. 33 (K being the critical point). The region to the right of the curve corresponds to liquid states, and that to the left to gaseous states. It should be remembered, however, that when there is a critical point the liquid and gaseous phases can, strictly speaking, be distinguished only when they are in equilibrium with each other.

A diagram of this type leads to the following curious effect. If a liquid whose composition is represented by the line AC (passing to the right of the point K) is heated in a closed vessel, then, after boiling begins (at the point B), the quantity of vapour will gradually increase as heating continues, but after a certain time it begins to decrease again, and the vapour disappears entirely at the point C. This is called *retrograde condensation*.

FIG. 33

FIG. 34

4. Two liquids which mix, but not in all proportions. The phase diagram is as shown in Fig. 34. At temperatures above that of the critical point K, the components mix in any proportion. Below this temperature the components do not mix in the proportions represented by points within the hatched region. In this region there is separation into two liquid mixtures whose concentrations are given by the points of intersection of the corresponding horizontal line with the equilibrium curve. Similar diagrams are possible with K a minimum, or with two critical points, an upper and a lower, so that the region of separation into two phases (two solutions) is bounded by a closed curve.

5. In the liquid (or gaseous) state the two components mix in any proportion, but in the solid (or liquid) state they do not mix in all proportions (limited miscibility). In this case there is a triple point. According as the temperature of the triple point lies below the pure-component phase equilibrium temperatures (the points A and C) or between them (it obviously cannot lie above them on the assumption made here that the components mix in any proportion in the higher phase), the phase diagram appears as in Fig. 35 or Fig. 36 respectively. For example, let the phase of unlimited

miscibility be a liquid, and that of limited miscibility be a solid. The region
above the curve *ABC* (Fig. 35) or *ADC* (Fig. 36) is the region of liquid states;
the regions bounded by the curves *ADF* and *CEG* (Fig. 35) or *ABF* and *CEG*
(Fig. 36) are the regions of homogeneous solid phases (solid solutions). At
the triple point (whose temperature is given by the line *DBE*) the liquid and
two solid solutions of different concentrations are in equilibrium. The point
B in Fig. 35 is called the *eutectic point*. A liquid mixture whose concentration

FIG. 35

FIG. 36

corresponds to this point freezes completely, without change of concentration, whereas at other concentrations a solid mixture freezes out with a
concentration different from that of the liquid. The regions *ADB* and *CBE*
(Fig. 35) and *ADB* and *CDE* (Fig. 36) correspond to separation into a liquid
phase and one of the solid phases; the regions *DEGF* (Fig. 35) and *BEGF*
(Fig. 36) correspond to separation into two solid phases.

   If, in a diagram such as Fig. 35, the components do not mix at all in the
solid state, the phase diagram takes the form shown in Fig. 37. In the

FIG. 37

FIG. 38

hatched regions above the line *ABC*, the mixed liquid phase is in equilibrium
with the solid phase of one of the pure substances, and below *ABC* we have
the two pure solid phases. When the temperature of the liquid mixture
decreases, one or the other of the pure substances freezes out according as
the concentration of the liquid lies to the right or the left of the eutectic
point. As the temperature decreases further, the composition of the liquid
varies along the curve *DB* or *EB*, and the liquid freezes completely at the
eutectic point *B*.

   6. In the liquid state the two components mix in any proportion, but in the
solid state they do not mix at all, forming only a chemical compound of

definite composition. The phase diagram is shown in Fig. 38. The straight line *DE* gives the composition of the chemical compound. There are two triple points, *B* and *G*, at which there is equilibrium between the liquid phase, the solid chemical compound, and the solid phase of one of the pure components. Between the points *B* and *G* lies a point of equal concentration, *D* (cf. Fig. 29). It is easy to see where separation occurs, and which are the resulting phases: in the region *DBE* they are a liquid phase and the solid chemical compound; below the line *CE*, the chemical compound and one of the solid pure substances, and so on. The freezing of the liquid terminates at one of the eutectic points *G* and *B*, according as the concentration of the liquid lies to the right or to the left of the line *DE*.

FIG. 39

FIG. 40

7. In the liquid state the two components mix in any proportion, but in the solid state they do not mix at all, forming only a chemical compound; this compound, however, decomposes at a certain temperature, before it melts. The straight line defining the composition of this compound cannot terminate at a point of equal concentration as in case 6, since it does not reach the melting point. It can therefore terminate at a triple point of the type shown in Fig. 30, § 97 (the point *A* in Fig. 39). In Fig. 39, which shows one possible form of the phase diagram for this case, it is easy to see which phases result from the separation at various points in the hatched region.

8. In the solid state the components do not mix at all, and in the liquid state they mix only in certain proportions. In this case there are two triple points, at which the liquid is in equilibrium with the two solid pure substances (the point *B* in Fig. 40) or one of the pure substances is in equilibrium with two mixed liquid phases of different concentrations (the point *D*). The regions not hatched in Fig. 40, above *ABC* and above *DE*, represent liquid states with various concentrations; the hatched region above *CD* is that of separation into two liquid phases, *DEF* is that of separation into a liquid and one of the solid pure substances, and so on.

§ 99. Intersection of singular curves on the equilibrium surface

The four kinds of line discussed in § 97 (critical lines, three-phase lines, lines of equal concentration and pure-substance lines) all lie on the same surface, the equilibrium surface. They will therefore in general intersect. Some properties of the points of intersection of such lines are described below.

It may be shown that no two critical lines can intersect, nor can two lines of equal concentration. We shall not pause to prove these statements here.

Let us now list (again without proof) the properties of the remaining points of intersection. All these properties follow almost immediately from the general properties of equilibrium curves given in § 97. The diagrams will show the projections of the intersecting lines on the PT-plane (see § 97); their form is, of course, chosen arbitrarily. A dotted line everywhere denotes a critical line; a continuous line, a line of phase equilibrium for a pure substance; a broken line, a line of equal concentration; and a dot-and-dash line, a three-phase line. The letters have the same significance as in Figs. 24 to 27 (§ 97).

At a point of intersection between a critical line and a pure-substance line (Fig. 41a) both lines terminate, and similarly for a critical line and a three-phase line (Fig. 41b). When a pure-substance line intersects a line of equal concentration, only the latter terminates (Fig. 41c), the two curves touching at the point of intersection. The same occurs when a line of equal concentration meets a critical line (Fig. 41d) or a three-phase line (Fig. 41e). In each case the line of equal concentration terminates at the point of intersection, the two curves touching at this point.

FIG. 41

The point of intersection of three-phase lines (Fig. 41f) is a quadruple point, i.e. a point where four phases are in equilibrium. Four three-phase lines meet at this point, corresponding to equilibrium between each three of the four phases.

Finally, the point where a pure-substance line intersects a three-phase line (Fig. 41g) must clearly be also a point of intersection between the three-phase line and all three pure-substance phase equilibrium lines (corresponding to equilibrium between each two of the three pure-substance phases).

§ 100. Gases and liquids

Let us now consider in more detail the equilibrium of liquid and gaseous phases consisting of two components.

When the temperature is sufficiently high (T large in comparison with the mean interaction energy of the molecules) all substances mix in any proportion. On the other hand, since a substance is a gas at such temperatures, we can say that all substances have unlimited miscibility in the gas phase (although when there are critical lines the difference between the liquid and the gas becomes to some extent arbitrary, and so likewise does the foregoing formulation).

In the liquid state, some substances mix in any proportion, others only in certain proportions (liquids of limited miscibility).

In the former case, when the two components mix in any proportion in both phases, the phase diagrams contain no triple points, since the system cannot consist of more than two phases (all liquid states are one phase, and the same applies to gaseous states). Let us consider the projection of the singular lines of the equilibrium surface on the *PT*-plane. There are two lines of phase equilibrium for the pure substances (i.e. for concentrations $x = 0$ and $x = 1$ in both phases). One of these lines is itself in the *PT*-plane, and the other is in a plane parallel to it, so that its projection has the same form as the line itself. Each of these lines terminates at a point which is a critical point for phases of the corresponding pure substance. A critical line begins and ends at these points (at a point of intersection of a critical line and a pure-substance line, both terminate; see § 99). Thus the projection of these various lines on the *PT*-plane has the form shown in Fig. 42; the notation is the same as in §§ 97 and 99. The letters g and l have a similar significance to a, b, c

FIG. 42

FIG. 43

in the diagrams in §§ 97 and 99: g denotes gas, and l liquid. The regions g and l contain projections of gaseous and liquid states respectively; the region g-l includes these and also states where separation into liquid and gas occurs; above the critical line, the difference between liquid and gas does not exist.

If there is also a line of equal concentration, the projection on the PT-plane is as shown in Fig. 43. The projection of the line of equal concentration lies either above the line from the origin O to B (as in Fig. 43), or below OC, but not between them. Only A, B, C are points of intersection of lines. The point D does not correspond to a true intersection of the pure-substance line with the critical line; these intersect only in projection. The letters l_1 and l_2 in the diagram denote liquid phases of different concentrations. Above the line of equal concentration there is only one liquid phase.[†]

All these properties of the projections of the singular lines on the PT-plane become obvious if we consider the phase diagrams corresponding to the cross-sections of the equilibrium surface by various planes of constant temperature (or pressure). For example, the cross-sections corresponding to pressures below that at B and to pressures between those of A and B in Fig. 42 give phase diagrams as shown in Figs. 31 and 33 respectively. Figure 44 shows cross-sections for various successive temperature ranges in Fig. 43 (T_A, T_B, T_C being the temperatures corresponding to the points A, B, C):

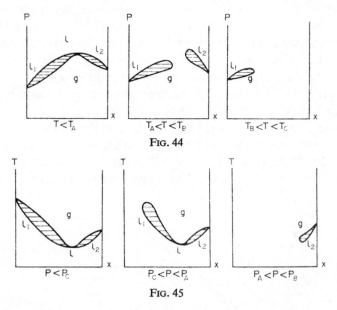

FIG. 44

FIG. 45

[†] Not being concerned with solid phases, we shall conventionally show lines in all the (P, T) diagrams as starting from the origin, as if solidification did not occur.

the region of separation into two phases "breaks up" at the point of equal concentration, and two critical points are formed; thereupon, first one and then the other hatched region shrink to a point on the ordinate axis and disappear. Figure 45 shows similar cross-sections for successive pressure ranges.

If the two components have limited miscibility in the liquid state, there is a three-phase line, which terminates at a point where it intersects a critical line starting from that point. Figures 46 and 47 show the two essentially different types of (P, T) projection that can occur in this case. They differ in that in Fig. 46 the projection of the three-phase line lies above both the pure-substance lines, while in Fig. 47 it lies between them; the three-phase line cannot lie below both the pure-substance lines, since in the gaseous state the two components mix in any proportion. In both cases there are two critical lines, one of which runs out towards high pressures.

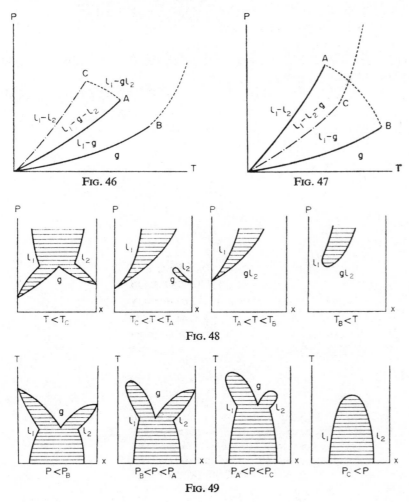

FIG. 46   FIG. 47

FIG. 48

FIG. 49

Figures 48 and 49 show a number of successive cross-sections by Px and Tx planes for the case shown in Fig. 46.

In conclusion, it should be emphasised that the examples of (P, T) diagrams discussed in this section are only the most typical ones for equilibrium of liquid and gaseous phases; they do not exhaust all theoretically possible forms.

CHAPTER X

CHEMICAL REACTIONS

§ 101. The condition for chemical equilibrium

A CHEMICAL reaction occurring in a mixture of reacting substances ultimately leads to the establishment of an equilibrium state in which the quantity of each of the substances that take part in the reaction no longer changes. This case of thermodynamic equilibrium is called *chemical equilibrium*. Any chemical reaction can take place, in general, in either direction; until equilibrium is reached, one direction predominates, but in equilibrium the two opposite reactions occur at rates such that the total numbers of particles of each of the reacting substances remain constant. The object of thermodynamics as applied to chemical reactions is to study only the chemical equilibrium, not the course of the reaction leading to that equilibrium.

It is important to note that the state of chemical equilibrium is independent of how and under what conditions the reaction occurred;[†] it depends only on the conditions under which the mixture of reacting substances exists in equilibrium. In deriving the condition for chemical equilibrium, we can therefore make any desired assumptions concerning the course of the reaction.

First of all, we shall describe the method to be used for expressing the reaction. Chemical reactions are commonly written as symbolic equations, which, if all the terms are taken to one side, have the form

$$\sum_i \nu_i A_i = 0, \tag{101.1}$$

where the A_i are the chemical symbols of the reacting substances, and the coefficients ν_i are positive or negative integers. For example, in the reaction $2H_2 + O_2 = 2H_2O$ or $2H_2 + O_2 - 2H_2O = 0$ the coefficients are $\nu_{H_2} = 2$, $\nu_{O_2} = 1$, $\nu_{H_2O} = -2$.

Let us assume that the reaction occurs at constant temperature and pressure. In such processes the thermodynamic potential of the system tends to a minimum. In equilibrium, therefore, the potential Φ must have its least

---

[†] In particular, it is independent of whether a catalyst took part in the reaction.

possible value (for given P and T). Let $N_1, N_2, \ldots$ be the numbers of particles of the various substances taking part in the reaction. Then the necessary
condition for Φ to be a minimum can be written as the vanishing of the total
derivative of Φ (for given P and T) with respect to one of the N_i, say N_1:

$$\frac{\partial \Phi}{\partial N_1} + \frac{\partial \Phi}{\partial N_2}\frac{dN_2}{dN_1} + \frac{\partial \Phi}{\partial N_3}\frac{dN_3}{dN_1} + \ldots = 0.$$

The changes in the numbers N_i during the reaction are related by the reaction
equation: it is clear that, if N_1 changes by ν_1, each of the other N_i will change
by ν_i, so that $dN_i = (\nu_i/\nu_1)\,dN_1$, $dN_i/dN_1 = \nu_i/\nu_1$. The foregoing equation
may therefore be written

$$\sum_i \frac{\partial \Phi}{\partial N_i}\frac{\nu_i}{\nu_1} = 0.$$

Finally, putting $\partial \Phi/\partial N_i = \mu_i$ and multiplying by ν_1, we have

$$\sum_i \nu_i \mu_i = 0. \tag{101.2}$$

This is the required condition for chemical equilibrium. In order to obtain
it, therefore, we must replace the symbols A_i by the corresponding chemical
potentials μ_i in the equation of the chemical reaction. When several different
reactions can occur in the mixture, the equilibrium condition will be a set of
several equations such as (101.2), each obtained by the above method from
the equation of the corresponding reaction.

It may be noted that the condition (101.2) retains its form even when the
reacting substances are distributed in the form of solutes in two different
phases in contact. This follows from the fact that in equilibrium the chemical
potentials of each substance in either phase must be equal, in accordance
with the conditions for phase equilibrium.

§ 102. The law of mass action

Let us apply the general condition for chemical equilibrium, derived in
§ 101, to reactions taking place in a gas mixture, assuming that the gas may be
regarded as an ideal one.

The chemical potential of each gas in the mixture is (see § 93)

$$\mu_i = T \log P_i + \chi_i(T), \tag{102.1}$$

where P_i is the partial pressure of the ith gas in the mixture; $P_i = c_iP$ if P
is the total pressure of the mixture and $c_i = N_i/N$ is the concentration of the
gas in question, defined as the ratio of the number N_i of molecules of that gas
to the total number $N = \Sigma N_i$ of molecules in the mixture.

It is now easy to write down the condition of chemical equilibrium for reactions in a mixture of gases. Substitution of (102.1) in (101.2) gives

$$\sum_i \nu_i \mu_i = T \sum_i \nu_i \log P_{0i} + \sum_i \nu_i \chi_i = 0,$$

where the P_{0i} are the partial pressures of the gases in a state of chemical equilibrium.

Using the notation

$$K_p(T) = e^{-\Sigma \nu_i \chi_i / T}, \qquad (102.2)$$

we thus have

$$\prod_i P_{0i}^{\nu_i} = K_p(T). \qquad (102.3)$$

Instead of P_{0i} we can substitute Pc_{0i}, where the c_{0i} are the concentrations of the gases in chemical equilibrium. Then

$$\prod_i c_{0i}^{\nu_i} = P^{-\Sigma \nu_i} K_p(T) \equiv K_c(P, T). \qquad (102.4)$$

The quantity on the right of (102.3) or (102.4) is a function only of temperature and pressure, and does not depend on the initial amounts of the reacting gases: this quantity is called the *chemical equilibrium constant*, and the law expressed by formula (102.3) or (102.4) is called the *law of mass action*.

The dependence of the gas reaction equilibrium constant on the pressure is entirely determined by the factor $P^{-\Sigma \nu_i}$ on the right-hand side of equation (102.4); if the quantities of reacting substances are expressed in terms of their partial pressures, the equilibrium constant is independent of pressure. The determination of its dependence on temperature, however, requires further assumptions concerning the properties of the gases.

For example, if the gases have constant specific heats, a comparison of the expression (102.1) with formula (43.3) for the thermodynamic potential of such a gas shows that the functions $\chi_i(T)$ are of the form

$$\chi_i(T) = \varepsilon_{0i} - c_{pi} T \log T - T \zeta_i, \qquad (102.5)$$

where c_{pi} is the specific heat and ζ_i the chemical constant of the gas. Substituting this expression in (102.2), we obtain for the equilibrium constant the formula

$$K_p(T) = e^{\Sigma \nu_i \zeta_i} T^{\Sigma c_{pi} \nu_i} e^{-\Sigma \nu_i \varepsilon_{0i} / T}, \qquad (102.6)$$

which is essentially an exponential function of temperature.

The law of mass action is valid also for reactions between solutes, provided that the solution may be regarded as weak. For the chemical potential of each solute has the form

$$\mu_i = T \log c_i + \psi_i(P, T). \qquad (102.7)$$

The concentration c_i is here defined as the ratio of the number of particles of the solute in question to the number of solvent particles ($c_i = n_i/N$). Substituting (102.7) in the equilibrium condition (101.2), we find in the same way

$$\prod_i c_{0i}^{\nu_i} = K(P, T), \qquad (102.8)$$

with the equilibrium constant

$$K(P, T) = e^{-\Sigma \nu_i \psi_i/T}. \qquad (102.9)$$

Unlike the case of gas reactions, the dependence of the equilibrium constant on the pressure here remains indeterminate.

If the reaction involves, as well as gases or solutes, substances in a pure condensed phase (i.e. not mixed with other substances), e.g. pure solids, then the equilibrium condition again leads to the law of mass action. Here, however, since the chemical potential of the pure phases depends only on the pressure and temperature, the left-hand side of the equation for the law of mass action will not involve the quantities of the pure phases, i.e. the product of the concentrations of the gases (or solutes) must be written as if the solids were absent. The latter affect only the dependence of the equilibrium constant on pressure and temperature.

If only gases and solids take part in the reaction, then, since the pressure of the gases is comparatively small, the chemical potential of the solids may be regarded as independent of the pressure, and the dependence of the equilibrium constant on the pressure remains the same as in (102.4). The sum $\Sigma \nu_i$ in the exponent must of course denote only the sum of the coefficients of the gaseous substances in the reaction equation.

Finally, the law of mass action is valid also for reactions in weak solutions where the solvent as well as the solutes takes part in the reactions. For, when its chemical potential is substituted in the condition for chemical equilibrium, the small terms which contain the concentration may be omitted, and the potential then reduces to a quantity which depends only on temperature and pressure. Thus we again obtain the equation of the law of mass action, and its left-hand side again involves only the concentrations of the reacting solutes, not that of the solvent.

PROBLEMS

PROBLEM 1. Find the degree of dissociation of a diatomic gas at high temperatures, if the gas molecule consists of identical atoms and has no spin or orbital angular momentum in the ground state.

SOLUTION. The reaction has the form $A_2 = 2A$. In these Problems, we shall use the suffixes 1 and 2 to denote quantities relating respectively to the atomic (A) and molecular (A_2) components of the mixture. The degree of dissociation is defined as the ratio $\alpha = N_1/2N_0$ of the number $\frac{1}{2}N_1$ of dissociated molecules to the total number of molecules (that would exist in a non-dissociating gas) $N_0 = N_2 + \frac{1}{2}N_1$. In accordance with the law of mass action (102.3), we have

$$\frac{P_2}{P_1^2} = \frac{N_2(N_1 + N_2)}{PN_1^2} = \frac{1 - \alpha^2}{4\alpha^2 P} = K_P(T),\tag{1}$$

whence

$$\alpha = [1 + 4PK_P(T)]^{-1/2}.$$

The equilibrium constant K_P is found by substituting in (102.6) the specific heats $c_{p1} = \frac{5}{2}$, $c_{p2} = \frac{9}{2}$ and the chemical constants

$$\zeta_1 = \log [g_1(m/2\pi\hbar^2)^{3/2}], \quad \zeta_2 = \log [(I/\hbar^6\omega)(m/\pi)^{3/2}]$$

(see (45.4), (46.4), (49.8)), where m is the mass of the atom A, and g_1 the statistical weight of the ground state of the atom A; at sufficiently high temperatures $g_1 = (2S + 1)(2L + 1)$, where S and L are the spin and orbital angular momentum of the atom.[†]
  The result is

$$K_P(T) = \frac{8I\pi^{3/2}}{g_1^2\omega m^{3/2}T^{1/2}} e^{\varepsilon_0/T},\tag{2}$$

where $\varepsilon = 2\varepsilon_{01} - \varepsilon_{02}$ is the dissociation energy of the molecule.

  PROBLEM 2. The same as Problem 1, but to find the specific heat.

  SOLUTION. The entropy of the gas may be calculated as the sum

$$S = N_1\left(c_{p1} + \frac{\varepsilon_{01} - \mu_1}{T}\right) + N_2\left(c_{p2} + \frac{\varepsilon_{02} - \mu_2}{T}\right)$$

$$= N_1\left(c_{p1} + \frac{\varepsilon_{01}}{T}\right) + N_2\left(c_{p2} + \frac{\varepsilon_{02}}{T}\right) - (N_1 + 2N_2)\frac{\mu_1}{T};$$

the entropy of each component is expressed in terms of its chemical potential according to (43.6) and (43.3), and the equilibrium equation $\mu_2 = 2\mu_1$ is used. Expressing N_1 and N_2 in terms of N_0 and α, writing the chemical potential as

$$\mu_1 = \varepsilon_{01} + T \log P_1 - c_{p1}T \log T - \zeta_1 T, \quad P_1 = 2\alpha P/(1 + \alpha),$$

and substituting the values of c_{p1} and c_{p2}, we obtain

$$S = N_0\left[-\frac{1 - \alpha}{T}\varepsilon_0 + 5 \log T + \frac{1}{2}\alpha - 2 \log \frac{\alpha}{1 + \alpha} + \text{constant}\right],\tag{3}$$

where the dissociation energy ε_0 is again used, and the constant represents the temperature-independent terms, which do not influence the required specific

[†] See the footnote to § 104.

heat $C_p = T(\partial S/\partial T)_P$. From (1) we calculate the derivative

$$\left(\frac{\partial \alpha}{\partial T}\right)_P = -\frac{1}{2}(1-\alpha^2)\alpha \frac{d \log K_P}{dT}$$

$$= \frac{(1-\alpha)^2 \alpha}{2T}\left(\frac{\varepsilon_0}{T}+\frac{1}{2}\right),$$

with K_P from (2). Now differentiating the entropy (3), we have finally

$$C_P = \frac{1}{2}N_0\left[9+\alpha+\alpha(1-\alpha^2)\left(\frac{\varepsilon_0}{T}+\frac{1}{2}\right)^2\right].$$

PROBLEM 3. Determine the dependence of the concentration of hydrogen dissolved as H atoms in a metal on the pressure of H_2 gas over the metal.

SOLUTION. Regarding the process as a chemical reaction $H_2 = 2H$, we can
write the equilibrium condition as $\mu_{H_2} = 2\mu_H$; μ_{H_2} is written as the chemical potential of an ideal gas, $\mu_{H_2} = T \log P + \chi(T)$, and μ_H as that of the solute in a solution,
$\mu_H = T \log c + \psi$. Since ψ depends only slightly on the pressure (cf. § 90), we find
$c = \text{constant} \times \sqrt{P}$.

§ 103. Heat of reaction

A chemical reaction is accompanied by the absorption or evolution of heat.
In the former case the reaction is said to be *endothermic*, and in the latter
case *exothermic*. It is evident that, if any particular reaction is exothermic,
the reverse reaction will be endothermic, and vice versa.

The amount of heat involved in a reaction depends on the conditions under
which the reaction occurs. Hence, for instance, we must distinguish the heats
of reaction at constant volume and at constant pressure (although the difference is usually quite small).

As in calculating the heat of solution (§ 91), we first find the maximum work
which can be obtained by means of the chemical reaction. We call a reaction
between one group of molecules as shown by the reaction equation an "elementary reaction", and calculate the change in the thermodynamic potential
of a mixture of reacting substances when a small number δn of elementary
reactions take place, assuming that the reaction occurs at constant temperature and pressure. We have

$$\delta \Phi = \sum_i \frac{\partial \Phi}{\partial N_i}\delta N_i = \sum_i \mu_i \delta N_i.$$

The change in the number of molecules of the ith substance after δn elementary reactions is clearly $\delta N_i = -\nu_i \delta n$. Thus

$$\delta \Phi = -\delta n \sum_i \nu_i \mu_i. \tag{103.1}$$

In equilibrium $\delta \Phi/\delta n$ is zero, as we should expect.

Formula (103.1) is the general expression for the minimum work which must be done in order to bring about δn elementary reactions. It is also the maximum work which can be obtained from that number of reactions occurring in the reverse direction.

Let us first suppose that the reaction is between gases. Using the expression (102.1) for μ_i, we find

$$\delta \Phi = -\delta n \left(T \sum_i v_i \log P_i + \sum_i v_i \chi_i \right),$$

or, in terms of the equilibrium constant,

$$\delta \Phi = T \, \delta n \left[-\sum_i v_i \log P_i + \log K_p(T) \right]$$

$$= T \, \delta n \left[-\sum_i v_i \log c_i + \log K_c(P, T) \right]. \tag{103.2}$$

For reactions in solution we similarly find, using the expressions (102.7) and (102.9),

$$\delta \Phi = T \, \delta n \left[-\sum_i v_i \log c_i + \log K(P, T) \right]. \tag{103.3}$$

The sign of $\delta \Phi$ determines the direction in which the reaction takes place: since Φ tends to a minimum, for $\delta \Phi < 0$ the reaction occurs in the forward direction (i.e. "from left to right" in the equation of the chemical reaction), while if $\delta \Phi > 0$ the reaction will actually go in the opposite direction in the mixture concerned. It may be noted, however, that the direction of the reaction is also evident directly from the law of mass action: we form the product $\Pi P_i^{v_i}$ for the mixture in question and compare it with the value of the equilibrium constant for the reaction. If, for instance, we find that $\Pi P_i^{v_i} > K_p$, this means that the reaction will occur in the forward direction, so as to reduce the partial pressures of the original substances (which have positive v_i in the reaction equation), and increase those of the reaction products (for which $v_i < 0$).

We can now determine also the heat absorbed (or evolved, according to sign), again for δn elementary reactions. Formula (91.4) shows that this heat δQ_p is, for a reaction at constant temperature and pressure,

$$\delta Q_p = -T^2 \left(\frac{\partial}{\partial T} \frac{\delta \Phi}{T} \right)_P.$$

For reactions between gases we have, substituting (103.2),

$$\delta Q_p = -T^2 \, \delta n \frac{\partial \log K_p(T)}{\partial T}. \tag{103.4}$$

Similarly, for solutions

$$\delta Q_p = -T^2 \, \delta n \, \frac{\partial \log K(P, T)}{\partial T}. \tag{103.5}$$

We may note that δQ_p is simply proportional to δn and does not depend on the values of the concentrations at any instant. These formulae are therefore valid for any δn, whether small or not.

If $Q_p > 0$, i.e. the reaction is endothermic, $\partial \log K/\partial T < 0$, and the equilibrium constant decreases with increasing temperature. On the other hand, for an exothermic reaction ($Q_p < 0$) the equilibrium constant increases with temperature. An increase in the equilibrium constant signifies that the chemical equilibrium is shifted back towards the re-formation of the initial substances—the reaction goes "from right to left", so as to increase the product $\Pi c_{0i}^{\nu_i}$. Conversely, a decrease in the equilibrium constant signifies a shift of the equilibrium towards formation of the reaction products. In other words, we can formulate the following rule: heating shifts the equilibrium in the direction of the endothermic process, and cooling in the direction of the exothermic process. This rule is entirely in agreement with Le Chatelier's principle.

For reactions between gases the heat of reaction at constant volume (and temperature) is also of interest. This quantity δQ_v is related in a simple manner to δQ_p. The quantity of heat absorbed in a process at constant volume is equal to the change in the energy of the system, whereas δQ_p is equal to the change in the heat function. Since $E = W - PV$, it is clear that $\delta Q_v = \delta Q_p - \delta(PV)$, or, substituting $PV = T \Sigma N_i$ and $\delta N_i = -\nu_i \, \delta n$,

$$\delta Q_v = \delta Q_p + T \, \delta n \sum_i \nu_i. \tag{103.6}$$

Finally, let us determine the change in volume of a mixture of reacting substances as a result of a reaction occurring at constant pressure (and temperature). For gases, the problem is trivial:

$$\delta V = (T/P) \, \delta N = -(T/P) \, \delta n \sum_i \nu_i. \tag{103.7}$$

In particular, reactions in which the total number of particles is unchanged ($\Sigma \nu_i = 0$) occur without change of volume.

For reactions in weak solutions we use the formula $\delta V = \partial \, \delta \Phi/\partial P$ and, substituting (103.3), we obtain

$$\delta V = T \, \delta n \, \frac{\partial \log K(P, T)}{\partial P}. \tag{103.8}$$

Thus a change in volume in the reaction is due to a pressure dependence of the equilibrium constant. In a similar way to the previous discussion of the temperature dependence, we easily deduce that an increase in pressure favours reactions in which the volume decreases (i.e. shifts the equilibrium in the direction of such reactions), and a decrease in pressure favours reactions which lead to an increase in volume, again in complete agreement with Le Chatelier's principle.

§ 104. Ionisation equilibrium

At sufficiently high temperatures, collisions between gas particles may cause their ionisation. The existence of such *thermal ionisation* leads to the establishment of an equilibrium in which certain fractions of the total number of gas particles are in various stages of ionisation. Let us consider thermal ionisation of a monatomic gas; this is the most interesting case, since chemical compounds are usually completely dissociated before the onset of thermal ionisation.

Thermodynamically, ionisation equilibrium is a particular case of chemical equilibrium corresponding to a series of simultaneously occurring "ionisation reactions", which may be written

$$A_0 = A_1 + e^-, \quad A_1 = A_2 + e^-, \ldots, \tag{104.1}$$

where the symbol A_0 denotes the neutral atom, A_1, A_2, ... the singly, doubly, etc., ionised atoms and e^- the electron. For these reactions the application of the law of mass action gives the set of equations

$$c_{n-1}/c_n c = P K_p^{(n)}(T) \quad (n = 1, 2, \ldots), \tag{104.2}$$

where c_0 is the concentration of neutral atoms, c_1, c_2, ... the concentrations of the various ions, and c the concentration of electrons (each defined as the ratio of the number of particles of the kind in question to the total number of particles, including electrons). To these equations we must add one which expresses the electrical neutrality of the gas as a whole:

$$c = c_1 + 2c_2 + 3c_3 + \ldots \, . \tag{104.3}$$

Equations (104.2) and (104.3) determine the concentrations of the various ions in ionisation equilibrium.

The equilibrium constants $K_p^{(n)}$ can be calculated without difficulty. All gases which take part in "reactions" (gases of neutral atoms, ions, or electrons) are monatomic and have constant specific heat $c_p = \frac{5}{2}$, and their chemical constants are $\zeta = \log[g(m/2\pi\hbar^2)^{3/2}]$, where m is the mass of a particle of the gas considered, and g the statistical weight of its ground state; for electrons, $g = 2$, while for atoms and ions $g = (2L+1)(2S+1)$, where

L and S are the orbital angular momentum and spin of the atom or ion.[†]
Substituting these values in formula (102.6), we obtain the following expression for the required equilibrium constants:

$$K_p^{(n)}(T) = \frac{g_{n-1}}{2g_n} \left(\frac{2\pi}{m}\right)^{3/2} \frac{\hbar^3}{T^{5/2}} e^{I_n/T} \qquad (104.4)$$

(M. N. Saha, 1921) where m is the electron mass and $I_n = \varepsilon_{0n} - \varepsilon_{0,n-1}$ the energy of the nth ionisation (nth ionisation potential) of the atom.

The degree of n-fold ionisation of the gas becomes of the order of unity as the temperature increases and the equilibrium constant $K_c^{(n)} = PK_p^{(n)}$ decreases to a value of the order of unity. It is very important to note that, despite the exponential dependence of the equilibrium constant on temperature, this stage is reached not when $T \sim I_n$ but at considerably lower temperatures. The reason is that the coefficient of the exponential $e^{I_n/T}$ is small: the quantity $(P/T)(\hbar^2/mT)^{3/2} = (N/V)(\hbar^2/mT)^{3/2}$ is in general very small, being for $T \sim I$ of the order of the ratio of the atomic volume to the volume V/N per atom in the gas.

Thus the gas will be considerably ionised even at temperatures which are small compared with the ionisation energy, but the number of excited atoms in the gas will still be small, since the excitation energy of the atom is in general of the same order as the ionisation energy.

When T becomes comparable with the ionisation energy, the gas is almost completely ionised. At temperatures of the order of the binding energy of the last electron in the atom, the gas may be regarded as consisting of electrons and bare nuclei only.

The binding energy I_1 of the first electron is usually much less than the subsequent ones I_n; there is therefore a range of temperatures in which the gas may be supposed to include only neutral atoms and singly charged ions. Defining the *degree of ionisation* α of the gas as the ratio of the number of ionised atoms to the total number of atoms, we have

$$c = c_1 = \alpha/(1+\alpha), \quad c_0 = (1-\alpha)/(1+\alpha),$$

and equation (104.2) gives $(1-\alpha^2)/\alpha^2 = PK_p^{(1)}$, whence

$$\alpha = 1/\sqrt{(1+PK_p^{(1)})}. \qquad (104.5)$$

This entirely determines the degree of ionisation as a function of pressure and temperature (in the temperature range considered).

[†] For reasons given below we may assume that all atoms and ions are in the ground state, even in a considerably ionised gas. If the atom (or ion) ground state has a fine structure, we assume that T is large compared with the intervals in this structure.

§ 105. Equilibrium with respect to pair production

At extremely high temperatures, comparable with the rest energy[†] mc^2 of the electron, collisions of particles in matter may be accompanied by the formation of electron–positron pairs. The number of particles itself then ceases to be a given quantity, and depends on the conditions of thermal equilibrium.

Pair production (and the reverse process, annihilation) can be regarded thermodynamically as a "chemical reaction" $e^+ + e^- = \gamma$, where the symbols e^+ and e^- denote a positron and an electron, and γ denotes one or more photons. The chemical potential of the photon gas is zero (§ 63). The condition of equilibrium for pair production is therefore

$$\mu^- + \mu^+ = 0, \tag{105.1}$$

where μ^- and μ^+ are the chemical potentials of the electron and positron gases. It should be emphasised that μ here denotes the relativistic expression for the chemical potential, including the rest energy of the particles (cf. § 27), which plays an important part in pair production.

Even at temperatures $T \sim mc^2$, the number of pairs formed per unit volume is very large in comparison with the atomic electron density (see the next footnote). We can therefore suppose with sufficient accuracy that the number of electrons is equal to the number of positrons. Then $\mu^- = \mu^+$, and the condition (105.1) gives $\mu^- = \mu^+ = 0$, i.e. in equilibrium the chemical potentials of the electrons and positrons must be zero.

Electrons and positrons obey Fermi statistics; their number is therefore obtained by integrating the distribution (56.3) with $\mu = 0$:

$$N^+ = N^- = \frac{V}{\pi^2 \hbar^3} \int_0^\infty \frac{p^2 \, dp}{e^{\varepsilon/T} + 1}, \tag{105.2}$$

where ε is determined from the relativistic expression $\varepsilon = c \sqrt{(p^2 + m^2 c^2)}$.

For $T \ll mc^2$, this number is exponentially small ($\propto e^{-mc^2/T}$). In the opposite case ($T \gg mc^2$) we can put $\varepsilon = cp$, and formula (105.2) gives

$$N^+ = N^- = \frac{V}{\pi^2} \left(\frac{T}{\hbar c} \right)^3 \int_0^\infty \frac{x^2 \, dx}{e^x + 1}.$$

[†] The energy $mc^2 = 0.51 \times 10^6$ eV, so that the temperature $mc^2/k = 6 \times 10^9$ degrees.

316 *Chemical Reactions*

The integral in this formula can be expressed in terms of the ζ function (see the second footnote to § 58), giving[†]

$$N^+ = N^- = \frac{3\zeta(3)}{2\pi^2}\left(\frac{T}{\hbar c}\right)^3 V = 0.183(T/\hbar c)^3 V. \tag{105.3}$$

The energy of the positron and electron gases is similarly

$$E^+ = E^- = \frac{VT}{\pi^2}\left(\frac{T}{\hbar c}\right)^3 \int_0^\infty \frac{x^3\,dx}{e^x+1} = 7\pi^2 VT^4/120(\hbar c)^3. \tag{105.4}$$

This quantity is $\frac{7}{8}$ of the energy of black-body radiation in the same volume.

PROBLEM

Determine the equilibrium density of electrons and positrons for $T \ll mc^2$.

SOLUTION. Using the expression (46.1a) for the chemical potential (to which mc^2 must be added), we obtain

$$n^+n^- = 4(mT/2\pi\hbar^2)^3\,e^{-2mc^2/T},$$

where $n^- = N^-/V$ and $n^+ = N^+/V$ are the electron and positron densities. If n_0 is the initial electron density (in the absence of pair production), then $n^- = n^+ + n_0$, and we find

$$n^+ = n^- - n_0 = -\tfrac{1}{2}n_0 + \left[\tfrac{1}{4}n_0^2 + \tfrac{1}{2}(mc/\hbar)^6\,(T/\pi mc^2)^3\,e^{-2mc^2/T}\right]^{1/2}.$$

[†] For $T \sim mc^2$ the volume per pair formed is $\sim (\hbar/mc)^3$. This volume is very small in comparison with the atomic volume (the cube of the Bohr radius, $(\hbar^2/me^2)^3$).

CHAPTER XI

PROPERTIES OF MATTER AT VERY HIGH DENSITY

§ 106. The equation of state of matter at high density

THE study of the properties of matter at extremely high density is of fundamental importance. Let us follow qualitatively the change in these properties
as the density is gradually increased.

When the volume per atom becomes less than the usual size of the atom,
the atoms lose their individuality, and so the substance is transformed into
a highly compressed plasma of electrons and nuclei. If the temperature of
the substance is not too high, the electron component of this plasma is a
degenerate Fermi gas. An unusual property of such a gas has been mentioned
at the end of § 57: it becomes more nearly ideal as the density increases.
Thus, when the substance is sufficiently compressed, the interaction of the
electrons with the nuclei (and with one another) becomes unimportant, and
the formulae for an ideal Fermi gas may be used. According to (57.9) this
occurs when $n_e \gg (m_e e^2/\hbar^2)^3 Z^2$ holds, where n_e is the number density of
electrons, m_e the electron mass, and Z some mean atomic number of the
substance. We therefore find for the total mass density of the substance the
inequality

$$\varrho \gg (m_e e^2/\hbar^2)^3 m' Z^2 \sim 20 Z^2 \text{ g/cm}^3, \qquad (106.1)$$

where m' is the mass per electron, so that $\varrho = n_e m'$.[†] The "gas of nuclei"
may still be far from degeneracy, because of the large mass of the nucleus,
but its contribution to the pressure of the substance, for example, is in any
case entirely negligible in comparison with that of the electron gas.

Thus the thermodynamic quantities for a substance under the conditions

---

[†] In all the numerical estimates given in this section it is assumed that the mean
atomic weight of the substance is twice its mean atomic number, so that m' is
twice the nucleon mass.

It may be mentioned that the degeneracy temperature of the electrons corresponding to a density $\varrho \sim 20 Z^2$ g/cm^3 is of the order of $10^6 Z^{4/3}$ degrees.

in question are given by the formulae derived in § 57, applied to the electron component. In particular, for the pressure we have[†]

$$P = \frac{(3\pi^2)^{2/3}}{5} \frac{\hbar^2}{m_e} \left(\frac{\varrho}{m'}\right)^{5/3}. \tag{106.2}$$

The condition (106.1) on the density gives for the pressure the numerical inequality $P \gg 5 \times 10^8 \, Z^{10/3}$ bar.

In the above formulae the electron gas is assumed non-relativistic. This implies that the Fermi limiting momentum p_F is small compared with mc (see § 61), giving the numerical inequalities

$$\varrho \ll 2 \times 10^6 \text{ g/cm}^3, \quad P \ll 10^{17} \text{ bar.}$$

When the density and pressure of the gas become comparable with these values, the electron gas becomes relativistic, and when the opposite inequalities hold we have the extreme relativistic case, where the equation of state is determined by formula (61.4):[‡]

$$P = \tfrac{1}{4}(3\pi^2)^{1/3} \hbar c (\varrho/m')^{4/3}. \tag{106.3}$$

A further increase in density leads to states where nuclear reactions consisting in the capture of electrons by nuclei (with emission of neutrinos) are thermodynamically favoured. Such a reaction decreases the charge on the nucleus (leaving its atomic weight constant), and this in general causes a decrease in the binding energy of the nucleus, i.e. a decrease in its mass defect. The energy required to bring about such a process is more than counterbalanced at sufficiently high densities by the decrease in the energy of the degenerate electron gas because of the smaller number of electrons.

It is not difficult to write down the thermodynamic conditions which govern the "chemical equilibrium" of the nuclear reaction mentioned, which may be symbolically written as

$$A_Z + e^- = A_{Z-1} + \nu,$$

where A_Z denotes a nucleus of atomic weight A and charge Z, e^- an electron and ν a neutrino. The neutrinos are not retained by matter and leave

---

[†] Numerically, this formula gives

$$P = 1.0 \times 10^{13} \, (\varrho/A')^{5/3} \text{ dyn/cm}^2 = 1.0 \times 10^7 \, (\varrho/A')^{5/3} \text{ bar,} \tag{106.2a}$$

where $A' = m/m_n$ is the atomic weight of the substance per electron (m_n being the nucleon mass); ϱ is measured in g/cm^3. The corrections to (106.2) because of the Coulomb interaction of the particles have been discussed in § 80.

[‡] With the notation of (106.2a),

$$P = 1.2 \times 10^9 \, (\varrho/A')^{4/3} \text{ bar.} \tag{106.3a}$$

the body; such a process must lead to a steady cooling of the body. Thus thermal equilibrium can be meaningfully considered in these conditions only if the temperature of the substance is taken as zero. The chemical potential of the neutrinos will not then appear in the equation of equilibrium. The chemical potential of the nuclei is mainly governed by their internal energy, which we denote by $-\varepsilon_{A,Z}$ (the term "binding energy" usually refers to the positive quantity $\varepsilon_{A,Z}$). Finally, let $\mu_e(n_e)$ denote the chemical potential of the electron gas as a function of the number density n_e of particles in it. Then the condition of chemical equilibrium takes the form $-\varepsilon_{A,Z}+\mu_e(n_e) = -\varepsilon_{A,Z-1}$, or, putting $\varepsilon_{A,Z}-\varepsilon_{A,Z-1} = \Delta$,

$$\mu_e(n_e) = \Delta.$$

Using formula (61.2) for the chemical potential of an extreme relativistic degenerate gas, we thus find

$$n_e = \Delta^3/3\pi^2(c\hbar)^3. \tag{106.4}$$

The equilibrium condition therefore gives a constant value of the electron density. This means that, as the density of the substance gradually increases, the nuclear reaction in question begins to occur when the electron density reaches the value (106.4). As the substance is further compressed, more and more nuclei will each capture an electron, so that the total number of electrons will decrease but their density will remain constant. Together with the electron density the pressure of the substance will also remain constant, being again determined mainly by the pressure of the electron gas: substitution of (106.4) in (106.3) gives

$$P = \Delta^4/12\pi^2(\hbar c)^3. \tag{106.5}$$

This will continue until each nucleus has captured an electron.

At still higher densities and pressures the nuclei will capture further electrons, the nuclear charge being thus reduced further. Ultimately the nuclei will contain so many neutrons that they become unstable and break up. At a density $\varrho \sim 3\times10^{11}$ g/cm³ (and pressure $P \sim 10^{24}$ bar) the neutrons begin to be more numerous than the electrons, and when $\varrho \sim 10^{12}$ g/cm³ the pressure due to the neutrons begins to predominate (F. Hund, 1936). This is the beginning of a density region in which matter may be regarded as essentially a degenerate neutron Fermi gas with a small number of electrons and various nuclei, whose concentrations are given by the equilibrium conditions for the corresponding nuclear reactions. The equation of state of matter in this range is

$$P = \frac{(3\pi^2)^{2/3}}{5}\frac{\hbar^2}{m_n^{8/3}}\varrho^{5/3} = 5.5\times10^3 \varrho^{5/3} \text{ bar,} \tag{106.6}$$

where m_n is the neutron mass.

Finally, at densities $\varrho \gg 6 \times 10^{15}$ g/cm³, the degenerate neutron gas becomes extreme-relativistic, and the equation of state is

$$P = \frac{(3\pi^2)^{1/3}}{4} \hbar c \left(\frac{\varrho}{m_n}\right)^{4/3} = 1.2 \times 10^9 \, \varrho^{4/3} \text{ bar.} \tag{106.7}$$

It should be remembered, however, that at densities of the order of that of nuclear matter the specifically nuclear forces (strong interaction of nucleons) become important. In this range of densities formula (106.7) can be only qualitative. In the present state of our knowledge concerning strong interactions we can draw no definite conclusions concerning the state of matter at densities considerably above the nuclear value. We shall merely mention that in this range other particles besides neutrons may be expected to appear. Since particles of each kind occupy a separate group of states, the conversion of neutrons into other particles may be thermodynamically favoured because of the decrease in the limiting energy of the Fermi distribution of neutrons.

§ 107. Equilibrium of bodies of large mass

Let us consider a body of very large mass, the parts of which are held together by gravitational attraction. Actual bodies of large mass that are known to us, namely the stars, continuously radiate energy and are certainly not in thermal equilibrium. It is, however, of fundamental interest to discuss an equilibrium body of large mass. We shall neglect the effect of temperature on the equation of state, i.e. consider a body at absolute zero—a "cold" body. Since in actual conditions the temperature of the outer surface is considerably lower than the internal temperature, a discussion of a body with a constant non-zero temperature is in any case devoid of physical meaning.

We shall further assume that the body is not rotating; then it will be spherical in equilibrium, and the density distribution will be symmetrical about the centre.

The equilibrium distribution of density (and of the other thermodynamic quantities) in the body will be determined by the following equations. The Newtonian gravitational potential ϕ satisfies the differential equation $\triangle\phi = 4\pi G\varrho$, where ϱ is the density of the substance and G the Newtonian constant of gravitation. In the case of spherical symmetry,

$$\frac{1}{r^2} \frac{d}{dr} \left(r^2 \frac{d\phi}{dr}\right) = 4\pi G\varrho. \tag{107.1}$$

Moreover, in thermal equilibrium the condition (25.2) must be satisfied. In the gravitational field the potential energy of a particle of the body of mass

m' is $m'\phi$, and so we have

$$\mu + m'\phi = \text{constant}, \tag{107.2}$$

where for brevity the suffix zero is omitted from the chemical potential of the substance in the absence of the field. Expressing ϕ in terms of μ by means of (107.2) and substituting in (107.1), we have

$$\frac{1}{r^2} \frac{d}{dr} \left(r^2 \frac{d\mu}{dr} \right) = -4\pi m' G \varrho. \tag{107.3}$$

As the mass of the gravitating body increases, so of course does its mean density, as the following calculations will confirm. When the total mass M of the body is sufficiently large, therefore, we can, as shown in § 106, regard the substance as a degenerate electron Fermi gas, initially non-relativistic and then at still greater masses relativistic.

The chemical potential of a non-relativistic degenerate electron gas is related to the density ϱ of the body by

$$\mu = \frac{(3\pi^2)^{2/3}}{2} \frac{\hbar^2}{m_e m'^{2/3}} \varrho^{2/3}; \tag{107.4}$$

see formula (57.3), with $\varrho = m'N/V$ (m' is the mass per electron and m_e the electron mass). Expressing ϱ in terms of μ and substituting in (107.3), we have[†]

$$\frac{1}{r^2} \frac{d}{dr} \left(r^2 \frac{d\mu}{dr} \right) = -\lambda \mu^{3/2}, \quad \lambda = \frac{2^{7/2} m_e^{3/2} m'^2 G}{3\pi \hbar^3}. \tag{107.5}$$

[†] It is easy to see that, for an electrically neutral gas consisting of electrons and atomic nuclei, the equilibrium condition can be written in the form (107.2) with the electron chemical potential as μ and the mass per electron as m'. For the derivation of this equilibrium condition (§ 25) involves considering the transport of an infinitesimal amount of substance from one place to another. In a gas consisting of both positively and negatively charged particles, such transport must be regarded as that of a certain quantity of neutral matter (i.e. electrons and nuclei together). The separation of the positive and negative charges is energetically very unfavourable, because of the resulting very large electric fields. We therefore obtain the equilibrium condition in the form

$$\mu_{nuc} + Z\mu_{el} + (m_{nuc} + Zm_{el})\phi = 0$$

(with Z electrons per nucleus). Owing to the large mass of the nuclei (compared with that of the electrons) their chemical potential is very small compared with μ_{el}. Neglecting μ_{nuc} and dividing the equation by Z, we obtain

$$\mu_{el} + m'\phi = 0.$$

As in § 106, it will be assumed in the numerical estimates that m' is twice the nucleon mass ($m' = 2m_n$).

The physically meaningful solutions of this equation must not have singularities at the origin: $\mu \rightarrow$ constant for $r \rightarrow 0$. This requirement necessarily imposes on the first derivative the condition

$$d\mu/dr = 0 \quad \text{for} \quad r = 0, \tag{107.6}$$

as follows immediately from equation (107.5) after integration over r:

$$\frac{d\mu}{dr} = -\frac{\lambda}{r^2} \int_0^r r^2 \mu^{3/2} \, dr.$$

A number of important results can be derived from equation (107.5) by simple dimensional considerations. The solution of (107.5) contains only two constants, λ and (for instance) the radius R of the body, a knowledge of which uniquely defines the solution. From these two quantities we can form only one quantity with the dimensions of length, the radius R itself, and one with the dimensions of energy, $1/\lambda^2 R^4$ (the constant λ having dimensions $\text{cm}^{-2} \text{ erg}^{-1/2}$). It is therefore clear that the function $\mu(r)$ must have the form

$$\mu(r) = \frac{1}{\lambda^2 R^4} f\left(\frac{r}{R}\right), \tag{107.7}$$

where f is some function of the dimensionless ratio r/R only. Since the density ϱ is proportional to $\mu^{3/2}$, the density distribution must be of the form

$$\varrho(r) = \frac{\text{constant}}{R^6} F\left(\frac{r}{R}\right).$$

Thus, when the size of the sphere varies, the density distribution in it remains similar in form, the density at corresponding points being inversely proportional to R^6. In particular, the mean density of the sphere is inversely proportional to R^6:

$$\bar{\varrho} \propto 1/R^6.$$

The total mass M of the body is therefore inversely proportional to the cube of the radius:

$$M \propto 1/R^3.$$

These two relations may also be written

$$R \propto M^{-1/3}, \quad \bar{\varrho} \propto M^2. \tag{107.8}$$

Thus the dimensions of an equilibrium sphere are inversely proportional to the cube root of its total mass, and the mean density is proportional to the square of the mass. The latter result confirms the assumption made above that the density of a gravitating body increases as its mass increases.

The fact that a gravitating sphere of non-relativistic degenerate Fermi gas can be in equilibrium for any total mass M can be seen *a priori* from the following qualitative argument. The total kinetic energy of the particles in such a gas is proportional to $N(N/V)^{2/3}$ (see (57.6)), or, what is the same thing, to $M^{5/3}/R^2$, and the gravitational energy of the gas as a whole is negative and proportional to M^2/R. The sum of two such expressions can have a minimum (as a function of R) for any M, and at the minimum $R \propto M^{-1/3}$.

Substituting (107.7) in (107.5) and using the dimensionless variable $\xi = r/R$, we find that the function $f(\xi)$ satisfies the equation

$$\frac{1}{\xi^2} \frac{d}{d\xi} \left(\xi^2 \frac{df}{d\xi} \right) = -f^{3/2} \qquad (107.9)$$

with the boundary conditions $f'(0) = 0$, $f(1) = 0$. This equation cannot be solved analytically, and must be integrated numerically. It may be mentioned that as a result we find $f(0) = 178.2$, $f'(1) = -132.4$.

Using these numerical values it is easy to determine the value of the constant MR^3. Multiplying equation (107.1) by $r^2\, dr$ and integrating from 0 to R, we obtain

$$GM = R^2[d\phi/dr]_{r=R} = -(R^2/m')[d\mu/dr]_{r=R} = -f'(1)/m'\lambda^2 R^3,$$

whence

$$MR^3 = 91.9\hbar^6/G^3 m_e^3 m'^5 = 2.2 \times 10^{13}\,(m_n/m')^5\,\odot\,\text{km}^3, \qquad (107.10)$$

where $\odot = 2 \times 10^{33}$ g is the Sun's mass. Finally, the ratio of the central density $\varrho(0)$ to the mean density $\bar{\varrho} = 3M/4\pi R^3$ is easily found to be

$$\varrho(0)/\bar{\varrho} = -f^{3/2}(0)/3f'(1) = 5.99. \qquad (107.11)$$

Curve 1 in Fig. 50 (p. 324) shows the ratio $\varrho(r)/\varrho(0)$ as a function of r/R.[†]

Let us now examine the equilibrium of a sphere consisting of a degenerate extreme-relativistic electron gas. The total kinetic energy of the particles of such a gas is proportional to $N(N/V)^{1/3}$ (see (61.3)), and hence to $M^{4/3}/R$; the gravitational energy is proportional to $-M^2/R$. Thus the two quantities depend on R in the same manner, and their sum will also be of the form constant/R. It follows that the body cannot be in equilibrium: if the constant is positive, the body will tend to expand until the gas becomes non-relativistic; if the constant is negative, a decrease of R to zero corresponds to a decreased total energy, i.e. the body will contract without limit. The body can be in

---

[†] In § 106 we have seen that matter may be regarded as a non-relativistic degenerate electron gas at densities $\varrho \gg 20Z^2$ g/cm^3. If this inequality is satisfied for the mean density of the sphere considered, its mass must satisfy the condition $M \gg 5 \times 10^{-3}\,Z\odot$. The corresponding radii are less than $5 \times 10^4\,Z^{-1/3}$ km.

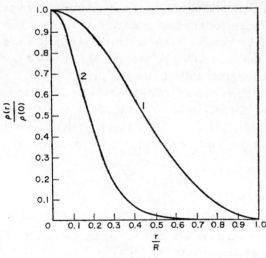

FIG. 50

equilibrium only in the special case where the constant is zero, and the equilibrium is then neutral, the value of R being arbitrary.

This qualitative argument is, of course, entirely confirmed by exact quantitative analysis. The chemical potential of the relativistic gas considered is related to the density by

$$\mu = (3\pi^2)^{1/3}\,\hbar c(\varrho/m')^{1/3} \tag{107.12}$$

(see (61.2)). Instead of (107.5) we now have

$$\frac{1}{r^2}\frac{d}{dr}\left(r^2\frac{d\mu}{dr}\right) = -\lambda\mu^3, \quad \lambda = \frac{4Gm'^2}{3\pi c^3\hbar^3}. \tag{107.13}$$

Since λ now has dimensions $\mathrm{erg}^{-2}\,\mathrm{cm}^{-2}$, we find that the chemical potential as a function of r must be of the form

$$\mu(r) = \frac{1}{R\sqrt{\lambda}}f\left(\frac{r}{R}\right), \tag{107.14}$$

and the density distribution

$$\varrho(r) = \frac{\text{constant}}{R^3}F\left(\frac{r}{R}\right).$$

Thus the mean density is inversely proportional to R^3, and the total mass $M \propto R^3\bar{\varrho}$ is independent of R:

$$\bar{\varrho} \propto 1/R^3, \quad M = \text{constant} \equiv M_0. \tag{107.15}$$

M_0 is the only value of the mass for which equilibrium is possible: for $M >$
M_0 the body will tend to contract indefinitely, and for $M < M_0$ it will expand.

For an exact calculation of the "critical mass" M_0, it is necessary to integrate numerically the equation

$$\frac{1}{\xi^2} \frac{d}{d\xi} \left(\xi^2 \frac{df}{d\xi} \right) = -f^3, \quad f'(0) = 0, \quad f(1) = 0, \qquad (107.16)$$

which is satisfied by the function $f(\xi)$ in (107.14). The result is $f(0) = 6.897$,
$f'(1) = -2.018$. For the total mass we find

$$GM_0 = R^2[d\phi/dr]_{r=R} = -f'(1)/m'\sqrt{\lambda},$$

whence

$$M_0 = \frac{3.1}{m'^2} \left(\frac{\hbar c}{G} \right)^{3/2} = 5.8(m_n/m')^2 \odot. \qquad (107.17)$$

Putting $m' = 2m_n$, we find $M_0 = 1.45 \odot$. Finally, the ratio of the central density to the mean density is $\varrho(0)/\bar{\varrho} = -f^3(0)/3f'(1) = 54.2$. Curve 2 in Fig. 50 shows $\varrho(r)/\varrho(0)$ in the extreme relativistic case as a function of r/R.[†]

The results obtained above concerning the relation between the mass and the radius of a "cold" spherical body in equilibrium can be represented by a single relation $M = M(R)$ for all radii R. For large R (and therefore for small densities), the electron gas may be regarded as non-relativistic, and the function $M(R)$ decreases as $1/R^3$. When R is sufficiently small, however, the density is so large that we have the extreme relativistic case, and the function $M(R)$ is almost a constant M_0; strictly $M(R) \to M_0$ when $R \to 0$. Figure 51 shows the curve $M = M(R)$ calculated with $m' = 2m_n$.[‡] It should be noted that the limiting value $1.45 \odot$ is reached only very gradually; this is because the density decreases rapidly away from the centre of the body, and so the extreme relativistic case may hold near the centre while the gas remains non-relativistic in a considerable part of the volume of the body. We may also mention that the initial part of the curve (R small) has no real physical significance: at sufficiently small radii the density becomes so large that nuclear reactions begin to occur. The pressure will then increase with density less

† The formal problem of the equilibrium of a gravitating gaseous sphere with a power-law dependence of P on ϱ was investigated by R. Emden (1907). The physical deduction of the existence of the limiting mass and its value (107.17) was made by S. Chandrasekhar (1931) and L. Landau (1932).

‡ The intermediate part of the curve is constructed by numerical integration of equation (107.3) with the exact relativistic equation of state for a degenerate gas; see § 61, Problem 3.

rapidly than $\varrho^{4/3}$, and for such an equation of state no equilibrium is possible.[†]

Finally, this curve also has no meaning for large values of R (and small M): as has already been mentioned (see the second footnote to this section), in this range the equation of state used above becomes invalid. Here it should be pointed out that there is an upper limit to the possible size of a "cold" body,

FIG. 51

since on the curve in Fig. 51 large dimensions of the body correspond to small masses and small densities, but when the density is sufficiently small the substance will be in the ordinary "atomic" state and will be solid at the low temperatures here considered. The dimensions of a body consisting of such a substance will obviously decrease as its mass decreases further, and not increase as shown in Fig. 51. The true curve $R = R(M)$ must therefore have a maximum for some value of M.

The order of magnitude of the maximum radius can easily be determined by noting that it must correspond to the density at which the interaction between electrons and nuclei becomes important, i.e. for $\varrho \sim (m_e e^2/\hbar^2)^3 m'Z^2$ (see 106.1)). Combining this with equation (107.10), we obtain

$$R_{\max} \sim \hbar^2/G^{1/2}em_e m'Z^{1/3} \sim 10^5 m_n/m'Z^{1/3} \text{ km.} \tag{107.18}$$

[†] If the chemical potential is proportional to a power of the density, $\mu \propto \varrho^n$ (and so $P \propto \varrho^{n+1}$), the internal energy of the body is proportional to $V\varrho^{n+1}$, i.e. to M^{n+1}/R^{3n}; the gravitational energy is again proportional to $-M^2/R$. It is easy to see that for $n < \frac{1}{3}$ the sum of two such expressions has an extremum as a function of R, but this extremum is a maximum, not a minimum.

§ 108. The energy of a gravitating body

The gravitational potential energy E_{gr} of a body is given by the integral

$$E_{gr} = \tfrac{1}{2} \int \varrho \phi \, dV, \tag{108.1}$$

taken over the whole volume of the body. It will, however, be more convenient for us to start from a different expression for this quantity, which may be found as follows. Let us imagine the body to be gradually "built up" from material brought from infinity. Let $M(r)$ be the mass of substance within a sphere of radius r. Let us suppose that a mass $M(r)$ with a certain value of r has already been brought from infinity; then the work required to add a further mass $dM(r)$ is equal to the potential energy of that mass (in the form of a spherical shell of radius r and thickness dr) in the field of the mass $M(r)$, i.e. $-GM(r) \, dM(r)/r$. The total gravitational energy of a sphere of radius R is therefore

$$E_{gr} = -G \int \frac{M(r) \, dM(r)}{r}. \tag{108.2}$$

Differentiation of the equilibrium condition (107.2) gives

$$v \frac{dP}{dr} + m' \frac{d\phi}{dr} = 0;$$

the differentiation must be at constant temperature, and $(\partial \mu / \partial P)_T = v$ is the volume per particle. The derivative $-d\phi/dr$ is the force of attraction on unit mass at a distance r from the centre, and equals $-GM(r)/r^2$. Using also the density $\varrho = m'/v$, we have

$$\frac{1}{\varrho} \frac{dP}{dr} = -\frac{GM(r)}{r^2}. \tag{108.3}$$

Expressing $GM(r)/r$ from this in terms of dP/dr and writing $dM(r) = \varrho(r) \cdot 4\pi r^2 \, dr$, we can put (108.2) in the form

$$E_{gr} = 4\pi \int_0^R r^3 \frac{dP}{dr} \, dr,$$

and then integrate by parts (bearing in mind that at the boundary of the body $P(R) = 0$ and that $r^3 P \to 0$ as $r \to 0$):

$$E_{gr} = -12\pi \int_0^R P r^2 \, dr = -3 \int P \, dV. \tag{108.4}$$

Thus the gravitational energy of an equilibrium body can be expressed as an integral of the pressure over the volume.

Let us apply this formula to the degenerate Fermi gases considered in § 107. We make the calculation for the general case, and take the chemical potential of the substance to be proportional to some power of its density:

$$\mu = K\varrho^{1/n}. \tag{108.5}$$

Since $d\mu = v\, dP = (m'/\varrho)\, dP$, we have

$$P = \frac{1}{n+1}\frac{K}{m'}\varrho^{1/n+1}. \tag{108.6}$$

In the equilibrium condition $\mu/m' + \phi = $ constant, the constant is just the potential at the boundary of the body, where μ vanishes; this potential is $-GM/R$ ($M = M(R)$ being the total mass of the body), and so we can write

$$\phi = -\frac{\mu}{m'} - \frac{GM}{R}.$$

We substitute this expression in the integral (108.1) which gives the gravitational energy, and use formulae (108.5), (108.6), obtaining

$$E_{\text{gr}} = -\frac{1}{2m'}\int \mu\varrho\, dV - \frac{GM}{2R}\int \varrho\, dV = -\frac{n+1}{2}\int P\, dV - \frac{GM^2}{2R}.$$

Finally, expressing the integral on the right in terms of E_{gr} by (108.4), we have

$$E_{\text{gr}} = -\frac{3}{5-n}\frac{GM^2}{R}. \tag{108.7}$$

Thus the gravitational energy of the body can be expressed by a simple formula in terms of its total mass and its radius.

A similar formula can also be obtained for the internal energy E of the body. The internal energy per particle is $\mu - Pv$ (for zero temperature and entropy); the energy per unit volume is therefore

$$\frac{1}{v}(\mu - Pv) = \frac{\varrho\mu}{m'} - P = nP,$$

where (108.5) and (108.6) have been used to derive the second equation. The internal energy of the whole body is therefore

$$E = n\int P\, dV = -\frac{1}{3}nE_{\text{gr}} = \frac{n}{5-n}\frac{GM^2}{R}. \tag{108.8}$$

Finally, the total energy of the body is

$$E_{\text{tot}} = E + E_{\text{gr}} = -\frac{3-n}{5-n}\frac{GM^2}{R}. \tag{108.9}$$

For a non-relativistic degenerate gas $n = \frac{3}{2}$, and so[†]

$$E_{gr} = -\frac{6}{7}\frac{GM^2}{R}, \quad E = \frac{3}{7}\frac{GM^2}{R}, \quad E_{tot} = -\frac{3}{7}\frac{GM^2}{R}. \quad (108.10)$$

In the extreme relativistic case, $n = 3$, so that

$$E_{gr} = -E = -\frac{3}{2}\frac{GM^2}{R}, \quad E_{tot} = 0. \quad (108.11)$$

The total energy is zero in this case, in accordance with the qualitative arguments given in § 107 concerning the equilibrium of such a body.[‡]

§ 109. Equilibrium of a neutron sphere

For a body of large mass there are two possible equilibrium states. One corresponds to the state of matter consisting of electrons and nuclei, as assumed
in the numerical estimates in § 107. The other corresponds to the "neutron"
state of matter, in which almost all the electrons have been captured by pro-
tons and the substance may be regarded as a neutron gas. When the body is
sufficiently massive, the second possibility must always become thermodynamically more favourable than the first (W. Baade and F. Zwicky, 1934).
Although the transformation of nuclei and electrons into free neutrons involves a considerable expenditure of energy, when the total mass of the body
is sufficiently great this is more than counterbalanced by the release of gravitational energy owing to the decrease in size and increase in density of the
body.

First of all, let us examine the conditions in which the neutron state of a
body can correspond to any thermodynamic equilibrium (which may be
metastable). To do this, we start from the equilibrium condition $\mu + m_n \phi =$
constant, where μ is the chemical potential (the thermodynamic potential per
neutron), m_n the neutron mass, and ϕ the gravitational potential.

Since the pressure must be zero at the boundary of the body, it is clear that
in an outer layer of the body the substance will be at low pressure and

---

[†] In this case $2E = -E_{gr}$, in agreement with the virial theorem of mechanics,
applied to a system of particles interacting according to Newton's law; see *Mechanics*, § 10.

[‡] In order to avoid misunderstanding, it may be mentioned that the energy E
(and therefore E_{tot} in (108.11)) includes the rest energy of the particles (which pro-
duce the pressure P). If E_{tot} is defined as the "binding energy" of the body (with
the energy of the matter dissipated through space taken as zero), the rest energy
of the particles must be subtracted from it.

density and will therefore consist of electrons and nuclei. Although the thickness of this "shell" may be comparable with the radius of the dense inner neutron "core", the density of the outer layer is much lower, and so its total mass may be regarded as small compared with the mass of the core.[†]

Let us compare the values of $\mu + m_n \phi$ at two points: in the dense core near its boundary and near the outer boundary of the shell. The gravitational potential at these points may be taken as $-GM/R$ and $-GM/R'$, where R and R' are the core and shell radii, and M the mass of the core, which in the approximation used here is equal to the total mass of the body. The chemical potential at both points is determined mainly by the internal energy (binding energy) of the corresponding particles, which is large compared with their thermal energy. The difference between the two chemical potentials may therefore be taken as simply equal to the difference between the rest energy of a neutral atom (i.e. a nucleus and Z electrons) per unit atomic weight and the rest energy of the neutron; let this quantity be denoted by Δ. Then, equating the values of $\mu + m_n \phi$ at the two points considered, we have

$$m_n MG \left(\frac{1}{R} - \frac{1}{R'} \right) = \Delta.$$

Hence, whatever the radius R', the mass and radius of the neutron core must satisfy the inequality

$$m_n MG/R > \Delta. \tag{109.1}$$

Applying the results of § 107 to a spherical body consisting of a degenerate (non-relativistic) neutron gas, we find that M and R are related by

$$MR^3 = 91.9\hbar^6/G^3 m_n^8 = 3.6 \times 10^3 \odot \text{ km}^3 \tag{109.2}$$

(formula (107.10) with m_e and m' replaced by m_n). Hence expressing M in terms of R and substituting in (109.1), we obtain an inequality for M, which in numerical form is $M > \sim 0.2\odot$. For example, with Δ for oxygen we get $M > 0.17\odot$, and for iron $M > 0.18\odot$. These masses correspond to radii $R < 26$ km.[‡]

This inequality gives a lower limit of mass, beyond which the neutron state of the body cannot be stable. It does not, however, ensure complete stability; the state may be metastable. To determine the limit of metastability, we must compare the total energies of the body in two states: the neutron state and the electron–nucleus state. The conversion of the whole mass M

[†] There is, of course, no sharp boundary between the "core" and the "shell", and the transition between them is continuous.

[‡] It must be emphasised that no literal astrophysical significance is to be attached to the numerical estimates in this section, which are based on simple assumptions about the structure of the body.

from the electron–nucleus state to the neutron state requires an expenditure of energy $M\Delta/m_n$ to counterbalance the binding energy of the nuclei. In the process, energy is released because of the contraction of the body; according to formula (108.10), this gain of energy is

$$\frac{3GM^2}{7}\left(\frac{1}{R_n}-\frac{1}{R_e}\right),$$

where R_n is the radius of the body in the neutron state, given by formula (109.2), and R_e its radius in the electron–nucleus state, given by (107.10). Since $R_e \gg R_n$, the quantity $1/R_e$ may be neglected, and we obtain the following sufficient condition for complete stability of the neutron state of the body (omitting the suffix in R_n):

$$3GMm_n/7R > \Delta. \tag{109.3}$$

Comparing this condition with (109.1) and using (109.2), we see that the lower limit of mass determined by the inequality (109.3) is greater by a factor $(7/3)^{3/4} = 1.89$ than that given by (109.2). Numerically, the limit of metastability of the neutron state is therefore at a mass $M \cong \frac{1}{3}\odot$ (and radius $R \cong 22$ km).[†]

Let us now consider the upper limit of the range of mass values for which a neutron body can be in equilibrium. If we were to use the results of § 107 (formula (107.17), with m_n in place of m'), the value obtained for this limit would be $6\,\odot$. In reality, however, these results are not applicable here, for the following reason. In a relativistic neutron gas, the kinetic energy of the particles is of the order of, or greater than, the rest energy, and the gravitational potential $\phi \sim c^2$.[‡] In consequence it is no longer valid to use the Newtonian gravitational theory, and the calculations must be based on the general theory of relativity; and, as we shall see later, we find that the extreme relativistic case is no longer reached. The calculations must therefore make use of the exact equation of state of a degenerate Fermi gas; see § 61, Problem 3.

The calculations are effected by numerical integration of the equations of a spherically symmetric static gravitational field, and the results are as follows.[§]

[†] The mean density of the body is then 1.4×10^{13} g/cm^3, and so the neutron gas may in fact still be regarded as non-relativistic, and the formulae used here are still valid.

[‡] In the relativistic *electron* gas, the kinetic energy of the particles is comparable with the rest energy of the electrons, but is still small in comparison with the rest energy of the nuclei, which contribute most of the mass of the substance.

[§] The details of the calculations are given in the original paper by J. R. Oppenheimer and G. M. Volkoff, *Physical Review* **55**, 374, 1939,

The limiting mass of a neutron sphere in equilibrium is found to be only $M_{max} = 0.76\odot$, and this value is reached at a finite radius $R_{min} = 9.4$ km. Figure 52 shows a graph of the relation obtained between the mass M and the radius R. Stable neutron spheres of larger mass or smaller radius cannot, therefore, exist. It should be mentioned that the mass M here denotes the product $M = Nm_n$, where N is the total number of particles (i.e. neutrons) in the sphere. This quantity is not equal to the gravitational mass M_{gr} of the body, which determines the gravitational field created by it in the surrounding space. Because of the "gravitational mass defect", in stable states we always have $M_{gr} < M$ (in particular, for $R = R_{min}$, $M_{gr} = 0.95M$).[†]

Fig. 52

The question arises of the behaviour of a spherical body of mass exceeding M_{max}. It is clear *a priori* that such a body must tend to contract indefinitely. The nature of this unrestrained *gravitational collapse* has been described in *Fields*, §§ 102–104.

It should be noted that the possibility in principle of gravitational collapse, which (for the model considered of a spherical body) is unavoidable for $M > M_{max}$, is not in fact restricted to large masses. A "collapsing" state exists for any mass, but for $M < M_{max}$ it is separated by a very high energy barrier from the static equilibrium state.[‡]

[†] The point $R = R_{min}$ in Fig. 52 is in fact a maximum on the curve $M = M(R)$. This curve continues beyond the maximum as an inward spiral which asymptotically approaches a centre. The parameter which increases monotonically along the curve is the density at the centre of the sphere, which tends to infinity for a sphere corresponding to the limiting point of the spiral (N. A. Dmitriev and S. A. Kholin, 1963). However, no part of the curve for $R < R_{min}$ corresponds to a stable state of the sphere. For an account of the investigation see N. A. Dmitriev and S. A. Kholin, *Voprosy kosmogonii* **9**, 254, 1963; B. K. Harrison, K. S. Thorne, M. Wakano and J. A. Wheeler, *Gravitation Theory and Gravitational Collapse*, University of Chicago Press, Chicago, 1965.

[‡] See Ya. B. Zel'dovich, *Soviet Physics JETP* **15**, 446, 1962.

CHAPTER XII

FLUCTUATIONS

§ 110. The Gaussian distribution

IT HAS already been stressed several times that the physical quantities which describe a macroscopic body in equilibrium are, almost always, very nearly equal to their mean values. Nevertheless, deviations from the mean values, though small, do occur (quantities are said to *fluctuate*), and the problem arises of finding the probability distribution of these deviations.

Let us consider some closed system, and let x be some physical quantity describing the system as a whole or some part of it (in the former case x must not, of course, be a quantity which is strictly constant for a closed system, such as its energy). In what follows it will be convenient to suppose that the mean value $\bar{x}$ has already been subtracted from x, and so we shall everywhere assume that $\bar{x} = 0$.

The discussion in § 7 has shown that, if the entropy of a system is formally regarded as a function of the exact values of the energies of the subsystems, the function e^S will give the probability distribution for these energies (formula (7.17)). It is easy to see, however, that the discussion made no use of any specific properties of the energy. Similar arguments will therefore show that the probability for a quantity x to have a value in the interval from x to $x+dx$ is proportional to $e^{S(x)}$, where $S(x)$ is the entropy formally regarded as a function of the exact value of x. Denoting this probability by $w(x)\,dx$, we have[†]

$$w(x) = \text{constant} \times e^{S(x)}. \tag{110.1}$$

Before proceeding to examine the consequences of this formula, let us consider its range of applicability. All the arguments leading to formula (110.1) tacitly assume that the quantity x behaves classically.[‡] We must therefore find a condition which ensures that quantum effects are negligible.

[†] This formula was first applied to the study of fluctuations by A. Einstein (1907).

[‡] This does not mean, of course, that the whole system must be a classical one. Variables other than x pertaining to the system may have quantum behaviour.

As we know from quantum mechanics, the relation $\Delta E \, \Delta x \sim \hbar \dot{x}$ exists between the quantum uncertainties of energy and of some quantity x, $\dot{x}$ being the classical rate of change of x (see *Quantum Mechanics*, § 16).

Let τ be a time[†] expressing the rate of change of the quantity x which is considered here and which has a non-equilibrium value; then $\dot{x} \sim x/\tau$, and so $\Delta E \, \Delta x \sim \hbar x/\tau$. It is clear that the quantity x may be said to have a definite value only if its quantum uncertainty is small: $\Delta x \ll x$, whence $\Delta E \gg \hbar/\tau$. Thus the quantum uncertainty of energy must be large in comparison with $\hbar/\tau$. The entropy of the system will then have uncertainty $\Delta S \gg \hbar/\tau T$.

If formula (110.1) is to be meaningful, it is clearly necessary for the uncertainty of entropy to be small compared with unity:

$$T \gg \hbar/\tau, \qquad \tau \gg \hbar/T. \tag{110.2}$$

This is the required condition. When the temperature is too low or when the quantity x varies too rapidly (τ is too small) the fluctuations cannot be treated thermodynamically, and the purely quantum fluctuations become of major importance.

Let us now return to formula (110.1). The entropy S has a maximum for $x = \bar{x} = 0$. Hence $\partial S/\partial x = 0$ and $\partial^2 S/\partial x^2 < 0$ for $x = 0$. In fluctuations the quantity x is very small. Expanding $S(x)$ in powers of x and retaining only terms of up to the second order, we obtain

$$S(x) = S(0) - \tfrac{1}{2}\beta x^2, \tag{110.3}$$

where β is a positive constant. Substitution in (110.1) gives the probability distribution in the form

$$w(x) \, dx = A e^{-\frac{1}{2}\beta x^2} \, dx.$$

The normalisation constant A is given by the condition $\int w(x) \, dx = 1$; although the expression for $w(x)$ is valid for small x, the integrand decreases so rapidly with increasing $|x|$ that the range of integration may be extended from $-\infty$ to ∞. The integration gives $A = \sqrt{(\beta/2\pi)}$.

Thus the probability distribution of the various values of the fluctuation x is given by the formula

$$w(x) \, dx = \sqrt{(\beta/2\pi)} e^{-\frac{1}{2}\beta x^2} \, dx. \tag{110.4}$$

[†] The time τ need not be the same as the relaxation time for equilibrium to be reached with respect to x, and may be less than this time if x approaches $\bar{x}$ in an oscillatory manner. For example, if we consider the variation of pressure in a small region of the body (with linear dimensions $\sim a$), τ will be of the order of the period of acoustic vibrations with wavelength $\lambda \sim a$, i.e. $\tau \sim a/c$, where c is the velocity of sound.

Such a distribution is called a *Gaussian distribution.* It has a maximum when $x = 0$ and decreases rapidly and symmetrically as $|x|$ increases on either side of the maximum.

The mean square fluctuation is

$$\langle x^2 \rangle = \int_{-\infty}^{\infty} x^2 w(x)\, dx = \frac{1}{\beta}.$$

(110.5)

Thus we can write the Gaussian distribution in the form

$$w(x)\, dx = \frac{1}{\sqrt{(2\pi \langle x^2 \rangle)}} \exp\left(-\frac{x^2}{2 \langle x^2 \rangle}\right) dx.$$

(110.6)

As we should expect, the smaller $\langle x^2 \rangle$, the sharper is the maximum of $w(x)$.

Knowing the mean square $\langle x^2 \rangle$, we can find the corresponding quantity for any function $\phi(x)$. Since x is small, we have[†]

$$\langle (\Delta\phi)^2 \rangle = [(d\phi/dx)^2]_{x=0} \langle x^2 \rangle.$$

(110.7)

§ 111. The Gaussian distribution for more than one variable

In § 110 we have discussed the probability of a deviation of any one thermo-dynamic quantity from its mean value, disregarding the values of other quantities.[‡] In a similar manner we can determine the probability of a simultaneous deviation of several thermodynamic quantities from their mean values. Let these deviations be denoted by $x_1, x_2, \ldots, x_n$.

We define the entropy $S(x_1, \ldots, x_n)$ as a function of the quantities $x_1, x_2, \ldots, x_n$ and write the probability distribution in the form $w\, dx_1 \ldots dx_n$, with w given by (110.1). Let S be expanded in powers of the x_i; as far as the second-order terms, the difference $S - S_0$ is a negative-definite quadratic form:

$$S - S_0 = -\tfrac{1}{2} \sum_{i,\, k=1}^{n} \beta_{ik} x_i x_k$$

(clearly $\beta_{ik} = \beta_{ki}$). In the rest of this section we shall omit the summation sign, and summation from 1 to n over all repeated suffixes will be implied. Thus we write

$$S - S_0 = -\tfrac{1}{2}\beta_{ik} x_i x_k.$$

(111.1)

[†] It is, of course, implied that the function $\phi(x)$ does not vary greatly for values of $x \sim \langle x^2 \rangle^{1/2}$, and that the derivative $d\phi/dx$ is not zero at $x = 0$.

[‡] This means that the function $S(x)$ used in § 110 was the greatest possible value of the entropy for the given non-equilibrium value of x.

Substituting this expression in (110.1), we obtain for the required probability distribution

$$w = Ae^{-\frac{1}{2}\beta_{ik}x_ix_k}. \tag{111.2}$$

The constant A is determined from the normalisation condition

$$\int w \, dx_1 \ldots dx_n = 1.$$

Here, for the same reason as in § 110, the integration over each x_i can be taken from $-\infty$ to ∞. To calculate this integral we proceed as follows. The linear transformation

$$x_i = a_{ik}x'_k \tag{111.3}$$

of $x_1, x_2, \ldots, x_n$ converts the quadratic form $\beta_{ik}x_ix_k$ into a sum of squares $x'_ix'_i$. In order that

$$\beta_{ik}x_ix_k = x'_ix'_i \equiv x'_ix'_k\delta_{ik}$$

should be valid, the transformation coefficients must satisfy the relations

$$\beta_{ik}a_{il}a_{km} = \delta_{lm}. \tag{111.4}$$

The determinant of the matrix of quantities on the left of this equation is the product of the determinant $\beta = |\beta_{ik}|$ and two determinants $a = |a_{ik}|$. The determinant $|\delta_{ik}| = 1$. The above relation therefore shows that

$$\beta a^2 = 1. \tag{111.5}$$

The Jacobian of the linear transformation from the variables x_i to the x'_i is a constant, the determinant a. After the transformation, therefore, the normalisation integral separates into a product of n identical integrals and, using (111.5), we find

$$Aa\left[\int_{-\infty}^{\infty} \exp\left(-\frac{1}{2}x'^2\right) dx'\right]^n = \frac{A}{\sqrt{\beta}}(2\pi)^{n/2} = 1.$$

Thus we obtain finally the Gaussian distribution for more than one variable in the form

$$w = \frac{\sqrt{\beta}}{(2\pi)^{n/2}} \exp\left(-\frac{1}{2}\beta_{ik}x_ix_k\right). \tag{111.6}$$

Let us define the quantities

$$X_i = -\partial S/\partial x_i = \beta_{ik}x_k, \tag{111.7}$$

which we refer to as *thermodynamically conjugate* to the x_i;[†] and let us determine the mean values of the products $x_i X_k$:

$$\langle x_i X_k \rangle = \frac{\sqrt{\beta}}{(2\pi)^{n/2}} \int \cdots \int x_i \beta_{kl} x_l \exp\left(-\tfrac{1}{2}\beta_{ik} x_i x_k\right) dx_1 \ldots dx_n.$$

To calculate the integral, let us assume for the moment that the mean values $\overline{x_i}$ are not zero but x_{i0}, say. Then in (111.6) x_i must be replaced by $x_i - x_{i0}$, and the definition of the mean gives

$$\overline{x_i} = \frac{\sqrt{\beta}}{(2\pi)^{n/2}} \int \cdots \int x_i \exp\left[-\tfrac{1}{2}\beta_{ik}(x_i - x_{i0})(x_k - x_{k0})\right] dx_1 \ldots dx_n = x_{i0}.$$

Differentiating this equation with respect to x_{k0} and then again putting all the x_{i0} equal to zero, we have δ_{ik} on the right and the required integral on the left.

Thus we find

$$\langle x_i X_k \rangle = \delta_{ik}. \tag{111.8}$$

Substituting (111.7) gives $\beta_{kl}\langle x_i x_l \rangle = \delta_{ik}$, whence

$$\langle x_i x_k \rangle = \beta^{-1}_{ik}, \tag{111.9}$$

where β^{-1}_{ik} is an element of the matrix inverse to β_{ik}.

Lastly, let us determine $\langle X_i X_k \rangle$. According to (111.7) and (111.8),

$$\langle X_i X_k \rangle = \beta_{il}\langle x_l X_k \rangle = \beta_{il}\delta_{lk},$$

i.e.

$$\langle X_i X_k \rangle = \beta_{ik}. \tag{111.10}$$

It is easy to determine also the mean square fluctuation of any function $\phi(x_1, \ldots, x_n)$. Since the deviations from the mean values are small, $\Delta\phi = (\partial\phi/\partial x_i)\Delta x_i$, where the $\partial\phi/\partial x_i$ denote the values of the derivatives for $x_1 = \ldots = x_n = 0$. Hence

$$\langle(\Delta\phi)^2\rangle = \frac{\partial\phi}{\partial x_i}\frac{\partial\phi}{\partial x_k}\langle x_i x_k \rangle = \frac{\partial\phi}{\partial x_i}\frac{\partial\phi}{\partial x_k}\beta^{-1}_{ik}. \tag{111.11}$$

If the fluctuations of any two of the x_i (x_1 and x_2, say) are statistically independent, the mean value $\langle x_1 x_2 \rangle$ is equal to the product of the mean

---

[†] Note that, with the linear relation (111.7), the conjugacy is reciprocal: if the same entropy is expressed in terms of the quantities X_i, then

$$x_i = -\partial S/\partial X_i, \tag{111.7a}$$

since, from (111.7),

$$dS = -X_k\, dx_k = -\beta_{ki}x_i\, dx_k = -x_i\, d(\beta_{ik}x_k) = -x_i\, dX_i.$$

values $\bar{x_1}$ and $\bar{x_2}$; since each of these is zero, so is $\langle x_1 x_2 \rangle$, and from (111.9) this implies that $\beta^{-1}_{12} = 0$. It is easy to see that for a Gaussian distribution the converse theorem is also valid: if $\langle x_1 x_2 \rangle = 0$ (i.e. $\beta^{-1}_{12} = 0$), the fluctuations of x_1 and x_2 are statistically independent. For the probability distribution w_{12} of the quantities x_1 and x_2 is obtained by integrating the distribution (111.6) over all the other x_i; the result is an expression of the form

$$w_{12} = \text{constant} \times \exp\left\{-\tfrac{1}{2}\beta'_{11}x_1^2 - \beta'_{12}x_1 x_2 - \tfrac{1}{2}\beta'_{22}x_2^2\right\}$$

(where the coefficients β'_{ik} are in general different from the corresponding β_{ik}). Applying formula (111.9) to this distribution, we find that $\langle x_1 x_2 \rangle = \beta'^{-1}_{12}$. If $\langle x_1 x_2 \rangle = 0$, then $\beta'^{-1}_{12} = 0$. But for a matrix of order two the vanishing of the inverse matrix element β'^{-1}_{12} implies that of the element β'_{12}.[†] Thus w_{12} separates into a product of two independent Gaussian distributions for the quantities x_1 and x_2, which are therefore statistically independent.

PROBLEM

Determine the mean value $\langle \exp(\alpha_i x_i) \rangle$, where the α_i are constants and the x_i are fluctuating quantities subject to the Gaussian distribution (111.2).

SOLUTION. We have to calculate the integral

$$\langle \exp(\alpha_i x_i) \rangle = A \int \exp\left(\alpha_i x_i - \tfrac{1}{2}\beta_{ik}x_i x_k\right) dx_1 \ldots dx_n.$$

By the transformation (111.3) the exponent in the integrand becomes

$$\alpha_i x_i - \tfrac{1}{2}\beta_{ik}x_i x_k = \alpha_i a_{ik}x'_k - \tfrac{1}{2}x'_k x'_k$$
$$= -\tfrac{1}{2}(x'_k - \alpha_i a_{ik})^2 + \tfrac{1}{2}\alpha_i a_{ik}\alpha_l a_{lk},$$

and the integration then gives

$$\langle \exp(\alpha_i x_i) \rangle = \exp\left(\tfrac{1}{2}\alpha_i \alpha_l a_{ik}a_{lk}\right).$$

According to (111.4), $a_{ik} = a^{-1}_{km}\beta^{-1}_{mi}$ and so $a_{ik}a_{lk} = \beta^{-1}_{li}$. Thus, using (111.9), we have finally

$$\langle \exp(\alpha_i x_i) \rangle = \exp\left(\tfrac{1}{2}\alpha_i \alpha_k \langle x_i x_k \rangle\right).$$

§ 112. Fluctuations of the fundamental thermodynamic quantities

Let us now calculate the mean square fluctuations of the fundamental thermodynamic quantities, pertaining to any small part of a body. This small part must still, of course, contain a sufficient number of particles. At very

---

[†] For a matrix of order two we have

$$\beta^{-1}_{12} = \beta_{12}/(\beta^2_{12} - \beta_{11}\beta_{22}).$$

low temperatures, however, this condition may be weaker than (110.2), which ensures that quantum fluctuations are absent, as assumed; in this case the minimum permissible dimensions of the parts of the body will be determined by the latter condition.[†] To avoid misunderstanding, it should be emphasised that the degree of importance of quantum fluctuations has no bearing on the influence of quantum effects on the thermodynamic quantities (or equation of state) of the substance: the fluctuations may be classical while at the same time the equation of state is given by the formulae of quantum mechanics.

For quantities such as energy and volume, which have a purely mechanical significance as well as a thermodynamic one, the concept of fluctuations is self-explanatory, but it needs more precise treatment for quantities such as entropy and temperature, whose definition necessarily involves considering the body over finite intervals of time. For example, let $S(E, V)$ be the equilibrium entropy of the body as a function of its (mean) energy and (mean) volume. By the fluctuation of entropy we shall mean the change in the function $S(E, V)$, formally regarded as a function of the exact (fluctuating) values of the energy and volume.

As we have seen in the preceding sections, the probability w of a fluctuation is proportional to e^{S_t}, where S_t is the total entropy of a closed system, i.e. of the body as a whole. We can equally well say that w is proportional to $e^{\Delta S_t}$, where ΔS_t is the change in entropy in the fluctuation. According to (20.8) we have $\Delta S_t = -R_{min}/T_0$, where R_{min} is the minimum work needed to carry out reversibly the given change in the thermodynamic quantities in the small part considered (relative to which the remainder of the body acts as a medium). Thus

$$w \propto e^{-R_{min}/T_0}. \tag{112.1}$$

T constant
P constant.

Here we substitute for R_{min} the expression

$$R_{min} = \Delta E - T_0 \Delta S + P_0 \Delta V,$$

where $\Delta E, \Delta S, \Delta V$ are the changes in the energy, entropy and volume of the small part of the body in the fluctuation, and T_0, P_0 the temperature and pressure of the medium, i.e. the equilibrium (mean) values of the temperature and pressure of the body. In what follows we shall omit the suffix zero from all quantities which are the coefficients of fluctuations; the equilibrium values will always be meant. Thus we have

probability of fluctuation.

$$w \propto \exp\left(-\frac{\Delta E - T\Delta S + P\Delta V}{T}\right). \tag{112.2}$$

[†] For example, for pressure fluctuations the condition $\tau \gg \hbar/T$ with $\tau \sim a/c$ (see the third footnote to § 110) gives $a \gg \hbar c/T$.

$\left(\frac{\partial S}{\partial V}\right)_T = \left(\frac{\partial P}{\partial T}\right)_V$

$\left(\begin{array}{cc} & T.\phi \\ V & \\ E & \phi \\ S & W & P \end{array}\right)$

F

340   *Fluctuations*

We may note that in this form the expression (112.2) is applicable to any fluctuations, small or large (a large fluctuation here meaning one in which, for example, ΔE is comparable with the energy of the small part of the body, though of course still small compared with the energy of the whole body). The application of formula (112.2) to small fluctuations (which are those generally occurring) leads to the following results.

Expanding ΔE in series, we obtain (cf. § 21)

$$\Delta E - T\,\Delta S + P\,\Delta V = \frac{1}{2}\left[\frac{\partial^2 E}{\partial S^2}(\Delta S)^2 + 2\frac{\partial^2 E}{\partial S\,\partial V}\Delta S\,\Delta V + \frac{\partial^2 E}{\partial V^2}(\Delta V)^2\right].$$

It is easily seen that this expression may be written as

$$\frac{1}{2}\left[\Delta S\Delta\left(\frac{\partial E}{\partial S}\right)_V + \Delta V\Delta\left(\frac{\partial E}{\partial V}\right)_S\right] = \frac{1}{2}(\Delta S\,\Delta T - \Delta P\,\Delta V).$$

Thus we obtain the fluctuation probability (112.2) in the form

$$w \propto \exp\left(\frac{\Delta P\,\Delta V - \Delta T\,\Delta S}{2T}\right). \qquad (112.3)$$

From this general formula we can find the fluctuations of various thermodynamic quantities. Let us first take V and T as independent variables. Then

$$\Delta S = \left(\frac{\partial S}{\partial T}\right)_V \Delta T + \left(\frac{\partial S}{\partial V}\right)_T \Delta V = \frac{C_v}{T}\Delta T + \left(\frac{\partial P}{\partial T}\right)_V \Delta V,$$

$$\Delta P = \left(\frac{\partial P}{\partial T}\right)_V \Delta T + \left(\frac{\partial P}{\partial V}\right)_T \Delta V;$$

see (16.3). Substituting these expressions in the exponent in formula (112.3), we find that the terms in $\Delta V\,\Delta T$ cancel, leaving

$$w \propto \exp\left\{-\frac{C_v}{2T^2}(\Delta T)^2 + \frac{1}{2T}\left(\frac{\partial P}{\partial V}\right)_T(\Delta V)^2\right\}. \qquad (112.4)$$

This expression separates into two factors, one depending only on ΔT and the other only on ΔV. In other words, the fluctuations of temperature and of volume are statistically independent, so that

$$\langle\Delta T\,\Delta V\rangle = 0. \qquad (112.5)$$

Comparing successively each of the two factors of (112.4) with the general formula (110.6) for the Gaussian distribution, we find the following expressions for the mean square fluctuations of temperature[†] and volume:

$$\langle(\Delta T)^2\rangle = T^2/C_v, \qquad (112.6)$$

$$\langle(\Delta V)^2\rangle = -T(\partial V/\partial P)_T. \qquad (112.7)$$

† If T is measured in degrees, $\langle(\Delta T)^2\rangle = kT^2/C_v.$

(handwritten marginal notes)

first order terms cancel

$\Delta S\,\Delta\left(\frac{\partial E}{\partial S}\right) =$

$\Delta S\,\Delta S\,\frac{\partial^2 E}{\partial S^2}$

$+\,\Delta S\,\Delta V\,\frac{\partial^2 E}{\partial V\,\partial S}$

$S = S(V,T)$

$P = P(V,T)$

$\frac{T}{T}\,\frac{\partial S}{\partial T}$

$\frac{1}{T}\,\frac{T\partial S}{\partial T} = \frac{1}{T}C_v$

$K = -\frac{1}{V}\left(\frac{\partial V}{\partial P}\right)_T$

These quantities are positive by virtue of the thermodynamic inequalities $C_v > 0$ and $(\partial P/\partial V)_T < 0$.

Let us now take P and S as the independent variables in (112.3). Then

$$\Delta V = \left(\frac{\partial V}{\partial P}\right)_S \Delta P + \left(\frac{\partial V}{\partial S}\right)_P \Delta S,$$

$$\Delta T = \left(\frac{\partial T}{\partial S}\right)_P \Delta S + \left(\frac{\partial T}{\partial P}\right)_S \Delta P = \frac{T}{C_p} \Delta S + \left(\frac{\partial T}{\partial P}\right)_S \Delta P.$$

From the formula $dW = T\,dS + V\,dP$,

$$(\partial V/\partial S)_P = \partial^2 W/\partial P\,\partial S = (\partial T/\partial P)_S,$$

and so

$$\Delta V = \left(\frac{\partial V}{\partial P}\right)_S \Delta P + \left(\frac{\partial T}{\partial P}\right)_S \Delta S.$$

Substitution of ΔV and ΔT in (112.3) gives

$$w \propto \exp\left\{\frac{1}{2T}\left(\frac{\partial V}{\partial P}\right)_S (\Delta P)^2 - \frac{1}{2C_p}(\Delta S)^2\right\}. \qquad (112.8)$$

As in (112.4), this expression is a product of factors, one depending only on ΔP and the other only on ΔS. In other words, the fluctuations of entropy and of pressure are statistically independent,[†] so that

$$\langle \Delta S\,\Delta P \rangle = 0. \qquad (112.9)$$

For the mean square fluctuations of entropy and pressure we find

$$\langle (\Delta S)^2 \rangle = C_p, \qquad (112.10)$$

$$\langle (\Delta P)^2 \rangle = -T(\partial P/\partial V)_S. \quad = T/\sqrt{V}\,K_s \qquad (112.11)$$

These formulae show that the mean square fluctuations of additive thermodynamic quantities (volume and entropy) are proportional to the dimensions (the volume) of those parts of the body to which they relate. Accordingly the root-mean-square fluctuations of these quantities are proportional to the square root of the volume, and the relative fluctuations are inversely proportional to this square root, in agreement with the general results in § 2 (formula (2.5)). But for quantities such as temperature and pressure the r.m.s. fluctuations themselves are inversely proportional to the square root of the volume.

Formula (112.7) determines the fluctuation of the volume of some part of the body, containing a certain number N of particles. Dividing both sides by

---

[†] The statistical independence of the pairs of quantities T, V and S, P is already obvious from the following argument. If we take $x_1 = \Delta S$ and $x_2 = \Delta V$ as the x_i in the formulae in § 111, the corresponding X_i will be $X_1 = \Delta T/T$, $X_2 = -\Delta P/T$ (see § 22). But $\langle x_i X_k \rangle = 0$ for $i \neq k$ according to the general formula (111.8), and this gives (112.5) and (112.9).

N^2, we find the volume fluctuation per particle:

$$\langle(\Delta(V/N))^2\rangle = -\frac{T}{N^2}\left(\frac{\partial V}{\partial P}\right)_T. \tag{112.12}$$

This fluctuation must obviously be independent of whether we consider it for constant volume or for constant number of particles. From (112.12) we can therefore find the fluctuation of the number of particles in a fixed volume in the body. Since V is then constant, we must put $\Delta(V/N) = V\,\Delta(1/N) = -(V/N^2)\,\Delta N$. Substitution in (112.12) gives

$$\langle(\Delta N)^2\rangle = -(TN^2/V^2)(\partial V/\partial P)_T. \tag{112.13}$$

For certain calculations it is convenient to write this formula in a different form. Since the derivative $(\partial V/\partial P)_T$ is regarded as taken with N constant, we write

$$-\frac{N^2}{V^2}\left(\frac{\partial V}{\partial P}\right)_{T,N} = N\left(\frac{\partial}{\partial P}\frac{N}{V}\right)_{T,N}.$$

The number of particles N, as a function of P, T and V, must be of the form $N = Vf(P, T)$, as shown by considerations of additivity (cf. § 24); that is, N/V is a function of P and T only, and it therefore does not matter whether N/V is differentiated at constant N or constant V. Hence we can write

$$N\left(\frac{\partial}{\partial P}\frac{N}{V}\right)_{T,N} = \frac{N}{V}\left(\frac{\partial N}{\partial P}\right)_{T,V}$$

$$= \left(\frac{\partial N}{\partial P}\right)_{T,V}\left(\frac{\partial P}{\partial\mu}\right)_{T,V} = \left(\frac{\partial N}{\partial\mu}\right)_{T,V},$$

where we have used the equation $N/V = (\partial P/\partial\mu)_{T,V}$, which follows from formula (24.14): $d\Omega = -V\,dP = -S\,dT - N\,d\mu$. Thus we have for the fluctuation of the number of particles the formula[†]

$$\langle(\Delta N)^2\rangle = T(\partial N/\partial\mu)_{T,V}. \tag{112.14}$$

[†] This formula can also be easily derived directly from the Gibbs distribution. According to the definition of the mean value,

$$\bar{N} = e^{\Omega/T}\sum_N Ne^{\mu N/T}\sum_n e^{-E_{nN}/T}.$$

Differentiation of this expression with respect to μ (for constant V and T) gives

$$\frac{\partial\bar{N}}{\partial\mu} = \frac{1}{T}e^{\Omega/T}\sum_N\left(N^2+N\frac{\partial\Omega}{\partial\mu}\right)e^{\mu N/T}\sum_n e^{-E_{nN}/T} = \frac{1}{T}\left(\langle N^2\rangle+\bar{N}\frac{\partial\Omega}{\partial\mu}\right).$$

But $\partial\Omega/\partial\mu = -\bar{N}$, and so

$$\frac{\partial\bar{N}}{\partial\mu} = \frac{1}{T}(\langle N^2\rangle-\bar{N}^2) = \frac{1}{T}\langle(\Delta N)^2\rangle,$$

which gives (112.14).

We could also use the Gibbs distribution to derive expressions for the fluctuations of the other thermodynamic quantities.

A body is characterised not only by the thermodynamic quantities considered above but also by the momentum **P** of its macroscopic motion relative to the medium. In a state of equilibrium there is no macroscopic motion, i.e. **P** = 0. Motion may, however, result from fluctuation; let us determine the probability of such a fluctuation. The minimum work $R_{\min}$ in this case is simply equal to the kinetic energy of the body: $R_{\min} = P^2/2M = \frac{1}{2}Mv^2$, where M is its mass and $\mathbf{v} = \mathbf{P}/M$ the velocity of the macroscopic motion. Thus the required probability is

$$w \propto e^{-Mv^2/2T}. \tag{112.15}$$

It may be noted that the fluctuations of velocity are statistically independent of those of the other thermodynamic quantities. The mean square fluctuation of each Cartesian component of the velocity is equal to

$$\langle(\Delta v_x)^2\rangle = T/M, \tag{112.16}$$

and is inversely proportional to the mass of the body.

The foregoing formulae show that the mean square fluctuations of such quantities as energy, volume, pressure and velocity vanish at absolute zero (as the first power of the temperature). This is a general property of all thermodynamic quantities which also have a purely mechanical significance, but is not in general true of such purely thermodynamic quantities as entropy and temperature.

Formula (112.6) for the fluctuations of temperature can also be interpreted from a different point of view. As we know, the concept of temperature may be introduced through the Gibbs distribution; it is then regarded as a parameter defining this distribution. As applied to an isolated body, the Gibbs distribution gives a complete description of the statistical properties, except for the inaccuracy that it leads to very small but non-zero fluctuations of the total energy of the body; these cannot in fact exist (see the end of § 28). Conversely, if the energy is regarded as a given quantity, we cannot assign a definite temperature to the body, and we must suppose that the temperature undergoes fluctuations in accordance with (112.6), where C_v denotes the specific heat of the body as a whole. This quantity (112.6) obviously describes the accuracy with which the temperature of an isolated body can be defined.

PROBLEMS

PROBLEM 1. Find the mean square fluctuation of the energy (using V and T as independent variables).

SOLUTION. We have

$$\Delta E = \left(\frac{\partial E}{\partial V}\right)_T \Delta V + \left(\frac{\partial E}{\partial T}\right)_V \Delta T = \left[T\left(\frac{\partial P}{\partial T}\right)_V - P\right]\Delta V + C_v \Delta T.$$

Squaring and averaging, we obtain

$$\langle(\varDelta E)^2\rangle = -\left[T\left(\frac{\partial P}{\partial T}\right)_V - P\right]^2 T\left(\frac{\partial V}{\partial P}\right)_T + C_v T^2.$$

PROBLEM. 2. Find $\langle(\varDelta W)^2\rangle$ (with variables P and S).

SOLUTION. $\langle(\varDelta W)^2\rangle = -TV^2(\partial P/\partial V)_S + T^2 C_p$.

PROBLEM 3. Find $\langle\varDelta T\,\varDelta P\rangle$ (with variables V and T).

SOLUTION. $\langle\varDelta T\,\varDelta P\rangle = (T^2/C_v)\,(\partial P/\partial T)_V$.

PROBLEM 4. Find $\langle\varDelta V\,\varDelta P\rangle$ (with variables V and T).

SOLUTION. $\langle\varDelta V\,\varDelta P\rangle = -T$.

PROBLEM 5. Find $\langle\varDelta S\,\varDelta V\rangle$ (with variables V and T).

SOLUTION. $\langle\varDelta S\,\varDelta V\rangle = (\partial V/\partial T)_P\,T$.

PROBLEM 6. Find $\langle\varDelta S\,\varDelta T\rangle$ (with variables V and T).

SOLUTION. $\langle\varDelta S\,\varDelta T\rangle = T$.

PROBLEM 7. Find the mean square fluctuation deviation of a simple pendulum suspended vertically.

SOLUTION. Let l be the length of the pendulum, m its mass, and ϕ the angle of deviation from the vertical. The work R_{min} is here just the mechanical work done against gravity in the deviation of the pendulum; for small ϕ, $R_{min} = \frac{1}{2}mg \cdot l\phi^2$. Hence

$$\langle\phi^2\rangle = T/mgl.$$

PROBLEM 8. Find the mean square fluctuation deviation of the points of a stretched string.

SOLUTION. Let l be the length of the string, and F its tension. Let us take a point at a distance x from one end of the string, and let y be its transverse displacement. To determine $\langle y^2\rangle$ we must consider the equilibrium form of the string when the displacement y of the point x is given; this consists of two straight segments from the fixed ends of the string to the point x, y. The work done in such a deformation of the string is

$$R_{min} = F[\sqrt{(x^2+y^2)}-x]+F[\sqrt{\{(l-x)^2+y^2\}}-(l-x)] \cong \frac{1}{2}Fy^2\left(\frac{1}{x}+\frac{1}{l-x}\right).$$

Thus the mean square is

$$\langle y^2\rangle = (T/Fl)\,x(l-x).$$

PROBLEM 9. Determine the mean value of the product of the fluctuation displacements of two different points of the string.

SOLUTION. Let y_1, y_2 be the transverse movements of points at distances x_1, x_2 from one end of the string (with $x_2 > x_1$). The equilibrium form for given y_1, y_2

consists of three straight segments, and the work is

$$R_{min} = \frac{1}{2} F\left(y_1^2 \frac{x_2}{x_1(x_2-x_1)} + y_2^2 \frac{l-x_1}{(l-x_2)(x_2-x_1)} - 2y_1 y_2 \frac{1}{x_2-x_1} \right);$$

from formula (111.8) we then have

$$\langle y_1 y_2 \rangle = (T/Fl) x_1 (l-x_2).$$

§ 113. Fluctuations in an ideal gas

The mean square fluctuation of the number of particles within some rela-
tively small volume in an ordinary ideal gas is found by substituting $V = NT/P$ in formula (112.13). This gives the simple result

$$\langle (\Delta N)^2 \rangle = N. \tag{113.1}$$

The relative fluctuation of the number of particles is therefore just the reciprocal square root of the mean number of particles:

$$\frac{\langle (\Delta N)^2 \rangle^{1/2}}{N} = \frac{1}{\sqrt{N}}.$$

In order to calculate the fluctuation of the number of particles in an ideal
Bose or Fermi gas, we must use formula (112.14), with the expression (56.5)
substituted for N as a function of μ, T, V, obtained by integrating the corresponding distribution function. We shall not pause to write out here the fairly
lengthy expressions which result, but simply note the following point. We
have seen that, in a Bose gas at temperatures $T < T_0$ (see § 62), the pressure
is independent of the volume, i.e. the compressibility becomes infinite.
According to formula (112.13) this would imply that the fluctuations of
number of particles also become infinite. This means that, in calculating
fluctuations in a Bose gas, the interaction between its particles cannot be
neglected at low temperatures, however weak this interaction may be. When
the interaction, which must exist in any actual gas, is taken into account,
the resulting fluctuations are finite.

Next, let us examine fluctuations in the distribution of the gas particles
over the various quantum states. We again consider the quantum states of
the particles (including different states of their translational motion); let n_k
be their occupation numbers.

Let us consider an assembly of n_k particles in the kth quantum state. Since
this set of particles is statistically independent of the remaining particles in
the gas (cf. § 37), we can apply formula (112.14) to it:

$$\langle (\Delta n_k)^2 \rangle = T \frac{\partial \overline{n_k}}{\partial \mu}. \tag{113.2}$$

$W_{mN} = \dfrac{e^{\frac{N\mu - E_{m,N}}{T}}}{\sum e^{\frac{\Omega}{T}}} = e^{\frac{\Omega + N\mu - E_m N}{T}}$

$\Omega = -T\ln Z$

$\bar{N} = \sum_{mN} N\, W_{mN}$

$\dfrac{\partial \bar{N}}{\partial \mu} = \sum_{mN} N\left(\dfrac{\partial \Omega}{\partial \mu} + N\right)\dfrac{1}{T} W_{mN}$

$-\bar{N}$

$= -\dfrac{\bar{N}^2}{T} + \dfrac{\overline{N^2}}{T}$

$= \dfrac{\langle N^2 \rangle - \langle N \rangle^2}{T}$

$= \dfrac{\langle (N - \langle N \rangle)^2 \rangle}{T}$

$= \dfrac{\langle (\Delta N)^2 \rangle}{T}$

$\langle (\Delta N)^2 \rangle = T\dfrac{\partial \bar{N}}{\partial \mu}$

346

Fluctuations

For a Fermi gas we must substitute

$$\bar{n}_k = \frac{1}{e^{(\varepsilon_k - \mu)/T} + 1}.$$

The differentiation gives

F.D.

$$\langle (\Delta n_k)^2 \rangle = \bar{n}_k(1 - \bar{n}_k). \tag{113.3}$$

Similarly, for a Bose gas

B.E.

$$\langle (\Delta n_k)^2 \rangle = \bar{n}_k(1 + \bar{n}_k). \tag{113.4}$$

For a Boltzmann gas the substitution $\bar{n}_k = e^{(\mu - \varepsilon_k)/T}$ naturally gives

$$\langle (\Delta n_k)^2 \rangle = \bar{n}_k, \tag{113.5}$$

which is obtained from both (113.3) and (113.4) when $\bar{n}_k \ll 1$. Summation of (113.3) or (113.4) over a group of G_j neighbouring states containing altogether $N_j = \Sigma n_k$ particles gives, by virtue of the statistical independence (already mentioned) of the fluctuations of the various n_k,

$$\langle (\Delta N_j)^2 \rangle = G_j \bar{n}_j(1 \mp \bar{n}_j) = \bar{N}_j(1 \mp \bar{N}_j/G_j), \tag{113.6}$$

where $\bar{n}_j$ is the common value of the neighbouring $\bar{n}_k$, and $\bar{N}_j = \bar{n}_j G_j$.

These formulae can be applied, in particular, to black-body radiation (an equilibrium Bose gas of photons), for which we must put $\mu = 0$ in (113.4). Let us consider the set of quantum states for the photon (in a volume V) with neighbouring frequencies in a small interval $\Delta\omega_j$. The number of such states is $G_j = V\omega_j^2\,\Delta\omega_j/\pi^2 c^3$; see (63.3). The total energy of the quanta in this frequency interval is $E_{\Delta\omega_j} = N_j\hbar\omega_j$. Multiplying (113.6) by $(\hbar\omega_j)^2$ and omitting the suffix j, we obtain the following expression for the fluctuation of the energy $E_{\Delta\omega}$ of black-body radiation in a given frequency interval $\Delta\omega$:

$$\langle (\Delta E_{\Delta\omega})^2 \rangle = \hbar\omega \cdot E_{\Delta\omega} + \pi^2 c^3 (E_{\Delta\omega})^2/V\omega^2\,\Delta\omega, \tag{113.7}$$

a relation first derived by A. Einstein (1909).

PROBLEM

Determine $\langle (\Delta N)^2 \rangle$ for an electron gas at temperatures much lower than the degeneracy temperature.

SOLUTION. In calculating $(\partial N/\partial\mu)_{T,V}$ we can use the expression (57.3) for μ at absolute zero. A simple calculation gives

$$\langle (\Delta N)^2 \rangle = \frac{3^{1/3} mT}{\pi^{4/3}\hbar^2}\left(\frac{N}{V}\right)^{1/3} V.$$

In general

$w(x) \sim e^{-\frac{R_{min}}{T}}$

$\propto e^{-\frac{\alpha x^2}{2T}}$

$\langle x^2 \rangle = \dfrac{T}{\alpha}$

R_{min} is min work to move x away from zero.

§ 114. Poisson's formula

Knowing the mean square fluctuation of the number of particles in a given volume of gas (113.1), we can write down the corresponding Gaussian probability distribution for fluctuations in this number of particles:

$$w(N)\,dN = \frac{1}{\sqrt{(2\pi\bar{N})}}\exp\left\{-\frac{(N-\bar{N})^2}{2\bar{N}}\right\}dN. \qquad (114.1)$$

This formula is, however, valid only for small fluctuations: the deviation $N-\bar{N}$ must be small compared with the number $\bar{N}$ itself.

If the volume selected in the gas is sufficiently small, the number of particles in it is small, and we may also consider large fluctuations, where $N-\bar{N}$ becomes comparable with $\bar{N}$. This problem is meaningful only for a Boltzmann gas, since in a Fermi or Bose gas the probability of such fluctuations can become appreciable only in volumes so small that quantum fluctuations become important.

The solution of this problem is most simply found as follows. Let V_0 and N_0 be the total volume of the gas and the number of particles in it, and V a part of the volume, small compared with V_0. Since the gas is uniform, the probability that any given particle is in the volume V is obviously just the ratio V/V_0, and the probability that N given particles are simultaneously present in it is $(V/V_0)^N$. Similarly, the probability that a particle is not in the volume V is $(V_0-V)/V_0$, and the same probability for N_0-N given particles simultaneously is $[(V_0-V)/V_0]^{N_0-N}$. The probability w_N that the volume V contains N molecules in all is therefore given by

$$w_N = \frac{N_0!}{N!\,(N_0-N)!}\left(\frac{V}{V_0}\right)^N\left(1-\frac{V}{V_0}\right)^{N_0-N}, \qquad (114.2)$$

where a factor has been included which gives the number of possible ways of choosing N out of N_0 particles.

In the case under consideration, $V \ll V_0$, and the number N, though it may differ considerably from its mean value $\bar{N}$, is of course assumed small compared with the total number N_0 of particles in the gas. Then we may put $N_0! \cong (N_0-N)!\,N_0^N$ and neglect N in the exponent N_0-N, obtaining

$$w_N = \frac{1}{N!}\left(\frac{N_0 V}{V_0}\right)^N\left(1-\frac{V}{V_0}\right)^{N_0}.$$

But $N_0 V/V_0$ is just the mean number $\bar{N}$ of particles in the volume V. Hence

$$w_N = \frac{\bar{N}^N}{N!}\left(1-\frac{\bar{N}}{N_0}\right)^{N_0}.$$

Finally, using the well-known formula

$$\lim_{n \to \infty} \left(1 - \frac{x}{n}\right)^n = e^{-x},$$

we replace $(1 - \bar{N}/N_0)^{N_0}$ with N_0 large by $e^{-\bar{N}}$ and obtain the required probability distribution in the form[†]

$$w_N = \frac{\bar{N}^N e^{-\bar{N}}}{N!}. \tag{114.3}$$

This is called *Poisson's formula.* It is easily seen to satisfy the normalisation condition

$$\sum_{N=0}^{\infty} w_N = 1.$$

From this distribution we can calculate the mean square fluctuation of the number of particles:

$$\langle N^2 \rangle = \sum_{N=0}^{\infty} N^2 w_N = e^{-\bar{N}} \sum_{N=1}^{\infty} \frac{\bar{N}^N N}{(N-1)!}$$

$$= e^{-\bar{N}} \sum_{N=2}^{\infty} \frac{\bar{N}^N}{(N-2)!} + e^{-\bar{N}} \sum_{N=1}^{\infty} \frac{\bar{N}^N}{(N-1)!} = \bar{N}^2 + \bar{N}.$$

Hence we find as before

$$\langle (\Delta N)^2 \rangle = \langle N^2 \rangle - \bar{N}^2 = \bar{N}, \tag{114.4}$$

and the mean square fluctuation of the number of particles is equal to $\bar{N}$ for any value of $\bar{N}$, and not only for large values.

We may note that formula (114.3) can also be derived directly from the Gibbs distribution. According to the latter, the distribution of N gas particles, considered simultaneously, among various quantum states is given by the expression $\exp\{(\Omega + \mu N - \Sigma\varepsilon_k)/T\}$, where $\Sigma\varepsilon_k$ is the sum of the energies of the individual particles. To derive the required probability w_N we must sum this expression over all particle states belonging to a given volume V. If we sum over the states of each particle independently, the result must be divided by $N!$ (cf. § 41), giving

$$w_N = \frac{e^{\Omega/T}}{N!} \left(\sum_k e^{(\mu-\varepsilon_k)/T}\right)^N = \frac{e^{\Omega/T}}{N!} \left(\sum_k \bar{n}_k\right)^N.$$

[†] For small fluctuations ($|N - \bar{N}| \ll \bar{N}$ with $\bar{N}$ large), this formula naturally becomes (114.1). This is easily seen by using Stirling's asymptotic expression for the factorial of a large number N, $N! = \sqrt{(2\pi N)} \cdot N^N e^{-N}$, and expanding $\log w_n$ in powers of $N - \bar{N}$.

The sum is just the mean number $\bar{N}$ of particles in the volume considered. Thus we find

$$w_N = \text{constant} \times \bar{N}^N / N!,$$

and the normalisation condition then shows that the constant is $e^{-\bar{N}}$,[†] giving again formula (114.3).

§ 115. Fluctuations in solutions

The fluctuations of thermodynamic quantities in solutions can be calculated by the same method as that used in § 112 for fluctuations in bodies consisting of identical particles. The calculations are considerably simplified by the following argument.

Let us take some small part of the solution, containing a given number N of solvent molecules, and try to calculate the mean fluctuation of the number n of solute molecules in that part of the solution or, what is the same thing, the fluctuation of the concentration $c = n/N$ in that part. To do so, we must consider the most complete equilibrium of the solution that is possible for a given non-equilibrium value of n; cf. the first footnote to § 111. Taking a given value of the concentration does not affect the establishment of equilibrium between the small part considered and the remainder of the solution as regards exchange of energy between them or change in their volumes. The former means (see § 9) that the temperature remains constant throughout the solution; the latter means that the pressure remains constant throughout the solution (§ 12). Thus to calculate the mean square $\langle (\Delta c)^2 \rangle$, it is sufficient to consider the fluctuations of concentration occurring at constant temperature and pressure.

This fact in itself signifies that the fluctuations of concentration are statistically independent of those of temperature and pressure, i.e.[‡]

$$\langle \Delta T \, \Delta c \rangle = 0, \quad \langle \Delta c \, \Delta P \rangle = 0. \tag{115.1}$$

[†] That is, $\Omega = -PV = -\bar{N}T$, in accordance with the equation of state of an ideal gas.

[‡] This may be more rigorously proved by the method indicated in the third footnote to § 112. Using the formula $dE = T \, dS - P \, dV + \mu' \, dn$ (with N constant), we can rewrite (96.1) as

$$dR_{\min} = (T - T_0) \, dS - (P - P_0) \, dV + (\mu' - \mu_0') \, dn.$$

Thus, if we take $x_1 = \Delta S, x_2 = \Delta V, x_3 = \Delta n$, then $X_1 = \Delta T/T, X_2 = -\Delta P/T, X_3 = \Delta \mu'/T$ are thermodynamically conjugate to them. The equations (115.1) then follow, since $\langle x_3 X_1 \rangle = 0, \langle x_3 X_2 \rangle = 0$.

The minimum work necessary to change the number n by Δn at constant pressure and temperature is, by (96.1), $R_{\min} = \Delta\Phi - \mu'\,\Delta n$, where μ' is the chemical potential of the solute. Expanding $\Delta\Phi$ in powers of Δn, we have

$$\Delta\Phi \cong \left(\frac{\partial\Phi}{\partial n}\right)_{P,T}\Delta n + \left(\frac{\partial^2\Phi}{\partial n^2}\right)_{P,T}\cdot\frac{1}{2}(\Delta n)^2 = \mu'\,\Delta n + \left(\frac{\partial\mu'}{\partial n}\right)_{P,T}\cdot\frac{1}{2}(\Delta n)^2,$$

and hence

$$R_{\min} = \frac{1}{2}\left(\frac{\partial\mu'}{\partial n}\right)_{P,T}(\Delta n)^2.$$

Substituting this expression in the general formula (112.1) and comparing with the Gaussian distribution formula (110.5), we obtain for the required mean square of the fluctuation of the number n

$$\langle(\Delta n)^2\rangle = \frac{T}{(\partial\mu'/\partial n)_{P,T}}, \tag{115.2}$$

or, dividing by N^2, for the mean square fluctuation of concentration

$$\langle(\Delta c)^2\rangle = \frac{T}{N(\partial\mu'/\partial c)_{P,T}}. \tag{115.3}$$

The latter expression is inversely proportional to the amount of matter (N) in the small part considered, as it should be (see the discussion following (112.11)).

For weak solutions, $\partial\mu'/\partial n = T/n$, and formula (115.2) gives

$$\langle(\Delta n)^2\rangle = n. \tag{115.4}$$

It should be noticed that there is a complete analogy (as was to be expected) with formula (113.1) for the fluctuations of the number of particles in an ideal gas.

§ 116. Spatial correlation of density fluctuations

The statement that in a homogeneous isotropic medium (gas or liquid) all positions of the particles in space are equally probable applies to any one particle on condition that all other particles can occupy arbitrary positions, and is not, of course, in contradiction with the fact that the interaction between different particles must cause some correlation in their positions. If we consider, say, two particles simultaneously, then for a given position of one particle the various positions of the other will not be equally probable.

Let $n(\mathbf{r})$ denote the exact (fluctuating) number density of particles. The product $n\,dV$ is the number of particles present in a volume element dV at a given instant. To represent the correlation between the positions of particles at two points in space, we use the spatial correlation function of the density fluctuations:

$$\langle \Delta n_1 \Delta n_2 \rangle = \overline{n_1 n_2} - \bar{n}^2, \tag{116.1}$$

where $\Delta n = n - \bar{n}$ and the suffixes 1 and 2 denote the values of $n(\mathbf{r})$ at two points $\mathbf{r}_1$ and $\mathbf{r}_2$. In a homogeneous isotropic medium, the correlation function depends only on the magnitude $r = |\mathbf{r}_2 - \mathbf{r}_1|$ of the distance between these points. When $r \to \infty$, the fluctuations at the points $\mathbf{r}_1$ and $\mathbf{r}_2$ become statistically independent, and the correlation function therefore tends to zero.

It is useful to clarify the significance of this correlation function in the following way. Since the volume dV is infinitesimal, it cannot contain more than one particle at a time; the probability of finding two particles in it at once is an infinitesimal quantity of a higher order. The mean number of particles $\bar{n} \, dV$ is therefore also the probability that a particle is present in dV. Let $\bar{n} w_{12}(r) \, dV_2$ denote the probability that there is a particle in the volume element dV_2 when there is one in dV_1 ($w_{12} \to 1$ as $r \to \infty$). It is evident from the foregoing that the mean value

$$\langle n_1 \, dV_1 \cdot n_2 \, dV_2 \rangle = \bar{n} \, dV_1 \cdot \bar{n} w_{12} \, dV_2.$$

Hence $\langle n_1 n_2 \rangle = w_{12} \bar{n}^2$. In this equation, valid when $\mathbf{r}_1 \neq \mathbf{r}_2$, we cannot, however, go to the limit $\mathbf{r}_2 \to \mathbf{r}_1$, since the derivation ignores the fact that, if the points 1 and 2 coincide, a particle in dV_1 is also in dV_2. A relation which takes account of this is clearly

$$\langle n_1 n_2 \rangle = \bar{n}^2 w_{12} + \bar{n} \delta(\mathbf{r}_2 - \mathbf{r}_1). \tag{116.2}$$

For let us take a small volume ΔV, multiply (116.2) by $dV_1 \, dV_2$, and integrate over ΔV. The term $\bar{n}^2 w_{12}$ then gives a second-order small quantity (proportional to $(\Delta V)^2$); the term containing the delta function gives a first-order quantity, $\bar{n} \Delta V$, as it should, since (to the first order) the small volume can only contain either no particle or one particle.

The delta-function term is conveniently separated also from the correlation function (116.1), by writing the latter as

$$\langle \Delta n_1 \Delta n_2 \rangle = \bar{n} \delta(\mathbf{r}_2 - \mathbf{r}_1) + \bar{n} v(r), \tag{116.3}$$

where

$$v(r) = \bar{n}[w_{12}(r) - 1]. \tag{116.4}$$

The term "correlation function" will be applied to both the original quantity $\langle \Delta n_1 \Delta n_2 \rangle$ and the function $v(r)$.[†]

[†] The function $v(r)$ differs from $\omega_{12}(r)$ in § 79 as regards the normalisation: $\bar{n}\omega_{12} = v$.

$\langle m_1 m_2 \rangle =$
$\langle (\bar{m}+\Delta m_1)(\bar{m}+\Delta m_2) \rangle$
$= \langle \bar{m}^2 + \Delta m_1 \bar{m}$
$+ \Delta m_2 \bar{m} + \Delta m_1 \Delta m_2 \rangle$
$= \bar{m}^2 + \langle \Delta m_1 \Delta m_2 \rangle$

$\langle \Delta N)^2 \rangle$
$= \langle (N-\bar{N})^2 \rangle$
$= \int \langle (m_1 - \bar{m})(m_2-m_2) \rangle dV_1 dV_2$
$= \int \langle \Delta n_1 \Delta m_2 \rangle dU_1 dU_2$
$= \int G(r_2-r_1) dV_1 dU_2$
$= V \int G(r) dr$

$\overline{\langle (\Delta N)^2 \rangle} = \dfrac{\bar{N}^2}{V} K_T$

$\int G(r) dr = \dfrac{\bar{N}^2}{V^2} K_T$

Let us now integrate (116.3) with respect to V_1 and V_2 over a finite volume V. With N the total number of particles in this volume (so that $\bar{n}V = \bar{N}$), we find

$$\langle (\Delta N)^2 \rangle = \bar{N} + \bar{n} \iint \nu(r)\, dV_1\, dV_2,$$

or changing from the integration over V_1 and V_2 to one over the coordinates of one particle and the relative coordinates $\mathbf{r} = \mathbf{r}_2 - \mathbf{r}_1$,

$$\int \nu\, dV = \frac{\langle (\Delta N)^2 \rangle}{\bar{N}} - 1. \tag{116.5}$$

Thus the integral of the correlation function over a certain volume is expressed in terms of the mean square fluctuation of the total number of particles in that volume. Using for the latter the thermodynamic formula (112.13), we can express the integral in terms of the thermodynamic quantities:

$$\int \nu\, dV = -\frac{TN}{V^2} \left(\frac{\partial V}{\partial P} \right)_T - 1. \tag{116.6}$$

In an ordinary (classical) ideal gas this gives $\int \nu\, dV = 0$, as it should: in such a gas there is no correlation between the positions of different particles, since they do not interact with one another, either directly or by exchange interaction as in a quantum ideal gas.

On the other hand, in a liquid (far from the critical point) the first term in (116.6) is small compared with unity, because the compressibility of a liquid is small. In this case the integral is close to -1.[†] The principal interaction forces between the liquid particles have a range of the order of the molecular dimensions a. When these forces are taken into account, the correlation function $\nu(r)$ decreases exponentially with increasing distance, the exponent being $\sim -r/a$.[‡]

Since the fluctuations of density and temperature are statistically independent, the temperature may be regarded as constant in considering the fluctuations of density. The total volume of the body is also constant, by definition. Under these conditions the minimum work needed to bring the body out of equilibrium is equal to the change ΔF_t in the total free energy of the body, so that the probability of the fluctuation may be written as

$$w \propto e^{-\Delta F_t/T}. \tag{116.7}$$

The change ΔF_t due to the density fluctuations can be put in the form

$$\Delta F_t = \tfrac{1}{2} \iint \phi(r)\, \Delta n_1\, \Delta n_2\, dV_1\, dV_2. \tag{116.8}$$

[†] The value -1 corresponds, as it were, to the mutual impenetrability of the liquid particles, regarded as closely packed solid spheres.

[‡] There are also, however, weak longer-range (van der Waals) interaction forces, which give rise to a term in the correlation function that decreases less rapidly (by a power law) with increasing distance (see Part 2).

We shall show how the correlation function $\nu(r)$ may be found from the function $\phi(r)$.[†]

Let us consider a body of large but finite volume V, and expand Δn as a Fourier series:

$$\Delta n = \sum_{\mathbf{k}} \Delta n_{\mathbf{k}} e^{i\mathbf{k}\cdot\mathbf{r}}, \quad \Delta n_{\mathbf{k}} = \frac{1}{V}\int \Delta n\cdot e^{-i\mathbf{k}\cdot\mathbf{r}}\, dV; \qquad (116.9)$$

since Δn is real, $\Delta n_{-\mathbf{k}} = \Delta n_{\mathbf{k}}^{*}$. When these expressions are substituted in (116.8) and the integration is carried out, all the terms containing products $\Delta n_{\mathbf{k}} \Delta n_{\mathbf{k}'} e^{i(\mathbf{k}+\mathbf{k}')\cdot\mathbf{r}}$ with $\mathbf{k}' \neq -\mathbf{k}$ give zero, and the result is

$$\Delta F_t = \tfrac{1}{2} V \sum_{\mathbf{k}} |\Delta n_{\mathbf{k}}|^2 \phi(k), \qquad (116.10)$$

where the same letter ϕ with the new argument k denotes the Fourier component of $\phi(r)$:

$$\phi(k) = \int \phi(r) e^{-i\mathbf{k}\cdot\mathbf{r}}\, dV. \qquad (116.11)$$

Since each term in the sum (116.10) depends on only one of the $\Delta n_{\mathbf{k}}$, the fluctuations of the different $\Delta n_{\mathbf{k}}$ are statistically independent. Each square $|\Delta n_{\mathbf{k}}|^2$ appears in the sum twice (for $\pm\mathbf{k}$), so that the probability distribution for its fluctuations is

$$w \propto \exp\left\{-\frac{V}{T}\phi(k)\,|\Delta n_{\mathbf{k}}|^2\right\}.$$

Lastly, since $|\Delta n_{\mathbf{k}}|^2$ is the sum of the squares of two independent quantities ($\Delta n_{\mathbf{k}}$ being complex), we find from this for the mean square fluctuation

$$\langle |\Delta n_{\mathbf{k}}|^2 \rangle = T/V\phi(k). \qquad (116.12)$$

Multiplying both sides of (116.3) by $\exp(-i\mathbf{k}\cdot\mathbf{r}) = \exp[-i\mathbf{k}\cdot(\mathbf{r}_2-\mathbf{r}_1)]$ and again integrating with respect to V_1 and V_2, we have

$$\langle |\Delta n_{\mathbf{k}}|^2 \rangle = \bar{n}[1+\nu(k)]/V, \quad \nu(k) = \int \nu(r) e^{-i\mathbf{k}\cdot\mathbf{r}}\, dV. \qquad (116.13)$$

Substitution of (116.12) finally gives the required result

$$\nu(k) = \frac{T}{\bar{n}\phi(k)} - 1. \qquad (116.14)$$

PROBLEM

Find the first term in an expansion of the correlation function for a rarefied gas in powers of N/V.

[†] In mathematical parlance, $\phi(r)$ is the second variational derivative of ΔF^t with respect to $n(\mathbf{r})$.

SOLUTION. We start from formula (79.2). In the first approximation, we can assume that all particles except for two specified ones are far apart and their interaction is negligible, so that the integration gives V^{N-2}. To the same accuracy, $F = F_{id}$. The result is

$$v(r) = \bar{n}[e^{-U(r)/T} - 1],$$

where $U(r)$ is the interaction energy for two gas particles.

It may be noted that substituting this expression in (79.1) gives as the energy of the gas

$$E = E_{id} + \frac{N^2}{2V^2} V \int U\left(1 + \frac{v}{\bar{n}}\right) dV$$

$$= E_{id} + \frac{N^2}{2V} \int U e^{-U/T} dV.$$

This result is, of course, in agreement with formulae (74.4) and (74.5) for the free energy of a slightly non-ideal gas.

§ 117. Correlation of density fluctuations in a degenerate gas

As already mentioned in § 116, in a classical ideal gas there is no correlation between the positions of the various particles. In quantum mechanics, however, such a correlation exists, because particles in an ideal gas interact indirectly, on account of the principle of symmetry of wave functions.[†]

The problem of determining the correlation function in a degenerate gas can be most simply solved by the second quantisation method, which has already been used in § 80 to calculate the energy of an electron gas. In this method, the particle number density corresponds to the operator

$$\hat{n}(\mathbf{r}) = \hat{\psi}^+(\mathbf{r})\,\hat{\psi}(\mathbf{r});$$

on substitution of the ψ-operators (80.5), it is expressed by the sum

$$\hat{n}(\mathbf{r}) = \sum_{\sigma, \sigma', \mathbf{p}, \mathbf{p}'} \hat{a}^+_{\mathbf{p}\sigma}\hat{a}_{\mathbf{p}'\sigma'}\psi^*_{\mathbf{p}\sigma}(\mathbf{r})\,\psi_{\mathbf{p}'\sigma'}(\mathbf{r}), \tag{117.1}$$

where the summation is taken over all values of the momenta $\mathbf{p}$ and $\mathbf{p}'$ (for free particles in the volume V) and over the spin components σ, σ'.[‡] However, since the spin wave functions for different values of σ are orthogonal, only

[†] The correlation of fluctuations in a Fermi gas was considered by V. S. Fursov (1937), and in a Bose gas by A. D. Galanin (1940).

[‡] It may be recalled that the wave functions of a particle having spin are spinors, and that the product of wave functions in (117.1) is in fact the "scalar product" of covariant and contravariant spinors with the appropriate summation over spinor indices; the latter should not be confused with the suffixes σ and σ' which denote the eigenvalues of the spin component in the states concerned.

the terms in the sum that have $\sigma = \sigma'$ are in fact non-zero. In the products $\psi_{p\sigma}^* \psi_{p'\sigma}$, the normalised spin factors give unity, so that the wave functions can be written simply as coordinate plane waves:

$$\psi_p = \frac{1}{\sqrt{V}} e^{i\mathbf{p}\cdot\mathbf{r}/\hbar}. \tag{117.2}$$

It is easily seen that the diagonal terms ($\mathbf{p} = \mathbf{p'}$) of the sum (117.1) give just the mean density $\bar{n}$: since the operator $\hat{a}_{p\sigma}^+ \hat{a}_{p\sigma}$ is simply the number $n_{p\sigma}$ of particles in the corresponding quantum state, the sum of these terms is

$$\frac{1}{V} \sum_{\sigma, p} n_{p\sigma} = \frac{N}{V} = \bar{n}.$$

We can therefore write

$$\Delta\hat{n} = \hat{n}(\mathbf{r}) - \bar{n}$$
$$= \sum_{\sigma, p, p'}{}' \hat{a}_{p\sigma}^+ \hat{a}_{p'\sigma} \psi_p^* \psi_{p'}, \tag{117.3}$$

where the prime to the summation sign denotes that the diagonal terms are to be omitted. Using this expression, we can easily calculate the required mean value $\langle \Delta n_1 \Delta n_2 \rangle$.

This mean value is calculated in two stages. First of all, the quantum averaging over the states of the particles is to be carried out. This amounts to taking the corresponding diagonal matrix element of the quantity concerned. Multiplying together the two operators (117.3), which belong to two different points $\mathbf{r}_1$ and $\mathbf{r}_2$, we obtain a sum of terms containing various products of the operators $\hat{a}_{p\sigma}$ and $\hat{a}_{p\sigma}^+$ taken four at a time. But among these products only those which contain two pairs of operators $\hat{a}_{p\sigma}$ and $\hat{a}_{p\sigma}^+$ with the same suffixes have non-zero diagonal matrix elements, i.e. the relevant terms are

$$\sum_{\sigma, p, p'}{}' \hat{a}_{p\sigma}^+ \hat{a}_{p'\sigma} \hat{a}_{p'\sigma}^+ \hat{a}_{p\sigma} \psi_p^*(\mathbf{r}_1)\, \psi_{p'}(\mathbf{r}_1)\, \psi_{p'}^*(\mathbf{r}_2)\, \psi_p(\mathbf{r}_2).$$

These terms form diagonal matrices with

$$\hat{a}_{p'\sigma} \hat{a}_{p'\sigma}^+ = 1 \mp n_{p'\sigma}, \quad \hat{a}_{p\sigma}^+ \hat{a}_{p\sigma} = n_{p\sigma};$$

here and henceforward, the upper sign refers to the case of Fermi statistics and the lower sign to that of Bose statistics. Substituting also the functions ψ_p (117.2), we obtain

$$\frac{1}{V^2} \sum_{\sigma, p, p'}{}' (1 \mp n_{p'\sigma}) n_{p\sigma} e^{i(\mathbf{p}-\mathbf{p'})\cdot(\mathbf{r}_2-\mathbf{r}_1)/\hbar}.$$

This expression must now be averaged in the statistical sense, i.e. over the equilibrium distribution of particles among the various quantum states. Since particles in different quantum states behave independently of one another, the numbers $n_{p\sigma}$ and $n_{p'\sigma}$ are averaged independently. The required mean value is thus found to be

$$\langle \Delta n_1 \Delta n_2 \rangle = \frac{1}{V^2} \sum_{\sigma, \mathbf{p}, \mathbf{p}'}{}' (1 \mp \overline{n_{\mathbf{p}'\sigma}})\overline{n_{\mathbf{p}\sigma}} e^{i(\mathbf{p}-\mathbf{p}')\cdot(\mathbf{r}_2-\mathbf{r}_1)/\hbar}. \tag{117.4}$$

We now change in the usual way from summation over $\mathbf{p}$ and $\mathbf{p}'$ to integration over $V\,d^3p\,V\,d^3p'/(2\pi\hbar)^6$; the restriction $\mathbf{p} \neq \mathbf{p}'$ then becomes unimportant. The integral separates into two parts, of which the first is

$$\sum_\sigma \int \overline{n_{\mathbf{p}\sigma}} e^{i(\mathbf{p}-\mathbf{p}')\cdot(\mathbf{r}_2-\mathbf{r}_1)/\hbar}\frac{d^3p\,d^3p'}{(2\pi\hbar)^6}.$$

The integration over $d^3p'/(2\pi\hbar)^3$ gives the delta function $\delta(\mathbf{r}_2-\mathbf{r}_1)$, which enables us to put $\mathbf{r}_2-\mathbf{r}_1 = 0$ in the remaining integrand; this leaves

$$\delta(\mathbf{r}_2-\mathbf{r}_1) \sum_\sigma \int \overline{n_{\mathbf{p}\sigma}}\frac{d^3p}{(2\pi\hbar)^3} = \bar{n}\delta(\mathbf{r}_2-\mathbf{r}_1).$$

This is just the first term in (116.3). The correlation function (the second term in (116.3)) is therefore

$$v(r) = \mp \frac{1}{\bar{n}} \sum_\sigma \left| \int e^{i\mathbf{p}\cdot\mathbf{r}/\hbar}\overline{n_{\mathbf{p}\sigma}}\frac{d^3p}{(2\pi\hbar)^3} \right|^2. \tag{117.5}$$

In a gas in equilibrium, the distribution of particles among quantum states is given by the Fermi or Bose distribution formula:

$$\overline{n_{\mathbf{p}\sigma}} \equiv \overline{n_{\mathbf{p}}} = [e^{(\varepsilon-\mu)/T}\pm 1]^{-1}. \tag{117.6}$$

These numbers are independent of σ; the summation over σ in (117.5) therefore gives simply a factor $g = 2s+1$ (where s is the spin of the particle). Thus we have finally the following formula for the correlation function:[†]

$$v(r) = \mp \frac{g}{\bar{n}} \left| \int \frac{e^{i\mathbf{p}\cdot\mathbf{r}/\hbar}}{e^{(\varepsilon-\mu)/T}\pm 1}\frac{d^3p}{(2\pi\hbar)^3} \right|^2, \tag{117.7}$$

or, after integration over the directions of $\mathbf{p}$,

$$v(r) = \mp \frac{g}{4\pi^4\bar{n}r^2\hbar^4} \left| \int_0^\infty \frac{\sin(pr/\hbar)p\,dp}{e^{(\varepsilon-\mu)/T}\pm 1} \right|^2. \tag{117.8}$$

We may also state a formula for the mean squares of the Fourier components of the density fluctuations, which is easily obtained by substituting $v(r)$ from (117.7) in the general formula (116.13) and integrating over the coordinates:[‡]

$$\langle |\Delta n_{\mathbf{k}}|^2 \rangle = \frac{g}{V} \int \overline{n_{\mathbf{p}}}(1 \mp \overline{n_{\mathbf{p}+\hbar\mathbf{k}}})\frac{d^3p}{(2\pi\hbar)^3}. \tag{117.9}$$

[†] For a Bose gas, this formula applies only to temperatures above the Bose–Einstein condensation point (see Problem 4).

[‡] The Fourier components $\Delta n_{\mathbf{k}}$ of the gas density fluctuations should not be confused with the particle quantum state occupation numbers $\bar{n}_{\mathbf{p}}$.

Formula (117.7) shows first of all that for a Fermi gas $v(r) < 0$, but for a Bose gas $v(r) > 0$. In other words, in a Bose gas the presence of a particle at some point in space increases the probability that another particle is near that point, i.e. the particles "attract" one another; in a Fermi gas, on the contrary, they "repel" one another (cf. the remark at the end of § 56).

In accordance with the discussion at the beginning of this section, if we go to the limit of classical mechanics, the correlation function tends to zero: as $\hbar \to 0$, the frequency of the oscillating factor $\exp(i\mathbf{p} \cdot \mathbf{r}/\hbar)$ in the integrand in (117.7) increases without limit, and the integral tends to zero.

As $r \to 0$, the function $v(r)$ tends to a constant limit:

$$v(0) = \mp \frac{g}{\bar{n}} \left| \int \bar{n}_p \frac{d^3 p}{(2\pi\hbar)^3} \right|^2 = \mp \frac{\bar{n}}{g}. \tag{117.10}$$

Let us apply the formula (117.8) to a Fermi gas at $T = 0$. In this case the distribution function is a step function: $\bar{n}_p = 1$ for $p < p_F$ and $\bar{n}_p = 0$ for $p > p_F$, where $p_F = \hbar(6\pi^2\bar{n}/g)^{1/3}$ is the limiting momentum. We therefore find

$$v(r) = -\frac{g}{4\pi^4\hbar^4\bar{n}r^2} \left| \int_0^{p_F} p \sin \frac{pr}{\hbar} \, dp \right|^2.$$

Let us take fairly large distances, assuming that $p_F r/\hbar \gg 1$. Accordingly, we retain only the term in the lowest power of $1/r$ in the integral:

$$v(r) = -\frac{3\hbar}{2\pi^2 p_F r^4} \cos^2 \frac{p_F r}{\hbar}.$$

The squared cosine varies rapidly over distances Δr that are small in comparison with the distances considered. Averaging over such a distance, we find

$$v(r) = -3\hbar/4\pi^2 p_F r^4. \tag{117.11}$$

PROBLEMS

PROBLEM 1. Determine the mean square of the Fourier component (with small wave numbers: $k \ll p_F/\hbar$) of the density fluctuations in a Fermi gas at $T = 0$.

SOLUTION. The integrand in (117.9) is non-zero (and equal to unity) only at points where $\bar{n}_p = 1$, $\bar{n}_{p+\hbar k} = 0$, i.e. at points in a sphere of radius p_F and also not in a similar sphere whose centre is shifted by $\hbar k$. A calculation of the volume of this region when $\hbar k \ll p_F$ gives

$$\langle |\Delta n_k|^2 \rangle = \pi g k p_F^2/(2\pi)^3 \hbar^2 V$$

$$= 3k\hbar\bar{n}/4p_F V.$$

PROBLEM 2. Determine the correlation function for a Fermi gas at temperatures small compared with the degeneracy temperature.

SOLUTION. In the integral in (117.8) we put $\mu \simeq \varepsilon_F = p_F^2/2m$ and transform it as follows:

$$I = \int_0^\infty \frac{p \sin (pr/\hbar)\,dp}{e^{(\varepsilon-\varepsilon_F)/T}+1} = -\hbar \frac{\partial}{\partial r} \int_0^\infty \frac{\cos (pr/\hbar)\,dp}{e^{(\varepsilon-\varepsilon_F)T}+1}.$$

Integration by parts is followed by a change to the variable $x = p_F(p-p_F)/mT$. Since T is small, the integrand decreases rapidly with increasing $|x|$, and the integral over x can therefore be extended from $-\infty$ to ∞:

$$I = -\hbar^2 \frac{\partial}{\partial r} \frac{1}{r} \int_{-\infty}^\infty \sin\left(\frac{p_F r}{\hbar}+\lambda x r\right) \frac{dx}{(e^x+1)(e^{-x}+1)}$$

$$= -\hbar^2 \frac{\partial}{\partial r}\left\{ \frac{\sin p_F r/\hbar}{r} \int_{-\infty}^\infty e^{i\lambda r x} \frac{dx}{(e^x+1)(e^{-x}+1)}\right\},$$

where $\lambda = mT/\hbar p_F$. The resulting integral is transformed by the substitution $1/(e^x+1) = u$ into Euler's beta integral, giving

$$I = \hbar^2 \frac{\partial}{\partial r}\left\{\frac{\pi\lambda}{\sinh \pi\lambda r} \sin \frac{p_F r}{\hbar}\right\}.$$

For distances $r \gg \hbar/p_F$, averaging the rapidly varying squared cosine, we have finally

$$\nu(r) = -\frac{3(mT)^2}{4\hbar p_F^3 r^2} \sinh^{-2}\left(\frac{\pi m T r}{\hbar p_F}\right)^2.$$

As $T \to 0$, this becomes (117.11). In the asymptotic region, where $p_F r/\hbar$ is large compared with ε_F/T as well as with unity, we find

$$\nu(r) = -\frac{3(mT)^2}{\hbar p_F^3 r^2} \exp\left(-\frac{2\pi m T r}{\hbar p_F}\right).$$

PROBLEM 3. Determine the correlation function for a Bose gas at large distances $(r \gg \hbar/\sqrt{(mT)})$ for temperatures above the point T_0 at which Bose–Einstein condensation begins, but close to this point.

SOLUTION. Near the point T_0, the chemical potential $|\mu|$ is small (see § 62, Problem). The integral in (117.7), which we denote by I, is then determined by small values of p: $\varepsilon/T \sim p^2/mT \sim |\mu|/T \ll 1$. Hence, expanding the integrand in powers of ε and μ, we find[†]

$$I \simeq T \int \frac{e^{i\mathbf{p}\cdot\mathbf{r}/\hbar}}{p^2/2m+|\mu|} \frac{d^3p}{(2\pi\hbar)^3} = \frac{mT}{2\pi\hbar^2 r} \exp\left\{-r\frac{(2m|\mu|)^{1/2}}{\hbar}\right\}.$$

[†] A Fourier transformation formula is used:

$$\int \frac{e^{-\varkappa r}}{r} e^{-i\mathbf{k}\cdot\mathbf{r}}\,dV = \frac{4\pi}{\varkappa^2+k^2}, \qquad \int \frac{e^{i\mathbf{k}\cdot\mathbf{r}}}{\varkappa^2+k^2} \frac{d^3k}{(2\pi)^3} = \frac{e^{-\varkappa r}}{4\pi r}.$$

This is most simply derived by noticing that the function $\phi = e^{-\varkappa r}/r$ satisfies

Finally

$$v(r) = \frac{gm^2T^2}{4\pi^2\bar{n}\hbar^4r^2} \exp\left\{-r\frac{2(2m\,|\mu\,|)^{1/2}}{\hbar}\right\}.$$

PROBLEM 4. Determine the correlation function for a Bose gas with $T < T_0$.

SOLUTION. For $T < T_0$, a finite fraction of the number of particles ($N_{\varepsilon=0}$) is in states with $\mathbf{p} = 0$ (the condensate). Returning to the expression (117.4), we must first (before changing from summation to integration) separate the terms with $\mathbf{p}$ or $\mathbf{p}' = 0$, noting that the number of particles in each quantum state with $\mathbf{p} = 0$ is $n_{\mathbf{p}=0} = N_{\varepsilon=0}/g$. The sum is then transformed as in the text, and instead of (117.7) we find

$$v(r) = \frac{2n_0}{\bar{n}}\,I + \frac{g}{\bar{n}}\,I^2, \quad I = \int e^{i\mathbf{p}\cdot\mathbf{r}/\hbar}\overline{n}_\mathbf{p}\,\frac{d^3p}{(2\pi\hbar)^3},$$

where $n_0 = N_{\varepsilon=0}/g$; $\overline{n}$ is given by the Bose distribution formula with $\mu = 0$:

$$\overline{n} = [e^{\varepsilon/T}-1]^{-1}.$$

At distances $r \gg \hbar/\sqrt{(mT)}$, the integral $I = mT/2\pi\hbar^2r$ (from Problem 3 with $\mu = 0$), so that

$$v(r) = \frac{mTn_0}{\pi\bar{n}\hbar^2r} + \frac{gm^2T^2}{4\pi^2\bar{n}\hbar^4r^2};$$

the second term may be neglected if T is not too close to T_0 (and so n_0 is not too small). In the opposite case, at distances $r \ll \hbar/\sqrt{(mT)}$, the integral

$$I \cong \int \overline{n}_\mathbf{p}\frac{d^3p}{(2\pi\hbar)^3} = \frac{\bar{n}-n_0}{g},$$

so that

$$v(r) \cong v(0) = (\bar{n}^2-n_0^2)/g\bar{n}.$$

Note that the integral $\int v\,dV$ for a Bose gas diverges if $T < T_0$, and the calculation from (116.5) would therefore lead to an infinite value of the fluctuation of the number of particles, in accordance with the comment in § 113.

§ 118. Correlations of fluctuations in time

Let us consider a physical quantity which describes a closed system in thermodynamic equilibrium, or part of such a system. (In the former case, the quantity must not be one that by definition remains constant for a closed

---

the differential equation

$$\Delta\phi - \varkappa^2\phi = -4\pi\,\delta(\mathbf{r}).$$

Multiplying this by $e^{-i\mathbf{k}\cdot\mathbf{r}}$ and integrating over all space (with twofold integration by parts in the first term), we obtain the required result.

system, such as the energy.) This quantity will undergo small variations in time, fluctuating about its mean value. Let $x(t)$ again denote the difference between the quantity and its mean value (so that $\bar{x} = 0$).

There is some correlation between the values of $x(t)$ at different instants; this means that the value of x at a given instant t affects the probabilities of its various values at a later instant t'. In the same way as for the spatial correlation discussed in the preceding sections, we can characterise the time correlation by the mean value of the product $\langle x(t)\,x(t')\rangle$. The averaging here is, as usual, understood in a statistical sense, i.e. as an averaging over the probabilities of all possible values of the quantity x at the times t and t'. As has been mentioned in § 1, this statistical averaging is equivalent to a time averaging—in this case over one of the times t and t' for a given difference $t'-t'$. The quantity thus obtained,

$$\phi(t'-t) = \langle x(t)\,x(t')\rangle, \tag{118.1}$$

depends only on the difference $t'-t$, and the definition may therefore also be written

$$\phi(t) = \langle x(0)\,x(t)\rangle. \tag{118.2}$$

As the time difference increases without limit, the correlation clearly tends to zero, and accordingly the function $\phi(t)$ likewise tends to zero. Note also that, because of the obvious symmetry of the definition (118.1) as regards the interchange of t and t', the function $\phi(t)$ is even:

$$\phi(t) = \phi(-t). \tag{118.3}$$

By regarding $x(t)$ as a function of time, we have implicitly assumed that it behaves classically. The definition given above can, however, be put in a form that is applicable to quantum variables also. To do this, we must consider in place of the quantity x its quantum-mechanical time-dependent (Heisenberg) operator $\hat{x}(t)$. The operators $\hat{x}(t)$ and $\hat{x}(t')$ relating to different instants do not in general commute, and the correlation function must now be defined as

$$\phi(t'-t) = \tfrac{1}{2}\langle \hat{x}(t)\,\hat{x}(t') + \hat{x}(t')\,\hat{x}(t)\rangle, \tag{118.4}$$

where the averaging is with respect to the exact quantum state.[†]

Let us assume that the quantity x is such that by specifying a definite value of it (considerably greater than the mean fluctuation $\langle x^2\rangle^{1/2}$) one could

[†] It may be again mentioned that, according to the basic principles of statistical physics, the result of the averaging does not depend on whether it is done mechanically with respect to the exact wave function of a stationary state of the system, or statistically by means of the Gibbs distribution. The only difference is that in the former case the result is expressed in terms of the energy of the body, and in the latter case as a function of its temperature.

characterise a definite state of partial equilibrium. That is, the relaxation time for the establishment of partial equilibrium for a given value of x is assumed to be much less than that required to reach the equilibrium value of x itself. This condition is satisfied for a wide class of quantities of physical interest. The fluctuations of such quantities will be said to be *quasi-stationary*.[†] In the rest of this section we shall consider fluctuations of this type and, moreover, shall assume the quantity x to be classical.[‡]

We shall also assume that, as complete equilibrium is approached, no other deviations from equilibrium arise in the system which would require the use of further quantities to describe them. In other words, at every instant the state of the non-equilibrium system must be entirely defined by the value of x; a more general case will be discussed in § 119.

Let the quantity x have at some instant t a value which is large compared with the mean fluctuation, i.e. the system be far from equilibrium. Then we can say that at subsequent instants the system will tend to reach the equilibrium state, and accordingly x will decrease. Under the assumptions made, its rate of change will be at every instant entirely defined by the value of x at that instant: $\dot{x} = \dot{x}(x)$. If x is still comparatively small, then $\dot{x}(x)$ can be expanded in powers of x, keeping only the linear term:

$$\mathrm{d}x/\mathrm{d}t = -\lambda x, \tag{118.5}$$

where λ is a positive constant; there is no zero-order term in this expansion, since the rate of change $\mathrm{d}x/\mathrm{d}t$ must be zero in complete equilibrium (i.e. at $x = 0$). Equation (118.5) is the linearised macroscopic "equation of motion" of the non-equilibrium system, describing its relaxation process (the physical nature of which depends entirely on the nature of the quantity x). The constant $1/\lambda$ determines the order of magnitude of the relaxation time for the establishment of complete equilibrium.

Returning to fluctuations in an equilibrium system, let us define a quantity $\xi_x(t)$ as the mean value of x at an instant $t > 0$ with the condition that it had some given value x at the prior instant $t = 0$; such a mean value is in general not zero. Evidently the correlation function $\phi(t)$ may be written in terms of $\xi_x(t)$ as

$$\phi(t) = \langle x\xi_x(t)\rangle, \tag{118.6}$$

where the averaging is only over the probabilities of the various values of x at the initial instant $t = 0$.

[†] This term is more appropriate than *thermodynamic*, as used in the previous edition of the book.

[‡] The final formulae for the quasi-stationary fluctuations of quantum variables are obtained from that for classical variables by a simple modification described in § 124 (see (124.19)).

For values of ξ_x large compared with the mean fluctuation, it follows from (118.5) that also

$$d\xi_x(t)/dt = -\lambda\xi_x(t), \qquad t > 0. \tag{118.7}$$

Taking into account the fact that $\xi_x(t)$ is an averaged quantity, we must expect that this equation is true even when ξ_x is not large. Integrating (118.7) we find, since $\xi_x(0) = x$ by definition,

$$\xi_x(t) = xe^{-\lambda t},$$

and finally, substituting in (118.6), we obtain a formula for the time correlation function:

$$\phi(t) = \langle x^2\rangle e^{-\lambda t}.$$

This formula as it stands is, however, valid only for $t > 0$, since in its derivation (equation (118.7)) we have assumed that the instant t is later than $t = 0$. On the other hand, since $\phi(t)$ is an even function, we can write the final formula as

$$\phi(t) = \langle x^2\rangle e^{-\lambda|t|} = (1/\beta)e^{-\lambda|t|}, \tag{118.8}$$

with $\langle x^2\rangle$ from (110.5), and it is then applicable for both positive and negative t. This function has two different derivatives at $t = 0$, the reason being that we are considering time intervals long compared with the time for the establishment of partial equilibrium (equilibrium with a given value of x). The consideration of short times, which is not possible within the "quasi-stationary" theory, would of course give $d\phi/dt = 0$ for $t = 0$, as must be true for any even function of t that has a continuous derivative.

The above theory can also be formulated in another way that may have certain advantages.

The equation $\dot{x} = -\lambda x$ for the quantity x itself (rather than its mean value ξ_x) is valid, as already mentioned, only when x is large compared with the mean fluctuation of x. For arbitrary values of x, we write

$$\dot{x} = -\lambda x + y, \tag{118.9}$$

thus defining a new quantity y. Although the magnitude of the oscillations of y does not change with time, when x is large (in the sense already defined) y is relatively small and may be neglected in equation (118.9). It is necessary to regard the quantity y thus defined (called the *random force*) as the source of the fluctuations of x. The correlation function of the random force, $\langle y(0)y(t)\rangle$, must be specified in such a way as to lead to the correct result (118.8) for $\langle x(0)x(t)\rangle$. To do so, we must put

$$\langle y(0)y(t)\rangle = 2\lambda\langle x^2\rangle \delta(t) = (2\lambda/\beta)\delta(t). \tag{118.10}$$

This is easily seen by writing the solution of equation (118.9):

$$x(t) = e^{-\lambda t} \int_{-\infty}^{t} y(\tau) e^{\lambda \tau} \, d\tau,$$

and averaging the product $x(0)\, x(t)$ after expressing it as a double integral.

The vanishing of the expression (118.10) when $t \neq 0$ signifies that the values of $y(t)$ at different instants are uncorrelated. In reality, of course, this statement is an approximation and signifies only that the values of $y(t)$ are correlated over time intervals of the order of the time for partial equilibrium to be established (equilibrium with a given value of x), which, as already mentioned, is regarded as negligibly small in the theory given here.

§ 119. Time correlations of the fluctuations of more than one variable

The results obtained in § 118 can be generalised to fluctuations in which several quantities $x_1, x_2, \ldots, x_n$ simultaneously deviate from their equilibrium values. We shall again suppose that the latter have been subtracted, so that each mean value $\bar{x}_i = 0$.

The correlation functions for the fluctuations of these quantities are defined (in the classical theory) as

$$\phi_{ik}(t' - t) = \langle x_i(t')\, x_k(t) \rangle. \tag{119.1}$$

By virtue of this definition, they have the obvious symmetry property

$$\phi_{ik}(t) = \phi_{ki}(-t). \tag{119.2}$$

There is, however, another symmetry property of the correlation functions, which has a profound physical significance. It arises from the symmetry under time reversal of the equations of mechanics which describe the motion of the particles in the body.[†] Because of this symmetry, it does not matter which of the quantities x_i and x_k is taken at the earlier or later instant in the averaging. Hence $\langle x_i(t')\, x_k(t) \rangle = \langle x_i(t)\, x_k(t') \rangle$, i.e.

$$\phi_{ik}(t) = \phi_{ik}(-t). \tag{119.3}$$

From (119.2) and (119.3) it also follows that $\phi_{ik}(t) = \phi_{ki}(t)$.

In this derivation we have tacitly assumed that the quantities x_i themselves are unaffected by time reversal. There are, however, quantities which change sign under time reversal, for instance quantities proportional to the velocities of some macroscopic motions. If both quantities x_i and x_k have this property,

[†] The system is assumed to be not in a magnetic field and not rotating as a whlole (see § 120).

the relation (119.3) is again valid, but if only one of the two changes sign the symmetry under time reversal implies that $\langle x_i(t') x_k(t) \rangle = - \langle x_i(t) x_k(t') \rangle$, i. e.

$$\phi_{ik}(t) = - \phi_{ik}(-t).\qquad(119.4)$$

Together with (119.2), this gives $\phi_{ik}(t) = - \phi_{ki}(t)$.

We shall now assume, as in § 118, that the fluctuations are quasi-stationary, i. e. that the set of values of $x_1, \ldots, x_n$ (outside the limits of their mean fluctuations) determines a macroscopic state of partial equilibrium. As complete equilibrium is approached, the x_i vary with time; we assume that the set of functions $x_i(t)$ completely describes this process, and that no other deviations from equilibrium occur in it. Then the rates of change of the x_i in any non-equilibrium state are functions of the values $x_1, \ldots, x_n$ in that state:

$$\dot{x}_i = \dot{x}_i(x_1, \ldots, x_n).\qquad(119.5)$$

If the system is in a state fairly close to complete equilibrium, i.e. if the x_i may be regarded as small, the functions (119.5) may be expanded in powers of the x_i as far as the first order, i.e. as the linear sums

$$\dot{x}_i = - \lambda_{ik} x_k\qquad(119.6)$$

with constant coefficients λ_{ik}; these expressions are the generalisation of (118.5).[†]

From this we can go to the equations for the correlation functions in the same way as in § 118. We define the mean values $\xi_i(t)$ of the quantities x_i at a time $t > 0$ for given values of all the $x_1, x_2, \ldots$ at the earlier time $t = 0$; these values themselves are, for brevity, omitted from the notation $\xi_i(t)$. These quantities satisfy the same equations (119.6):

$$\xi_i = - \lambda_{ik} \xi_k,\qquad(119.7)$$

and do so for arbitrarily small values of $\xi_i(t)$ as well as for those which are large compared with the mean fluctuations. The correlation functions are obtained from the $\xi_i(t)$ by multiplying by $x_l = x_l(0)$ and averaging with respect to the probabilities of the various values of x_l: $\phi_{il}(t) = \langle \xi_i(t) x_l \rangle$. This operation performed on equation (119.7) gives

$$\frac{d}{dt} \phi_{il}(t) = - \lambda_{ik} \phi_{kl}(t), \qquad (t > 0)\qquad(119.8)$$

the suffix l in these equations being a free suffix.

As already mentioned, equations (119.6) are just the linearised macroscopic "equations of motion" of the non-equilibrium system, describing its

---

† As in § 111, summation from 1 to n over repeated Latin suffixes is implied.

$P[x(t) \mid x(0)]$

relaxation process. We see that the set of equations for the correlation functions of the fluctuations is obtained by simply replacing the quantities $x_i(t)$ in these "equations of motion" by functions $\phi_{il}(t)$ with a free suffix l that takes all values from 1 to n. The equations thus obtained refer to times $t > 0$, and are to be integrated with the "initial condition"

$$\phi_{ik}(0) = \langle x_i(0)\, x_k(0) \rangle \equiv \langle x_i x_k \rangle = \beta^{-1}_{ik}\,; \tag{119.9}$$

the mean values $\langle x_i x_k \rangle$ must be equal to their known values (111.9). For times $t < 0$, the correlation functions are then found directly from their symmetry properties.

§ 120. The symmetry of the kinetic coefficients

Let us return to the macroscopic equations (119.6), which describe the relaxation of a slightly non-equilibrium system:[†]

$$\dot{x}_i = -\lambda_{ik} x_k\,. \tag{120.1}$$

These equations have a deep-lying internal symmetry, which becomes explicit only when the right-hand sides are expressed not in terms of the macroscopic quantities x_i themselves (whose rates of change are on the left-hand sides) but in terms of the thermodynamically conjugate quantities

generalized force

$$X_i = -\partial S/\partial x_i\,, \tag{120.2}$$

which have already been used in § 111. In a state of equilibrium, the entropy of the system is a maximum, so that all the $X_i = 0$. If $x_1, x_2, \ldots$ are not zero but comparatively small (i.e. in slightly non-equilibrium states of the system), the X_i may be expressed as linear functions:

$$X_i = \beta_{ik} x_k\,. \tag{120.3}$$

positive

The constant coefficients β_{ik} are the first derivatives of the X_i, i.e. the second derivatives of S, and therefore

$$\beta_{ik} = \beta_{ki}\,. \tag{120.4}$$

[†] In particular applications, cases occur where the complete equilibrium which is being approached depends on external parameters (such as volume or external field) which themselves vary slowly with time; the equilibrium (mean) values of the quantities considered therefore vary also. If this variation is sufficiently slow, we can again use the relations derived here, except that the mean values $\bar{x}_i$ cannot be regarded as always equal to zero. If they are denoted by $x_i^{(0)}$, then (120.1) must be replaced by

$$\dot{x}_i = -\lambda_{ik}(x_k - x_k^{(0)})\,. \tag{120.1a}$$

$$S = S_0 - \frac{\beta_{ik}\, x_i\, x_k}{2}$$

If we express the quantities x_i in terms of the X_i from (120.3) and substitute in (120.1), we obtain the relaxation equations in the form

$$\dot{x}_i = -\gamma_{ik} X_k, \tag{120.5}$$

where

$$\gamma_{ik} = \lambda_{il}\beta^{-1}_{lk} \tag{120.6}$$

are new constants called *kinetic coefficients*. We shall now prove the *principle of the symmetry of the kinetic coefficients* or *Onsager's principle* (L. Onsager, 1931), according to which

$$\gamma_{ik} = \gamma_{ki}. \tag{120.7}$$

The proof is based on the fact mentioned in § 119 that similar equations (120.1) and (120.5) are satisfied by the quantities that describe fluctuations in an equilibrium system. We define mean values $\xi_i(t)$ of the fluctuating quantities x_i, and mean values $\Xi_i(t)$ of the X_i, at a time t, for given values of $x_1, x_2, \ldots$ at $t = 0$; then

$$\dot{\xi}_i = -\gamma_{ik}\Xi_k. \qquad (t > 0) \tag{120.8}$$

We now make use of the symmetry of the fluctuations (in an equilibrium system) under time reversal, expressed by the relation (119.3), which may be written

$$\langle x_i(t)\, x_k(0)\rangle = \langle x_i(0)\, x_k(t)\rangle, \tag{120.9}$$

or, with the $\xi_i(t)$,

$$\langle \xi_i(t) x_k\rangle = \langle x_i \xi_k(t)\rangle, \tag{120.10}$$

where the averaging is with respect to the probabilities of the various values of all the x_i at $t = 0$. We differentiate this equation with respect to t and substitute the derivatives $\dot{\xi}_i$ from (120.8):

$$\gamma_{il}\langle \Xi_l(t)x_k\rangle = \gamma_{kl}\langle x_i\Xi_l(t)\rangle.$$

When $t = 0$, the Ξ_l are evidently equal to $X_l(0)$; hence, putting $t = 0$ in the above equation, we get

$$\gamma_{il}\langle X_l x_k\rangle = \gamma_{kl}\langle X_l x_i\rangle,$$

where the two factors in the averaged products are taken at the same instant. From (111.8), such mean values $\langle X_l x_k\rangle = \delta_{lk}$, and we arrive at the required result (120.7).[†]

[†] Here it would be incorrect to use instead of (120.9) the relation (119.2), which gives $\langle x_i(0)\, x_k(t)\rangle = \langle x_i(-t)\, x_k(0)\rangle$. It can be shown that differentiating this equation with respect to t and then putting $t = 0$ gives (with the use of (120.9)) $\langle \dot{x}_i x_k\rangle = 0$. In reality, however, the functions $\phi_{ik}(t)$ (like $\phi(t)$ in § 118) have in this approximation two different derivatives at $t = 0$, according as $t \to +0$ or $t \to -0$.

The following two comments should, however, be made regarding this result. The proof depends on the symmetry of the equations of motion with respect to time, and the formulation of this symmetry is somewhat altered for fluctuations in a uniformly rotating body and for bodies in an external magnetic field: in these cases the symmetry under time reversal holds only if the sign of the angular velocity of rotation Ω or of the magnetic field $\mathbf{H}$ is simultaneously changed. Thus in these cases the kinetic coefficients depend on Ω or $\mathbf{H}$ as parameter, and we have the relations

$$\gamma_{ik}(\Omega) = \gamma_{ki}(-\Omega), \quad \gamma_{ik}(\mathbf{H}) = \gamma_{ki}(-\mathbf{H}). \tag{120.11}$$

Moreover, it has been assumed in the derivation that the quantities x_i and x_k are unaffected by time reversal. The relation (120.9), and therefore the result (120.7), remain valid if both quantities change sign under time reversal (are both proportional to the velocities of some macroscopic motions). But if one of x_i and x_k changes sign and the other remains unchanged, the derivation must start from (119.4) instead of (119.3), and the principle of the symmetry of the kinetic coefficients is formulated as

$$\gamma_{ik} = -\gamma_{ki}. \tag{120.12}$$

Exactly similar results are valid for the kinetic coefficients ζ_{ik} which appear in the relaxation equations when these are put in the form thermodynamically conjugate to equations (120.5):

$$\dot{X}_i = -\zeta_{ik}x_k, \quad \zeta_{ik} = \beta_{ik}\lambda_{lk}. \tag{120.13}$$

The coefficients ζ_{ik} have symmetry properties similar to those of the γ_{ik}. This can be seen by a corresponding derivation, but is in any case evident because of the reciprocal relation between the quantities x_i and X_i; see the second footnote to § 111.

If all the quantities $x_1, \ldots, x_n$ behave in the same way under time reversal, so that the matrix of the γ_{ik} is entirely symmetrical, the rates $\dot{x}_i$ can be represented as derivatives

$$\dot{x}_i = -\partial f/\partial X_i, \quad f = \tfrac{1}{2}\gamma_{ik}X_iX_k, \tag{120.14}$$

of a quadratic function of the X_i with coefficients γ_{ik}.

The importance of the function f is due to the fact that it determines the rate of change of the entropy of the system:

$$\dot{S} = (\partial S/\partial x_i)\dot{x}_i = -X_i\dot{x}_i = X_i \, \partial f/\partial X_i,$$

and, since f is a quadratic function of the X_i, Euler's theorem gives

$$\dot{S} = 2f. \tag{120.15}$$

SP - M

As equilibrium is approached, the entropy increases to a maximum. The quadratic form f must therefore be positive-definite, and this imposes certain conditions on the coefficients γ_{ik}. The function f may also be expressed in terms of the x_i, and its derivatives then give the rates $\dot{X}_i$:

$$\dot{X}_i = -\partial f/\partial x_i, \qquad f = \tfrac{1}{2}\zeta_{ik}x_i x_k. \qquad (120.16)$$

Here, of course, we have as before $\dot{S} = -x_i \dot{X}_i = 2f$.

For a system consisting of a body in an external medium we can transform (120.15) by using the fact that the change in entropy of a closed system in a deviation from equilibrium is $-R_{min}/T_0$, where R_{min} is the minimum work needed to bring the system from the equilibrium state into the one considered (see (20.8)).[†] Putting also $R_{min} = \Delta E - T_0\,\Delta S + P_0\,\Delta V$ (where E, S and V relate to the body, and T_0, P_0 are the temperature and pressure of the medium), we obtain

$$\dot{E} - T_0\dot{S} + P_0\dot{V} = -2fT_0. \qquad (120.17)$$

In particular, if the deviation from equilibrium occurs with the temperature and pressure of the body constant and equal to T_0 and P_0,

$$\dot{\Phi} = -2fT, \qquad (120.18)$$

and at constant temperature and volume

$$\dot{F} = -2fT. \qquad (120.19)$$

§ 121. The dissipative function

The macroscopic motion of bodies surrounded by an external medium is in general accompanied by irreversible frictional processes, which ultimately bring the motion to a stop. The kinetic energy of the bodies is thereby converted into heat and is said to be *dissipated*.

A purely mechanical treatment of such a motion is clearly impossible: since the energy of macroscopic motion is converted into the energy of thermal motion of the molecules of the body and the medium, such a treatment would require the setting up of the equations of motion for all these molecules. The problem of setting up equations of motion in the medium which contain only the "macroscopic" coordinates of the bodies is therefore a problem of statistical physics.

[†] Owing to this relation between the change of the entropy and R_{min}, the X_i can also be defined as

$$X_i = (1/T_0)\,\partial R_{min}/\partial x_i, \qquad (120.2a)$$

which is sometimes more convenient ta an (120.2); cf. (22.7).

This problem, however, cannot be solved in a general form. Since the internal motion of the atoms in a body depends not only on the motion of the body at a given instant but also on the previous history of the motion, the equations of motion will in general contain not only the macroscopic co-ordinates $Q_1, Q_2, \ldots, Q_s$ of the bodies and their first and second time derivatives but also all the higher-order derivatives (more precisely, the functions $Q_i(t)$ are subject to some integral operator). The Lagrangian for the macroscopic motion of the system does not then exist, of course, and the equations of motion will be entirely different in different cases.

The equations of motion can be derived in a general form if it may be assumed that the state of the system at a given instant is completely determined by the values of the coordinates Q_i and velocities $\dot{Q}_i$, and that the higher-order derivatives may be neglected; a more precise criterion of small-ness has to be established in each particular case. We shall further suppose that the velocities $\dot{Q}_i$ are themselves so small that their higher powers may be neglected, and finally that the motion consists of small oscillations about certain equilibrium positions. The latter is the case usually met with in this connection. We shall assume the coordinates Q_i to be chosen so that $Q_i = 0$ in the equilibrium position. Then the kinetic energy $K(\dot{Q}_i)$ of the system will be a quadratic function of the velocities $\dot{Q}_i$ and independent of the coordinates Q_i themselves; the potential energy $U(Q_i)$ due to the external forces will be a quadratic function of the coordinates Q_i.

We define the generalised momenta P_i in the usual way:

$$P_i = \partial K(\dot{Q}_i)/\partial \dot{Q}_i. \tag{121.1}$$

These equations define the momenta as linear combinations of the velocities; using them to express the velocities in terms of the momenta and substituting in the kinetic energy, we obtain the latter as a quadratic function of the momenta, with

$$\dot{Q}_i = \partial K(P_i)/\partial P_i. \tag{121.2}$$

If the dissipative processes are entirely neglected, the equations of motion will be as in ordinary mechanics, according to which the time derivatives of the momenta are equal to the corresponding generalised forces:

$$\dot{P}_i = -\partial U/\partial Q_i. \tag{121.3}$$

First of all, let us note that equations (121.2), (121.3) are in formal agreement with the principle of the symmetry of the kinetic coefficients, if the quantities $x_1, x_2, \ldots, x_{2s}$ defined in § 120 are taken as the coordinates Q_i and momenta P_i. For the minimum work needed to bring the bodies from a state of rest in their equilibrium positions to the positions Q_i with momenta P_i is $R_{min} = K(P_i) + U(Q_i)$. The quantities $X_1, X_2, \ldots, X_{2s}$ will

$$\dot{X}_1 = \gamma_{11} \bar{X}_1 + \gamma_2 \bar{X}_2$$

$$\dot{Q} = \gamma_{11}\frac{\Delta T}{T^2} + \gamma_{12}\frac{\Delta V}{T}$$

$$I = \gamma_{21}\frac{\Delta T}{T^2} + \delta_{12}\frac{\Delta V}{T}$$

$$\Delta T$$
$$e$$
$$\Delta Q \underset{K}{\underbrace{\qquad}}$$

$$\dot{Q} = K A \Delta T$$
$$\ell$$

$$\boxed{\gamma_{11} = \frac{K A T^2}{\ell}}$$

$$I = \frac{V}{r^2}$$

$$\frac{\sigma A}{\ell} = R$$

$$\gamma_{22} = \frac{\ell T}{\sigma A R}$$

$$\gamma_{12} = \gamma_{21} = \gamma$$

$$S = S_0 - \frac{\beta_{ik} X_i X_k}{2}$$

$$\dot{X}_i = - \lambda_{ij} X_j$$

therefore be the derivatives (see the previous footnote)

$$X_{Q_i} = (1/T)\, \partial R_{\min}/\partial Q_i = (1/T)\, \partial U/\partial Q_i,$$
$$X_{P_i} = (1/T)\, \partial R_{\min}/\partial P_i = (1/T)\, \partial K/\partial P_i$$

and equations (121.2), (121.3) will correspond to the relations (120.5) with $\gamma_{Q_iP_i} = -T = -\gamma_{P_iQ_i}$, in accordance with the rule (120.12); this is a case where one quantity (Q_i) remains invariant under time reversal but the other (P_i) changes sign.

In accordance with the general relations (120.5), we can now write the equations of motion allowing for dissipative processes, by adding to the right-hand sides of equations (121.2), (121.3) certain linear combinations of the quantities X_{Q_i}, X_{P_i}, such that the required symmetry of the kinetic coefficients is maintained. It is easy to see, however, that equations (121.2) must be left unchanged, since they are simply a consequence of the definition (121.1) of the momenta and do not depend on the presence or absence of dissipative processes. This shows that the terms added to equations (121.3) can only be linear combinations of the quantities X_{P_i} (i.e. of the derivatives $\partial K/\partial P_i$), since otherwise the symmetry of the kinetic coefficients would be violated.

Thus we have a set of equations of the form

$$\dot{P}_i = -\frac{\partial U}{\partial Q_i} - \sum_{k=1}^{s} \gamma_{ik} \frac{\partial K}{\partial P_k},$$

where the constant coefficients γ_{ik} are related by

$$\gamma_{ik} = \gamma_{ki}. \tag{121.4}$$

Putting $\partial K/\partial P_k = \dot{Q}_k$, we have finally

$$\dot{P}_i = -\frac{\partial U}{\partial Q_i} - \sum_{k=1}^{s} \gamma_{ik}\dot{Q}_k. \tag{121.5}$$

These are the required equations of motion. We see that the presence of dissipative processes leads, in this approximation, to the appearance of *frictional forces* which are linear functions of the velocities. Owing to the relation (121.4) these forces can be written as the derivatives, with respect to the corresponding velocities, of the quadratic function

$$f = \tfrac{1}{2} \sum_{i,k} \gamma_{ik}\dot{Q}_i\dot{Q}_k, \tag{121.6}$$

which is called the *dissipative function*. Then

$$\dot{P}_i = -\frac{\partial U}{\partial Q_i} - \frac{\partial f}{\partial \dot{Q}_i}. \tag{121.7}$$

Using the Lagrangian $L = K - U$, we can write these equations of motion as

$$\frac{d}{dt}\left(\frac{\partial L}{\partial \dot{Q}_i}\right) - \frac{\partial L}{\partial Q_i} = -\frac{\partial f}{\partial \dot{Q}_i},$$

(121.8)

which differs from the usual form of Lagrange's equations by the presence of the derivative of the dissipative function on the right.

The existence of friction causes a decrease in the total mechanical energy $K + U$ of the moving bodies. In accordance with the general results of § 120, the rate of decrease is determined by the dissipative function. This will be proved afresh here, since there is some difference in the notation used in the present section. We have

$$\frac{d}{dt}(K+U) = \sum_{i=1}^{s}\left(\frac{\partial K}{\partial P_i}\dot{P}_i + \frac{\partial U}{\partial Q_i}\dot{Q}_i\right) = \sum_i \dot{Q}_i\left(\dot{P}_i + \frac{\partial U}{\partial Q_i}\right),$$

or, substituting (121.7) and using the fact that the dissipative function is quadratic,

$$\frac{d}{dt}(K+U) = -\sum_i \dot{Q}_i \frac{\partial f}{\partial \dot{Q}_i} = -2f,$$

(121.9)

as it should be.

Finally, we may mention that, when there is an external magnetic field, the equations of motion again take the form (121.5), but (121.4) is replaced by

$$\gamma_{ik}(\mathbf{H}) = \gamma_{ki}(-\mathbf{H}).$$

As a result, there is no dissipative function whose derivatives determine the frictional forces. The equations of motion therefore cannot be written in the form (121.7).

§ 122. Spectral resolution of fluctuations

The spectral resolution of a fluctuating quantity $x(t)$ is defined by the usual Fourier expansion formulae:

$$x_\omega = \int_{-\infty}^{\infty} x(t)e^{i\omega t}\, dt,$$

(122.1)

and conversely

$$x(t) = \int_{-\infty}^{\infty} x_\omega e^{-i\omega t}\frac{d\omega}{2\pi}.$$

(122.2)

It should be noted that the integral (122.1) in fact diverges, since $x(t)$ does not tend to zero as $|t| \to \infty$. This is unimportant, however, as regards the formal derivations below, whose purpose is to calculate the mean squares, which are certainly finite.[†]

Substituting (122.2) in the definition (118.1) of the correlation function, we obtain

$$\phi(t'-t) = \langle x(t') x(t) \rangle$$

$$= \int_{-\infty}^{\infty} \int_{-\infty}^{\infty} \langle x_\omega x_{\omega'} \rangle e^{-i(\omega t + \omega' t')} \frac{d\omega\, d\omega'}{(2\pi)^2}. \qquad (122.3)$$

If this integral is to be a function of the difference $t - t'$ only, the integrand must contain a delta function of $\omega + \omega'$, i.e. we must have

$$\langle x_\omega x_{\omega'} \rangle = 2\pi (x^2)_\omega\, \delta(\omega + \omega'). \qquad (122.4)$$

This relation is to be regarded as the definition of the quantity symbolically denoted by $(x^2)_\omega$. Although the quantities x_ω are complex, $(x^2)_\omega$ is clearly real. The expression (122.4) is zero except when $\omega' = -\omega$, and is symmetrical with respect to interchange of ω and ω'; hence $(x^2)_\omega = (x^2)_{-\omega}$; and a change in the sign of ω is equivalent to taking the complex conjugate.

Substituting (122.4) in (122.3) and eliminating the delta function by integrating with respect to ω, we find

$$\phi(t) = \int_{-\infty}^{\infty} (x^2)_\omega\, e^{-i\omega t}\, \frac{d\omega}{2\pi}. \qquad (122.5)$$

In particular, $\phi(0)$ is the mean square of the fluctuating quantity:

$$\langle x^2 \rangle = \int_{-\infty}^{\infty} (x^2)_\omega\, \frac{d\omega}{2\pi} = \int_0^{\infty} 2(x^2)_\omega\, \frac{d\omega}{2\pi}. \qquad (122.6)$$

We see that the spectral density of the mean square fluctuation is just the quantity $(x^2)_\omega$, or $2(x^2)_\omega$ if the integral is taken only over positive frequencies. The same quantity is also, by (122.5), the Fourier component of the correlation function. Conversely

$$S_x(\omega) = (x^2)_\omega = \int_{-\infty}^{\infty} \phi(t) e^{i\omega t}\, dt. \qquad (122.7)$$

power spectrum

In the formulae given above, $x(t)$ has been regarded as a classical quantity. For a quantum variable, the expansions (122.1) and (122.2) must relate to

[†] We follow S. M. Rytov in the treatment of the spectral resolution of fluctuations.

$$S_x(q,\omega) = \int dt\, d^3r\, \langle x(r,t) x(0,0) \rangle\, e^{i(q \cdot r - \omega t)}$$

Normalization ??

a time–dependent operator $\hat{x}(t)$, and the spectral density $(x^2)_\omega$ is defined by

$$\tfrac{1}{2}\langle \hat{x}_\omega \hat{x}_{\omega'} + \hat{x}_{\omega'} \hat{x}_\omega \rangle = 2\pi (x^2)_\omega \, \delta(\omega + \omega') \qquad (122.8)$$

instead of (122.4).

The expression (118.8) was obtained for the correlation function of quasi-stationary fluctuations of a single quantity. An elementary integration gives as its spectral resolution

$$(x^2)_\omega = \frac{1}{\beta} \left[\frac{1}{\lambda - i\omega} + \frac{1}{\lambda + i\omega} \right] = \frac{2\lambda}{\beta(\omega^2 + \lambda^2)} . \qquad (122.9)$$

In accordance with the physical significance of the approximation, which corresponds to quasi-stationary fluctuations, this expression is valid only for frequencies small compared with the reciprocal of the time for partial equilibrium to be established.

In terms of the random force $y(t)$ defined at the end of § 118, the time dependence of the fluctuating quantity x is described by the equation $\dot{x} = -\lambda x + y$. Multiplying by $e^{i\omega t}$ and integrating with respect to t from $-\infty$ to ∞ (the term $\dot{x}e^{i\omega t}$ being integrated by parts[†]), we obtain $(\lambda - i\omega)x_\omega = y_\omega$. Hence it is evident that we must put

$$(y^2)_\omega = (\omega^2 + \lambda^2)(x^2)_\omega = 2\lambda/\beta. \qquad (122.10)$$

This expression can also, of course, be obtained directly from (118.10). The presence of $\delta(t)$ in (118.10) corresponds to the fact that $(y^2)_\omega$ is independent of ω in (122.10).

The above formulae can be immediately generalised to the simultaneous fluctuations of several thermodynamic variables. The corresponding correlation functions $\phi_{ik}(t)$ have been determined in § 119. The components of their spectral resolution are

$$(x_i x_k)_\omega = \int_{-\infty}^{\infty} \phi_{ik}(t) e^{i\omega t} \, dt$$

$$\equiv \int_{-\infty}^{\infty} \langle x_i(t) x_k(0) \rangle e^{i\omega t} \, dt, \qquad (122.11)$$

and instead of (122.4) we have

$$\langle x_{i\omega} x_{k\omega'} \rangle = 2\pi (x_i x_k)_\omega \, \delta(\omega + \omega'); \qquad (122.12)$$

the order of the factors in $(x_i x_k)_\omega$ is significant.

[†] The terms in $x(\pm \infty)$ are to be omitted; they occur because, as mentioned above, the integrals (122.1) are actually divergent. Formally, they are still unimportant in the calculation of the average $\langle y_\omega y_{\omega'} \rangle$, since they are finite when $\omega' = -\omega$ and can be omitted in comparison with the principal (delta function) term.

A change in the sign of the time is equivalent to the change $\omega \rightarrow -\omega$ in the spectral resolution, which in turn implies taking the complex conjugate of $(x_i x_k)_\omega$. The equation $\phi_{ik}(t) = \phi_{ki}(-t)$ (119.2) therefore shows that

$$(x_i x_k)_\omega = (x_k x_i)_{-\omega} = (x_k x_i)_\omega^*. \tag{122.13}$$

The symmetry of the fluctuations under time reversal, expressed by equation (119.3) or (119.4), is written in terms of the spectral resolution as

$$(x_i x_k)_\omega = \pm (x_i x_k)_{-\omega} = \pm (x_i x_k)_\omega^*, \tag{122.14}$$

where the $+$ and $-$ signs respectively relate to cases where x_i and x_k behave similarly or differently under time reversal; in the former case, therefore, $(x_i x_k)_\omega$ is real and symmetrical in the suffixes i and k, while in the latter case it is imaginary and antisymmetrical.

The equations (119.8) are satisfied by the correlation functions of quasi-stationary fluctuations. They are easily solved by means of the spectral resolution.

Since the equations (119.8) refer only to times $t > 0$, we shall apply to them a "one-sided" Fourier transformation, multiplying by $e^{i\omega t}$ and integrating with respect to t from 0 to ∞. The term $e^{i\omega t}\dot\phi_{il}(t)$ is integrated by parts; since $\phi_{il}(\infty) = 0$, we have

$$-\phi_{il}(0) - i\omega (x_i x_l)_\omega^{(+)} = -\lambda_{ik}(x_k x_l)_\omega^{(+)},$$

with the notation

$$(x_i x_l)_\omega^{(+)} = \int_0^\infty \phi_{il}(t) e^{i\omega t} \, dt. \tag{122.15}$$

The value of $\phi_{il}(0)$ is determined by the "initial condition" (119.9); hence

$$(\lambda_{ik} - i\omega \delta_{ik})(x_k x_l)_\omega^{(+)} = \beta^{-1}_{il},$$

or

$$(\zeta_{ik} - i\omega \beta_{ik})(x_k x_l)_\omega^{(+)} = \delta_{il},$$

where the coefficients λ_{ik} have been replaced by the kinetic coefficients $\zeta_{ik} = \beta_{il}\lambda_{lk}$ (see (120.13)), which are more convenient because of their symmetry. The solution of these algebraic equations is

$$(x_k x_l)_\omega^{(+)} = (\zeta - i\omega \beta)^{-1}_{kl},$$

where the index -1 denotes the inverse matrix.

On the other hand, the desired components of the spectral resolution (122.11) are expressed in terms of the components of the "one-sided" expansion (122.15) by

$$(x_i x_k)_\omega = (x_i x_k)_\omega^{(+)} + (x_k x_i)_\omega^{(+)*} ; \tag{122.16}$$

this is easily seen by writing the integral from $-\infty$ to ∞ as a sum of two integrals from $-\infty$ to 0 and from 0 to ∞, replacing t by $-t$ in the first, and using the symmetry property (119.2). Thus we have finally

$$(x_i x_k)_\omega = (\zeta - i\omega\beta)^{-1}_{ik} + (\zeta + i\omega\beta)^{-1}_{ki}. \tag{122.17}$$

By the symmetry properties of the matrices ζ_{ik} and β_{ik}, the quantities (122.17) necessarily have the properties (122.13) and (122.14).[†]

The results obtained above can be put in a different form by using in the relaxation equations "random forces" in the same way as was done at the end of § 118 for a single fluctuating quantity. The correlation properties of these forces take an especially simple form if they are brought into the equations written in terms of thermodynamically conjugate quantities, as in (120.5) or (120.13). For example, with random forces Y_i, the equations (120.13) become

$$\dot{X}_i = -\zeta_{ik}x_k + Y_i; \tag{122.18}$$

the quantities Y_i may be neglected when the x_i exceed their mean fluctuations. As in the derivation of (122.10), a simple calculation gives for the spectral resolution of the correlation functions of the random forces

$$(Y_i Y_k)_\omega = \zeta_{ik} + \zeta_{ki}. \tag{122.19}$$

As in (122.10), these are independent of the frequency.

If random forces y_i are used in equation (120.5),

$$\dot{x}_i = -\gamma_{ik}X_k + y_i, \tag{122.20}$$

their correlation function is given by the analogous formula

$$(y_i y_k)_\omega = \gamma_{ik} + \gamma_{ki}. \tag{122.21}$$

This is obvious without further calculations if we again use the reciprocity of the relation between the x_i and the X_i (see the second footnote to § 111). The advantage of formulae (122.19) and (122.21) is that they involve the elements of the matrices ζ_{ik} and γ_{ik} themselves, and not of their inverses.[‡]

[†] The matrix of the quantities β_{ik} is always symmetrical. But, if certain x_i and x_k behave differently under time reversal, the corresponding $\beta_{ik} = 0$. This follows because β_{ik} is the coefficient of the product $x_i x_k$ in the quadratic form (111.1) which determines the change of entropy in a deviation from equilibrium. Since the entropy is invariant under time reversal, while the product $x_i x_k$ changes sign, the entropy cannot contain any such term, and we must have $\beta_{ik} = 0$.

[‡] Since the expressions (122.19) and (122.21) are independent of the frequency, it follows (as with (122.10) for a single fluctuating quantity) that the correlation functions $\langle Y_i(t) Y_k(0) \rangle$ and $\langle y_i(t) y_k(0) \rangle$ themselves contain a delta function of time:

$$\langle y_i(t) y_k(0) \rangle = (\gamma_{ik} + \gamma_{ki}) \delta(t). \tag{122.21a}$$

As an example of the use of the above formulae, let us consider fluctuations of a one-dimensional oscillator, i.e. a body which is at rest in the equilibrium position ($Q = 0$) but capable of executing small oscillations in some macroscopic coordinate Q. Because of fluctuations, the coordinate Q will in fact undergo deviations from the value $Q = 0$. The mean square of this deviation is determined directly from the coefficient in the quasi-elastic force which acts on the body during a deviation.

We write the potential energy of the oscillator in the form $U = \frac{1}{2}m\omega_0^2 Q^2$, where m is the "mass" (i.e. the coefficient of proportionality between the generalised momentum P and the velocity $\dot{Q}$: $P = m\dot{Q}$), and ω_0 the natural frequency of the oscillator (in the absence of friction). Then the mean square fluctuation is (cf. § 112, Problem 7)

$$\langle Q^2 \rangle = T/m\omega_0^2. \tag{122.22}$$

The spectral resolution of the fluctuations in the coordinate will be obtained for the general case where the oscillations are accompanied by friction.

The equations of motion of an oscillator with friction are

$$\dot{Q} = P/m, \tag{122.23}$$

$$\dot{P} = -m\omega_0^2 Q - \gamma P/m, \tag{122.24}$$

where $-\gamma P/m = -\gamma\dot{Q}$ is the frictional force. As shown in § 121, if Q and P are taken as x_1 and x_2, the corresponding X_1 and X_2 are $m\omega_0^2 Q/T$ and P/mT. Equations (122.23) and (122.24) then represent the relations $\dot{x}_i = -\gamma_{ik}X_k$, so that

$$\gamma_{11} = 0, \quad \gamma_{12} = -\gamma_{21} = -T, \quad \gamma_{22} = \gamma T.$$

In order to apply these equations to the fluctuations, we write (122.24) as

$$\dot{P} = -m\omega_0^2 Q - \gamma P/m + y, \tag{122.25}$$

with the random force y on the right; equation (122.23), which is the definition of the momentum, must be left unchanged. According to (122.21) we have immediately the spectral density of the fluctuations of the random force

$$(y^2)_\omega = 2\gamma_{22} = 2\gamma T. \tag{122.26}$$

Finally, in order to derive the required $(Q^2)_\omega$, we substitute $P = m\dot{Q}$ in (122.25), obtaining

$$m\ddot{Q} + \gamma\dot{Q} + m\omega_0^2 Q = y. \tag{122.27}$$

Multiplying by $e^{i\omega t}$ and integrating over time, we find

$$(-m\omega^2 - i\omega\gamma + m\omega_0^2)Q_\omega = y_\omega,$$

and hence finally

$$(Q^2)_\omega = 2\gamma T/[m^2(\omega^2 - \omega_0^2)^2 + \omega^2\gamma^2]. \tag{122.28}$$

§ 123. The generalised susceptibility

It is not possible to derive a general formula for the spectral distribution of arbitrary fluctuations analogous to formula (122.9) for the quasi-stationary fluctuations, but in many cases it is possible to relate the properties of fluctuations to quantities describing the behaviour of the body under certain external interactions. These may be fluctuations of either classical or quantum quantities.

Physical quantities of this type have the property that for each of them there exists an external interaction described by the presence, in the Hamiltonian of the body, of a perturbing operator of the type

$$\hat{V} = -\hat{x}f(t), \tag{123.1}$$

where $\hat{x}$ is the operator of the physical quantity concerned, and the perturbing *generalised force f* is a given function of time.

The quantum mean value $\bar{x}$ is not zero when such a perturbation is present (whereas $\bar{x} = 0$ in the equilibrium state in the absence of the perturbation), and it can be written in the form $\hat{\alpha}f$, where $\hat{\alpha}$ is a linear integral operator whose effect on the function $f(t)$ is given by a formula of the type

$$\bar{x}(t) = \hat{\alpha}f = \int_0^\infty \alpha(\tau) f(t-\tau)\, d\tau, \tag{123.2}$$

$\alpha(\tau)$ being a function of time which depends on the properties of the body. The value of $\bar{x}$ at time t can, of course, depend only on the values of the force f at previous (not subsequent) times; the expression (123.2) satisfies this requirement. The quantity $\bar{x}(t)$ is called the *response* of the system to the perturbation.

Any perturbation depending on time can be reduced by means of a Fourier expansion to a set of monochromatic components with a time dependence $e^{-i\omega t}$. Substituting in (123.2) f and $\bar{x}$ in the forms $f_\omega e^{-i\omega t}$ and $\bar{x}_\omega e^{-i\omega t}$, we find as the relation between the Fourier components of the force and the response

$$\bar{x}_\omega = \alpha(\omega)f_\omega, \tag{123.3}$$

where the function $\alpha(\omega)$ is given by

$$\alpha(\omega) = \int_0^\infty \alpha(t)e^{i\omega t}\, dt. \tag{123.4}$$

If this function is specified, the behaviour of the body under the perturbation in question is completely determined. We shall call[†] $\alpha(\omega)$ the *generalised*

† For example, f may be an electric field and x the electric dipole moment acquired by the body in that field. Then α is the electric polarisability of the body.

susceptibility. It plays a fundamental part in the theory described below, since, as we shall see, the fluctuations of the quantity x can be expressed in terms of it.[†]

The function $\alpha(\omega)$ is in general complex. Let its real and imaginary parts be denoted by α' and α'':

$$\alpha(\omega) = \alpha'(\omega) + i\alpha''(\omega). \tag{123.5}$$

The definition (123.4) shows immediately that

$$\alpha(-\omega) = \alpha^*(\omega). \tag{123.6}$$

Separating the real and imaginary parts, we find

$$\alpha'(-\omega) = \alpha'(\omega), \quad \alpha''(-\omega) = -\alpha''(\omega), \tag{123.7}$$

i.e. $\alpha'(\omega)$ is an even function of the frequency, and α'' an odd function. When $\omega = 0$ the function $\alpha''(\omega)$ changes sign, passing through zero (or in some cases through infinity).

It should be emphasised that the property (123.6) simply expresses the fact that the response $\bar{x}$ must be real for any real force f. If the function f is purely monochromatic and is given by the real expression

$$f(t) = \operatorname{re} f_0 e^{-i\omega t} = \tfrac{1}{2}(f_0 e^{-i\omega t} + f_0^* e^{i\omega t}), \tag{123.8}$$

then by applying the operator α to each of the two terms we obtain

$$\bar{x} = \tfrac{1}{2}[\alpha(\omega)f_0 e^{-i\omega t} + \alpha(-\omega)f_0^* e^{i\omega t}]; \tag{123.9}$$

the condition for this expression to be real is just (123.6).

As $\omega \to \infty$, the function $\alpha(\omega)$ tends to a real finite limit α_∞. For definiteness we shall suppose below that this limit is zero; a non-zero α_∞ requires only some obvious slight changes in some of the formulae.

The change in state of the body as a result of the "force" f is accompanied by absorption (dissipation) of energy; the source of this energy is the external interaction, and after absorption in the body it is converted into heat. This dissipation also can be expressed in terms of the quantity α. To do so, we use the equation $dE/dt = \overline{\partial H/\partial t}$, which states that the time derivative of the mean energy of the body is equal to the mean value of the partial derivative of the Hamiltonian of the body with respect to time (see § 11). Since only the perturbation $\hat{V}$ in the Hamiltonian depends explicitly on the time, we have

$$dE/dt = -\bar{x}\, df/dt. \tag{123.10}$$

[†] The quantity $\alpha(\omega)$ thus defined is more convenient than the quantity $Z(\omega) = -1/i\omega\alpha(\omega)$, called the *generalised impedance*, which is the coefficient in the relation $f_\omega = Z(\omega)\dot{x}_\omega$.

This relation is of importance in applications of the theory under discussion. If we know the expression for the change in energy in a particular process, a comparison with (123.10) will show which quantity is to be interpreted as the "force" f with respect to a given variable x.

Substituting $\bar{x}$ and f from (123.8) and (123.9) in (123.10) and averaging over time, we find the mean energy Q dissipated per unit time in the system under a monochromatic perturbation. The terms containing $e^{\pm 2i\omega t}$ vanish on averaging, and we obtain[†]

$$Q = \tfrac{1}{4}i\omega(\alpha^* - \alpha)\,|f_0|^2 = \tfrac{1}{2}\omega\alpha''(\omega)\,|f_0|^2. \tag{123.11}$$

From this we see that the imaginary part of the susceptibility determines the dissipation of energy. Since any actual process is always accompanied by some dissipation ($Q > 0$), we reach the important conclusion that, for all positive values of the variable ω, the function α'' is positive and not zero.

It is possible to derive some very general relations concerning the function $\alpha(\omega)$ by using the methods of the theory of functions of a complex variable. We regard ω as a complex variable ($\omega = \omega' + i\omega''$) and consider the properties of the function $\alpha(\omega)$ in the upper half of the ω-plane. From the definition (123.4) and the fact that $\alpha(t)$ is finite for all positive t, it follows that $\alpha(\omega)$ is a one-valued regular function everywhere in the upper half-plane. For, when $\omega'' > 0$, the integrand in (123.4) includes the exponentially decreasing factor $e^{-t\omega''}$ and, since the function $\alpha(t)$ is finite throughout the range of integration, the integral converges. The function $\alpha(\omega)$ has no singularity on the real axis ($\omega'' = 0$), except possibly at the origin.[‡] It is useful to notice that the conclusion that $\alpha(\omega)$ is regular in the upper half-plane is, physically, a consequence of the causality principle. Owing to this principle, the integration in (123.2) is taken only over times previous to t, and the range of

---

[†] If the function $f(t)$ is not purely monochromatic but the perturbation acts over a finite time ($f \to 0$ as $|t| \to \infty$), the total energy dissipation in the whole time is expressed in terms of the Fourier component of the perturbation by the integral

$$\int_{-\infty}^{\infty} Q\,dt = -\int_{-\infty}^{\infty} i\omega\alpha(\omega)\,|f_\omega|^2\,\frac{d\omega}{2\pi}$$

$$= \int_{0}^{\infty} 2\omega\alpha''(\omega)\,|f_\omega|^2\,\frac{d\omega}{2\pi}.$$

[‡] In the lower half-plane, the definition (123.4) is invalid, since the integral diverges. Hence the function $\alpha(\omega)$ can be defined in the lower half-plane only as the analytical continuation of the expression (123.4) from the upper half-plane, and in general has singularities in this region.

integration in (123.4) therefore extends from 0 to ∞ rather than from $-\infty$ to ∞.

It is evident also from the definition (123.4) that

$$\alpha(-\omega^*) = \alpha^*(\omega). \tag{123.12}$$

This generalises the relation (123.6) for real ω. In particular, for purely imaginary ω we have $\alpha(i\omega'') = \alpha^*(i\omega'')$, i.e. the function $\alpha(\omega)$ is real on the imaginary axis.

We shall prove the following theorem. The function $\alpha(\omega)$ does not take real values at any finite point in the upper half-plane except on the imaginary axis, where it decreases monotonically from a positive value $\alpha_0 > 0$ at $\omega = i0$ to zero at $\omega = i\infty$. Hence, in particular, it will follow that the function $\alpha(\omega)$ has no zeros in the upper half-plane.

To prove the theorem[†] we use a theorem from the theory of functions of a complex variable, according to which the integral

$$\frac{1}{2\pi i} \int \frac{d\alpha(\omega)}{d\omega} \frac{d\omega}{\alpha(\omega)-a}, \tag{123.13}$$

taken round some closed contour C, is equal to the difference between the number of zeros and the number of poles of the function $\alpha(\omega)-a$ in the region bounded by the contour. Let a be a real number and let C be taken as a contour consisting of the real axis and an infinite semicircle in the upper half-plane (Fig. 53). Let us first suppose that α_0 is finite. Since in the upper

Fig. 53

half-plane the function $\alpha(\omega)$ has no pole, the same is true of $\alpha(\omega)-a$, and the integral in question gives simply the number of zeros of the difference $\alpha-a$, i.e. the number of points at which $\alpha(\omega)$ takes the real value a.

[†] The proof given here is due to N. N. Meĭman.

To calculate the integral, we write it as

$$\frac{1}{2\pi i} \int_{C'} \frac{d\alpha}{\alpha - a},$$

the integration being round a contour C' in the plane of the complex variable α which is the map of the contour C in the ω-plane. The whole infinite semicircle is mapped on to the point $\alpha = 0$, and the origin ($\omega = 0$) is mapped on to another real point α_0. The right and left halves of the real axis of ω are mapped in the α-plane on to some very complicated (generally self-intersecting) curves which are entirely in the upper and lower half-planes respectively. It is important to note that these curves nowhere meet the real axis (except at $\alpha = 0$ and $\alpha = \alpha_0$), since α does not take real values for any real finite ω except $\omega = 0$. Because of this property of the contour C', the total change of the argument of the complex number $\alpha - a$ on passing round C' is 2π (if a lies between 0 and α_0 as shown in Fig. 53) or zero (if a lies outside that range), whatever the number of self-intersections of the contour. Hence it follows that the expression (123.13) is equal to 1 if $0 < a < \alpha_0$ and zero for any other value of a.

Thus we conclude that the function $\alpha(\omega)$ takes, in the upper half-plane of ω, each real value of a in this range once only, and values outside this range not at all. Hence we can deduce first of all that on the imaginary axis, where the function $\alpha(\omega)$ is real, it cannot have either a maximum or a minimum, since otherwise it would take some values at least twice. Consequently, $\alpha(\omega)$ varies monotonically on the imaginary axis, taking on that axis, and nowhere else, all real values from α_0 to zero once only.

If $\alpha_0 = \infty$ (i.e. $\alpha(\omega)$ has a pole at the point $\omega = 0$), the above proof is affected only in that on passing along the real axis (in the ω-plane) it is necessary to avoid the origin by means of an infinitesimal semicircle above it. The change in the contour C' in Fig. 53 can be regarded as the result of moving α_0 to infinity. The function $\alpha(\omega)$ then decreases monotonically from ∞ to 0 on the imaginary axis.

Let us now derive a formula relating the real and imaginary parts of the function $\alpha(\omega)$. To do so, we choose some real positive value ω_0 of ω and integrate the expression $\alpha/(\omega - \omega_0)$ round the contour shown in Fig. 54. This

FIG. 54

contour includes the whole of the real axis, indented upwards at the point $\omega = \omega_0 > 0$ (and also at the point $\omega = 0$ if the latter is a pole of the function $\alpha(\omega)$), and is completed by an infinite semicircle. At infinity, $\alpha \to 0$, and the function $\alpha/(\omega - \omega_0)$ therefore tends to zero more rapidly than $1/\omega$. The integral

$$\int_C \frac{\alpha(\omega)}{\omega - \omega_0} \, d\omega$$

consequently converges; and since $\alpha(\omega)$ is regular in the upper half-plane, and the point $\omega = \omega_0$ has been excluded from the region of integration, the function $\alpha/(\omega - \omega_0)$ is analytic everywhere inside the contour C, and the integral is therefore zero.

The integral along the infinite semicircle is also zero. The point ω_0 is avoided by means of an infinitesimal semicircle whose radius ϱ tends to zero. The direction of integration is clockwise, and the contribution to the integral is $-i\pi\alpha(\omega_0)$. If α_0 is finite, the indentation at the origin is unnecessary, and the integration along the whole real axis therefore gives

$$\lim_{\varrho \to 0} \left\{ \int_{-\infty}^{\omega_0 - \varrho} \frac{\alpha}{\omega - \omega_0} \, d\omega + \int_{\omega_0 + \varrho}^{\infty} \frac{\alpha}{\omega - \omega_0} \, d\omega \right\} - i\pi\alpha(\omega_0) = 0.$$

The first term is the principal value of the integral from $-\infty$ to ∞. Indicating this in the usual way, we have

$$i\pi\alpha(\omega_0) = P \int_{-\infty}^{\infty} \frac{\alpha}{\omega - \omega_0} \, d\omega. \tag{123.14}$$

Here the variable of integration ω takes only real values. We replace it by ξ, call the given real value ω instead of ω_0, and write the function $\alpha(\omega)$ of the real variable ω in the form $\alpha = \alpha' + i\alpha''$. Taking the real and imaginary parts of (123.14), we obtain the following two formulae:

$$\alpha'(\omega) = \frac{1}{\pi} P \int_{-\infty}^{\infty} \frac{\alpha''(\xi)}{\xi - \omega} \, d\xi, \tag{123.15}$$

$$\alpha''(\omega) = -\frac{1}{\pi} P \int_{-\infty}^{\infty} \frac{\alpha'(\xi)}{\xi - \omega} \, d\xi, \tag{123.16}$$

called the *dispersion relations* and first derived by H. A. Kramers and R. de L. Kronig (1927). It should be emphasised that the only essential property of the function $\alpha(\omega)$ used in the proof is that it is regular in the upper half-plane.[†] Hence we can say that Kramers and Kronig's formulae, like this property of $\alpha(\omega)$, are a direct consequence of the causality principle.

Using the fact that $\alpha''(\xi)$ is an odd function, we can rewrite (123.15) as

$$\alpha'(\omega) = \frac{1}{\pi} P \int_0^\infty \frac{\alpha''(\xi)}{\xi-\omega}\,d\xi + \frac{1}{\pi} P \int_0^\infty \frac{\alpha''(\xi)}{\xi+\omega}\,d\xi,$$

or

$$\alpha'(\omega) = \frac{2}{\pi} P \int_0^\infty \frac{\xi\alpha''(\xi)}{\xi^2-\omega^2}\,d\xi. \tag{123.17}$$

If the function $\alpha(\omega)$ has a pole at the point $\omega = 0$, near which $\alpha = iA/\omega$, the semicircle avoiding this pole gives a further real term $-A/\omega$, which must be added to the left-hand side of equation (123.14). Thus formula (123.16) becomes

$$\alpha''(\omega) = -\frac{1}{\pi} P \int_{-\infty}^\infty \frac{\alpha'(\xi)}{\xi-\omega}\,d\xi + \frac{A}{\omega}, \tag{123.18}$$

but (123.15) and (123.17) remain unchanged.

We may also derive a formula which expresses the values of $\alpha(\omega)$ on the positive imaginary axis in terms of the values of $\alpha''(\omega)$ on the real axis. To do so, we consider the integral

$$\int \frac{\omega\alpha(\omega)}{\omega^2+\omega_0^2}\,d\omega$$

taken along a contour consisting of the real axis and an infinite semicircle in the upper half-plane (ω_0 being a real number). This integral can be expressed in terms of the residue of the integrand at the pole $\omega = i\omega_0$. The integral along the infinite semicircle is zero, and so we have

$$\int_{-\infty}^\infty \frac{\omega\alpha(\omega)}{\omega^2+\omega_0^2}\,d\omega = i\pi\alpha(i\omega_0).$$

On the left-hand side the real part of the integral is zero, since the integrand is an odd function. Replacing ω by ξ and ω_0 by ω, we have finally

$$\alpha(i\omega) = \frac{2}{\pi} \int_0^\infty \frac{\xi\alpha''(\xi)}{\omega^2+\xi^2}\,d\xi. \tag{123.19}$$

[†] The property $\alpha \to 0$ as $\omega \to \infty$ is not essential: if the limit α_∞ were other than zero, we should simply take $\alpha-\alpha_\infty$ in place of α, with corresponding obvious changes in formulae (123.15), (123.16). See also § 126, Problem.

Integration of this with respect to ω gives

$$\int\limits_0^\infty \alpha(i\omega)\,d\omega = \int\limits_0^\infty \alpha''(\omega)\,d\omega. \qquad (123.20)$$

§ 124. The fluctuation–dissipation theorem

Let us now proceed to calculations which are designed to relate the fluctuations of a quantity x to the generalised susceptibility defined in § 123.

Let a body to which the quantity x refers be in a particular (nth) stationary state. The mean value (122.8) is calculated as the corresponding diagonal matrix element

$$\tfrac{1}{2}(\hat{x}_\omega \hat{x}_{\omega'} + \hat{x}_{\omega'}\hat{x}_\omega)_{nn} = \tfrac{1}{2}\sum_m [(x_\omega)_{nm}\,(x_{\omega'})_{mn} + (x_{\omega'})_{nm}\,(x_\omega)_{mn}], \qquad (124.1)$$

where the summation is over the whole spectrum of energy levels; since the operator $\hat{x}_\omega$ is complex, the two terms in the brackets are not equal.

The time dependence of the operator $\hat{x}(t)$ means that its matrix elements must be calculated by means of the time-dependent wave functions. We therefore have

$$(x_\omega)_{nm} = \int\limits_{-\infty}^\infty x_{nm}e^{i(\omega_{nm}+\omega)t}\,dt = 2\pi x_{nm}\,\delta(\omega_{nm}+\omega), \qquad (124.2)$$

where x_{nm} is the ordinary time-independent matrix element of the operator $\hat{x}$, expressed in terms of the coordinates of the particles of the body, and $\omega_{nm} = (E_n - E_m)/\hbar$ is the frequency of the transition between the states n and m. Thus

$$\tfrac{1}{2}(\hat{x}_\omega \hat{x}_{\omega'} + \hat{x}_{\omega'}\hat{x}_\omega)_{nn} = 2\pi^2 \sum_m |x_{nm}|^2\,[\delta(\omega_{nm}+\omega)\,\delta(\omega_{mn}+\omega')$$

$$+ \delta(\omega_{nm}+\omega')\,\delta(\omega_{mn}+\omega)],$$

where we have used the fact that $x_{nm} = x_{mn}^*$, since x is real. The products of delta functions in the brackets can clearly be written as

$$\delta(\omega_{nm}+\omega)\,\delta(\omega+\omega') + \delta(\omega_{mn}+\omega)\,\delta(\omega+\omega').$$

A comparison with (122.8) then gives

$$(x^2)_\omega = \pi \sum_m |x_{nm}|^2\,[\delta(\omega+\omega_{nm}) + \delta(\omega+\omega_{mn})]. \qquad (124.3)$$

The following comment may be made concerning the way in which this expression is written. Although the energy levels of a macroscopic body are, strictly speaking, discrete, they are so close together that in practice they form a continuous spectrum. Formula (124.3) may be written without the delta functions if it is averaged over small frequency intervals (which never-theless contain many levels). If $\Gamma(E)$ is the number of energy levels less than E, then

$$(x^2)_\omega = \pi\hbar \,|x_{nm}|^2 \left[\frac{d\Gamma}{dE_m} + \frac{d\Gamma}{dE_m'} \right], \qquad (124.4)$$

where $E_m = E_n + \hbar\omega$, $E_m' = E_n - \hbar\omega$.

Let us now assume that the body is subject to a periodic perturbation (with frequency ω), described by the operator

$$\hat{V} = -f\hat{x} = -\tfrac{1}{2}(f_0 e^{-i\omega t} + f_0^* e^{i\omega t})\hat{x}. \qquad (124.5)$$

Under the effect of the perturbation the system makes transitions, and the probability (per unit time) of the transition $n \to m$ is given by

$$w_{mn} = \frac{\pi\,|f_0|^2}{2\hbar^2} |x_{mn}|^2 \{\delta(\omega + \omega_{mn}) + \delta(\omega + \omega_{nm})\}; \qquad (124.6)$$

see *Quantum Mechanics*, § 42. The two terms in this formula correspond to those in (124.5). In each transition the system absorbs or emits a quantum $\hbar\omega_{mn}$. The sum

$$Q = \sum_m w_{mn}\hbar\omega_{mn}$$

is the mean energy absorbed by the body per unit time; this energy is supplied by the external perturbation, and after absorption in the body it is dissipated there. Substitution of (124.6) gives

$$Q = \frac{\pi}{2\hbar}\,|f_0|^2 \sum_m |x_{nm}|^2 \{\delta(\omega + \omega_{mn}) + \delta(\omega + \omega_{nm})\}\omega_{mn}$$

or, since the delta functions are zero except when their argument is zero,

$$Q = \frac{\pi}{2\hbar}\,\omega\,|f_0|^2 \sum_m |x_{nm}|^2 \{\delta(\omega + \omega_{nm}) - \delta(\omega + \omega_{mr})\}. \qquad (124.7)$$

Comparison of (124.7) and (123.11) gives

$$\alpha''(\omega) = \frac{\pi}{\hbar} \sum_m |x_{nm}|^2 \{\delta(\omega + \omega_{nm}) - \delta(\omega + \omega_{mn})\}. \qquad (124.8)$$

The quantities $(x^2)_\omega$ and α'' thus calculated are related in a simple manner but the relation appears only when these quantities are expressed in terms of the temperature of the body. To do this, we average by means of the Gibbs distribution (cf. the first footnote to § 118). For $(x^2)_\omega$ we have

$$(x^2)_\omega = \pi \sum_{n,\,m} \varrho_n\,|x_{nm}|^2 \{\delta(\omega + \omega_{nm}) + \delta(\omega + \omega_{mn})\},$$

where for brevity we have put $\varrho_n = e^{(F-E_n)/T}$, E_n denoting the energy levels of the body and F its free energy. Since the summation is now over both suffixes m and n, these can be interchanged. If this is done in the second term, we obtain

$$(x^2)_\omega = \pi \sum_{m,n} (\varrho_n + \varrho_m)\,|x_{nm}|^2\,\delta(\omega+\omega_{nm})$$

$$= \pi \sum_{m,n} \varrho_n(1+e^{\hbar\omega_{nm}/T})\,|x_{nm}|^2\,\delta(\omega+\omega_{nm})$$

or, because of the delta function in the summand,

$$(x^2)_\omega = \pi(1+e^{-\hbar\omega/T}) \sum_{m,n} \varrho_n\,|x_{nm}|^2\,\delta(\omega+\omega_{nm}).$$

In an exactly similar manner we obtain

$$\alpha'' = \frac{\pi}{\hbar}(1-e^{-\hbar\omega/T}) \sum_{m,n} \varrho_n\,|x_{nm}|^2\,\delta(\omega+\omega_{nm}).$$

A comparison of these two expressions gives

$$(x^2)_\omega = \hbar\alpha'' \coth\frac{\hbar\omega}{2T} = 2\hbar\alpha'' \left\{\frac{1}{2} + \frac{1}{e^{\hbar\omega/T}-1}\right\}. \qquad (124.9)$$

The mean square of the fluctuating quantity itself is given by the integral

$$\langle x^2 \rangle = \frac{\hbar}{\pi} \int_0^\infty \alpha''(\omega) \coth\frac{\hbar\omega}{2T}\,d\omega. \qquad (124.10)$$

These important formulae constitute the *fluctuation–dissipation theorem* (FDT), established by H. B. Callen and T. A. Welton (1951). They relate the fluctuations of physical quantities to the dissipative properties of the system when it is subject to an external interaction. It should be noted that the factor in the braces in (124.9) is the mean energy (in units of $\hbar\omega$) of an oscillator at temperature T; the term $\frac{1}{2}$ corresponds to the zero-point oscillations.

As at the end of §118 the results obtained above can be written in a different form by formally regarding the spontaneous fluctuations of the quantity x as due to the action of some fictitious random forces. It is convenient to write the formulae in terms of the Fourier components x_ω and f_ω as if x were a classical quantity. The relation between them is

$$x_\omega = \alpha(\omega)f_\omega, \qquad (124.11)$$

similarly to (123.3), and the mean square fluctuations can then be written in the form

$$\langle x_\omega x_{\omega'} \rangle = \alpha(\omega)\,\alpha(\omega')\,\langle f_\omega f_{\omega'} \rangle,$$

or with the spectral densities of fluctuations, by the definition (122.4):

$$(x^2)_\omega = \alpha(\omega)\,\alpha(-\omega)\,(f^2)_\omega$$
$$= |\alpha(\omega)|^2\,(f^2)_\omega.$$

The spectral density of the mean square of the random force is accordingly, from (124.9),

$$(f^2)_\omega = \frac{\hbar\alpha''(\omega)}{|\alpha(\omega)|^2}\coth\frac{\hbar\omega}{2T}. \tag{124.12}$$

This treatment may offer certain advantages in particular applications of the theory.

The derivation of the FDT is based on regarding the external interaction (124.5) as a small perturbation; this also ensures that the response of the system is linear (there is a linear relation between $\bar{x}$ and the force f). It should be emphasised, however, that this does not imply any physical restrictions on the permissible values of the mean fluctuation of x itself. The smallness of the interaction can always be achieved by making sufficiently small the auxiliary quantity f, which does not appear in the final formulation of the FDT. Thus, for the class of physical quantities x under consideration, the properties of their fluctuations (in a system in thermodynamic equilibrium) are entirely determined by the response of the system to an infinitesimal external interaction.

At temperatures $T \gg \hbar\omega$ we have $\coth(\hbar\omega/2T) \approx 2T/\hbar\omega$, and formula (124.9) becomes

$$(x^2)_\omega = (2T/\omega)\,\alpha''(\omega). \tag{124.13}$$

The constant $\hbar$ no longer appears, in accordance with the fact that under these conditions the fluctuations are classical.

If the inequality $T \gg \hbar\omega$ is valid for all frequencies of importance (those for which $\alpha''(\omega)$ is significantly different from zero), then we can take the classical limit in the integral formula (124.10) also:

$$\langle x^2 \rangle = \frac{2T}{\pi}\int\limits_0^\infty \frac{\alpha''(\omega)}{\omega}\,d\omega.$$

But from (123.17) this integral can be expressed in terms of the static value $\alpha'(0) = \alpha(0)$, and hence[†]

$$\langle x^2 \rangle = T\alpha(0). \tag{124.14}$$

[†] This expression can also be derived directly from the Gibbs distribution in classical statistics. Let $x = x(q, p)$ be a classical quantity. With a term $-xf$ (f constant) in the energy of the system, the mean value of x is

$$\bar{x} = \int x(q, p)\exp\left\{\frac{F - E(q, p) + x(q, p)f}{T}\right\}\,dq\,dp.$$

By definition, $\alpha(0) = d\bar{x}/df$ when $f \to 0$; differentiating the above expression, we

Lastly, let us consider the relationship between these results and the theory of quasi-stationary fluctuations (§ 118). First of all, we note that, if x is such that its fluctuations are small in the sense implied in § 110 (i.e. the expansion (110.3) of the entropy is permissible), the mean square $\langle x^2 \rangle = 1/\beta$. A comparison with (124.14) shows that for such a quantity

$$\alpha(0) = 1/\beta T. \qquad (124.15)$$

Next, let x be a quantity of the class whose fluctuations are quasi-stationary. We assume that the body is subject to the action of a static force f. This causes a displacement of the equilibrium state, in which $\bar{x}$ is now not zero but $\alpha(0)f = f/\beta T$. The macroscopic equation of the relaxation of a system far from equilibrium will then have the form

$$\dot{x} = -\lambda(x - f/\beta T), \qquad (124.16)$$

which differs from $\dot{x} = -\lambda x$ (118.5) in that $\dot{x}$ is zero not when $x = 0$ but when $x = f/\beta T$.

Equation (124.16) may be regarded as valid also when the body is subject to the action of a time-dependent perturbation, provided that the period of variation of the force $f(t)$ is large compared with the time for the establishment of the partial equilibrium corresponding to any given value of x. If $f(t)$ is a periodic function of time with frequency ω, the macroscopic value of $x(t)$ will vary with the same frequency. Substituting in equation (124.16) $f(t)$ and $x(t)$ in the form (123.8), (123.9) and separating the terms in $e^{-i\omega t}$ and $e^{i\omega t}$, we obtain

$$-i\omega\alpha(\omega)f_0 = -\lambda\alpha(\omega)f_0 + (\lambda/\beta T)f_0,$$

whence

$$\alpha(\omega) = \lambda/\beta T(\lambda - i\omega). \qquad (124.17)$$

According to the FDT (124.9), we then have

$$(x^2)_\omega = \frac{2\lambda}{\beta(\lambda^2 + \omega^2)} \frac{\hbar\omega}{2T} \coth \frac{\hbar\omega}{2T}. \qquad (124.18)$$

---

find

$$\alpha(0) = \frac{1}{T} \int x^2 \exp\left(\frac{F-E}{T}\right) dq \, dp = \frac{1}{T}\langle x^2 \rangle;$$

the free energy F also depends on f, but the term in the derivative $\partial F/\partial f$ disappears when we put $f = 0$, i.e. $\bar{x} = 0$.

This result generalises formula (122.9), which relates to fluctuations of a classical quantity. The expression (124.18) differs from (122.9) by a factor

$$\frac{\hbar\omega}{2T} \coth \frac{\hbar\omega}{2T},\tag{124.19}$$

which tends to unity in the classical limit ($\hbar\omega \ll T$).

Equation (124.16) can also be viewed in another way: not as the macroscopic equation of motion of a system far from equilibrium (under an external interaction), but as the equation for the fluctuations of the quantity $x(t)$ in an equilibrium closed system under the action of the random force f. In this interpretation, it corresponds to equation (118.9), and the two definitions of the random force differ only by a factor: $y = \lambda f/T\beta$. The spectral density $(y^2)_\omega$ is found by substituting (124.17) in (124.12):

$$(y^2)_\omega = \frac{2\lambda}{\beta}\frac{\hbar\omega}{2T} \coth \frac{\hbar\omega}{2T},\tag{124.20}$$

which differs from the previous expression (122.10) by the same factor (124.19).

§ 125. The fluctuation-dissipation theorem for more than one variable

The FDT can easily be generalised to the case where several fluctuating quantities x_i are considered simultaneously. The generalised susceptibilities in such a case are determined from the response of the system to a perturbation having the form

$$\hat{V} = -\hat{x}_i f_i(t),\tag{125.1}$$

and are the coefficients in the linear relation between the Fourier components of the mean values $\overline{x_i}(t)$ and the generalised forces $f_i(t)$:

$$\overline{x_{i\omega}} = \alpha_{ik}(\omega)f_{k\omega}.\tag{125.2}$$

The change in the energy of the system is expressed in terms of the external perturbation by

$$\dot{E} = -\dot{f}_i\overline{x}_i.\tag{125.3}$$

This formula, like (123.10), is generally used in particular applications of the theory to establish the actual relationship between the quantities x_i and f_i.

The spectral densities of fluctuations are defined in terms of the mean values of the symmetrised operator products:

$$\tfrac{1}{2}\langle \hat{x}_{i\omega}\hat{x}_{k\omega'} + \hat{x}_{k\omega'}\hat{x}_{i\omega}\rangle = 2\pi(x_ix_k)_\omega\,\delta(\omega+\omega'),\tag{125.4}$$

which generalise the expression (122.8). A calculation of this mean value as the diagonal (nn) matrix element, as in the derivation of (124.3), gives the result

$$(x_i x_k)_\omega = \pi \sum_m \{(x_i)_{nm} (x_k)_{mn} \delta(\omega + \omega_{nm})$$

$$+ (x_k)_{nm} (x_i)_{mn} \delta(\omega + \omega_{mn})\}. \tag{125.5}$$

Let a periodic perturbation with

$$f_i(t) = \tfrac{1}{2}(f_{0i} e^{-i\omega t} + f_{0i}^* e^{i\omega t}) \tag{125.6}$$

act on the system. The response is

$$\overline{x_i}(t) = \tfrac{1}{2}[\alpha_{ik}(\omega) f_{0k} e^{-i\omega t} + \alpha_{ik}^*(\omega) f_{0k}^* e^{i\omega t}]. \tag{125.7}$$

Substituting these two equations in (125.3) and averaging over the period of the perturbation, we obtain instead of (123.11) the following expression for the energy dissipation:

$$Q = \tfrac{1}{4} i\omega (\alpha_{ik}^* - \alpha_{ki}) f_{0i} f_{0k}^*. \tag{125.8}$$

On the other hand, a calculation similar to the derivation of (124.7) gives

$$Q = \frac{\pi}{2\hbar} \omega \sum_m f_{0i} f_{0k}^* [(x_i)_{mn} (x_k)_{nm} \delta(\omega + \omega_{nm})$$

$$- (x_i)_{nm} (x_k)_{mn} \delta(\omega + \omega_{mn})],$$

and a comparison with (125.8) shows that

$$\alpha_{ik}^* - \alpha_{ki} = -\frac{2\pi i}{\hbar} \sum_m [(x_i)_{mn} (x_k)_{nm} \delta(\omega + \omega_{nm})$$

$$- (x_i)_{nm} (x_k)_{mn} \delta(\omega + \omega_{mn})]. \tag{125.9}$$

Finally, averaging (125.5) and (125.9) over the Gibbs distribution as in § 124, we find the following generalisation of the FDT (124.9):

$$(x_i x_k)_\omega = \tfrac{1}{2} i\hbar (\alpha_{ki}^* - \alpha_{ik}) \coth (\hbar\omega/2T). \tag{125.10}$$

As in formulae (124.11) and (124.12), we can express (125.10) in terms of fictitious random forces whose action would give a result equivalent to the spontaneous fluctuations of the x_i. To do so, we write

$$x_{i\omega} = \alpha_{ik} f_{k\omega}, \qquad f_{i\omega} = \alpha^{-1}_{ik} x_{k\omega} \tag{125.11}$$

and

$$(f_i f_k)_\omega = \alpha^{-1}_{il} \alpha^{-1}_{km} (x_l x_m)_\omega.$$

Substitution of (125.10) gives

$$(f_i f_k)_\omega = \tfrac{1}{2} i\hbar (\alpha^{-1}_{ik} - \alpha^{-1*}_{ki}) \coth (\hbar\omega/2T). \tag{125.12}$$

From these results we can draw some conclusions concerning the symmetry properties of the generalised susceptibilities $\alpha_{ik}(\omega)$ (H. B. Callen, M. L. Barasch, J.L. Jackson and R.F. Greene, 1952). Let us first suppose that the quantities x_i and x_k are invariant under time reversal; then their operators $\hat{x}_i$ and $\hat{x}_k$ are real. We shall also suppose that the body has no magnetic structure (see the first footnote to § 128) and is not in an external magnetic field; then the wave functions of its stationary states are also real,[†] and consequently so are the matrix elements of the quantities x. Since the matrices x_{nm} are Hermitian, we have $x_{nm} = x_{mn}^* = x_{mn}$. Then the right-hand side of (125.9), and consequently the left-hand side, are symmetrical in the suffixes i and k. Hence $\alpha_{ik}^* - \alpha_{ki} = \alpha_{ki}^* - \alpha_{ik}$ or $\alpha_{ik} + \alpha_{ik}^* = \alpha_{ki} + \alpha_{ki}^*$, i.e. we conclude that the real part of α_{ik} is symmetric.

But the real and imaginary parts (α_{ik}' and α_{ik}'') of each α_{ik} are related by linear integral equations, namely Kramers and Kronig's formulae. Hence the symmetry of α_{ik}' implies that of α_{ik}'' and therefore that of α_{ik} itself. Our final result is therefore

$$\alpha_{ik}(\omega) = \alpha_{ki}(\omega). \tag{125.13}$$

These relationships are somewhat modified if the body is in an external magnetic field **H**. The wave functions of a system in a magnetic field are not real, but have the property $\psi^*(\mathbf{H}) = \psi(-\mathbf{H})$. Accordingly the matrix elements of the quantities x are such that $x_{nn}(\mathbf{H}) = x_{mn}(-\mathbf{H})$, and the expression on the right of (125.9) is unchanged, when the suffixes i and k are transposed, only if the sign of **H** is simultaneously changed. We therefore obtain the relation

$$\alpha_{ik}^*(\mathbf{H}) - \alpha_{ki}(\mathbf{H}) = \alpha_{ki}^*(-\mathbf{H}) - \alpha_{ik}(-\mathbf{H}).$$

Another relation is given by Kramers and Kronig's formula (123.14), according to which

$$\alpha_{ki} = i\hat{J}(\alpha_{ki}),$$

where $\hat{J}$ is a real linear operator. Adding this to the Hermitian conjugate equation $\alpha_{ik}^* = -i\hat{J}(\alpha_{ik}^*)$, we obtain

$$\alpha_{ik}^* + \alpha_{ki} = -i\hat{J}(\alpha_{ik}^* - \alpha_{ki});$$

here all the α_{ik} are, of course, taken for a fixed value of **H**. Hence we see that, if the difference $\alpha_{ik}^* - \alpha_{ki}$ has a particular symmetry property, then so has

---

[†] The exact energy levels of a system of interacting particles can be degenerate only with respect to the directions of the total angular momentum of the system. This source of degeneracy can be eliminated by assuming the body to be enclosed in a vessel with immovable walls. The energy levels of the body will not then be degenerate, and so the corresponding exact wave functions can be taken as real.

the sum $\alpha_{ik}^* + \alpha_{ki}$, and therefore α_{ik} itself. Thus

$$\alpha_{ik}(\omega; \mathbf{H}) = \alpha_{ki}(\omega; -\mathbf{H}). \tag{125.14}$$

Finally, let the quantities x include some which change sign under time reversal. The operator corresponding to such a quantity is purely imaginary, and so $x_{nm} = x_{mn}^* = -x_{mn}$. If the two quantities x_i, x_k are both of this kind, the derivation of (125.13) is unaffected, but if only one of them changes sign under time reversal, the right-hand side of equation (125.9) changes sign when the suffixes i, k are interchanged. Accordingly (125.13) becomes

$$\alpha_{ik}(\omega) = -\alpha_{ki}(\omega). \tag{125.15}$$

For a body in a magnetic field we have

$$\alpha_{ik}(\omega; \mathbf{H}) = -\alpha_{ki}(\omega; -\mathbf{H}). \tag{125.16}$$

All these relations can, of course, be derived also from formula (125.10) as a consequence of the symmetry of the fluctuations with respect ot time. For example, if the two quantities x_i and x_k behave in the same way under time reversal, then the symmetry mentioned implies that $(x_i x_k)_\omega$ is real and symmetric in the suffixes i and k (see § 122). Then the right-hand side of (125.10) must also be symmetric in these suffixes, and we return to the result (125.13). This derivation of the symmetry properties of the generalised susceptibilities is analogous to that of the principle of the symmetry of the kinetic coefficients in § 120; we shall see below that formulae (125.13)–(125.16) may be regarded as a generalisation of that principle.

The relation between the generalised susceptibilities and the kinetic coefficients is shown by comparing the FDT with the theory of quasi-stationary fluctuations of several variables. We shall give the relevant formulae without reiterating all the arguments similar to those at the end of § 124 for the case of a single variable.

The static values of the susceptibilities are related to the entropy expansion coefficients β_{ik} by

$$T\alpha_{ik}(0) = \beta_{ik}^{-1}.$$

The displacement of the equilibrium state when static forces f_k act on the system is therefore given by

$$\bar{x}_i = \alpha_{ik}(0)f_k = \beta_{ik}^{-1}f_k/T, \quad \bar{X}_i = \beta_{ik}\bar{x}_k = f_i/T.$$

The macroscopic equations of motion of a non-equilibrium system under the action of quasi-static forces $f_k(t)$ may be written

$$\dot{x}_i = -\gamma_{ik}(X_k - f_k/T), \tag{125.17}$$

which differs from (120.5) in that X_k is replaced by $X_k - f_k/T$.

Substituting in (125.17) $x_i(t)$ and $f_i(t)$ as the periodic functions (125.6), (125.7) (with the X_k written as the linear combinations $X_k = \beta_{kl}x_l$), we obtain

$$-i\omega\alpha_{im}f_{0m} = -\gamma_{ik}\beta_{kl}\alpha_{lm}f_{0m} + \frac{1}{T}\gamma_{im}f_{0m},$$

whence, since the f_{0m} are arbitrary, we have the relations

$$-i\omega\alpha_{im} + \gamma_{ik}\beta_{kl}\alpha_{lm} = \frac{1}{T}\gamma_{im}$$

or

$$\alpha_{ik} = \frac{1}{T}(\beta - i\omega\gamma^{-1})^{-1}_{ik}. \tag{125.18}$$

These are the required relations between the α_{ik} and the kinetic coefficients γ_{ik}.

The quantities β_{ik} are by definition symmetric with respect to their suffixes (being the derivatives $-\partial^2 S/\partial x_i\,\partial x_k$). Hence the symmetry of the α_{ik} implies that of the γ_{ik}, i.e. the ordinary principle of the symmetry of the kinetic coefficients.

Regarding the f_k in equations (125.17) as random forces, and substituting (125.18) in (125.12), we obtain

$$(f_i f_k)_\omega = \frac{\hbar\omega T}{2}(\gamma^{-1}_{ik} + \gamma^{-1}_{ki}) \coth \frac{\hbar\omega}{2T}.$$

If the random forces y_i are defined as in (122.20), then $y_i = \gamma_{ik}f_k/T$, and their spectral distribution is

$$(y_i y_k)_\omega = \frac{\hbar\omega}{2T}(\gamma_{ik} + \gamma_{ki}) \coth \frac{\hbar\omega}{2T}. \tag{125.19}$$

This relation differs from formula (122.21) by the same factor (124.19), which tends to unity in the classical limit.

§ 126. The operator form of the generalised susceptibility

The FDT can also be regarded from the opposite side, taking equation (124.9) from right to left and writing $(x^2)_\omega$ explicitly as the Fourier component of the correlation function:

$$\alpha''(\omega) = \frac{1}{2\hbar} \tanh \frac{\hbar\omega}{2T} \int_{-\infty}^{\infty} e^{i\omega t}\langle \hat{x}(0)\,\hat{x}(t) + \hat{x}(t)\,\hat{x}(0)\rangle\, dt. \tag{126.1}$$

In this form, the equation in principle allows the function $\alpha''(\omega)$ to be calculated from the microscopic properties of the system. Its disadvantage is that it gives directly only the imaginary part of $\alpha(\omega)$, not the whole function. A similar formula can be obtained that does not suffer from this disadvantage. To derive it, we make a direct quantum-mechanical calculation of the mean value $\bar{x}$ in the perturbed system, with the perturbation operator (124.5).[†]

Let $\Psi_n^{(0)}$ be the wave functions of the unperturbed system. Following the general method (see *Quantum Mechanics*, § 40), we seek the wave functions of the perturbed system in the first approximation as

$$\Psi_n = \Psi_n^{(0)} + \sum_m a_{mn}\Psi_m^{(0)}, \qquad (126.2)$$

where the coefficients a_{mn} satisfy the equations

$$i\hbar\frac{da_{mn}}{dt} = V_{mn}e^{i\omega_{mn}t} = -\frac{1}{2}x_{mn}e^{i\omega_{mn}t}(f_0e^{-i\omega t} + f_0^*e^{i\omega t}).$$

In solving this, we must assume that the perturbation is "adiabatically" applied until the time t from $t = -\infty$ (cf. *Quantum Mechanics*, § 43); this means that we must put $\omega \to \omega \mp i0$ in the factors $e^{\pm i\omega t}$ (where $i0$ denotes $i\delta$ with $\delta \to +0$). Then

$$a_{mn} = \frac{1}{2\hbar}x_{mn}e^{i\omega_{mn}t}\left[\frac{f_0e^{-i\omega t}}{\omega_{mn} - \omega - i0} + \frac{f_0^*e^{i\omega t}}{\omega_{mn} + \omega + i0}\right]. \qquad (126.3)$$

With the function Ψ_n thus found, we calculate the mean value $\bar{x}$ as the corresponding diagonal matrix element of the operator $\hat{x}$. In the same approximation,

$$\bar{x} = \int \Psi_n^*\hat{x}\Psi_n \, dq$$

$$= \sum_m (a_{mn}x_{nm}e^{i\omega_{nm}t} + a_{mn}^*x_{mn}e^{i\omega_{mn}t})$$

$$= \frac{1}{2\hbar}\sum_m x_{mn}x_{nm}\left[\frac{1}{\omega_{mn} - \omega - i0} + \frac{1}{\omega_{mn} + \omega + i0}\right]f_0e^{-i\omega t}$$

+complex conjugate.

Comparison of this result with the definition (123.9) shows that

$$\alpha(\omega) = \frac{1}{\hbar}\sum_m |x_{mn}|^2\left[\frac{1}{\omega_{mn} - \omega - i0} + \frac{[1}{\omega_{mn} + \omega + i0}\right]. \qquad (126.4)$$

[†] This is more direct than using Kramers and Kronig's relations to find $\alpha'(\omega)$, and thence $\alpha(\omega)$, from $\alpha''(\omega)$.

The real and imaginary parts of this expression can be separated by means of the formula

$$\frac{1}{x \pm i0} = P\frac{1}{x} \mp i\pi\,\delta(x);$$ (126.5)

see *Quantum Mechanics*, (43.10). For $\alpha''(\omega)$ we recover, of course, the previous result (124.8).

It is easy to see that the expression (126.4) is the Fourier transform of the function

$$\alpha(t) = \begin{cases} (i/\hbar)\langle\hat{x}(t)\,\hat{x}(0)-\hat{x}(0)\,\hat{x}(t)\rangle, & t > 0, \\ 0, & t < 0 \end{cases}$$ (126.6)

(as with the correlation function, this mean value depends, of course, only on the difference between the times at which the two operators $\hat{x}(t)$ are taken). For, calculating the function (126.6) as the diagonal matrix element with respect to the nth stationary state of the (unperturbed) system, we have for $t > 0$

$$\alpha(t) = \frac{i}{\hbar}\sum_m [x_{nm}(t)\,x_{mn}(0)-x_{nm}(0)\,x_{mn}(t)]$$

$$= \frac{i}{\hbar}\sum_m |x_{nm}|^2\,[e^{i\omega_{nm}t}-e^{i\omega_{mn}t}],$$

where the change to time-independent matrix elements is made by the usual rule:

$$x_{nm}(t) = x_{nm}e^{i\omega_{nm}t}.$$

Since the function $\alpha(t)$ is zero except when $t > 0$, its Fourier transform is calculated as[†]

$$\int_0^\infty e^{i\omega t}\,\mathrm{d}t = \frac{i}{\omega+i0},$$ (126.7)

and is the same as (126.4).

Thus we have as the final result

$$\alpha(\omega) = \frac{i}{\hbar}\int_0^\infty e^{i\omega t}\langle\hat{x}(t)\,\hat{x}(0)-\hat{x}(0)\,\hat{x}(t)\rangle\,\mathrm{d}t$$ (126.8)

(R. Kubo, 1956). Being valid for averaging over any given stationary state of the system, this formula also remains unchanged after averaging over the Gibbs distribution.

[†] The integral is calculated by deflecting the contour of integration (in the complex t-plane) upwards or downwards according to the sign of ω, i.e. replacing t by $t(1+i\delta\,\mathrm{sgn}\,\omega)$ and then making $\delta \to +0$.

The generalised susceptibilities $\alpha_{ik}(\omega)$ which determine the response of the system to a perturbation affecting several quantities x_i are given by a similar formula,

$$\alpha_{ik}(\omega) = \frac{i}{\hbar} \int_0^\infty e^{i\omega t} \langle x_i(t) x_k(0) - x_k(0) x_i(t) \rangle \, dt. \qquad (126.9)$$

PROBLEM

Determine the asymptotic form of $\alpha(\omega)$ for $\omega \to \infty$, if $\alpha(\infty) = 0$.

SOLUTION. When $\omega \to \infty$, small values of t are important in (126.8). Putting $\hat{x}(t) \cong \hat{x}(0) + t\dot{\hat{x}}(0)$, we find

$$\alpha(\omega) \cong \frac{i}{\hbar} \langle \dot{\hat{x}} \hat{x} - \hat{x} \dot{\hat{x}} \rangle \int_0^\infty t e^{i\omega t} \, dt,$$

the uniform argument $t = 0$ of the operators being omitted. The integral is calculated by differentiating (126.7) with respect to ω, and gives

$$\alpha(\omega) \cong -\frac{i}{\hbar \omega^2} \langle \dot{\hat{x}} \hat{x} - \hat{x} \dot{\hat{x}} \rangle; \qquad (1)$$

this formula is valid if the mean value of the commutator is other than zero.

The expression (1), being an even function of ω, is real, so that it is the asymptotic form of $\alpha'(\omega)$. From (123.15) we have with $\omega \to \infty$

$$\alpha'(\omega) \cong -\frac{2}{\pi \omega^2} \int_0^\infty \xi \alpha''(\xi) \, d\xi,$$

using the fact that $\alpha''(\xi)$ is odd. Comparison with (1) gives the following "sum rule" for $\alpha''(\omega)$;

$$\int_0^\infty \omega \alpha''(\omega) \, d\omega = \frac{i\pi}{2\hbar} \langle \dot{\hat{x}} \hat{x} - \hat{x} \dot{\hat{x}} \rangle. \qquad (2)$$

§ 127. Fluctuations in the curvature of long molecules

In ordinary molecules, the strong interaction between atoms reduces the thermal motion within molecules to small oscillations of the atoms about their equilibrium positions, which have practically no effect on the shape of the molecule. Molecules consisting of very long chains of atoms (e.g. long polymer hydrocarbon chains) behave quite differently, however. The great length of the molecule, together with the relative weakness of the forces tending to preserve the equilibrium straight shape of the molecule, means that the fluctuation curvature of the molecule may become very large and even cause the molecule to coil up. The great length of the molecule enables us to consider it as a kind of macroscopic linear system, and statistical methods may be used in order to calculate the mean values of quantities describing its curvature (S. E. Bresler and Ya. I. Frenkel', 1939).[†]

We shall consider molecules having a uniform structure along their length. Being concerned only with their shape, we can regard such molecules as uniform continuous threads. The shape of a thread is defined by specifying at each point in it the "curvature vector" ϱ, which is along the principal normal to the curve and equal in magnitude to the reciprocal of the radius of curvature.

The curvature of the molecule is in general small at each point; since the molecule is of great length this does not, of course, exclude very large relative movements of distant points of it. For small values of the vector ϱ, the free energy per unit length of the curved molecule can be expanded in powers of the components of this vector. Since the free energy is a minimum in the equilibrium position (the straight shape, with $\varrho = 0$ at every point) linear terms do not appear in the expansion, and we have

$$F = F_0 + \tfrac{1}{2} \sum_{i,\,k} a_{ik} \varrho_i \varrho_k, \tag{127.1}$$

where the values of the coefficients a_{ik} represent the properties of the straight molecule (its resistance to curvature) and are constant along its length, since the molecule is assumed homogeneous.

The vector ϱ is in the plane normal to the line of the molecule at the point considered, and has two independent components in that plane. Accordingly, the set of constants a_{ik} forms a symmetrical tensor of dimension two and rank two in this plane. We refer this to its principal axes, and denote its principal values by a_1, a_2; the thread which represents the molecule need not be axially symmetrical in its properties, and so a_1 and a_2 need not be equal. The expression (127.1) then becomes

$$F = F_0 + \tfrac{1}{2}(a_1 \varrho_1^2 + a_2 \varrho_2^2),$$

where ϱ_1 and ϱ_2 are the components of ϱ in the direction of the corresponding principal axes.

[†] In the theory given here, the molecule is regarded as an isolated system, its interaction with neighbouring molecules being neglected. In a solid or liquid substance this interaction may, of course, have a considerable effect on the shape of the molecules. Although the applicability of the results to actual substances is therefore very limited, their derivation is of considerable methodological interest.

Finally, integrating over the whole length of the molecule, we find the total change in its free energy due to a slight curvature:

$$\Delta F_t = \tfrac{1}{2} \int (a_1 \varrho_1^2 + a_2 \varrho_2^2)\, dl, \tag{127.2}$$

where l is a coordinate along the thread. It is clear that a_1 and a_2 are necessarily positive.

Let t_a and t_b be unit vectors along the tangents at two points a and b on the thread separated by a section of length l, and let $\theta = \theta(l)$ denote the angle between these tangents, i.e. $t_a \cdot t_b = \cos \theta$.

Let us first consider a curvature so slight that the angle θ is small even for distant points. We draw two planes through the vector t_a and the two principal axes of the tensor a_{ik} in the normal plane (at the point a). For small θ, the square of this angle may be written

$$\theta^2 = \theta_1^2 + \theta_2^2, \tag{127.3}$$

where θ_1 and θ_2 are the angles of rotation of the vector t_b relative to t_a in these two planes. The components of the curvature vector are related to the functions $\theta_1(l)$ and $\theta_2(l)$ by $\varrho_1 = d\theta_1(l)/dl$, $\varrho_2 = d\theta_2(l)/dl$, and the change in the free energy due to the curvature of the molecule may be written

$$\Delta F_t = \frac{1}{2} \int \left[a_1 \left(\frac{d\theta_1}{dl} \right)^2 + a_2 \left(\frac{d\theta_2}{dl} \right)^2 \right] dl. \tag{127.4}$$

In calculating the probability of a fluctuation with given values of $\theta_1(l) = \theta_1$ and $\theta_2(l) = \theta_2$ for a particular l, we must consider the most complete equilibrium possible for given θ_1 and θ_2 (see the first footnote to § 111). That is, we must determine the minimum possible value of the free energy for given θ_1 and θ_2. An integral of the form

$$\int_0^l \left(\frac{d\theta_1}{dl} \right)^2 dl$$

for given values of the function $\theta_1(l)$ at both limits ($\theta_1(0) = 0$, $\theta_1(l) = \theta_1$) has a minimum value if $\theta_1(l)$ varies linearly. Then

$$\Delta F_t = \frac{a_1 \theta_1^2}{2l} + \frac{a_2 \theta_2^2}{2l};$$

the fluctuation probability $w \propto e^{-\Delta F_t/T}$ (see (116.7)), and so we obtain for the mean squares of the two angles

$$\langle \theta_1^2 \rangle = lT/a_1, \quad \langle \theta_2^2 \rangle = lT/a_2.$$

The mean square of the angle $\theta(l)$ under consideration is

$$\langle \theta^2 \rangle = lT \left(\frac{1}{a_1} + \frac{1}{a_2} \right). \tag{127.5}$$

In this approximation it is, as we should expect, proportional to the length of the section of the molecule between the two points.

Curvature with large values of the angles $\theta(l)$ may now be treated as follows. The angles between the tangents $\mathbf{t}_a$, $\mathbf{t}_b$, $\mathbf{t}_c$ at three points a, b, c on the thread are related by the trigonometrical formula

$$\cos \theta_{ac} = \cos \theta_{ab} \cos \theta_{bc} - \sin \theta_{ab} \sin \theta_{bc} \cos \phi,$$

where ϕ is the angle between the planes $(\mathbf{t}_a, \mathbf{t}_b)$ and $(\mathbf{t}_b, \mathbf{t}_c)$. Averaging this formula and bearing in mind that, in the approximation considered, the fluctuations of curvature of the sections ab and bc of the molecule (for a given direction of the tangent $\mathbf{t}_b$ at the middle point) are statistically independent, we obtain

$$\langle \cos \theta_{ac} \rangle = \langle \cos \theta_{ab} \cos \theta_{bc} \rangle$$
$$= \langle \cos \theta_{ab} \rangle \langle \cos \theta_{bc} \rangle;$$

the term containing $\cos \phi$ gives zero on averaging.

This relation shows that the mean value $\langle \cos \theta(l) \rangle$ must be a multiplicative function of the length l of the section of the molecule between two given points. But for small $\theta(l)$ we must have, according to (127.5),

$$\langle \cos \theta(l) \rangle \cong 1 - \tfrac{1}{2} \langle \theta^2 \rangle = 1 - lT/a,$$

with the notation $2/a = 1/a_1 + 1/a_2$. The function which satisfies both these conditions is

$$\langle \cos \theta \rangle = e^{-lT/a}, \tag{127.6}$$

and this is the required formula. For large distances l, the mean value $\langle \cos \theta \rangle \cong 0$, in accordance with the statistical independence of the directions of sufficiently distant parts of the molecule.

By means of formula (127.6) it is easy to determine the mean square of the distance R (measured in a straight line) between the two ends of the molecule. If $\mathbf{t}(l)$ is a unit vector along the tangent at an arbitrary point in the molecule, the radius vector between its ends is

$$\mathbf{R} = \int_0^L \mathbf{t}(l) \, dl,$$

where L is the total length of the molecule. Writing the square of this integral as a double integral and averaging, we obtain

$$\langle R^2 \rangle = \int_0^L \int_0^L \mathbf{t}(l_1) \cdot \mathbf{t}(l_2) \, dl_1 \, dl_2 = \int_0^L \int_0^L e^{-T |l_1 - l_2|/a} \, dl_1 \, dl_2 .$$

The calculation of the integral gives the final formula

$$\langle R^2 \rangle = 2\left(\frac{a}{T}\right)^2 \left(\frac{LT}{a} - 1 + e^{-LT/a}\right). \tag{127.7}$$

For low temperatures ($LT \ll a$) this becomes

$$\langle R^2 \rangle = L^2(1 - LT/3a); \tag{127.8}$$

as $T \rightarrow 0$ the mean square $\langle R^2 \rangle$ tends to the square of the total length of the molecule, as it should. If $LT \gg a$ (high temperatures or great lengths L),

$$\langle R^2 \rangle = 2La/T. \tag{127.9}$$

Then $\langle R^2 \rangle$ is proportional to the length of the molecule, and as L increases the ratio $\langle R^2 \rangle/L^2$ tends to zero.

CHAPTER XIII

THE SYMMETRY OF CRYSTALS

§ 128. Symmetry elements of a crystal lattice

THE most usual properties of symmetry of macroscopic bodies relate to the symmetry of the configuration of particles in them.

Atoms and molecules in motion do not occupy precisely defined places in a body, and for an exact statistical description of their arrangement we must use a *density function* $\varrho(x, y, z)$, which gives the probability of various configurations of the particles: $\varrho\, dV$ is the probability that an individual particle is in the volume element dV. The symmetry properties of the configuration of the particles are determined by the coordinate transformations (translations, rotations and reflections) which leave the function $\varrho(x, y, z)$ invariant. The set of all such *symmetry transformations* for a given body forms what is called its *symmetry group*.

If the body consists of different kinds of atom, the function ϱ must be determined for each kind of atom separately; this, however, is unimportant here, since all these functions in an actual body will in practice possess the same symmetry. We could also use the function ϱ defined as the total electron density due to all the atoms at each point in the body.[†]

The highest symmetry is that of *isotropic* bodies (bodies whose properties are the same in all directions), which include gases, liquids and amorphous solids. It is evident that in such a body all positions in space of any given particle must be equally probable, i.e. we must have $\varrho = $ constant.

In *anisotropic* crystalline solids, on the other hand, the density function is not simply a constant. In this case it is a triply periodic function (with periods equal to those of the crystal lattice) and has sharp maxima at the lattice points. Besides translational symmetry, the lattice (i.e. the function $\varrho(x, y, z)$) also has, in general, symmetry under certain rotations and reflections. The

---

[†] Moving electrons can cause not only a mean charge density $e\varrho$ but also a mean current density $\mathbf{j}(x, y, z)$. Bodies in which there are non-zero currents are those having a "magnetic structure", and the symmetry of the vector function $\mathbf{j}(x, y, z)$ determines the symmetry of that structure. This is discussed in *Electrodynamics*, § 28.

lattice points which can be made to coincide by any symmetry transformation are said to be *equivalent*.

Proceeding to study the symmetry of a crystal lattice, we must first of all ascertain which elements can contribute to this symmetry.

The symmetry of a crystal lattice is based on its spatial periodicity, the property of being unchanged by a parallel displacement or *translation* through certain distances in certain directions;[†] translational symmetry will be further discussed in § 129.

As well as translational symmetry, the lattice may also be symmetrical under certain rotations and reflections; the corresponding symmetry elements (*axes of symmetry*, *planes of symmetry*, and *rotary-reflection axes*) are the same as those which can occur in symmetrical bodies of finite size (see *Quantum Mechanics*, § 91).

In addition, however, crystal lattices can also possess symmetry elements consisting of combinations of parallel translations with rotations and reflections. Let us first consider combinations of translations with the axes of symmetry. The combination of an axis of symmetry with a translation in a direction perpendicular to the axis does not give a new type of symmetry element. It is easy to see that a rotation through a certain angle followed by a translation perpendicular to the axis is equivalent to a rotation through the same angle about an axis parallel to the first. The combination of a rotation about an axis and a translation along that axis leads to a new type of symmetry element, a *screw axis*. The lattice has a screw axis of order n if it is unchanged by rotation through an angle $2\pi/n$ about the axis, accompanied by translation through a certain distance d along the axis.

After n rotations, with accompanying translations, about a screw axis of order n, the lattice is simply shifted along the axis by a distance nd. Thus, when there is a screw axis, the lattice must certainly also have a simple periodicity along this axis with a period not exceeding nd. This means that screw axes of order n can be correlated only with translations through distances $d = pa/n \, (p = 1, 2, \ldots, n-1)$, where a is the smallest period of the lattice in the direction of the axis. For example, a screw axis of order 2 can be of only one type, the translation being through half a period; screw axes of order 3 can be correlated with translations by $\frac{1}{3}$ or $\frac{2}{3}$ period, and so on.

Similarly, we can combine translations with planes of symmetry. Reflection in a plane together with translation in a direction perpendicular to the plane does not give a new type of symmetry element, since such a transformation is easily seen to be equivalent to a reflection in a plane parallel to the first. The combination of a reflection with a translation along a direction lying in

---

[†] Here the crystal lattice must be regarded as infinite, ignoring the faces of the crystal.

the reflection plane leads to a new type of symmetry element, a *glide-reflection plane* or *glide plane*. The lattice has a glide-reflection plane if it is unchanged by a reflection in this plane, accompanied by a translation through a certain distance d in a certain direction lying in this plane.

A twofold reflection in a glide-reflection plane amounts to a translation through a distance $2d$. It is therefore clear that a lattice can have only glide-reflection planes such that the translation distance $d = \frac{1}{2}a$, where a is the smallest period of the lattice in the direction of the translation.

The combination of rotary-reflection axes with translations does not lead to new types of symmetry element, since in this case any translation can be resolved into two parts, one perpendicular to the axis and the other parallel to it and therefore perpendicular to the reflection plane. Thus a rotary-reflection transformation followed by a translation is always equivalent to another rotary-reflection transformation about an axis parallel to the first.

§ 129. The Bravais lattice

The translational periods of a lattice can be represented by vectors **a** whose directions are those of the respective translations and whose magnitudes are equal to the distances concerned. The lattice has an infinity of different lattice vectors. These vectors are not all independent, however: one can always choose three basic lattice vectors (corresponding to the three dimensions of space) which do not lie in one plane, and then any other lattice vector can be represented as a sum of three vectors each an integral multiple of one of the basic vectors. If the basic vectors are denoted by a_1, a_2, a_3, an arbitrary lattice vector **a** will be of the form

$$\mathbf{a} = n_1\mathbf{a}_1 + n_2\mathbf{a}_2 + n_3\mathbf{a}_3, \tag{129.1}$$

where n_1, n_2, n_3 are any positive or negative integers or zero.

The choice of the basic lattice vectors is not unique. On the contrary, they may be chosen in an infinity of ways. Let $\mathbf{a}_1$, $\mathbf{a}_2$, $\mathbf{a}_3$ be basic lattice vectors, and let us replace them by other vectors $\mathbf{a}_1'$, $\mathbf{a}_2'$, $\mathbf{a}_3'$, defined by the formulae

$$\mathbf{a}_i' = \sum_k \alpha_{ik}\mathbf{a}_k \quad (i, k = 1, 2, 3), \tag{129.2}$$

where the α_{ik} are some integers. If the new periods $\mathbf{a}_i'$ are also basic lattice vectors, then, in particular, the vectors $\mathbf{a}_i$ must be expressible in terms of the $\mathbf{a}_i'$ as linear functions with integral coefficients; then any other lattice vector can also be expressed in terms of the $\mathbf{a}_i'$. In other words, if we express the $\mathbf{a}_i$ in terms of the $\mathbf{a}_i'$ in accordance with (129.2), the resulting formulae must be of the type

$$\mathbf{a}_i = \sum_k \beta_{ik} \mathbf{a}'_k ,$$

with the β_{ik} again integral. The determinant $|\beta_{ik}|$ is the reciprocal of the determinant $|\alpha_{ik}|$, and since both are integers it follows that the equation

$$|\alpha_{ik}| = \pm 1 \tag{129.3}$$

is a necessary and sufficient condition for the $\mathbf{a}'_i$ to be basic lattice vectors.

Let us choose a lattice point and mark off from it three basic lattice vectors. The parallelepiped formed by the three vectors is called a *unit cell* of the lattice. The whole lattice can then be regarded as a regular assembly of such parallelepipeds. All the unit cells are, of course, identical in their properties; they have the same shape and size, and each contains the same number of atoms of each kind identically arranged.

It is evident that identical atoms will be found at every vertex of every unit cell. All these vertices, therefore, are equivalent lattice points, and each can be brought to the position of any other by translation through a lattice vector. A set of all such equivalent points which can be brought into coincidence by a translation forms what is called a *Bravais lattice* of the crystal. This clearly does not include every point of the crystal lattice; indeed, in general it does not even include all equivalent points, since the lattice may contain equivalent points which can be made to coincide only by transformations involving rotations or reflections.

The Bravais lattice can be constructed by selecting any crystal lattice point and performing all possible translations. By taking initially some other point not in the first Bravais lattice we should obtain another Bravais lattice displaced relative to the first. It is therefore clear that the crystal lattice in general consists of several interpenetrating Bravais lattices, each corresponding to atoms of a particular type and position. All these lattices, regarded as sets of points (i.e. purely geometrically), are completely identical.

Let us return now to the unit cells. Because the choice of the basic lattice vectors is arbitrary, that of the unit cell is also not unique. The unit cell can be constructed from any basic vectors. The resulting cells are, of course, of varying shapes, but their volumes are all equal. This is most simply seen as follows. It is clear from the above discussion that each unit cell contains one point belonging to each of the Bravais lattices that can be constructed in the crystal concerned. Consequently, the number of unit cells in a given volume is always equal to the number of atoms of a particular type and position, i.e. is independent of the choice of cell. The volume of each cell is therefore the same, and equal to the total volume divided by the number of cells.

§ 130. **Crystal systems**

Let us now consider the possible types of symmetry of the Bravais lattices.

First, we shall prove a general theorem concerning the symmetry of crystal lattices with respect to rotations. Let us see which axes of symmetry the lattice can have. Let A (Fig. 55) be a point of a Bravais lattice, lying on an axis of symmetry perpendicular to the plane of the diagram. If B is another point separated from A by one of the possible translations, a similar axis of symmetry must pass through B.

Fig. 55

Let us now perform a rotation through an angle $\phi = 2\pi/n$ about the axis through A, where n is the order of the axis. Then the point B and the axis through it will move to B'. Similarly, a rotation about B carries A into A'. From their construction, the points A' and B' belong to the same Bravais lattice, and so can be made to coincide by a translation. The distance $A'B'$ must therefore also be a translational period of the lattice. If a is the shortest period in the direction concerned, the distance $A'B'$ must therefore be equal to pa with p integral. It is seen from the figure that this gives

$$a + 2a \sin\left(\phi - \tfrac{1}{2}\pi\right) = a - 2a \cos \phi$$
$$= pa,$$

or $\cos \phi = \tfrac{1}{2}(1-p)$.

Since $|\cos \phi| \leqslant 1$, p can be 3, 2, 1 or 0. These values correspond to $\phi = 2\pi/n$ with $n = 2, 3, 4$ or 6. Thus the crystal lattice can have axes of symmetry only of order 2, 3, 4 and 6.

Let us now examine the possible types of symmetry of the Bravais lattice under rotations and reflections. These types of symmetry are called *crystal systems*, and each corresponds to a certain set of axes and planes of symmetry, i.e. is a point group.

It is easy to see that every point of a Bravais lattice is a centre of symmetry thereof. For to each atom in a Bravais lattice there corresponds another atom collinear with that atom and with the lattice point considered, and such that the two atoms are equidistant from this lattice point. If the centre of symmetry is the only symmetry element of the Bravais lattice (apart from translations), we have

1. The *triclinic system*. This system, the least symmetrical of all, corresponds to the point group C_i. The points of a triclinic Bravais lattice lie at
the vertices of equal parallelepipeds with edges of arbitrary lengths and arbitrary angles between edges. Such a parallelepiped is shown in Fig. 56.

The Bravais lattices are customarily denoted by special symbols; that of the
triclinic system is denoted by Γ_t.

2. The *monoclinic system* is next in degree of symmetry. Its symmetry elements are a second-order axis and a plane of symmetry perpendicular to this
axis, forming the point group C_{2h}. This is the symmetry of a right parallelepiped with a base of any shape. The Bravais lattice for this system can be
constructed in two ways. In one, called the simple monoclinic Bravais lattice
(Γ_m), the lattice points are at the vertices of right parallelepipeds with the *ac*
face an arbitrary parallelogram (Fig. 56). In the other, the base-centred lattice (Γ_m^b), the lattice points are not only at the vertices but also at the centres
of opposite rectangular faces of the parallelepipeds.

FIG. 56

3. The *orthorhombic system* corresponds to the point group D_{2h}. This is the symmetry of a rectangular parallelepiped with edges of any length. The system has four types of Bravais lattice. In the simple orthorhombic lattice (Γ_o), the lattice points are at the vertices of rectangular parallelepipeds. In the base-centred lattice (Γ_o^b), there are in addition lattice points at the centres of two opposite faces of each parallelepiped. In the body-centred lattice (Γ_o^v), the points are at the vertices and centres of the parallelepipeds; finally, in the face-centred lattice (Γ_o^f), the points are at the vertices and at the centre of each face.

4. The *tetragonal system* represents the point group D_{4h}; this is the symmetry of a right square prism. The Bravais lattice for this system can be constructed in two ways, giving the simple and body-centred tetragonal Bravais lattices (Γ_q and Γ_q^v), whose points lie respectively at the vertices and at the vertices and centres of right square prisms.

5. The *rhombohedral* or *trigonal system* corresponds to the point group D_{3d}: this is the symmetry of a rhombohedron (a solid formed from a cube by stretching or compressing it along a spatial diagonal). In the only Bravais lattice possible in this system (Γ_{rh}) the lattice points are at the vertices of rhombohedra.

6. The *hexagonal system* corresponds to the point group D_{6h}; this is the symmetry of a regular hexagonal prism. The Bravais lattice for this system (Γ_h) can be constructed in only one way; its lattice points are at the vertices of regular hexagonal prisms and at the centres of their hexagonal bases. It is useful to mention the following difference between the rhombohedral and hexagonal Bravais lattices. In both, the lattice points lie in planes perpendicular to the axis of order 3 or 6, and form a network of equilateral triangles; but in the hexagonal lattice the points are directly superimposed in successive such planes (in the direction of the C_6 axis); these planes are shown in plan in Fig. 57. In the rhombohedral lattice, on the other hand, the points in each plane lie above the centres of the triangles formed by the points in the previous plane, as shown by the circles and crosses in Fig. 57.

7. The *cubic system* corresponds to the point group O_h; this is the symmetry of a cube. This system has three types of Bravais lattice: the simple cubic (Γ_c), the body-centred cubic (Γ_c^v) and the face-centred cubic (Γ_c^f).

In the sequence of systems: triclinic, monoclinic, orthorhombic, tetragonal, cubic, each has higher symmetry than those which precede it, i.e. each contains all the symmetry elements which appear in the preceding ones. The rhombohedral system is similarly of higher symmetry than the monoclinic, while at the same time it is of lower symmetry than the cubic and hexagonal systems: its symmetry elements are present in both of the latter, which are the two systems of highest symmetry.

We may also mention the following fact. It might appear at first sight that further types of Bravais lattice beyond the fourteen listed above are possible. For instance, if we add to the simple tetragonal lattice a point at the centre of each opposite square base of the prisms, the lattice would again be of tetragonal symmetry. However, it is easy to see that this would not give a new Bravais lattice. For, on joining the points of such a lattice in the manner indicated in Fig. 58 by the broken lines, we see that the new lattice is again a simple tetragonal one. The same is easily found to be true in all similar cases.

FIG. 57   FIG. 58   FIG. 59

The Bravais lattice parallelepipeds shown in Fig. 56 themselves have all the symmetry elements of the system to which they belong. However, it must be remembered that, for all the Bravais lattices except the simple ones, these parallelepipeds are not unit cells: the lattice vectors from which they are constructed are not basic ones. As the basic lattice vectors in the face-centred Bravais lattices we can take the vectors from any vertex of the parallelepiped to the centres of the faces, in the body-centred lattices from a vertex to the centres of the parallelepipeds, and so on. Fig. 59 shows the unit cells for the cubic lattices Γ_c^f and Γ_c^v; these cells are rhombohedra and do not themselves possess all the symmetry elements of the cubic system. It is evident that the volume v_f of the face-centred Bravais parallelepiped is four times that of the unit cell: $v_f = 4v$. The volumes of the body-centred and base-centred parallelepipeds are twice that of the unit cell: $v_v = 2v$, $v_b = 2v$.

In order to define completely the triclinic Bravais lattice, it is necessary to specify six quantities: the lengths of the edges of its parallelepipeds and the angles between the edges. In the monoclinic system four quantities are sufficient, since two of the angles between the edges are always right angles. Similarly, we easily find that the Bravais lattices of the various systems are defined by the following numbers of quantities (lengths of edges of parallelepipeds or angles between edges): triclinic 6, monoclinic 4, orthorhombic 3, tetragonal 2, rhombohedral 2, hexagonal 2, cubic 1.

§ 131. Crystal classes

In many effects which may be called macroscopic, a crystal behaves as a homogeneous and continuous body. The macroscopic properties of the crystal depend only on the direction considered in it. For example, the properties of the passage of light through a crystal depend only on the direction of the light ray; the thermal expansion of a crystal is in general different in different directions; finally, the elastic deformations of a crystal under various external forces also depend on direction.

On the other hand, the symmetry of crystals brings about an equivalence of various directions in them. All macroscopic properties of a crystal will be exactly the same in such directions. We can therefore say that the macroscopic properties of the crystal are determined by the symmetry of directions in it. For instance, if the crystal has a centre of symmetry, every direction in it will be equivalent to the opposite direction.

Translational symmetry of the lattice does not lead to equivalence of directions, since parallel displacements do not affect directions. For the same reason, the difference between screw axes and simple axes of symmetry, and between simple planes of symmetry and glide-reflection planes, does not affect the symmetry of directions.

Thus the symmetry of directions, and therefore that of the macroscopic properties of the crystal, are determined by its axes and planes of symmetry, with screw axes and glide planes regarded as ordinary axes and planes. Such sets of symmetry elements are called *crystal classes*.

As we already know, an actual crystal may be regarded as a set of several interpenetrating identical Bravais lattices. Because of this superposition of the Bravais lattices, the symmetry of an actual crystal is in general different from that of the corresponding Bravais lattice.

In particular, the set of symmetry elements forming the class of a given crystal is in general different from its system. It is evident that the addition of further points to a Bravais lattice can only eliminate some of its axes or planes of symmetry, not introduce new ones. Thus the crystal class contains fewer (or at most the same number of) symmetry elements than the corresponding system, i.e. the set of axes and planes of symmetry of the Bravais lattice of the crystal in question.

From this we can derive a method of finding all the classes belonging to a given system. To do so, we must find all the point groups which contain some or all of the symmetry elements of the system. It may happen, however, that a point group thus obtained comprises symmetry elements present in more than one system. For example, we have seen in § 130 that all Bravais lattices have a centre of symmetry. The point group C_i is therefore present in all systems. Nevertheless, the distribution of crystal classes among systems is usually physically unique: each class must be assigned to the system of lowest symmetry among those which contain it. For example, the class C_i must be assigned to the triclinic system, which has no symmetry element except a centre of inversion. With this method of assigning the classes, a crystal having a certain Bravais lattice will never be placed in a class which could be constructed from a Bravais lattice of a system of lower symmetry—with one exception (see below).

The necessity of satisfying this condition is physically evident: it is physically most improbable that the atoms in a crystal which belong to its Bravais lattice should be arranged more symmetrically than is required by the symmetry of the crystal. Moreover, even if such a configuration were to occur by chance, any external perturbation, even a weak one (heating, for example), would be sufficient to destroy this configuration, since it is not imposed by the symmetry of the crystal. For instance, if a cubic Bravais lattice were to occur in a crystal belonging to a class for which the tetragonal system was sufficient, even a slight interaction would be capable of lengthening or shortening one of the edges of the cubic cell, converting it into a right square prism.

From this example we see the importance of the fact that the Bravais lattice of a system of higher symmetry can be converted to that of a system of lower symmetry by means of an arbitrarily small deformation. There is one exceptional case, however, where such a transformation is not possible: a hexagonal Bravais lattice cannot be converted by any infinitesimal deformation into the lattice of the rhombohedral system, which is of lower symmetry. For we see from Fig. 57 that, to transform the hexagonal into the rhombohedral lattice, it is necessary to move the lattice points in alternate layers by a finite amount from the vertices to the centres of the triangles. In consequence, all the classes of the rhombohedral system can be obtained with either a hexagonal or a rhombohedral Bravais lattice.[†]

Thus, to find all the crystal classes, we must first look for the point groups of the triclinic system, which has the lowest symmetry, and then go on in turn to systems of higher symmetry, omitting those of their point groups (i.e. classes) which have already been assigned to systems of lower symmetry. It is found that there are altogether 32 classes; a list of these arranged according to systems is as follows:

System	Classes		
Triclinic	C_1,	C_i	
Monoclinic	C_s,	C_2,	C_{2h}

[†] Crystals of rhombohedral classes with a hexagonal Bravais lattice are usually assigned to the rhombohedral system.

System	Classes				
Orthorhombic	C_{2v},	D_2,	D_{2h}		
Tetragonal	S_4,	D_{2d},	C_4,	C_{4h},	
	C_{4v},	D_4,	D_{4h}		
Rhombohedral	C_3,	S_6,	C_{3v},	D_3,	D_{3d}
Hexagonal	C_{3h},	D_{3h},	C_6,	C_{6h},	
	C_{6v},	D_6,	D_{6h}		
Cubic	T,	T_h,	T_d,	O,	O_h

In each of these sets of classes the last is the one of highest symmetry, and contains all the symmetry elements of the corresponding system. The classes whose symmetry is equal to that of the system are called *holohedral* classes. Those whose number of different symmetry transformations (rotations and reflections, including the identical transformation), is less than for a holohedral class by a factor of two or four are called *hemihedral* and *tetartohedral* classes respectively. For example, in the cubic system the class O_h is holohedral, O, T_h and T_d are hemihedral, and T is tetartohedral.

§ 132. Space groups

Having studied the symmetry of the Bravais lattices and the symmetry of directions in the crystal, we can, finally, go on to consider the complete actual symmetry of crystal lattices. This symmetry may be termed microscopic, in contradistinction to the macroscopic symmetry of crystals discussed in § 131. The microscopic symmetry determines those properties of a crystal which depend on the arrangement of the atoms in its lattice (e.g. the scattering of X-rays by the crystal).

The set of (actual) symmetry elements of the crystal lattice is called its *space group*. The lattice always has a certain translational symmetry, and may also have simple, rotary-reflection and screw axes of symmetry and simple and glide-reflection planes of symmetry. The translational symmetry of the lattice is entirely determined by its Bravais lattice, since by the definition of the latter the crystal lattice can have no translational periods except those of its Bravais lattice. Hence, to determine the space group of a crystal, it is sufficient to find the Bravais lattice and to enumerate the symmetry elements which involve rotations and reflections, including of course the relative position of these axes and planes of symmetry. It must also be remembered that the translational symmetry of the crystal lattice means that, if the lattice possesses an axis or plane of symmetry, there exists an infinity of parallel axes or planes which are carried into one another by displacements through the lattice vectors. Finally, in addition to these axes (or planes) of symmetry separated by lattice vectors, the simultaneous presence of translational symmetry and the axes (or planes) of symmetry results in the existence of other axes (or planes) which cannot be made to coincide with the former by a translation through any lattice vector. For example, the presence of a plane of symmetry involves not only planes parallel to it at distances equal to the lattice vector but also planes of symmetry which bisect each lattice vector: it is easily seen that reflection in any plane followed by translation through a distance d in a direction perpendicular to the plane is equivalent to reflection in a plane parallel to the first and at a distance $\frac{1}{2}d$ from it.

The possible space groups can be divided among the crystal classes, each space group being assigned to the class where the set of axes and planes of symmetry is the same as in the space group when no distinction is made in the latter between simple and screw axes and between simple and glide planes. Altogether 230 different space groups[†] are possible; they were first found by E. S. Fedorov (1895). The space groups are distributed among classes as shown in Table 1.

Table 1

Class	Number of groups	Class	Number of groups
C_1	1	S_6	2
C_i	1	C_{3v}	6
C_s	4	D_3	7
C_2	3	D_{3d}	6
C_{2h}	6	C_{3h}	1
C_{2v}	22	C_6	6
D_2	9	C_{6h}	2
D_{2h}	28	D_{3h}	4
S_4	2	C_{6v}	4
C_4	6	D_6	6
C_{4h}	6	D_{6h}	4
D_{2d}	12	T	5
C_{4v}	12	T_h	7
D_4	10	T_d	6
D_{4h}	20	O	8
C_3	4	O_h	10

[†] Including eleven pairs that are the same except for the direction of rotation about their screw axes.

We shall not pause here to enumerate the symmetry elements of all the space groups, which would be a very lengthy process. They may be found in manuals of crystallography.[†]

The space groups that contain no screw axes and glide planes are said to be *symmorphic*; there are 73 of these. The remaining 157 space groups contain such symmetry elements. The crystal lattices that belong to non-symmorphic space groups must certainly contain at least two identical atoms in each unit cell: since a rotation about a screw axis or a reflection in a glide plane involves a translation through a fraction of a basic lattice vector, the Bravais lattice points are not carried into one another by such transformations, and the crystal lattice must therefore consist of at least two interpenetrating Bravais lattices occupied by identical atoms.

§ 133. The reciprocal lattice

All physical quantities which describe the properties of a crystal lattice have the same periodicity as the lattice itself. Such quantities are, for example, the charge density due to the electrons in the atoms in the lattice, the probability of finding an atom at a particular point in the lattice, and so on.

Let the function $U(\mathbf{r})$ be any such quantity. Its periodicity implies that

$$U(\mathbf{r}+n_1\mathbf{a}_1+n_2\mathbf{a}_2+n_3\mathbf{a}_3) = U(\mathbf{r}) \tag{133.1}$$

for any integral n_1, n_2, n_3 ($\mathbf{a}_1$, $\mathbf{a}_2$, $\mathbf{a}_3$ being the basic vectors of the lattice).

Let us expand the periodic function $U(\mathbf{r})$ as a triple Fourier series, which may be written

$$U = \sum_{\mathbf{b}} U_{\mathbf{b}} e^{i\mathbf{b}\cdot\mathbf{r}}, \tag{133.2}$$

where the summation is over all possible values of the vector $\mathbf{b}$. These are determined from the requirement that the function U, when put in the form of the series (133.2), satisfies the periodicity condition (133.1). This means that the exponential factors must be left unchanged when $\mathbf{r}$ is replaced by $\mathbf{r}+\mathbf{a}$, $\mathbf{a}$ being any lattice vector. For this to be so it is necessary that the scalar product $\mathbf{a}\cdot\mathbf{b}$ should always be an integral multiple of 2π. Taking $\mathbf{a}$ successively as the basic vectors $\mathbf{a}_1$, $\mathbf{a}_2$, $\mathbf{a}_3$, we must therefore have $\mathbf{a}_1\cdot\mathbf{b} = 2\pi p_1$, $\mathbf{a}_2\cdot\mathbf{b} = 2\pi p_2$, $\mathbf{a}_3\cdot\mathbf{b} = 2\pi p_3$, where p_1, p_2, p_3 are positive or negative integers or zero. The solution of these three equations has the form

$$\mathbf{b} = p_1\mathbf{b}_1+p_2\mathbf{b}_2+p_3\mathbf{b}_3, \tag{133.3}$$

[†] A full account of the space groups is given, for example, by G.Yu. Lyubarskiĭ, *The Application of Group Theory in Physics*, Pergamon, Oxford, 1960, and in the *International Tables for X-Ray Crystallography*, Vol. I, Kynoch Press, Birmingham, 1952. The latter also lists the equivalent points for each space group.

where the vectors $\mathbf{b}_i$ are given in terms of the $\mathbf{a}_i$ by

$$\mathbf{b}_1 = 2\pi\mathbf{a}_2\times\mathbf{a}_3/v, \quad \mathbf{b}_2 = 2\pi\mathbf{a}_3\times\mathbf{a}_1/v, \quad \mathbf{b}_3 = 2\pi\mathbf{a}_1\times\mathbf{a}_2/v,$$

$$v = \mathbf{a}_1\cdot\mathbf{a}_2\times\mathbf{a}_3. \tag{133.4}$$

We have thus determined the possible values of the vector $\mathbf{b}$. The summation in (133.2) is taken over all integral values of p_1, p_2, p_3.

Geometrically, the product $v = \mathbf{a}_1\cdot\mathbf{a}_2\times\mathbf{a}_3$ represents the volume of the parallelepiped formed by the vectors $\mathbf{a}_1, \mathbf{a}_2, \mathbf{a}_3$, i.e. the volume of the unit cell; the products $\mathbf{a}_1\times\mathbf{a}_2$, etc., represent the areas of the three faces of this cell. The vectors $\mathbf{b}_i$ therefore have the dimensions of reciprocal length, and in magnitude are equal to 2π times the reciprocal altitudes of the parallelepiped formed by the vectors $\mathbf{a}_1, \mathbf{a}_2, \mathbf{a}_3$.

From (133.4) it is seen that $\mathbf{b}_i$ and $\mathbf{a}_i$ are related by

$$\mathbf{a}_i\cdot\mathbf{b}_k = 0 \quad \text{if} \quad i \neq k,$$
$$\phantom{\mathbf{a}_i\cdot\mathbf{b}_k} = 2\pi \quad \text{if} \quad i = k. \tag{133.5}$$

Hence the vector $\mathbf{b}_1$ is perpendicular to $\mathbf{a}_2$ and $\mathbf{a}_3$, and similarly for $\mathbf{b}_2$ and $\mathbf{b}_3$.

Having defined the vectors $\mathbf{b}_i$, we can formally construct a lattice with $\mathbf{b}_1, \mathbf{b}_2, \mathbf{b}_3$ as basic vectors. This is called the *reciprocal lattice*, and the vectors $\mathbf{b}_1, \mathbf{b}_2, \mathbf{b}_3$ are called the (basic) vectors of the reciprocal lattice.[†]

Let us calculate the volume of the unit cell of the reciprocal lattice. This is $v' = \mathbf{b}_1\cdot\mathbf{b}_2\times\mathbf{b}_3$. Substitution of the expressions (133.4) gives

$$v' = \frac{(2\pi)^3}{v^3}\,\mathbf{a}_2\times\mathbf{a}_3\cdot(\mathbf{a}_3\times\mathbf{a}_1)\times(\mathbf{a}_1\times\mathbf{a}_2)$$

$$= \frac{(2\pi)^3}{v^3}\,(\mathbf{a}_2\times\mathbf{a}_3\cdot\mathbf{a}_1)\,(\mathbf{a}_3\times\mathbf{a}_1\cdot\mathbf{a}_2)$$

$$= (2\pi)^3/v. \tag{133.6}$$

It is evident that the reciprocal lattice cell corresponding to a triclinic Bravais lattice will also be an arbitrary parallelepiped. Similarly, the reciprocal lattices of the simple Bravais lattices of the other systems are also simple lattices of the same system; for example, the reciprocal lattice of a simple cubic Bravais lattice also has a simple cubic cell. It is also easy to see by a straightforward construction that the reciprocal lattices of the face-centred Bravais lattices (orthorhombic, tetragonal and cubic) are body-centred lattices of the corresponding systems; the volume of the Bravais parallelepiped in the reciprocal lattice is $v'_v = (2\pi)^3\,8/v_f$, where v_f is the volume of the Bravais parallelepiped in the original lattice. Conversely, the body-centred

---

[†] The definition (133.4), which is now customary in the physics literature, differs by factors of 2π from that usual in pure crystallography.

original lattices correspond to face-centred reciprocal lattices, again with $v_f' = (2\pi^3)8/v_v$. Lastly, base-centred lattices have reciprocal lattices which are also base-centred, with $v_b' = (2\pi)^3 \, 4/v_b$.

An equation of the form $\mathbf{b} \cdot \mathbf{r} = $ constant, where $\mathbf{b}$ is a given vector, represents a plane perpendicular to the vector $\mathbf{b}$ and at a distance from the origin equal to the constant divided by b. Let us take the origin at any of the Bravais lattice points, and let $\mathbf{b} = p_1\mathbf{b}_1 + p_2\mathbf{b}_2 + p_3\mathbf{b}_3$ be any vector of the reciprocal lattice (p_1, p_2, p_3 being integers). Also writing $\mathbf{r}$ in the form $\mathbf{a} = n_1\mathbf{a}_1 + n_2\mathbf{a}_2 + n_3\mathbf{a}_3$, we obtain the equation of a plane:

$$\mathbf{b} \cdot \mathbf{a}/2\pi = n_1 p_1 + n_2 p_2 + n_3 p_3 = m, \tag{133.7}$$

where m is a given constant. If this equation represents a plane containing an infinity of Bravais lattice points (called a *crystal plane*), it must be satisfied by a set of integers n_1, n_2, n_3. For this to be so, the constant m must clearly be an integer also. For given p_1, p_2, p_3, when the constant m takes various integral values, equation (133.7) successively defines an infinity of crystal planes which are all parallel. A particular family of parallel crystal planes thus defined corresponds to each reciprocal lattice vector.

The numbers p_1, p_2, p_3 in (133.7) can always be taken as mutually prime, i.e. as having no common divisor except unity. If there were such a divisor, both sides of the equation could be divided by it, leaving an equation of the same form. The numbers p_1, p_2, p_3 are called the *Miller indices* of the family of crystal planes in question and are written as $(p_1 p_2 p_3)$.

The plane (133.7) intersects the coordinate axes (taken along the basic lattice vectors $\mathbf{a}_1, \mathbf{a}_2, \mathbf{a}_3$) at the points $ma_1/p_1, ma_2/p_2, ma_3/p_3$. The ratio of the intercepts (measured in units of a_1, a_2, a_3 respectively) is $1/p_1 : 1/p_2 : 1/p_3$, i.e. they are in inverse proportion to the Miller indices. For instance, the Miller indices of planes parallel to the coordinate planes (i.e. having intercepts in the ratio $\infty : \infty : 1$) are (100), (010), (001) for the three coordinate planes respectively. Planes parallel to the diagonal plane of the basic parallelepiped of the lattice have indices (111), and so on.

It is easy to find the distance between two successive planes of the same family. The distance of the plane (133.7) from the origin is $2\pi m/b$, where b is the length of the reciprocal lattice vector concerned. The distance of the next plane from the origin is $2\pi(m+1)/b$, and the distance d between these two planes is

$$d = 2\pi/b. \tag{133.8}$$

The following formula is useful in applications:

$$\sum_{\mathbf{b}} e^{i\mathbf{b}\cdot\mathbf{r}} = v \sum_{\mathbf{a}} \delta(\mathbf{r}-\mathbf{a}), \tag{133.9}$$

where the summations on the left and right are taken over all vectors of the reciprocal and original lattices respectively. The sum on the right is a function of **r** that is periodic in the original lattice, and the left-hand side expresses it as a Fourier series.[†] The corresponding formula

$$\sum_{a} e^{i\mathbf{k}\cdot\mathbf{a}} = v' \sum_{b} \delta(\mathbf{k}-\mathbf{b}) \qquad (133.10)$$

follows immediately from (133.9), because of the symmetrical relationship between the two lattices.

§ 134. Irreducible representations of space groups

The physical applications of the theory of symmetry generally involve using the mathematical formalism of what are called *representations* of groups. In this section we shall discuss the question of the classification and method of constructing the irreducible representations of the space groups.[‡]

Let us first summarise in more mathematical terms the information given in the preceding sections about the structure of space groups.

Each space group contains a sub-group of translations comprising an infinity of all possible parallel displacements which leave the crystal lattice unchanged; this sub-group is the mathematical expression of the Bravais lattice of the crystal. The complete space group is obtained from this sub-group by adding n symmetry elements involving rotations and reflections, where n is the number of symmetry transformations of the corresponding crystal class; we shall call these the *rotational* elements. Every element of the space group may be represented as the product of one of the translations and one of the rotational elements.[§]

If the space group does not contain screw axes and glide planes (a symmorphic group), the rotational elements can be taken simply as the n symmetry transformations (rotations and reflections) of the crystal class. In non-symmorphic groups, the rotational elements are rotations and reflections combined with a simultaneous translation through a certain fraction of one of the basic vectors of the lattice. In such space groups, the rotational ele-

---

[†] The divergence of the sum when $\mathbf{r} = \mathbf{a}$ occurs because the lattice is infinite. In considering a lattice of large but finite volume, the sum when $\mathbf{r} = \mathbf{a}$ is to be equated to the number of lattice cells N.

[‡] The reader is assumed familiar with group theory to the extent given, for example, in *Quantum Mechanics*, Chapter XII.

[§] The translation sub-group is Abelian (all its elements commute) and is a normal divisor of the complete space group: all elements of the group that are conjugate to translations are also translations. (Two elements A and B are said to be conjugate if $A = C^{-1}BC$ with C a third element of the group.)

ments of symmetry are "interlinked" with translations and do not themselves form a sub-group; for example, a repeated reflection in a glide plane is not an identical transformation but a translation through one of the basic vectors of the lattice.

In order to characterise explicitly the elements of the space group, it is convenient to denote them by symbols $(P|\mathbf{t})$, where P is any rotation or reflection and $\mathbf{t}$ the vector of a simultaneous translation; the effect of this element on the radius vector $\mathbf{r}$ of any point is shown by $(P|\mathbf{t})\mathbf{r} = P\mathbf{r} + \mathbf{t}$. The multiplication of elements follows the obvious rule

$$(P'|\mathbf{t}')(P|\mathbf{t}) = (P'P|P'\mathbf{t}+\mathbf{t}'). \tag{134.1}$$

The element inverse to $(P|\mathbf{t})$ is

$$(P|\mathbf{t})^{-1} = (P^{-1}|-P^{-1}\mathbf{t}); \tag{134.2}$$

on multiplication by $(P|\mathbf{t})$ it gives the unit element $(E|0)$ of the group, where E denotes the identical rotation.

In particular, the pure translations are represented by the symbol $(E|\mathbf{a})$, where $\mathbf{a}$ is any of the lattice vectors. The rotational elements in symmorphic groups, chosen as described above, are of the form $(P|0)$. In non-symmorphic groups, the rotational elements have the form $(P|\boldsymbol{\tau})$, where $\boldsymbol{\tau}$ is the fraction of a lattice vector through which translation takes place along the screw axis or glide plane. In the former case, the set of rotational transformations $(P|0)$ is itself a sub-group of the space group. In the latter case, the elements $(P|\boldsymbol{\tau})$ themselves do not form a sub-group, since a repeated application of them gives not the identical transformation but a translation through one of the basic lattice vectors. The rotations and reflections P as such (i.e. if the differences between simple and screw axes, and between simple and glide planes, are ignored) always form a group, however, namely the point symmetry group which defines the crystal class; this point group may conveniently be called in this respect the *direction group* of the lattice.[†]

Let us now construct the irreducible representations of the space groups.[‡]

Any such representation can be given by a set of functions of the form

$$\phi_{\mathbf{k}\alpha} = u_{\mathbf{k}\alpha}e^{i\mathbf{k}\cdot\mathbf{r}}, \tag{134.3}$$

[†] In every case, the relation between the space group and the direction group may be formulated in the following way, using the group-theory standpoint. Let all the elements of the space group be distributed into n affine classes, each containing an infinity of products of one rotational element with all possible translations, i.e. all elements of the form $(P|\boldsymbol{\tau}+\mathbf{a})$ with given P and $\boldsymbol{\tau}$. If we now regard the whole of each of these classes as an element in a new group, we get a *factor group* of the original space group, which is isomorphous with the direction group.

[‡] The arguments below are due to F. Seitz (1936).

where the **k** are constant wave vectors, the $u_{k\alpha}$ are functions invariant under translations, and the suffix $\alpha = 1, 2, \ldots$ labels functions with the same **k**.

As a result of a parallel displacement $\mathbf{r} \rightarrow \mathbf{r} + \mathbf{a}$ (where **a** is any vector of the lattice), the functions (134.3) are multiplied by constants $e^{i\mathbf{k}\cdot\mathbf{a}}$. In other words, the matrices of translations are diagonal in the representation given by the functions (134.3). It is evident that two vectors **k** which differ by **b** (where **b** is any vector of the reciprocal lattice) will give the same law of transformation of the functions $\phi_{k\alpha}$ by translations: since $\mathbf{a}\cdot\mathbf{b}$ is an integer times 2π, $e^{i\mathbf{a}\cdot\mathbf{b}} = 1$. Such vectors **k** will be said to be *equivalent*. If we imagine the vectors **k** drawn from a vertex of a reciprocal lattice cell to various points, the non-equivalent vectors will correspond to the points in one unit cell.

By the application of a rotational element of symmetry $(P\,|\,\tau)$, the function $\phi_{k\alpha}$ is transformed into a linear combination of the functions $\phi_{k'\alpha}$ with various values of α and a vector $\mathbf{k}'$ which is obtained from **k** by means of the rotation or reflection in question, performed in the reciprocal lattice: $\mathbf{k}' = P\mathbf{k}$.[†] The set of all (non-equivalent) vectors **k** which can be obtained from one another by the application of all n rotational elements of the group is called the *star* of the wave vector **k**. In the general case of arbitrary **k** the star contains n vectors (*rays*). The functions $\phi_{k\alpha}$ which form the basis of an irreducible representation must always include functions having all the different rays of the star of **k**: since functions with non-equivalent **k** are multiplied by different factors under translations, no choice of linear combinations of them can bring about a decrease in the number of functions which are transformed into combinations of one another.

For certain values of **k** the number of rays in its star may be less than n, since it may happen that some of the rotational elements of symmetry leave **k** unchanged or transform it into an equivalent vector. For example, if the vector **k** is along an axis of symmetry, it is unchanged by rotations about this axis; a vector **k** from a vertex to the centre of the unit cell, i.e. $\mathbf{k} = \frac{1}{2}\mathbf{b}_i$, where $\mathbf{b}_i$ is one of the basic vectors of the reciprocal lattice, is transformed by inversion into the equivalent vector $-\mathbf{k} = -\frac{1}{2}\mathbf{b}_i = \mathbf{k} - \mathbf{b}_i$.

The set of rotational elements of symmetry (regarded as all being simple rotations or reflections P) which appear in a given space group and which do not alter the vector **k** (or which transform it into an equivalent vector) is called the *proper symmetry group* of the vector **k**, or simply the group of **k**; it is one of the ordinary point symmetry groups.

Let us first consider the simple case of symmorphic space groups. The base functions of an irreducible representation of such a group can be written

---

[†] In transforming the vector **k** in the reciprocal lattice all axes and planes of symmetry must, of course, be treated as simple ones, i.e. only the direction group is to be considered.

as products

$$\phi_{k\alpha} = u_\alpha \psi_k, \tag{134.4}$$

where the u_α are invariant under translations and the ψ_k are linear combinations of the expressions $e^{i\mathbf{k}\cdot\mathbf{r}}$ (with equivalent $\mathbf{k}$) invariant with respect to the transformations in the proper symmetry group of the vector $\mathbf{k}$; in (134.4) this vector takes all the values in its star. In translations the functions u_α are unchanged, but the functions ψ_k, and therefore the $\phi_{k\alpha}$, are multiplied by $e^{i\mathbf{k}\cdot\mathbf{a}}$. In rotations and reflections belonging to the group of $\mathbf{k}$, the functions ψ_k are unchanged but the functions u_α are transformed into combinations of one another. Thus the functions u_α give one of the irreducible representations of the point group, these being called in this connection *small representations*. Finally, rotational elements which are not in the group of $\mathbf{k}$ transform sets of functions (134.4) with non-equivalent $\mathbf{k}$ into combinations of one another. The dimension of the representation of the space group thus constructed is equal to the number of rays in the star of $\mathbf{k}$ multiplied by the dimension of the small representation.

Thus the problem of finding all irreducible representations of symmorphic space groups reduces entirely to the classification of the vectors $\mathbf{k}$ with respect to their proper symmetry and the known problem of discovering the irreducible representations of finite point groups.

Let us now consider space groups which have screw axes or glide planes. The presence of such elements of symmetry is still unimportant if the wave vector $\mathbf{k}$ is such that it remains unchanged (i.e. is not transformed into an equivalent vector) under all the transformations in its group.[†] In this case the corresponding irreducible representations are again given by functions of the form (134.4), in which the u_α form the basis of a representation of the point group of the vector $\mathbf{k}$. The only difference from the previous case will be that under rotational transformations the functions $\psi_k = e^{i\mathbf{k}\cdot\mathbf{r}}$ in (134.4) will not remain unchanged, but will be multiplied by $e^{i\mathbf{k}\cdot\boldsymbol{\tau}}$.

Functions of the form (134.4) become inapplicable, however, if there are several equivalent vectors $\mathbf{k}$ transformed into one another by transformations in their proper symmetry group. In a rotational transformation with a simultaneous translation $\boldsymbol{\tau}$, functions $e^{i\mathbf{k}\cdot\mathbf{r}}$ with equivalent but different values of $\mathbf{k}$ are multiplied by different factors (since $\mathbf{b}\cdot\boldsymbol{\tau}/2\pi$ is not integral), and therefore their linear combinations ψ_k will not be transformed into combinations of one another.

In such cases it is no longer possible to consider the rotational elements and the translations separately, but of the infinity of translations it is sufficient to consider a finite number only. These cases arise for vectors $\mathbf{k}$ from a

---

[†] This always includes, in particular, the vector $\mathbf{k} = 0$ and a vector in a general position in which the unit element of its group is the identical transformation.

vertex of a unit cell of the reciprocal lattice to various special points within
the cell, all or some of whose coordinates are simple rational fractions of the
basic vectors b_1, b_2, b_3.[†] We shall call the *extended group* of the wave vector
the group consisting of the rotational elements (together with the relevant
translations through fractions of a period τ) and those translations for
which $\mathbf{k} \cdot \mathbf{a}/2\pi$ is a rational fraction (< 1); the remaining translations
are again regarded as identical transformations. The functions $\phi_{k\alpha}$ which
give irreducible representations of the finite group thus formed (small
representations), together with the corresponding functions $\phi_{k'\alpha}$ for other
rays in the star of $\mathbf{k}$, give an irreducible representation of the space group.
The dimension of the small representations in these groups may be as high
as six (for the crystal class O_h).[‡]

This procedure will now be demonstrated for a specific example.

Let us consider the space group (D_{2h}^2) which corresponds to the simple
orthorhombic Bravais lattice and contains the following rotational elements:[§]

$$(E\,|\,0), \quad (C_2^x\,|\,0), \quad (C_2^y\,|\,0), \quad (C_2^z\,|\,0),$$
$$(I\,|\,\tau), \quad (\sigma_x\,|\,\tau), \quad (\sigma_y\,|\,\tau), \quad (\sigma_z\,|\,\tau),$$

the x, y and z axes being taken along the three basic vectors of the lattice;
$\tau = \frac{1}{2}(a_1 + a_2 + a_3)$; the axes of symmetry C_2 are simple axes but the planes σ
perpendicular to them are glide planes.

[†] In practice these fractions are usually $\frac{1}{2}$, $\frac{1}{3}$ or $\frac{2}{3}$, the last two values occurring for groups of the rhombohedral and hexagonal systems.

[‡] If the representations of the extended group of the wave vector are regarded
as representations of the unextended group (one of the point groups), then the
relations between the matrices $\hat{G}$ that represent the elements G of the group will be
different from those between these elements themselves: if $G_1 G_2 = G_3$, the corresponding matrices of the representation will in general be related not by a simi-
lar equation $\hat{G}_1 \hat{G}_2 = \hat{G}_3$ (as in ordinary representations), but by $\hat{G}_1 \hat{G}_2 = \omega_{12} \hat{G}_3$,
where ω_{12} is a phase factor equal to unity only in modulus: $|\omega_{12}| = 1$. Such rep-
resentations are said to be *projective*. All the essentially different projective representations can be listed once and for all in each point group, and thereafter used
as small representations for constructing irreducible representations of space
groups.

An account of the theory of projective representations, and tables of them for the
crystallographic point groups, will be found in G. L. Bir and G. E. Pikus, *Symmetry
and Deformation Effects in Semiconductors* (*Simmetriya i deformatsionnye éffekty v
poluprovodnikakh*), Nauka, Moscow, 1972.

There exist also complete tables of irreducible representations of space groups,
which are given by O. V. Kovalev, *Irreducible Representations of the Space Groups*,
Gordon & Breach, New York, 1965; C. J. Bradley and A. P. Cracknell, *The Mathematical Theory of Symmetry in Solids*, Clarendon Press, Oxford, 1972.

[§] The space groups are customarily denoted by the symbol for the crystal class,
with an index which arbitrarily numbers the group within the class.

Let us take, for example, the vector

$$\mathbf{k} = (\tfrac{1}{2}, 0, 0);\qquad\qquad(134.5)$$

the three numbers in parentheses give the components of the vector along the reciprocal lattice axes, measured in units of the edge lengths ($b_i = 2\pi/a_i$) of the reciprocal lattice cell. Its proper symmetry includes all the axes and planes of the point group $\mathbf{D}_{2h}$, and so this vector is its own star. The extended group is got by adding the translation $(E|\mathbf{a}_1)$ for which $\mathbf{k}\cdot\mathbf{a}/2\pi = \tfrac{1}{2}$. Thus we obtain a group of 16 elements in 10 classes as shown in the upper line in Table 2. The fact that the elements $(C_2^y|0)$ and $(C_2^y|\mathbf{a}_1)$, for example, are conjugate (i.e. belong to the same class) may be seen as follows. We have

$$(I|\boldsymbol{\tau})^{-1}(C_2^y|0)(I|\boldsymbol{\tau}) = (I|-\boldsymbol{\tau})(C_2^y|0)(I|\boldsymbol{\tau})$$
$$= (I|-\boldsymbol{\tau})(C_2^y I|C_2^y\boldsymbol{\tau})$$
$$= (C_2^y|-\boldsymbol{\tau}+C_2^y\boldsymbol{\tau}).$$

But

$$C_2^y\boldsymbol{\tau} = \tfrac{1}{2}(-\mathbf{a}_1+\mathbf{a}_2-\mathbf{a}_3),$$
$$-\boldsymbol{\tau}+C_2^y\boldsymbol{\tau} = -\mathbf{a}_1-\mathbf{a}_3$$
$$= \mathbf{a}_1-(2\mathbf{a}_1+\mathbf{a}_3),$$

and, since translations through $\mathbf{a}_3$ and $2\mathbf{a}_1$ must be regarded as identical transformations, we have

$$(I|\boldsymbol{\tau})^{-1}(C_2^y|0)(I|\boldsymbol{\tau}) = (C_2^y|\mathbf{a}_1).$$

Table 2

	$(C_2^z\|0)$	$(C_2^z\|0)$	$(I\|\boldsymbol{\tau})$	$(\sigma_x\|\boldsymbol{\tau})$	$(\sigma_y\|\boldsymbol{\tau})$	$(\sigma_z\|\boldsymbol{\tau})$
$(E\|0)\ (E\|\mathbf{a}_1)\ (C_2^z\|0)\ (C_2^x\|\mathbf{a}_1)$						
	$(C_2^y\|\mathbf{a}_1)$	$(C_2^z\|\mathbf{a}_1)$	$(I\|\boldsymbol{\tau}+\mathbf{a}_1)$	$(\sigma_x\|\boldsymbol{\tau}+\mathbf{a}_1)$	$(\sigma_y\|\boldsymbol{\tau}+\mathbf{a}_1)$	$(\sigma_z\|\boldsymbol{\tau}+\mathbf{a}_1)$
Γ_1	2	-2	2	-2	0	0	0	0	0	0
Γ_2	2	-2	-2	2	0	0	0	0	0	0

From the numbers of elements and classes in the group we find that it has eight one-dimensional and two two-dimensional irreducible representations ($8\cdot 1^2+2\cdot 2^2 = 16$). All the one-dimensional representations are obtained from representations of the point group $\mathbf{D}_{2h}$, the translation $(E|\mathbf{a}_1)$ being assigned the character 1. These representations, however, occur here as "spurious" representations and must be rejected. They do not solve the problem in question: their base functions are invariant under all translations, whereas the functions $e^{i\mathbf{k}\cdot\mathbf{r}}$ with given $\mathbf{k}$ are certainly not invariant with respect to the translation $(E|\mathbf{a}_1)$. Thus there remain only two irreducible representations, whose characters are shown in Table 2. The base functions of these representations can be taken as

$$\Gamma_1 : \cos \pi x, \quad \sin \pi x,$$
$$\Gamma_2 : \cos \pi x \sin 2\pi y, \quad \sin \pi x \sin 2\pi y;$$

the coordinates x, y, z are measured in units of the respective basic vectors $\mathbf{a}_1$, $\mathbf{a}_2$, $\mathbf{a}_3$.

Let us also consider the representations corresponding to the star of two vectors

$$\mathbf{k} = \left(\tfrac{1}{2}, 0, \varkappa\right), \quad \left(\tfrac{1}{2}, 0, -\varkappa\right) \tag{134.6}$$

with proper symmetry C_{2v} (the axis C_2 being along the z-axis); $\varkappa$ is an arbitrary number between 0 and 1 $\left(\text{other than } \tfrac{1}{2}\right)$. The extended group of $\mathbf{k}$ contains eight elements in five classes (Table 3). (The dependence of the base functions of the representations of this group on z reduces to a common factor $e^{2\pi i \varkappa z}$ or $e^{-2\pi i \varkappa z}$, which is invariant under all transformations of the group; it is therefore unnecessary to extend the group by translations along the z-axis.) There are four one-dimensional and one two-dimensional irreducible representation of this group. The one-dimensional representations must be rejected for the same reason as previously, leaving only one representation, whose characters are shown in Table 3. Its base functions can be taken as

$$e^{\pm 2\pi i \varkappa z} \cos \pi x, \quad e^{\pm 2\pi i \varkappa z} \sin \pi x,$$

with the plus and minus signs in the exponent for the two vectors (134.6) respectively; the complete irreducible representation of the whole space group is four-dimensional, and is given by all four of these functions together.

Table 3

		$(C_2^z \mid 0)$	$(\sigma_x \mid \boldsymbol{\tau})$	$(\sigma_y \mid \boldsymbol{\tau})$
$(E \mid 0)$	$(E \mid \mathbf{a}_1)$			
		$(C_2^z \mid \mathbf{a}_1)$	$(\sigma_x \mid \boldsymbol{\tau} + \mathbf{a}_1)$	$(\sigma_y \mid \boldsymbol{\tau} + \mathbf{a}_1)$
2	-2	0	0	0

§ 135. Symmetry under time reversal

In physical applications of the theory of symmetry groups, a further condition is usually imposed on their representations, namely that the base functions should be real (or rather reducible to a real form). This arises in consequence of the symmetry under time reversal. In quantum mechanics, this symmetry has the result that complex-conjugate wave functions must correspond to the same energy level of the quantum system, and must therefore appear among the base functions of the same *physically irreducible* representation (cf. *Quantum Mechanics*, § 96). In classical theory, however, this symmetry is expressed by the invariance of the equations of motion under the change $t \to -t$; the equation contains time derivatives of an even order, the second. Consequently, the equations for the displacements u_s of the atoms in the lattice remain real when their solution is sought in the complex form (69.6) ($\propto e^{-i\omega t}$); the amplitudes of these expressions can therefore be taken to be real.[†]

Real base functions, of course, remain real under the action of any element of symmetry; thus all the group representation matrices are real. If, however, any irreducible representation does not satisfy the condition, it must be combined with the complex-conjugate representation to form one physically irreducible representation of twice the dimension. Let us consider from this standpoint the cases that can occur for representations of space groups (C. Herring, 1937).

The simplest case in this respect is the one where the stars of the wave vectors $\mathbf{k}$ and $-\mathbf{k}$ do not coincide. Then the irreducible representations constructed from each star are certainly complex. For example, for the star of $\mathbf{k}$ the base functions of the representations are multiplied by factors $e^{i\mathbf{k}\cdot\mathbf{a}}$ under translations $(E|\mathbf{a})$, and these factors include none that are complex conjugates; it is therefore clear that no choice of linear combinations of these functions can bring the matrices of the transformations to a real form. On the other hand, by taking the complex conjugates of these functions we obtain the complex-conjugate representation belonging to the star of the vector $-\mathbf{k}$. By combining these two representations, we arrive at a real representation. Thus, to find a physically irreducible representation, we must include $-\mathbf{k}$ together with $\mathbf{k}$ in the star of each wave vector. That is, to obtain the whole of the required star we must apply to some original $\mathbf{k}$ all the elements of the direction group plus the centre of symmetry.

If, however, the star of the wave vector already contains all the necessary values of $\mathbf{k}$, this does *not* guarantee that the irreducible representations constructed from them are real, as can be shown by a simple example.

Let us consider the symmorphic space group S_4^1, belonging to the crystal class S_4 and having a simple tetragonal Bravais lattice. In this group, we take the representations corresponding to the star of two vectors

$$\mathbf{k} = (0, 0, \varkappa), \quad (0, 0, -\varkappa), \tag{135.1}$$

where the z-axis is along the symmetry axis S_4, and $\varkappa$ is an arbitrary number between 0 and 1 $\left(\text{other than } \frac{1}{2}\right)$. The proper symmetry of these vectors is C_2,

---

[†] This will not be so in the presence of a magnetic field or in a crystal having a magnetic structure.

a point group having two one-dimensional representations with the following characters:

$$
\begin{array}{ccc}
 & E & C_2 \\
A & 1 & 1 \\
B & 1 & -1
\end{array}
$$

Taking the first of these as the small representation, we get a two-dimensional representation of the whole space group, whose basis can be taken as the complex conjugate functions $\exp(\pm 2\pi i \varkappa z)$; this representation is therefore real. When B is the small representation, we have a two-dimensional representation of the whole group, with base functions

$$
\exp(2\pi i \varkappa z)\cos 2\pi x, \quad \exp(-2\pi i \varkappa z)\sin 2\pi x.
$$

The characters of the rotational elements of the group in this representation are

$$
\begin{array}{cccc}
(E|0) & (S_4|0) & (C_2|0) & (S_4^3|0) \\
2 & 0 & -2 & 0
\end{array}
$$

and those of the translations are

$$
\begin{array}{ccc}
(E|\mathbf{a}_1) & (E|\mathbf{a}_2) & (E|\mathbf{a}_3) \\
2 & 2 & 2\cos 2\pi x
\end{array}
$$

All these characters are real, but the representation is nevertheless complex: its base functions cannot be brought into a real form. The physically irreducible representation is found by adding to these functions their complex conjugates, and in this case therefore by combining two complex-conjugate but equivalent (with the same characters) representations.[†]

In the example considered, the symmetry under time reversal doubles the dimension of the physically irreducible representation for values of the wave vector on a straight line (the axis of symmetry) in **k**-space. There are also cases where this doubling occurs for values of **k** occupying a whole plane in **k**-space, this plane being perpendicular to a second-order screw axis.

Let us consider, for example, the non-symmorphic space group C_2^2, which belongs to the crystal class C_2 and has a simple monoclinic Bravais lattice. The second-order axis, which we take as the z-axis, is here a screw axis, with a translation through half a lattice vector: $(C_2|\frac{1}{2}\mathbf{a}_3)$. In this group, we take the star of two wave vectors:

$$
\mathbf{k} = (\varkappa, \lambda, \tfrac{1}{2}), \quad (-\varkappa, -\lambda, \tfrac{1}{2}), \tag{135.2}
$$

[†] This kind of situation did not arise in point groups, for which all the irreducible representations with real characters are real.

where $\varkappa$ and λ are arbitrary numbers between 0 and $\frac{1}{2}$; the x- and y-axes are oblique, in a plane perpendicular to the axis of symmetry. The star includes $\mathbf{k}$ and $-\mathbf{k}$, since the vectors $\left(-\varkappa, -\lambda, -\frac{1}{2}\right)$ and $\left(-\varkappa, -\lambda, \frac{1}{2}\right)$ are equivalent. This star corresponds to two equivalent (with the same real characters) two-dimensional irreducible representations of the group, given respectively by the base functions

$$e^{\pm 2\pi (\varkappa x + \lambda y)} e^{inz}$$

and their complex conjugates. A physically irreducible representation is obtained by combining these two complex-conjugate representations. Its four base functions fall into two pairs each corresponding to one of the two wave vectors in the star:

$$e^{2\pi i(\varkappa x + \lambda y)} e^{\pm inz} \quad \text{and} \quad e^{-2\pi i(\varkappa x + \lambda y)} e^{\pm inz}.$$

If the irreducible representation is found together with its base functions, there is no difficulty in seeing whether it is real or complex. Nevertheless, in more complicated cases (and for the investigation of certain general problems) it is useful to have a criterion for answering this question from the characters of the small representation. Such a criterion can be obtained from the following general theorem in the theory of group representations.[†]

For each of the irreducible representations of the group, the following sum can have one of three values:

$$\frac{1}{g} \sum_G \chi(G^2) = +1 \text{ (a)}, \ 0 \text{ (b)}, \ -1 \text{ (c)}. \tag{135.3}$$

The summation is over all the elements in the group, and g is the order of the group. Correspondingly the representation is (a) real, (b) complex with non-equivalent complex-conjugate representations (they have complex-conjugate characters), (c) complex with equivalent complex-conjugate representations (they have equal real characters).

We shall outline the way in which this criterion can be converted for application to space groups, but not give the details. According to the method described in § 134 for the construction of irreducible representations of space groups, their characters may be expressed as

$$\chi[(P|\tau + \mathbf{a})] = \sum_i \chi_{\mathbf{k}_i}[(P|\tau)] \exp(i\mathbf{k}_i \cdot \mathbf{a}), \tag{135.4}$$

where $\chi_{\mathbf{k}}[(P|\mathbf{k})]$ are the characters of the rotational elements of the group in the small representation, and the summation is over those rays $\mathbf{k}_1, \mathbf{k}_2, \ldots$ of

† It is proved, for instance, in the books mentioned in the footnotes to §§ 132 and 134.

the star of the wave vector for which P is an element of the symmetry group. Applying this formula to the element

$$(P|\tau+\mathbf{a})^2 = (P^2|\tau+P\tau+\mathbf{a}+P\mathbf{a})$$
$$= (P|\tau)^2(E|\mathbf{a}+P\mathbf{a}),$$

we have

$$\chi[(P|\tau+\mathbf{a})^2] = \sum_i \chi_{\mathbf{k}_i}[(P|\tau)^2]e^{i\mathbf{a}\cdot(\mathbf{k}_i+P^{-1}\mathbf{k}_i)},$$

with the substitution $\mathbf{k}_i\cdot P\mathbf{a} = \mathbf{a}\cdot P^{-1}\mathbf{k}_i$ in the exponent. These characters are to be summed over all translations and all rotational elements $(P|\tau)$. The sum

$$\sum_{\mathbf{a}} \exp\{i\mathbf{a}\cdot(\mathbf{k}_i+P^{-1}\mathbf{k}_i)\}$$

is zero except when $\mathbf{k}_i+P^{-1}\mathbf{k}_i = 0$ or $\mathbf{b}$. Lastly, because all rays in the star are of equal significance, in the sum over i which is finally to be calculated all the terms are the same.

We thus obtain the *Herring criterion*:

$$\frac{1}{n_{\mathbf{k}}}\Sigma\chi_{\mathbf{k}}[(P|\tau)^2] = +1 \text{ (a)}, \ 0 \text{ (b)}, \ -1 \text{ (c)}, \qquad (135.5)$$

where $\chi_{\mathbf{k}}$ are the characters of the small representation and the summation is over those rotational elements $(P|\tau)$ of the space group which convert $\mathbf{k}$ into a vector equivalent to $-\mathbf{k}$: $P\mathbf{k} = -\mathbf{k}+\mathbf{b}$;[†] $n_{\mathbf{k}}$ is the number of rotational elements in the proper symmetry of the wave vector.

In particular, if the space group contains no rotational elements having the specified property, the sum (135.5) has no terms, and case (b) occurs, in accordance with our previous discussion of the case where the stars of $\mathbf{k}$ and $-\mathbf{k}$ are not the same.

In the example of the group S_4^1 discussed above, the elements $(S_4|0)$ and $(S_4^3|0)$ have the desired property; their squares are the element $(C_2|0)$. The sum (135.5) is therefore

$$\tfrac{1}{2}\{\chi_{\mathbf{k}}[(S_4|0)^2]+\chi_{\mathbf{k}}[(S_4^3|0)^2]\} = \chi_{\mathbf{k}}[(C_2|0)];$$

it is 1 for the small representation A and -1 for the small representation B, for which we thus have cases (a) and (c), again in accordance with previous results.

[†] Here $(P|\tau)^2$ does not alter the vector $\mathbf{k}$ (or converts it into an equivalent vector), and is therefore certainly part of the proper symmetry group of the vector $\mathbf{k}$.

§ 136. Symmetry properties of normal vibrations of a crystal lattice

One physical application of the mathematical theory of the representations of space groups is the classification of the normal vibrations of a lattice by means of their symmetry properties.[†]

It may be recalled that a lattice with ν atoms in the unit cell has for any given wave vector $\mathbf{k}$ a total of 3ν normal vibrations, each with its frequency $\omega(\mathbf{k})$. That is, over the whole range of variation of $\mathbf{k}$ the dispersion relation $\omega = \omega(\mathbf{k})$ of the vibrations has 3ν branches $\omega_\alpha(\mathbf{k})$; each of the $\omega_\alpha(\mathbf{k})$ takes values in a certain finite range, the phonon *energy band*. All the essentially different values of the wave vector lie in one unit cell of the reciprocal lattice; if we consider the whole infinite reciprocal lattice, the functions $\omega_\alpha(\mathbf{k})$ are periodic in it:

$$\omega_\alpha(\mathbf{k}+\mathbf{b}) = \omega_\alpha(\mathbf{k}). \tag{136.1}$$

The physical principles of the classification of lattice vibrations by means of the irreducible representations of the symmetry group are the same as for the corresponding classification of finite symmetrical systems (polyatomic molecules; see *Quantum Mechanics*, § 100). The normal coordinates of the vibrations, which act as the basis of an irreducible representation of the symmetry group of the lattice, all belong to one frequency.

Each irreducible representation of the space group is specified, first of all, by its star of wave vectors. From this it follows immediately that the frequency is the same for all normal vibrations that differ only by values of $\mathbf{k}$ from the same star. Thus each function $\omega_\alpha(\mathbf{k})$ has the complete symmetry of directions for the crystal class concerned. As mentioned in § 135, the symmetry under time reversal means that the star of $\mathbf{k}$ must be supplemented by all vectors $-\mathbf{k}$ (if $\mathbf{k}$ and $-\mathbf{k}$ themselves do not coincide); thus we always have[‡]

$$\omega_\alpha(-\mathbf{k}) = \omega_\alpha(\mathbf{k}). \tag{136.2}$$

For a given value of $\mathbf{k}$ (i.e. for one ray of the star), the normal coordinates are distributed among the bases of the small representations corresponding to various frequencies. If the dimension f of the small representation is more than one, there is degeneracy for the given value of $\mathbf{k}$, the frequency in f branches being the same.

[†] Representations of space groups were first applied to the study of the physical properties of crystal lattices by F. Hund (1936) and by L. P. Bouckaert, R. Smoluchowski and E.P. Wigner (1936).

[‡] From the physical point of view, the relation between the transformation $\mathbf{k} \rightarrow -\mathbf{k}$ for the lattice vibrations and time reversal is obvious: a change in the sign of the time reverses the direction of wave propagation (or, in terms of the phonon picture, changes the sign of the phonon momentum $\mathbf{p} = \hbar\mathbf{k}$).

When the vector **k** has a general position in the reciprocal lattice, it has no proper symmetry (its group contains only the unit element, i.e. the identical transformation); all $3v$ values of $\omega_\alpha(\mathbf{k})$ are in general different. Degeneracy may occur if the proper symmetry of the wave vector is so high that its group has irreducible representations with dimension $f > 1$. With reference to the spatial symmetry only, this may take place either at isolated points in the lattice or on entire straight lines (axes of symmetry) in it. The symmetry under time reversal may also cause (twofold) degeneracy on entire planes in **k**-space (F. Hund, 1936; C. Herring, 1937). According to the discussion in § 135, this degeneracy can occur on planes perpendicular to a second-order screw axis; see the example of representations associated with the star (135.2).[†]

In order to classify the normal vibrations of a specific crystal lattice, we must first find the complete vibrational representation of the space group given by all the vibrational coordinates (atomic displacement vectors) together. This representation is reducible; by decomposing it into irreducible parts, we find the degree of degeneracy of the frequencies and the symmetry properties of the corresponding vibrations. It may happen that one representation appears several times in the vibrational representation: this means that there are several different frequencies with the same multiplicity corresponding to vibrations of the same symmetry.

This procedure is analogous to the method of classifying molecular vibrations (*Quantum Mechanics*, § 100). There is an important difference, however, in that the lattice vibrations are further characterised by the parameter **k**, which takes a continuous series of values, and the classification must be carried out for each value (or class of values) of the wave vector separately. If **k** is specified, the star of the irreducible representation of the space group is determined. It is therefore sufficient in practice to find the vibrational small representation and decompose it into irreducible small representations, i.e. irreducible representations of the symmetry group of the vector **k**.

The classification of the lattice vibrations in the limit as $\mathbf{k} \rightarrow 0$ is especially simple. When $\mathbf{k} = 0$, the irreducible small representations for all the space groups (whether or not symmorphic) are the same as the irreducible representations of the point symmetry group of the lattice, namely its crystal class. To find the vibrational representation (D_{vib}) we must consider only

---

[†] Besides degeneracies due to the lattice symmetry, there can also be degeneracy for "accidental" values of **k**; the existence of such degeneracies could be theoretically predicted only by actually solving the equations of motion of the atoms in a specific lattice. For an analysis of possible cases, see C. Herring, *Physical Review* **52**, 365, 1937 (reprinted in R.S. Knox and A. Gold, *Symmetry in the Solid State*, Benjamin, New York, 1964).

the atoms in one unit cell, i.e. regard all translationally equivalent atoms[†] as one. Without repeating the whole discussion of the atomic vibrations in the molecule for this case, we can formulate the following rule for finding the characters of the vibrational representation of the lattice when $\mathbf{k} = 0$. The characters of a rotation $C(\phi)$ through an angle ϕ about an axis of symmetry, or $S(\phi)$ about a rotary-reflection axis, are

$$\chi_{\text{vib}}(C) = \nu_C \chi_v(C), \qquad \chi_{\text{vib}}(S) = \nu_S \chi_v(S), \tag{136.3}$$

where

$$\chi_v(C) = 1 + 2 \cos \phi, \qquad \chi_v(S) = -1 + 2 \cos \phi$$

are the characters of the representation given by the three components of a (polar) vector; ν_C and ν_S are the numbers of atoms which remain fixed or move to translationally equivalent places in the transformation.[‡] The same formulae give the characters for reflection in a plane (transformation $S(0)$) and for inversion in a centre of symmetry (transformation $S(\pi)$). A rotation about a screw axis or a reflection in a glide plane certainly brings all the atoms into translationally non-equivalent positions, and for them we therefore always have $\chi_{\text{vib}} = 0$.

These rules may be illustrated by an example.[§] The diamond lattice belongs to the non-symmorphic space group $\boldsymbol{O}_h^7$. It has a face-centred cubic Bravais lattice with two like atoms in the unit cell at the vertices (000) and at the points $\left(\frac{1}{4} \frac{1}{4} \frac{1}{4}\right)$ on the spatial diagonals of the cubic cells.[||] Half the rotational elements of the group $\boldsymbol{O}_h^7$ coincide with rotations and reflections in the point group $\boldsymbol{T}_d$. These transformations leave both atoms fixed or move them to translationally equivalent positions. The characters of the vibrational representation for these elements are therefore $\chi_{\text{vib}} = 2\chi_v$. The remaining

---

[†] These are the occupied sites in any one Bravais lattice.

[‡] For a molecule, a subtraction is needed in the characters of the vibrational representation in order to eliminate the coordinates that correspond to the displacement or rotation of the molecule as a whole. For a lattice, the number (six) of these degrees of freedom is negligible compared with the total number of degrees of freedom, and such a subtraction is not necessary.

[§] To avoid misunderstanding, we may note that the classification of the limiting frequencies of the optical branches of the vibrations by means of the crystallographic symmetry only of the lattice is not valid for ionic crystals. The long-wavelength optical vibrations of the ionic lattice are accompanied by the occurrence of macroscopic polarisation of the crystal and an associated macroscopic electric field, which in general modifies (and reduces) the symmetry of the vibrations.

[||] The coordinates of the atoms are specified with respect to the edges of the cubic cell and in units of the length of these edges. The volume of a face-centred cubic cell is four times that of the unit cell. The basic lattice vectors go from the vertex to the points $\left(\frac{1}{2} \frac{1}{2} 0\right)$, $\left(\frac{1}{2} 0 \frac{1}{2}\right)$, $\left(0 \frac{1}{2} \frac{1}{2}\right)$, which are the centres of the faces of the cubic cell.

rotational elements of the group O_h^7 are screw rotations and reflections in glide planes, obtained by combining the elements of the group T_d with inversion $(I \mid \tau)$, where $\tau_i^i = \left(\frac{1}{2} \frac{1}{2} \frac{1}{2} \right)$; these elements move the atom at (000) to the translationally non-equivalent point $\left(\frac{1}{4} \frac{1}{4} \frac{1}{4} \right)$, so that their characters are $\chi_{\mathrm{vib}} = 0$. The decomposition of the resulting vibrational representation into irreducible representations of the point group O_h is $D_{\mathrm{vib}} = F_{2g} + F_{2u}$.[†] The coordinates of the acoustic vibrations which describe the movement of the cell as a whole when $\mathbf{k} = 0$ are transformed as the components of a vector, and therefore correspond to the representation F_{2u} by which such components are transformed in the group O_h. The representation F_{2g} corresponds to the triply degenerate limiting frequency of the optical vibrations.[‡]

The degeneracy of the optical vibrations is generally removed on leaving the point $\mathbf{k} = 0$. According to the symmetry, the amount of splitting may vary near $\mathbf{k} = 0$ as a homogeneous function of the first or second order in the components of the vector $\mathbf{k}$. The appropriate criterion is easily ascertained in terms of quantum perturbation theory. The Hamiltonian of lattice vibrations with a small wave vector $\mathbf{k} \equiv \delta \mathbf{k}$ has the form $\hat{H}_0 + \hat{\boldsymbol{\gamma}} \cdot \delta \mathbf{k}$, where $\hat{H}_0$ is the Hamiltonian of vibrations with $\mathbf{k} = 0$, and $\hat{\boldsymbol{\gamma}}$ is a vector operator; the term $\hat{\boldsymbol{\gamma}} \cdot \delta \mathbf{k}$ acts as the perturbation which causes the splitting. The amount of splitting is of the first order in $\delta \mathbf{k}$ if $\hat{\boldsymbol{\gamma}}$ has non-zero matrix elements for transitions between states belonging to the same degenerate vibrational frequency; otherwise, the splitting is of the second order in $\delta \mathbf{k}$. Here it must be noted that the operator $\hat{\boldsymbol{\gamma}}$ changes sign under time reversal, since the wave vector $\delta \mathbf{k}$ changes sign while the product $\hat{\boldsymbol{\gamma}} \cdot \delta \mathbf{k}$, like any Hamiltonian, must be invariant under time reversal. Thus the solution of the problem reduces to finding the selection rules for frequency-diagonal matrix elements of a vector operator that changes sign under time reversal (see *Quantum Mechanics*, § 97). If the degenerate frequency corresponds to an irreducible representation D, these rules depend on the decomposition of the antisymmetric part of its direct product with itself, $\{D^2\}$; non-zero matrix elements exist if this expansion contains parts by which the components of a vector are transformed.

[†] The point group O_h may be regarded as the direct product $O \times C_i$ or $T_d \times C_i$; here we use the second form. Accordingly, the irreducible representations of the group O_h are constructed from representations of the group T_d. In particular, the representations F_{2g} and F_{2u} of O_h are obtained from the representation F_2 of T_d, and differ in being respectively even and odd under inversion (see *Quantum Mechanics*, § 95).

[‡] The limiting frequency of the acoustic vibrations is always degenerate: the macroscopic nature of these vibrations leads to a single value $\omega = 0$ for all three branches, even if this is not required by symmetry. In this sense the degeneracy is "accidental".

The splitting will certainly be of the second order in $\delta\mathbf{k}$ if the point symmetry group of the lattice (the crystal class) contains a centre of inversion; this is evident, since the squared basis of the representation $\{D^2\}$ must be even under inversion, whereas the components of a vector change sign. If the crystal class does not contain an inversion, there are two possibilities. For example, in the crystal class O, the antisymmetric products for the two-dimensional irreducible representation E and the three-dimensional representations F_1 and F_2 are[†]

$$\{E^2\} = A_2, \quad \{F_1^2\} = \{F_2^2\} = F_1 + F_2.$$

The components of a vector are transformed by F_1; hence the splitting of the doubly degenerate frequency will be of the second order in $\delta\mathbf{k}$, and that of the triply degenerate ones will be of the first order.

Let us now consider vibrations with non-zero wave vector. For symmorphic space groups, they are classified in the same way as in the case $\mathbf{k} = 0$ described above. The irreducible small representations are here the same as the irreducible representations of the point symmetry group of the vector $\mathbf{k}$, and to find the vibrational small representation we must again consider only the atoms in one unit cell.

The procedure may be demonstrated for the optical vibrations of the diamond lattice. The face-centred Bravais lattice of this structure corresponds to a body-centred cubic reciprocal lattice. At the point $\mathbf{k} = 0$ (the vertex of the cubic cell) the proper symmetry of the wave vector is O_h, and there is (as shown above) one triply degenerate frequency of optical vibrations, corresponding to the representation F_{2g}; the characters of this representation are[‡]

	E	8C₃	3C₂	6σ′	6S₄	I	8S₆	3σ	6C₂′	6C₄
F_{2g}	3	0	−1	1	−1	3	0	−1	1	−1

Let us now see how this frequency is split when we leave the point $\mathbf{k} = 0$.

A displacement along the spatial diagonal of the cubic cell gives the vector $\mathbf{k}$ the proper symmetry C_{3v}. For this group the representation given by the same three vibrational coordinates is reducible:

	E	2C₃	3σ′
	3	0	1	$= E + A_1$,

[†] The notation for the irreducible representations of point groups is as in *Quantum Mechanics*, § 95.

[‡] The symmetry elements that appear in the point group T_d are put first, followed by the elements obtained on multiplication of these by the inversion I. The elements $3C_2$ are rotations through π about axes along the edges of the cubic cell; $6C_2'$ are rotations through π about the diagonals of the faces; $6\sigma'$ are reflections in planes through opposite edges; 3σ are reflections in the planes of the faces.

i.e. the triply degenerate frequency splits into one doubly degenerate and one non-degenerate frequency. A similar splitting occurs in a displacement along an edge of the cubic cell, where the proper symmetry of the wave vector is C_{4v}:

$$
\begin{array}{ccccc}
E & C_2 & 3C_4 & 2\sigma & 2\sigma' \\
3 & -1 & -1 & -1 & 1 = E + B_2.
\end{array}
$$

In a displacement along a diagonal of a face of the cubic cell, the proper symmetry of the vector $\mathbf{k}$ is lowered to C_{2v}, and the frequency splitting is complete:

$$
\begin{array}{cccc}
E & C_2' & \sigma & \sigma' \\
3 & 1 & -1 & 1 = A_1 + A_2 + B_2.
\end{array}
$$

For crystal lattices of non-symmorphic space groups, the procedure for classifying the normal vibrations is more laborious, and we shall not pause to discuss it here.[†]

§ 137. Structures periodic in one and two dimensions

A characteristic feature of solid crystals is that the density function $\varrho(x, y, z)$ is periodic in three dimensions and extends to unlimited distances. Let us consider the possibility of the existence in Nature of bodies whose density functions are periodic in only one or two dimensions (R. E. Peierls, 1934; L. D. Landau, 1937).

For example, a body with $\varrho = \varrho(x)$ could be regarded as consisting of parallel planes regularly arranged and lying perpendicular to the x-axis, with the atoms randomly distributed in each plane. When $\varrho = \varrho(x, y)$, the atoms would be randomly distributed along lines parallel to the z-axis, but these lines themselves would be regularly arranged.

To discuss this question, let us consider the displacements undergone by small parts of the body as a result of thermal fluctuations. It is clear that, if such displacements increase without limit as the size of the body increases, there will necessarily be a "smoothing-out" of the function ϱ, in contradiction with hypothesis. In other words, only those structures can occur for which the mean displacement remains finite when the dimensions of the body become arbitrarily large.

Let us first confirm that this condition is satisfied in an ordinary crystal. Let $\mathbf{u}(x, y, z)$ denote the vector of the fluctuation displacement of a small region with coordinates x, y, z and let $\mathbf{u}$ be represented as a Fourier series:

$$
\mathbf{u} = \sum_{\mathbf{k}} \mathbf{u_k} e^{i\mathbf{k}\cdot\mathbf{r}} ; \tag{137.1}
$$

[†] Examples of such groups are given in the book by Bir and Pikus quoted in § 134.

the components of the vector **k** take both positive and negative values, and the coefficients $\mathbf{u_k}$ are related by $\mathbf{u_{-k}} = \mathbf{u_k^*}$, since **u** is real. The series (137.1) will include only terms with not too large wave numbers, $k \lesssim 1/d$, where d is the linear dimension of the region undergoing displacement. We shall consider the fluctuations at constant temperature; then their probability is given by

$$w \propto \exp(-\Delta F_t/T), \tag{137.2}$$

where

$$\Delta F_t = \int (F - \bar{F}) \, dV \tag{137.3}$$

is the change in the total free energy of the body in the fluctuation, and F now denotes the free energy per unit volume of the body (cf. (116.7)).

To calculate ΔF_t, we must expand $F - \bar{F}$ in powers of the displacement. The expansion will involve not the function $\mathbf{u}(x, y, z)$ itself but only its derivatives, since $F - \bar{F}$ must vanish when $\mathbf{u} = $ constant, corresponding to a simple displacement of the body as a whole. It is evident that the terms in the expansion which are linear in these derivatives must be absent, since otherwise F could not have a minimum for $\mathbf{u} = 0$. Next, owing to the smallness of the wave numbers k, we need go only as far as the terms quadratic in the first derivatives of **u** in the expansion of the free energy, neglecting the terms containing the higher-order derivatives. Hence we find that ΔF_t has the form

$$\Delta F_t = \tfrac{1}{2} V \sum_{\mathbf{k}} u_{ik} u_{lk}^* \phi_{il}(k_x, k_y, k_z), \tag{137.4}$$

where the elements of the real tensor ϕ_{il} (i and l being the tensor indices, over which summation is implied) are quadratic functions of the components of the vector **k**.[†]

According to (111.9), we hence find for the mean square fluctuations of the Fourier components of the displacement vector

$$\langle u_{ik} u_{lk}^* \rangle = (T/V) \phi_{il}^{-1}(k_x, k_y, k_z), \quad \langle u_{ik} u_{lk'} \rangle = 0 \quad \text{for} \quad \mathbf{k'} \neq -\mathbf{k}, \tag{137.5}$$

where ϕ_{il}^{-1} are the components of the tensor inverse to ϕ_{il}.[‡] For clarity, this expression may be written

$$\langle u_{ik} u_{lk}^* \rangle = \frac{T}{V} \frac{A_{il}(\mathbf{n})}{k^2}, \tag{137.6}$$

[†] The terms containing products $u_{ik} u_{lk'}$ exp $[i(\mathbf{k}+\mathbf{k'})\cdot\mathbf{r}]$ with $\mathbf{k'} \neq -\mathbf{k}$ disappear on integration over the volume.

[‡] To determine the constant numerical coefficient in (137.5), we have to note that each product $u_{ik} u_{lk}^*$ appears twice (for $\pm\mathbf{k}$) in (137.4), giving 2 re $(u_{ik} u_{lk}^*)$, and the real part of the product $u_{ik} u_{lk}^*$ is itself the sum of two independent products.

where the A_{ll} depend only on the direction of the vector $\mathbf{k}$ ($\mathbf{n} = \mathbf{k}/k$). The mean values $\langle u_i u_l \rangle$ are found from (137.6) by summation over $\mathbf{k}$; changing in the usual manner from this to integration, we get, for example, the mean square displacement vector

$$\langle \mathbf{u}^2 \rangle = T \int \frac{A_{ll}(\mathbf{n})}{k^2} \frac{d^3k}{(2\pi)^3}$$

$$= T \int A_{ll}(\mathbf{n}) \frac{dk\, do}{(2\pi)^3}. \qquad (137.7)$$

The integral converges at the lower limit ($\mathbf{k} \to 0$) linearly in $\mathbf{k}$.[†] Thus the mean square of the fluctuation displacement is, as it should be, a finite quantity independent of the size of the body.

Next, let us consider a body with density function $\varrho = \varrho(x)$. Since $\varrho =$ constant along the y- and z-axes in such a body, no displacement along these axes can "smooth out" the density function, and such displacements are consequently of no interest here. We need therefore consider only a displacement u_x. Moreover, it is easy to see that the first derivatives $\partial u_x/\partial y$, $\partial u_x/\partial z$ cannot appear in the expansion of the free energy, since, if the body is rigidly rotated about the y- or z-axis, these derivatives change, whereas the free energy must obviously remain constant. Thus in the expansion of $F-\bar{F}$ we have to consider the following terms quadratic in the displacement:

$$\left(\frac{\partial u_x}{\partial x}\right)^2, \quad \frac{\partial u_x}{\partial x}\left(\frac{\partial^2 u_x}{\partial y^2} + \frac{\partial^2 u_x}{\partial z^2}\right), \quad \left(\frac{\partial^2 u_x}{\partial y^2} + \frac{\partial^2 u_x}{\partial z^2}\right)^2;$$

the derivatives with respect to y and z appear symmetrically, owing to the complete symmetry in the yz-plane. Substitution in (137.3) leads to terms of the types

$$|u_{x\mathbf{k}}|^2 k_x^2, \quad |u_{x\mathbf{k}}|^2 k_x\varkappa^2, \quad |u_{x\mathbf{k}}|^2 \varkappa^4,$$

where $\varkappa^2 = k_y^2 + k_z^2$. Although the two latter expressions include powers of the wave vector components higher than the first expression, they may be of the same order of magnitude, since nothing is known *a priori* concerning the relative magnitude of k_x and $\varkappa$.

Thus the change in the free energy will be of the form

$$\Delta F_t = \tfrac{1}{2}V \sum_{\mathbf{k}} |u_{x\mathbf{k}}|^2\, \phi(k_x, \varkappa^2), \qquad (137.8)$$

where ϕ is a quadratic function of the variables k_x and $\varkappa^2$. Instead of (137.7), we now have

$$\langle u_x^2 \rangle = T \int \frac{1}{\phi(k_x, \varkappa^2)} \frac{d^3k}{(2\pi)^3} = \frac{T}{8\pi^2} \int \frac{dk_x\, d(\varkappa^2)}{\phi(k_x, \varkappa^2)}. \qquad (137.9)$$

[†] The integrand as written is valid only when k is not too large.

This integral is logarithmically divergent when $k \to 0$. The divergence of the mean square displacement implies that a point to which a particular value of $\varrho(x)$ corresponds may be displaced through very large distances; in other words, the density $\varrho(x)$ is "smoothed out" through the whole body, so that no function $\varrho(x)$ is possible except the trivial case $\varrho = $ constant.

Similar arguments for a body with $\varrho = \varrho(x, y)$ give the following expression for the mean squares of the displacement:

$$\langle u_x^2 \rangle, \langle u_y^2 \rangle = \frac{T}{(2\pi)^3} \int \frac{dk_x \, dk_y \, dk_z}{\phi(k_x, k_y, k_z^2)}, \qquad (137.10)$$

where ϕ is again a quadratic function of its arguments. This integral is easily seen to converge at the lower limit, so that the mean fluctuational displacement remains finite. Thus bodies having such a structure could in theory exist, but it is not known whether they do in fact exist in Nature.

So far in this section we have been concerned with three-dimensional bodies, and only the arrangement of the atoms was assumed to have a two-dimensional (or one-dimensional) ordering. Let us now consider the possibility of an ordered configuration of atoms in two-dimensional systems where the atoms occupy only a certain surface.[†] The two-dimensional analogue of ordinary solid crystals would be a film in which the atoms are regularly arranged at the points of a plane lattice. This configuration could be described by a density function $\varrho(x, y)$ (which now has a different meaning from that used above, since only atoms on one surface $z = $ constant are considered). It is easy to see, however, that the thermal fluctuations "smooth out" such a crystal, so that $\varrho = $ constant is the only possibility: the mean values of the products of components of the fluctuational displacement $\mathbf{u}$ (in the xy-plane) are again determined by expressions of the form (137.6) and (137.7), except that the integration is now taken over two-dimensional $\mathbf{k}$-space:

$$\langle u_i u_l \rangle = T \int \frac{A_{il}(\mathbf{n})}{k^2} \frac{dk_x \, dk_y}{(2\pi)^2}, \qquad (137.11)$$

and the integral diverges logarithmically as $k \to 0$.

Here, however, the following reservation is necessary. The result obtained, strictly speaking, means only that the fluctuational displacement becomes infinite when the size (area) of the two-dimensional system increases without limit (so that the wave number may be arbitrarily small). But, because of the slow (logarithmic) divergence of the integral, the size of the film for which the

[†] Such are monomolecular adsorbed films at the interface between two isotropic phases; see § 159.

fluctuations are still small may be very great.[†] In such cases, a film of finite size might have practically "solid-crystal" properties and be approximately describable as a two-dimensional lattice. We shall see in § 138 that such properties of two-dimensional systems become more marked at lower temperatures.

§ 138. The correlation function in two-dimensional systems

The expression (137.11) gives the mean square fluctuational displacement at any specified point in a two-dimensional crystal system. A fuller understanding of the properties of such systems can be achieved by considering the correlation function between the fluctuations at different points in the system.

Let us note first of all that, when $T = 0$, a two-dimensional lattice of *any* size could exist: the divergence of the integral (137.11) is due to the thermal fluctuations ($T \neq 0$). Let $\varrho_0(\mathbf{r})$ be the density function of the system at $T = 0$.[‡] We can determine the correlation function of the density fluctuations for non-zero but sufficiently low temperatures (much less than the Debye temperature). Under these conditions, only long-wavelength vibrations are excited in the lattice, i.e. the variation of the density function is determined mainly by the long-wavelength fluctuations.

Let the atoms at the points $\mathbf{r}$ in the lattice undergo fluctuational displacements $\mathbf{u}(\mathbf{r})$. If the function $\mathbf{u}(\mathbf{r})$ varies only slightly over distances of the order of a lattice constant (corresponding to the fluctuations with small wave numbers that we are concerned with here), the change in the density at each point in space may be regarded as simply the result of a shift of the lattice by an amount equal to the local value of the displacement vector. The fluctuating density is then written $\varrho(\mathbf{r}) = \varrho_0[\mathbf{r} - \mathbf{u}(\mathbf{r})]$, and the correlation between its fluctuations at different points $\mathbf{r}_1$ and $\mathbf{r}_2$ is determined by the mean value

$$\langle \varrho(\mathbf{r}_1)\, \varrho(\mathbf{r}_2) \rangle = \langle \varrho_0[\mathbf{r}_1 - \mathbf{u}(\mathbf{r}_1)]\, \varrho_0[\mathbf{r}_2 - \mathbf{u}(\mathbf{r}_2)] \rangle. \tag{138.1}$$

The periodic function $\varrho_0(\mathbf{r})$ can be expanded as a Fourier series (cf. (133.2)):

$$\varrho_0(\mathbf{r}) = \bar{\varrho} + \sum_{\mathbf{b} \neq 0} \varrho_b e^{i\mathbf{b}\cdot\mathbf{r}}, \tag{138.2}$$

where the $\mathbf{b}$ are the vectors of the (plane) reciprocal lattice and the constant term $\bar{\varrho}$ has been separated from the sum. When these series are substituted

---

[†] The same applies to three-dimensional bodies with one-dimensional periodicity, for which the integral (137.9) diverges logarithmically.

[‡] Here and in the rest of § 138, $\mathbf{r} = (x, y)$ is the two-dimensional radius vector in the plane of the system.

in (138.1) and the result is averaged, the terms containing the products $\varrho_b \varrho_{b'}$ with $\mathbf{b}' \neq -\mathbf{b}$ disappear, as will be shown below. The product with $\mathbf{b}' = -\mathbf{b}$ gives a contribution to (138.1)

$$|\varrho_b|^2 \exp\left[i\mathbf{b}\cdot(\mathbf{r}_1-\mathbf{r}_2)\right] \langle \exp\left[-i\mathbf{b}\cdot(\mathbf{u}_1-\mathbf{u}_2)\right]\rangle; \tag{138.3}$$

for brevity, we write $\mathbf{u}(\mathbf{r}_1) = \mathbf{u}_1$, $\mathbf{u}(\mathbf{r}_2) = \mathbf{u}_2$.

The probability distribution for the fluctuations of the displacement vector is given by formula (137.2), in which ΔF_t is a quadratic functional of $\mathbf{u}(\mathbf{r})$. If the values of $\mathbf{u}(\mathbf{r})$ for different (discrete) points in space are regarded as different fluctuating quantities x_a ($a = 1, 2, \ldots$), this means that the probability distribution for them is Gaussian. We can then use in the averaging in (138.3) the formula

$$\langle \exp\left(\alpha_a x_a\right)\rangle = \exp\left(\tfrac{1}{2}\alpha_a\alpha_b\langle x_a x_b\rangle\right)$$

(see § 111, Problem), obtaining

$$\langle \exp\left[-i\mathbf{b}\cdot(\mathbf{u}_1-\mathbf{u}_2)\right]\rangle = \exp\left(-\tfrac{1}{2}b_i b_l \chi_{il}\right), \tag{138.4}$$

where

$$\chi_{il}(\mathbf{r}) = \langle (u_{i1}-u_{i2})(u_{l1}-u_{l2})\rangle$$
$$= 2\langle u_i u_l\rangle - \langle u_{i1}u_{l2}\rangle - \langle u_{i2}u_{l1}\rangle$$

($\mathbf{r} = \mathbf{r}_1 - \mathbf{r}_2$). It remains to substitute $\mathbf{u}_1$ and $\mathbf{u}_2$ as the expansions (137.1). Noting that the mean values $\langle u_{ik}u_{l\mathbf{k}'}\rangle$ are zero when $\mathbf{k}' \neq -\mathbf{k}$, and are given by (137.11) when $\mathbf{k}' = -\mathbf{k}$, we find

$$\chi_{il}(\mathbf{r}) = T\int \frac{A_{il}(\mathbf{n})}{k^2}\cdot 2(1-\cos\mathbf{k}\cdot\mathbf{r})\frac{dk_x\,dk_y}{(2\pi)^2}. \tag{138.5}$$

The integral converges for small k, since the factor $1 - \cos\mathbf{k}\cdot\mathbf{r} \propto k^2$ when $k \to 0$.[†] For large k, the integral diverges logarithmically. This divergence is actually due only to the fact that the approximations used are invalid for large k: when $k \gtrsim k_{max}$, $\hbar c k_{max} \sim T$ (where c is the velocity of sound; see § 110), the fluctuations cease to be classical. (At low temperatures this condition is violated sooner than $k \ll 1/a$, where a is the lattice constant.) Noting also that for large k the term containing the rapidly oscillating factor $\cos\mathbf{k}\cdot\mathbf{r}$ in the integrand may be omitted, we find

$$\chi_{il}(\mathbf{r}) = \frac{T}{\pi}\overline{A_{il}}\log k_{max}r; \tag{138.6}$$

† If we look into the origin of this factor, we see that it arises from the fact that $\mathbf{b}' = -\mathbf{b}$ in (138.3). It is easy to see that when $\mathbf{b}' \neq -\mathbf{b}$ there will not be cancellation in the integrand and the integral will diverge. Since these integrals occur in the exponent (cf. (138.4)), their divergence means that the corresponding contributions to the correlation function are zero.

the bar over A_{il} denotes averaging over the directions of the vector $\mathbf{k}$ in the plane.

The required correlation function is now obtained by substituting (138.6) in (138.3) and (138.4) and summing over $\mathbf{b}$; the asymptotic law of decrease of this function with increasing distance r is given by the least rapidly decreasing term in the sum:

$$\langle \varrho(\mathbf{r}_1)\,\varrho(\mathbf{r}_2) \rangle - \bar{\varrho}^2 \propto \frac{1}{r^{T\alpha_b}} \cos \mathbf{b} \cdot \mathbf{r},$$

$$\alpha_{\mathbf{b}} = b_i b_l \overline{A_{il}}/2\pi, \tag{138.7}$$

where $\mathbf{b}$ is to be taken as the reciprocal lattice basic vector for which α_b has its least value.

Thus, in a two-dimensional lattice, the correlation function does tend to zero as $r \to \infty$ (unlike a three-dimensional lattice, where it tends to a finite limit), but only according to a power law, and more slowly at low temperatures.[†] Note that the correlation function (138.7) depends on the direction of $\mathbf{r}$, and the system therefore remains anisotropic.

Similar but slightly more laborious calculations give a law of the same type for the correlation function in a three-dimensional system with density function $\varrho(x)$. For comparison, it may be recalled that in an ordinary liquid the correlation function decreases much more rapidly, in fact exponentially (see § 116).

§ 139. Symmetry with respect to orientation of molecules

The condition $\varrho = $ constant is necessary but certainly not sufficient for a body to be isotropic. This is clear from the following example. Let us imagine a body consisting of elongated molecules, all positions in space of a molecule as a whole (i.e. of its centre of mass) being equally probable, but the axes of the molecules being predominantly oriented in one direction. Such a body is obviously anisotropic, despite the fact that $\varrho = $ constant for each atom present in the molecule.

The property whose symmetry is here under consideration may be formulated in terms of a mutual correlation between the positions of the different atoms. Let $\varrho_{12}\,dV_2$ be the probability of finding an atom 2 in the volume element dV_2 for a given position of atom 1 (atoms of different types usually being involved); ϱ_{12} is a function of the coordinates $\mathbf{r}_1$ and $\mathbf{r}_2$ of the two atoms, and the symmetry properties of this function determine the symmetry of the body (in which $\varrho = $ constant).

[†] A correlation function of this form was found by T. M. Rice (1965) for a two-dimensional superconductor, and by B. Jancovici (1967) and V. L. Berezinskii (1971) for a two-dimensional lattice.

The fact that the density function ϱ is constant signifies that a relative displacement of parts of the body (without change of volume) does not affect the equilibrium state of the body, i.e. does not change its thermodynamic quantities. This is precisely the characteristic property of liquids (and gases). We must therefore regard bodies with $\varrho = $ constant and an anisotropic correlation function ϱ_{12} as forming a distinct category, *liquid crystals* (anisotropic fluids). These include bodies whose molecules are anisotropically distributed in respect of their spatial orientation.

Two types are possible as regards the symmetry of this distribution. For one type (*nematic* liquid crystals) the correlation function depends only on the difference $\mathbf{r}_{12} = \mathbf{r}_1 - \mathbf{r}_2$; when the length of this vector varies without change in its direction, no periodicity of the correlation function occurs, though it may undergo oscillations which become less as r_{12} increases. The function thus has no translational symmetry, and its symmetry group must consist of various rotations and reflections, i.e. is a point group. Geometrically, it may be any point group, with symmetry axes of any order, but apparently all the known nematic liquid crystals have an axis of complete axial symmetry, and the two directions along this axis are equivalent. The point groups $\boldsymbol{C_{\infty h}}$, $\boldsymbol{D_\infty}$ and $\boldsymbol{D_{\infty h}}$ have these properties.[†] We shall see in § 140, however, that the symmetry $\boldsymbol{D_\infty}$ (which has no plane of symmetry) makes the state of the liquid crystal unstable, and there consequently must appear a certain "secondary" periodic structure typical of liquid crystals of the other (*cholesteric*) type.

Besides these two types, there exist other anisotropic liquids with various layered structures, which are usually classed together as *smectic* liquid crystals. It seems that at least some of these have a density function $\varrho(x)$ that is periodic in one direction only. Such substances may be regarded as consisting of freely moving plane layers at equal distances apart. In each layer the molecules have an ordered orientation, but the configuration of their centres of mass is random.

It has been shown in § 137 that structures with a one-dimensional periodicity of the density function are smoothed out by thermal fluctuations. The divergence of these fluctuations is, however, only logarithmic. Although this excludes the possibility of a one-dimensional periodicity extended to arbitrarily large distances, it does not exlude its existence over comparatively small but nevertheless macroscopic regions of space, as already noted at the end of § 137.

Finally, it may be mentioned that in ordinary isotropic liquids also there are two different types of symmetry. If the liquid consists of a substance which does not have stereoisomers, it is completely symmetrical not only

[†] In the other groups with axial symmetry ($\boldsymbol{C_\infty}$ and $\boldsymbol{C_{\infty v}}$) the two directions along the axis are not equivalent. Such liquid crystals would in general be pyroelectric.

under a rotation through any angle about any axis but also under a reflection in any plane, i.e. its symmetry group is the complete group of rotations about a point, together with a centre of symmetry (group K_h). If the substance has two stereoisomeric forms, however, and the liquid contains different numbers of molecules of the two isomers, it will not possess a centre of symmetry and therefore will not allow reflections in planes. Its symmetry group is just the complete group of rotations about a point (group K).

§ 140. Nematic and cholesteric liquid crystals

The orientational symmetry of nematic liquid crystals is uniaxial: at every point in the liquid there is only one preferred direction of orientation of the molecules, namely that of axial symmetry. The macroscopic state of such a body can therefore be described by specifying one unit vector $\mathbf{n}(\mathbf{r})$ at each point, having this preferred direction; it is called the *director*. In a complete equilibrium state the body is homogeneous, i.e. $\mathbf{n} = $ constant. Inhomogeneous distributions $\mathbf{n}(\mathbf{r})$ describe various deformational states of the liquid crystal.

In a macroscopic deformation, $\mathbf{n}(\mathbf{r})$ varies slowly through the body; the characteristic dimensions of the deformation are large compared with molecular dimensions. The derivatives of $\mathbf{n}(\mathbf{r})$ with respect to the coordinates are therefore small quantities of an order that increases with that of the derivative. Representing the total free energy of the deformed liquid crystal (at a given temperature) as an integral $F_t = \int F \, dV$, we can expand the free energy density F in powers of the derivatives of $\mathbf{n}(\mathbf{r})$ (C. W. Oseen, 1933; F. C. Frank, 1958).

The expansion of the scalar quantity F can contain only scalar combinations of the components of the vector $\mathbf{n}$ and its derivatives. There are only two scalar combinations linear in the first derivatives: the true scalar div $\mathbf{n}$ and the pseudoscalar $\mathbf{n} \cdot \text{curl } \mathbf{n}$. The former, on integration over the volume, becomes an integral over the surface of the body, and is therefore unimportant in considering the bulk properties of the substance.

True scalars quadratic in the first derivatives can be found by taking the rank-four tensor

$$\frac{\partial n_k}{\partial x_i} \frac{\partial n_l}{\partial x_m}$$

and forming invariants from it by contraction with respect to pairs of suffixes or multiplication by components of the vector $\mathbf{n}$. It must be noted here that $\mathbf{n}$ is a unit vector, so that

$$\frac{\partial}{\partial x_i} \mathbf{n}^2 = 2n_k \frac{\partial n_k}{\partial x_i} = 0.$$

In this way we find the invariants

$$[(\mathbf{n}\cdot\text{grad})\mathbf{n}]^2, \quad \frac{\partial n_k}{\partial x_i}\frac{\partial n_k}{\partial x_i}, \quad (\text{div }\mathbf{n})^2, \quad \frac{\partial n_k}{\partial x_i}\frac{\partial n_i}{\partial x_k}.$$

The last two, however, differ only by a divergence:

$$\frac{\partial n_i}{\partial x_i}\frac{\partial n_k}{\partial x_k} - \frac{\partial n_k}{\partial x_i}\frac{\partial n_i}{\partial x_k} = \frac{\partial}{\partial x_i}\left(n_i\frac{\partial n_k}{\partial x_k} - n_k\frac{\partial n_i}{\partial x_k}\right),$$

so that their contributions to the total free energy differ only by an integral over the surface of the body, which is of no interest here (J. L. Ericksen, 1962). The invariant[†]

$$\frac{\partial n_k}{\partial x_i}\frac{\partial n_k}{\partial x_i} = (\mathbf{n}\cdot\text{curl }\mathbf{n})^2 + (\text{div }\mathbf{n})^2,$$

and $(\mathbf{n}\cdot\text{curl }\mathbf{n})^2$ may therefore be taken as the independent invariant. Lastly, there is a pseudoscalar quadratic in the first derivatives: $(\mathbf{n}\cdot\text{curl }\mathbf{n})\,\text{div }\mathbf{n}$.[‡]

Scalars linear in the second derivatives are of the same order of smallness as the above, but integration by parts reduces all of them to forms quadratic in the first derivatives.

Thus we have the following expression for the free energy density of a liquid crystal:

$$F = F_0 + b\mathbf{n}\cdot\text{curl }\mathbf{n} + \tfrac{1}{2}a_1(\text{div }\mathbf{n})^2 + \tfrac{1}{2}a_2(\mathbf{n}\cdot\text{curl }\mathbf{n})^2$$
$$+ \tfrac{1}{2}a_3[(\mathbf{n}\cdot\text{grad})\mathbf{n}]^2 + a_{12}(\mathbf{n}\cdot\text{curl }\mathbf{n})\,\text{div }\mathbf{n}, \qquad (140.1)$$

where b, a_1, a_2, a_3 and a_{12} are (temperature-dependent) constants.

As already mentioned in § 139, the directions of $\mathbf{n}$ and $-\mathbf{n}$ are equivalent in all known liquid crystals of these types; to meet this condition, we must put $a_{12} = 0$. Moreover, if the symmetry elements of the crystal include planes, $b = 0$: since $\mathbf{n}\cdot\text{curl }\mathbf{n}$ is a pseudoscalar and the free energy is a true scalar, the coefficient b must be a pseudoscalar; but a medium having planes of symmetry cannot have pseudoscalar characteristics, since reflection in the plane would give $b = -b$. Thus the free energy of a nematic liquid crystal is

$$F = F_0 + \tfrac{1}{2}a_1(\text{div }\mathbf{n})^2 + \tfrac{1}{2}a_2(\mathbf{n}\cdot\text{curl }\mathbf{n})^2 + \tfrac{1}{2}a_3[(\mathbf{n}\cdot\text{grad})\mathbf{n}]^2. \quad (140.2)$$

All three coefficients a_1, a_2, a_3 must be positive. Then $\mathbf{n} = \text{constant}$ corresponds to the equilibrium state.

[†] This is easily proved by expressing in components and taking the z-axis in the direction of $\mathbf{n}$ at the relevant point in space (then $\partial n_z/\partial x_i = 0$).

[‡] The product $(\mathbf{n}\cdot\text{grad})\mathbf{n}\cdot\text{curl }\mathbf{n} = 0$, since from grad $\mathbf{n}^2 = 0$ it follows that $(\mathbf{n}\cdot\text{grad})\mathbf{n} = -\mathbf{n}\times\text{curl }\mathbf{n}$.

If the liquid crystal does not have a plane of symmetry, then $b \neq 0.$[†] We can rewrite (140.1) (with $a_{12} = 0$) as

$$F = F_0 + \tfrac{1}{2}a_1(\text{div } \mathbf{n})^2 + \tfrac{1}{2}a_2(\mathbf{n}\cdot\text{curl } \mathbf{n} + q_0)^2 + \tfrac{1}{2}a_3[(\mathbf{n}\cdot\text{grad})\mathbf{n}]^2, \quad (140.3)$$

where $q_0 = b/a_2$ (and the constant $-b^2/2a_2$ is included in F_0). The equilibrium state of such a substance corresponds to a distribution of directions of the director such that

$$\text{div } \mathbf{n} = 0, \quad (\mathbf{n}\cdot\text{grad})\mathbf{n} = 0, \quad \mathbf{n}\cdot\text{curl } \mathbf{n} = -q_0.$$

These equations have the solutions

$$n_x = 0, \quad n_y = \cos q_0 x, \quad n_z = \sin q_0 x. \quad (140.4)$$

This structure (corresponding to cholesteric liquid crystals) can be imagined as the result of a uniform twisting about the x-axis of a nematic medium originally oriented with $\mathbf{n} = $ constant in one direction in the yz-plane. The orientational symmetry of a cholesteric crystal is periodic in one direction (the x-axis) in space, so that the correlation function $\varrho_{12} = \varrho_{12}(x, \mathbf{r}_{12})$. The vector $\mathbf{n}$ reverts to its original value after each interval of length $2\pi/q_0$ along the x-axis; but since the directions of $\mathbf{n}$ and $-\mathbf{n}$ are physically equivalent, the actual period of repetition of the structure is π/q_0. A structure of this type is usually said to be *helicoidal*.

The above theory is, of course, valid only if the period of the helicoidal structure is large in comparison with molecular dimensions. This condition is in fact satisfied in cholesteric liquid crystals (the period $\pi/q_0 \sim 10^{-5}$ cm).

§ 141. Fluctuations in liquid crystals

Let us consider the fluctuations of the direction of the director $\mathbf{n}$ in a nematic liquid crystal (P. G. de Gennes, 1968).

We express $\mathbf{n}$ as $\mathbf{n}_0 + \boldsymbol{\nu}$, where $\mathbf{n}_0 \equiv \bar{\mathbf{n}}$ is the equilibrium direction, constant throughout the volume, and $\boldsymbol{\nu} \equiv \varDelta\mathbf{n}$ is the fluctuation from this value. Since $\mathbf{n}^2 = \mathbf{n}_0^2 = 1$, $\mathbf{n}_0\cdot\boldsymbol{\nu} \cong 0$, i.e. the vector $\boldsymbol{\nu}$ is perpendicular to $\mathbf{n}_0$. Accordingly, the correlation function of the fluctuations,

$$\langle \nu_\alpha(\mathbf{r}_1)\, \nu_\beta(\mathbf{r}_2) \rangle, \quad (141.1)$$

is a two-dimensional tensor in the plane perpendicular to $\mathbf{n}_0$ (α and β being vector suffixes in that plane). In a homogeneous but anisotropic liquid, this

---

[†] This symmetry will always occur for a liquid crystal consisting of one stereoisomer of a substance that has molecules with mirror-type asymmetry (such as all the known cholesteric liquid crystals). Crystals consisting of two different stereoisomers of the same substance differ in the sign of the constant b.

function depends not only on the magnitude of the vector $\mathbf{r} = \mathbf{r}_2 - \mathbf{r}_1$ but also on its direction.

A magnetic field has a strong influence on the fluctuations of the director. This is due to the presence, in the free energy density of a liquid crystal, of a further term

$$F_{\text{magn}} = -\tfrac{1}{2}\chi_a(\mathbf{n}\cdot\mathbf{H})^2, \tag{141.2}$$

which depends on the vector $\mathbf{n}$ itself, not on its derivatives as in (140.2).[†] If $\chi_a > 0$, the equilibrium direction of $\mathbf{n}$ is the same as that of the field, but if $\chi_a < 0$ it is in the plane perpendicular to the field. We shall take the case where $\chi_a > 0$, so that $\mathbf{n}_0 \parallel \mathbf{H}$. Then $(\mathbf{n}\cdot\mathbf{H})^2 \cong H^2(1 - \nu^2)$; omitting the term independent of ν, we can write

$$F_{\text{magn}} = \tfrac{1}{2}\chi_a H^2\nu^2. \tag{141.3}$$

Taking F from (140.2) and (141.3), and retaining only quantities of the second order in ν, we get the following expression for the change in the total free energy in a fluctuation:

$$\Delta F_t = \tfrac{1}{2}\int \{a_1(\text{div } \nu)^2 + a_2(\text{curl}_x \nu)^2 + a_3(\partial\nu/\partial x)^2 + \chi_a H^2\nu^2\}\,dV, \tag{141.4}$$

with the x-axis in the direction of $\mathbf{n}_0$. It must be emphasised that, by using the expression (140.2) for the energy of the deformed crystal, we thereby limit the treatment to fluctuations with wavelengths much longer than molecular dimensions.

We now proceed as in § 116, expressing the fluctuating quantity $\nu(\mathbf{r})$ as a Fourier series in the volume V:

$$\nu = \sum_{\mathbf{k}} \nu_{\mathbf{k}} e^{i\mathbf{k}\cdot\mathbf{r}}, \quad \nu_{-\mathbf{k}} = \nu_{\mathbf{k}}^*. \tag{141.5}$$

After substitution of this series, the expression (141.4) forms a sum of terms $(\Delta F_t)_{\mathbf{k}}$, each of which depends only on the component $\nu_{\mathbf{k}}$ with a particular value of $\mathbf{k}$. Taking the xy-plane to contain the direction of $\mathbf{k}$ (and of $\mathbf{H}$), we get

$$(\Delta F_t)_{\mathbf{k}} = \tfrac{1}{2}V\{(a_1 k_y^2 + a_3 k_x^2 + \chi_a H^2)\,|\nu_{y\mathbf{k}}|^2 + (a_2 k_y^2 + a_3 k_x^2 + \chi_a H^2)\,|\nu_{z\mathbf{k}}|^2\}.$$

Hence (cf. § 116) we find for the mean square fluctuations

$$\left.\begin{aligned}
\langle|\nu_{y\mathbf{k}}|^2\rangle &= T/V(a_1 k_y^2 + a_3 k_x^2 + \chi_a H^2), \\
\langle|\nu_{z\mathbf{k}}|^2\rangle &= T/V(a_2 k_y^2 + a_3 k_x^2 + \chi_a H^2), \\
\langle\nu_{y\mathbf{k}}\nu_{z\mathbf{k}}\rangle &= 0.
\end{aligned}\right\} \tag{141.6}$$

[†] In a uniaxial anisotropic medium, the magnetic susceptibility is a tensor $\chi_{ik} = \chi_0\delta_{ik} + \chi_a n_i n_k$, and the magnetisation of the substance gives rise to a free energy term $-\tfrac{1}{2}\chi_{ik} H_i H_k$. The quantity (141.2) is the part of this contribution that depends on $\mathbf{n}$.

We see that, in the absence of the field, the fluctuations of the Fourier components ν_k increase without limit as $\mathbf{k} \to 0$ (the integrals over d^3k, which give the mean square of the vector ν itself, remain finite, however). The application of the magnetic field suppresses the fluctuations that have wave numbers $k \lesssim H(\chi_a/a)^{1/2}$ (where a is the order of magnitude of the coefficients a_1, a_2, a_3).[†]

The correlation function (141.1) can be calculated from (141.6), using the formula

$$\langle \nu_\alpha(\mathbf{r}_1)\, \nu_\beta(\mathbf{r}_2) \rangle = \int e^{i\mathbf{k}\cdot\mathbf{r}} \langle \nu_{\alpha k} \nu_{\beta k}^* \rangle V \, d^3k/(2\pi)^3 ; \qquad (141.7)$$

cf. (116.13). We shall not pause to give the fairly lengthy formula resulting from the integration,[‡] but only mention that, in the absence of the field, the correlation function decreases as $1/r$ with increasing distance $r = |\mathbf{r}_2 - \mathbf{r}_1|$. When the field is present, the decrease becomes exponential, with correlation radius $r_c \sim (a/\chi_a)^{1/2} H^{-1}$.

The fluctuations of the direction of the director in a cholesteric liquid crystal can be treated similarly, and only a brief commentary will be given here.

In a cholesteric medium, we can distinguish the fluctuations of the local direction of the helicoidal structure axis and those of the phase (the angle of rotation of the vector $\mathbf{n}$ about this axis). Fluctuations of the first type are finite, but the mean square fluctuation of the phase is (in the absence of a magnetic field) logarithmically divergent as $\mathbf{k} \to 0$. In this respect, the fluctuations in a medium with one-dimensional periodicity of the orientational structure are analogous to those in a medium with one-dimensional periodicity of the particle configuration (§ 137). Strictly speaking, such periodicity is consequently impossible in a medium of arbitrarily great extent, but, on account of the large value of the helicoidal structure period in cholesteric

---

[†] This type of fluctuation is analogous to the behaviour of the density fluctuations in an ordinary liquid near its critical point and that of the order parameter fluctuations near a phase transition point of the second kind (see § 146 and 152 below). Whereas in these latter cases the factor that suppresses the fluctuations is the "distance" from the points concerned, here it is the external magnetic field, which is independent of the temperature. Note that it is precisely the increase of the fluctuations of $\mathbf{n}$ when $\mathbf{k}$ is small that enables us to treat these fluctuations independently of those of other quantities. In this connection, it is important that we are not considering the neighbourhood of phase transition points of the second kind. Near such points, the fluctuations of other characteristics of the transition also increase, and the fluctuations of $\mathbf{n}$ cannot in general be dealt with independently of the others. It should also be emphasised that the increase of the fluctuations does not lead to any limitations on the range of applicability of (141.6), whereas the validity of formula (146.8), for example, is subject to the inequality (146.15).

[‡] In deriving this, the expressions (141.6) must, of course, be rewritten in a form that does not involve a particular choice of the coordinate axes.

liquid crystals, the divergence of the fluctuations would occur only for such huge dimensions that the whole question is a purely academic one.

We may add a few words about fluctuations in smectic liquid crystals consisting of regularly arranged plane layers. As already mentioned in § 139, such a structure is smoothed out by thermal fluctuations, and can therefore occur only in restricted volumes. It is interesting to note, however, that these fluctuations are suppressed by a magnetic field. The origin of this effect is as follows.

In each layer, the molecules are oriented in an ordered manner, with a preferred direction given by the director $\mathbf{n}$; let this be normal to the surface of the layer. When fluctuations occur, the surface of the layers is deformed and the director is rotated. Let $\mathbf{u}$ be the displacement vector of points in a layer, and $\boldsymbol{\nu}$ again the change in the director ($\mathbf{n} = \mathbf{n}_0 + \boldsymbol{\nu}$). In long-wavelength deformations, the layer may be regarded as a geometrical surface, and the small quantities $\boldsymbol{\nu}$ and $\mathbf{u}$ are then related by $\boldsymbol{\nu} = -\operatorname{grad}(\mathbf{u}\cdot\mathbf{n}_0)$ (the change of direction of the normal to the surface); for their Fourier components we have $\boldsymbol{\nu}_{\mathbf{k}} = -i\varkappa(\mathbf{u}_{\mathbf{k}}\cdot\mathbf{n}_0)$, where $\varkappa$ is the component of $\mathbf{k}$ in the plane of the layer. In the presence of the magnetic field, the change in direction of the director makes an additional contribution (141.3), proportional to ν^2, in $\varDelta F_t$. This in turn has the result that the integral (137.9) which gives the mean square of the fluctuational displacement acquires a further term $\sim \varkappa^2$ in the denominator of the integrand (together with a term in $\varkappa^4$), and in consequence the integral no longer diverges.

Lastly, let us consider fluctuations in two-dimensional liquid-crystal systems (films). In such a system, the orientation of the molecules is specified by a director $\mathbf{n}$ lying in the plane of the film. If we consider its fluctuations (with wave vectors $\mathbf{k}$ in the plane of the film), we get an expression analogous to (141.6): in the absence of a field, $\langle \nu_{\mathbf{k}}^2 \rangle \propto 1/\phi(k_x, k_y)$, where $\phi(k_x, k_y)$ is a quadratic function of the components of $\mathbf{k}$. But to find the total fluctuation $\langle \nu^2 \rangle$ this expression must now be integrated over $d^2k \propto k\,dk$, and the integral diverges logarithmically. Thus the thermal fluctuations smooth out such a two-dimensional liquid-crystal structure. As with a two-dimensional solid crystal (§ 137), however, the logarithmic divergence does not exclude the existence of such a structure in regions of finite size.

[handwritten, top left] First order

[handwritten graphs at top: φ vs T with φ₁, φ₂, labeled $S=\left(\frac{\partial \phi}{\partial T}\right)$, T_c; S vs T graph; ΔS vs T graph with T_c]

[handwritten left margin] characteristics

i) two free energies

ii) discontinuous struc. changes

iii) coexisting phases.

iv) discontinuous changes in S, E, V, etc.

[handwritten mid-left] 2nd order

[handwritten graph: φ vs T, with T_c, T]

CHAPTER XIV

PHASE TRANSITIONS OF THE SECOND KIND AND CRITICAL PHENOMENA

§ 142. Phase transitions of the second kind

IT HAS already been mentioned in § 83 that the transition between phases of different symmetry (crystal and liquid; different crystal modifications) cannot occur in a continuous manner such as is possible for a liquid and a gas. In every state the body has either one symmetry or the other, and therefore we can always assign it to one of the two phases.

The transition between different crystal modifications is usually effected by means of a phase transition in which there is a sudden rearrangement of the crystal lattice and the state of the body changes discontinuously. As well as such discontinuous transitions, however, another type of transition involving a change of symmetry is also possible.

[handwritten left, lower]

i) One free energy
   φ continuous
   $\frac{d\phi}{dT}$ continuous
   $\frac{d^2\phi}{dT^2}$ discontinuous

ii) Symmetry change.

iii) kink in S, E, V,

iv) No coexisting phases.

[handwritten near figure] $S = \frac{d\phi}{dT}$

[handwritten] $C = \frac{d\phi}{dT^2}$

[handwritten graphs of S vs T and C vs T with T_c]

● Ba   ○ O   ⊕ Ti

FIG. 60

To elucidate the nature of these transitions, let us consider a specific example. At high temperatures, $BaTiO_3$ has a cubic lattice whose unit cell is as shown in Fig. 60 (the barium atoms are at the vertices, the oxygen atoms at the centres of the faces, and the titanium atoms at the centres of the cells). As the temperature decreases below a certain value, the titanium and oxygen atoms begin to move relative to the barium atoms parallel to an edge of the cube. It is clear that, as soon as this movement begins, the symmetry of the lattice is affected, and it becomes tetragonal instead of cubic.

This example is typical in that there is no discontinuous change in state of

446

[handwritten bottom]

First order ⟹ first derivative of F is discont.

2nd order ⟹ 2nd deriv. is disconti.

the body. The configuration of atoms in the crystal[†] changes continuously. However, an arbitrarily small displacement of the atoms from their original symmetrical positions is sufficient to change the symmetry of the lattice. The resulting transition from one crystal modification to another is called a *phase transition of the second kind*, in contrast to ordinary phase transitions, which in this case are said to be of the first kind.[‡]

Thus a phase transition of the second kind is continuous in the sense that the state of the body changes continuously. It should be emphasised, however, that the symmetry, of course, changes discontinuously at the transition point, and at any instant we can say to which of the two phases the body belongs. But whereas at a phase transition point of the first kind bodies in two different states are in equilibrium, the states of the two phases are the same at a transition point of the second kind.

As well as cases where the change in symmetry of the body occurs by a displacement of the atoms (as in the example given above), the change in symmetry in a phase transition of the second kind may result from a change in the *ordering* of the crystal. It has already been mentioned in § 64 that the concept of ordering arises if the number of lattice points that can be occupied by atoms of a given kind exceeds the number of such atoms. We shall use the word "own" for the places occupied by atoms of the kind in question in a completely ordered crystal, in contrast to the "other" places which are taken by some of the atoms when the crystal becomes disordered. In many cases, which will be of interest in connection with transitions of the second kind, it is found that the "own" and "other" lattice sites are geometrically identical and differ only in that they have different probabilities of containing atoms of the kind in question.[§] If now these probabilities become equal (they will not be unity, of course), all such sites become equivalent, and therefore new symmetry elements appear, i.e. the symmetry of the lattice is increased. Such a crystal will be said to be *disordered*.

The foregoing may be illustrated by an example. The completely ordered alloy CuZn has a cubic lattice with the zinc atoms at the vertices, say, and the copper atoms at the centres of the cubic cells (Fig. 61a; a simple cubic Bravais lattice). When the alloy becomes disordered, copper and zinc atoms

[†] To simplify the discussion, we shall conventionally speak of the configuration of the atoms or its symmetry as if the atoms were at rest. In reality we should speak of the probability distribution for various configurations of the atoms in space, and of the symmetry of this distribution.

[‡] Phase transition points of the second kind are also called *Curie points*, especially when they are associated with a change in the magnetic structure of the body.

[§] We may note that in this case it can always be assumed that the probability of finding an atom at one of its own sites is greater than at one of the other sites simply because, if it were not, we could transpose the nomenclature of the sites.

change places, i.e. non-zero probabilities of finding atoms of either kind exist at every lattice site. Until the probabilities of finding copper (or zinc) atoms at the vertices and at the centres of the cells become equal (that is, while the crystal is not completely ordered), these sites remain non-equiv-alent, and the symmetry of the lattice is unchanged. But when the probabil-ities become equal, all sites become equivalent, and the symmetry of the crystal is raised: a new lattice vector appears, from a vertex to the centre of a cell, and the crystal acquires a body-centred cubic Bravais lattice (Fig. 61b).[†]

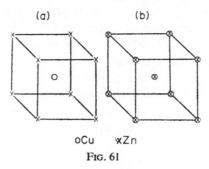

$$(a) \qquad (b)$$

oCu   ⅹZn

FIG. 61

So far we have discussed only transitions between different crystal modi-fications, but phase transitions of the second kind need not necessarily in-volve a change in symmetry of the configuration of atoms in the lattice. A transition of the second kind can also bring about a transformation be-tween two phases differing in some other property of symmetry, as for exam-ple at the Curie points of ferromagnetic or antiferromagnetic substances. In this case there is a change in symmetry of the arrangement of the elemen-tary magnetic moments in the body, or more precisely a disappearance of the currents $\mathbf{j}$ in it; see the first footnote to § 128. Other phase transitions of the second kind are the transition of a metal to the superconducting state (in the absence of a magnetic field) and that of liquid helium to the super-fluid state. In both these cases the state of the body changes continuously, but it acquires a qualitatively new property at the transition point.

Since the states of the two phases are the same at a transition point of the second kind, it is clear that the symmetry of the body at the transition point itself must contain all the symmetry elements of both phases. It will be

---

[†] Cases are in principle possible where the occurrence of ordering does not cause a change in the symmetry of the crystal. A phase transition of the second kind is then impossible: even if the transition from the ordered to the disordered state of the crystal were to occur continuously, there would still be no discontinuity of specific heat (see below). In such cases a phase transition of the first kind is, of course, possible.

shown below that the symmetry at the transition point itself is the same as the symmetry everywhere on one side of that point, i.e. the symmetry of one of the phases. Thus the change in symmetry of the body in a phase transition of the second kind has the following very important general property: the symmetry of one phase is higher than that of the other.[†] It should be emphasised that in a phase transition of the first kind the change in symmetry of the body is subject to no restriction, and the symmetries of the two phases may be unrelated.

In the great majority of the known instances of phase transitions of the second kind, the more symmetrical phase corresponds to higher temperatures and the less symmetrical one to lower temperatures. In particular, a transition of the second kind from an ordered to a disordered state always occurs with increasing temperature. This is not a law of thermodynamics, however, and exceptions are therefore possible.[‡]

For brevity we shall arbitrarily call the more symmetrical phase simply the symmetrical one, and the less symmetrical phase the unsymmetrical one.

To describe quantitatively the change in the structure of the body when it passes through the phase transition point, we can define a quantity η, called the *order parameter*, in such a way that it takes non-zero (positive or negative) values in the unsymmetrical phase and is zero in the symmetrical phase. For instance, in transitions where the atoms are displaced from their positions in the symmetrical phase, η may be taken as the amount of this displacement. For transitions with a change in the ordering of the crystal (e.g. in the CuZn alloy mentioned above), the parameter η may be defined as

$$\eta = (w_{Cu} - w_{Zn})/(w_{Cu} + w_{Zn}),$$

where w_{Cu} and w_{Zn} are the probabilities of finding a copper atom and a zinc atom respectively at any given lattice site. For magnetic transitions, η may be taken as the macroscopic magnetic moment per unit volume of a ferromagnet or the magnetic moment of the sub-lattice for an antiferromagnet.

It must again be emphasised that the symmetry of the body is changed (namely, increased) only when η becomes exactly zero; any non-zero value of the order parameter, however small, brings about a lowering of the

---

[†] It will be recalled that the term "higher symmetry" refers to a symmetry which includes all the symmetry elements (rotations, reflections and translational periods) of the lower symmetry, together with additional elements.

The condition mentioned is necessary but not sufficient for a phase transition of the second kind to be possible; we shall see later that the possible changes of symmetry in such a transition are subject to further restrictions.

[‡] One exception, for example, is the "lower Curie point" of Rochelle salt, below which the crystal is orthorhombic, but above which it is monoclinic.

symmetry. A passage through a phase transition point of the second kind has a continuous change of η to zero.

The absence of any discontinuous change of state at a phase transition point of the second kind has the result that the thermodynamic functions of the state of the body (its entropy, energy, volume, etc.) vary continuously as the transition point is passed. Hence a phase transition of the second kind, unlike one of the first kind, is not accompanied by evolution or absorption of heat. We shall see, however, that the derivatives of these thermodynamic quantities (i.e. the specific heat of the body, the thermal expansion coefficient, the compressibility, etc.) are discontinuous at a transition point of the second kind.

We must remember that mathematically a phase transition point of the second kind is a singularity of the thermodynamic quantities, and in particular of the thermodynamic potential Φ; the nature of this singularity is discussed in §§ 148 and 149. In order to see this, let us first recall that a phase transition point of the first kind is not a singularity; it is a point at which the thermodynamic potentials $\Phi_1(P, T)$ and $\Phi_2(P, T)$ of the two phases are equal, and each of the functions Φ_1 and Φ_2 on either side of the transition point corresponds to an equilibrium (though possibly metastable) state of the body. In a phase transition of the second kind, however, the thermodynamic potential of each phase, if formally regarded on the far side of the transition point, corresponds to no equilibrium state, i.e. to no minimum of Φ; we shall see in § 143 that the thermodynamic potential of the symmetrical phase would indeed correspond to a maximum of Φ beyond the transition point.

This last result implies that superheating and supercooling effects are impossible in phase transitions of the second kind (whereas they can occur in ordinary phase transitions). In this case neither phase can exist beyond the transition point (here we ignore, of course, the time needed to establish the equilibrium distribution of atoms, which in solid crystals may be considerable).

PROBLEM

Let c be the concentration of atoms of one component of a binary solid solution, and c_0 the concentration of these atoms' "own" sites. If $c \neq c_0$ the crystal cannot be completely ordered. Assuming the difference $c - c_0$ small and the crystal almost completely ordered, determine the concentration λ of atoms at "other" sites, expressing it in terms of the value λ_0 which it would have at $c = c_0$ for given P and T (C. Wagner and W. Schottky, 1930).

SOLUTION. Considering throughout only the atoms of one component, we use the concentration λ of atoms at other sites and the concentration λ' of their own sites not occupied by these atoms; concentrations are defined with respect to the total number of all atoms in the crystal. Clearly

$$c - c_0 = \lambda - \lambda'. \tag{1}$$

We shall regard the crystal as a "solution" of atoms at other sites and of own sites not occupied by atoms, the "solvent" being represented by atoms at their "own" sites. The transition of atoms from "other" to their "own" sites can then be regarded as a "chemical reaction" between the "solutes" (with small concentrations λ and λ') to form the "solvent" (with concentration $\cong 1$). Applying to this "reaction" the law of mass action, we obtain $\lambda\lambda' = K$, where K depends only on P and T. For $c = c_0$ we must have $\lambda = \lambda' \equiv \lambda_0$; hence $K = \lambda_0^2$, and so

$$\lambda\lambda' = \lambda_0^2. \tag{2}$$

From (1) and (2) we find the required concentrations:

$$\lambda = \tfrac{1}{2}[(c-c_0)+\sqrt{\{(c-c_0)^2+4\lambda_0^2\}}],$$
$$\lambda' = \tfrac{1}{2}[-(c-c_0)+\sqrt{\{(c-c_0)^2+4\lambda_0^2\}}].$$

§ 143. The discontinuity of specific heat

The quantitative theory of phase transitions of the second kind is based on a consideration of the thermodynamic quantities of the body for given deviations from the symmetrical state (i. e. for given values of the order parameter η); we can represent the thermodynamic potential of the body, for example, as a function of P, T and η. Here it must of course be remembered that in the function $\Phi(P,T,\eta)$ the variable η is in one sense not on the same footing as the variables P and T; whereas the pressure and temperature can be specified arbitrarily, the value of η which actually occurs must itself be determined from the condition of thermal equilibrium, i.e. the condition that Φ is a minimum (for given P and T).

The continuity of the change of state in a phase transition of the second kind is expressed mathematically by the fact that the quantity η takes arbitrarily small values near the transition point. Considering the neighbourhood of this point. we expand $\Phi(P, T, \eta)$ in powers of η:

$$\Phi(P, T, \eta) = \Phi_0 + \alpha\eta + A\eta^2 + C\eta^3 + B\eta^4 + \dots, \tag{143.1}$$

where the coefficients $\alpha, A, B, C, \dots$ are functions of P and T.

It must emphasised, however, that writing Φ as the regular expansion (143.1) does not take account of the fact already mentioned that the transition point is a singularity of the thermodynamic potential; the same is true of the later expansion of the coefficients in (143.1) in powers of the temperature. This section and §§ 144–146 are concerned with a theory[†] based on the

---

[†] This theory is due to L. D. Landau (1937), who also first showed the general relation between phase transitions of the second kind and the change in the symmetry of the body.

validity of such expansions; the conditions for it to be applicable will be considered in § 146.

It can be shown (see § 144) that, if the states with $\eta = 0$ and $\eta \neq 0$ are of different symmetry (as we assume), the first-order term in the expansion (143.1) is identically zero: $\alpha \equiv 0$. The coefficient $A(P, T)$ in the second-order term is easily seen to vanish at the transition point, since in the symmetrical phase the value $\eta = 0$ must correspond to a minimum of Φ, and for this to be so it is evident that $A > 0$ is necessary, while on the other side of the transition point, in the unsymmetrical phase, non-zero values of η must correspond to the stable state (i. e. to the minimum of Φ), and this is possible only if $A < 0$; Fig. 62 shows the form of the function $\Phi(\eta)$ for $A < 0$ and $A > 0$. Since A is positive on one side of the transition point and negative on the other, it must vanish at the transition point itself.

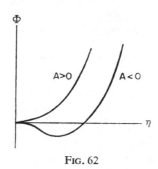

FIG. 62

But if the transition point itself is a stable state, i. e. if Φ as a function of η is a minimum at $\eta = 0$, it is necessary that the third-order term should be zero and the fourth-order term positive there:

$$A_c(P, T) = 0, \quad C_c(P, T) = 0, \quad B_c(P, T) > 0, \qquad (143.2)$$

where the suffix c refers to the transition point.

The coefficient B, being positive at the transition point, is of course also positive in the neighbourhood of that point.

Two cases can occur. In one, the third-order term is identically zero owing to the symmetry of the body: $C(P, T) \equiv 0$. Then there remains at the transition point only the one condition $A(P, T) = 0$, which determines P as a function of T or vice versa. Thus in the PT-plane there is a line of phase transition points of the second kind.[†]

[†] This condition, however, needs to be made more precise; see the sixth footnote to § 145.

If, however, C is not identically zero, the transition points are determined by the two equations $A(P,T) = 0, C(P,T) = 0$. In this case, therefore, the continuous phase transitions can occur only at isolated points.[†]

The most interesting case is, of course, that where there is a line of continuous-transition points. In what follows we shall take the discussion of phase transitions of the second kind to refer only to this case, which will now be considered. It includes, in particular, transitions due to the appearance or disappearance of a magnetic structure. This is because of the symmetry under time reversal. The thermodynamic potential of the body cannot be altered by this transformation, whereas the magnetic moment (which here acts as the order parameter) changes sign. It is therefore clear that in such cases the expansion of Φ contains no odd-order terms.

We shall therefore suppose that $C \equiv 0$ and the expansion of the thermodynamic potential has the form

$$\Phi(P, T, \eta) = \Phi_0(P, T) + A(P, T)\eta^2 + B(P, T)\eta^4. \tag{143.3}$$

Here $B > 0$, while the coefficient $A > 0$ in the symmetrical phase and $A < 0$ in the unsymmetrical phase; the transition points are determined by the equation $A(P, T) = 0$.

In the theory given here, it is assumed that the function $A(P, T)$ has no singularity at the transition point, so that it can be expanded near this point in integral powers of the "distance" from the transition:

$$A(P, T) = a(P)(T - T_c), \tag{143.4}$$

where $T_c = T_c(P)$ is the transition temperature. The coefficient $B(P, T)$ may be replaced by $B(P, T_c)$. The expansion of the thermodynamic potential therefore becomes

$$\Phi(P, T) = \Phi_0(P, T) + a(P)(T - T_c)\eta^2 + B(P)\eta^4, \tag{143.5}$$

with $B(P) > 0$.

The dependence of η on the temperature near the transition point, in the unsymmetrical phase, is determined from the condition for Φ to be a minimum as a function of η. Equating the derivative $\partial\Phi/\partial\eta$ to zero, we obtain $\eta(A + 2B\eta^2) = 0$, and hence

$$\eta^2 = -A/2B = a(T_c - T)/2B; \tag{143.6}$$

[†] It can be shown (L. Landau, *Zhurnal éksperimental'noĭ i teoreticheskoĭ fiziki* **7**, 627, 1937; translation in *Collected Papers of L. D. Landau*, p.209, Pergamon, Oxford, 1965) that there is always a third-order term in the expansion for the transition between an isotropic liquid and a solid crystal.

the solution $\eta = 0$ with $A < 0$ corresponds to a maximum of Φ, not a minimum. It should be noted that the configuration of the two phases on the temperature scale depends on the sign of a: for $a > 0$ and $a < 0$, the unsymmetrical phase corresponds to the temperatures $T < T_c$ and $T > T_c$ respectively. We shall assume below the particular case where the symmetrical phase is at $T > T_c$, as happens in the great majority of cases, and accordingly that $a > 0$.

Neglecting higher powers of η_i, we find for the entropy

$$S = -\partial\Phi/\partial T = S_0 - (\partial A/\partial T)\eta^2 ;$$

the term containing the temperature derivative of η is zero, because $\partial\Phi/\partial\eta = 0$. In the symmetrical phase, $\eta = 0$ and $S = S_0$; in the unsymmetrical phase,

$$S = S_0 + (a^2/2B)(T-T_c). \tag{143.7}$$

At the transition point itself, this expression becomes S_0, and the entropy is therefore continuous, as it should be.

Finally, let us determine the specific heats $C_p = T(\partial S/\partial T)_P$ of the two phases at the transition point. For the unsymmetrical phase we have, differentiating (143.7),

$$C_p = C_{p0} + a^2 T_c/2B. \tag{143.8}$$

For the symmetrical phase $S = S_0$, and therefore $C_p = C_{p0}$. Thus the specific heat is discontinuous at a phase transition point of the second kind. Since $B > 0$, $C_p > C_{p0}$ at the transition point, i.e. the specific heat increases in going from the symmetrical to the unsymmetrical phase, whatever their respective positions on the temperature scale.

Other quantities besides C_p are discontinuous: C_v, the thermal expansion coefficient, the compressibility, etc. There is no difficulty in deriving relations between the discontinuities of all these quantities. First of all we note that the volume and the entropy are continuous at the transition point, i.e. their discontinuities ΔV and ΔS are zero:

$$\Delta V = 0, \quad \Delta S = 0.$$

We differentiate these equations with respect to temperature along the curve of transition points, i.e. assuming the pressure to be the function of temperature given by this curve. The result is

$$\Delta(\partial V/\partial T)_P + (dP/dT)\Delta(\partial V/\partial P)_T = 0, \tag{143.9}$$
$$\Delta C_p/T - (dP/dT)\Delta(\partial V/\partial T)_P = 0, \tag{143.10}$$

since $(\partial S/\partial P)_T = -(\partial V/\partial T)_P$. These two equations relate the discontinuities of the specific heat C_p, the thermal expansion coefficient and the compressibility at a phase transition point of the second kind (W. H. Keesom and P. Ehrenfest, 1933).

Differentiating along the curve of transition points the equations $\Delta S = 0$ and $\Delta P = 0$ (the pressure is, of course, unchanged in the transition), but with temperature and volume as independent variables, we find

$$\Delta(\partial P/\partial T)_V + (dV/dT)\,\Delta(\partial P/\partial V)_T = 0, \qquad (143.11)$$

$$\Delta C_v/T + (dV/dT)\,\Delta(\partial P/\partial T)_V = 0. \qquad (143.12)$$

It may be noted that

$$\Delta C_p = T(dP/dT)^2\,\Delta(-\partial V/\partial P)_T, \qquad (143.13)$$

$$\Delta C_v = -T(dV/dT)^2\,\Delta(-\partial V/\partial P)_T^{-1}, \qquad (143.14)$$

so that the discontinuities of the specific heat and the compressibility have the same sign. From the previous statement about the discontinuity of the specific heat, it follows that the compressibility decreases discontinuously on going from the unsymmetrical to the symmetrical phase.

To conclude this section, let us return to the first part and consider the meaning of the function $\Phi(P, T, \eta)$.

The formal introduction of this function for any values of η does not, in general, require the possible existence of actual macroscopic states (partial equilibria) corresponding to such values. We must emphasise, however, that such states do in fact exist near a phase transition point of the second kind: as the transition point is approached, the minimum of Φ as a function of η becomes steadily flatter. This means that the "restoring force" that tends to bring the body to the state having the equilibrium value of η becomes steadily weaker, so that the relaxation time for the establishment of equilibrium with respect to the order parameter increases without limit (and certainly becomes much longer than the time for equalisation of pressure throughout the body).

PROBLEM

Find the relation between the discontinuities of specific heat and heat of solution in a transition of the second kind in a solution (I. M. Lifshitz, 1950).

SOLUTION. The heat of solution per molecule of solute is given by $q = \partial W/\partial n - w_0'$, where W is the heat function of the solution and w_0' the heat function per particle of the pure solute. Since w_0' is not affected by the phase transition in solution, we have for the discontinuity of q

$$\Delta q = \Delta(\partial W/\partial n) = \Delta\frac{\partial}{\partial n}\left(\Phi - T\frac{\partial\Phi}{\partial T}\right) = -T\,\Delta(\partial^2\Phi/\partial n\,\partial T),$$

where we have used the fact that the chemical potential $\mu' = \partial\Phi/\partial n$ is continuous at the transition. On the other hand, differentiation of the equation $\Delta(\partial\Phi/\partial T) = 0$

(continuity of entropy) along the curve of the transition temperature as a function of the concentration c at constant pressure gives

$$\frac{dT_c}{dc}\Delta\frac{\partial^2\Phi}{\partial T^2}+N\Delta\frac{\partial^2\Phi}{\partial n\,\partial T} = 0.$$

Hence we have the required relation

$$N\Delta q = (dT_c/dc)\,\Delta C_p.$$

We may note that in the derivation of this relation no assumption has been made concerning the concentration of the solution.

§ 144. Effect of an external field on a phase transition

Let us now consider how the properties of a phase transition change when the body is subjected to an external field whose action depends on the value of the parameter η. Without specifying the physical nature of this field, we may formulate in general terms the assumptions made about it. These amount to asserting that the application of the field is described by the appearance in the Hamiltonian of the body of a perturbing operator having the form

$$\hat{H}_h = -\hat{\eta}hV, \tag{144.1}$$

which is linear in the field "strength" h and in the operator $\hat{\eta}$ of the quantity η; V is the volume of the body.[†] If the thermodynamic potential is defined as a function of P, T and h, the mean (equilibrium) value of η is given by

$$V\bar{\eta} = -\partial\Phi(P, T, h)/\partial h, \tag{144.2}$$

according to the theorem of differentiation with respect to a parameter; cf. (11.4) and (15.11).

In order to ensure that this relation is satisfied in the Landau theory, we must add to the expansion (143.5) a term $-\eta hV$:

$$\Phi(P, T, \eta) = \Phi_0(P, T) + at\eta^2 + B\eta^4 - \eta hV, \tag{144.3}$$

with the notation $t = T-T_c(P)$.[‡]

[†] For example, in a ferromagnet near its Curie point (the transition to the paramagnetic phase), the parameter η is the macroscopic magnetic moment per unit volume, and the field h is the magnetic field; in a ferroelectric, η is the electric dipole moment per unit volume of the body, and h the electric field. In other cases, the field h need not have a direct physical significance, but its formal introduction helps towards a fuller understanding of the properties of the phase transition.

[‡] In the Landau theory, the equilibrium value of $\eta(P, T)$ is determined by minimising this expansion, i.e. by the condition $\partial\Phi(P, T, \eta)/\partial\eta = 0$. The relation (144.2) is then, of course, satisfied:

$$\frac{\partial\Phi}{\partial h} = \left(\frac{\partial\Phi}{\partial h}\right)_\eta + \left(\frac{\partial\Phi}{\partial\eta}\right)_h\frac{\partial\eta}{\partial h} = \left(\frac{\partial\Phi}{\partial h}\right)_\eta = -V\eta.$$

We may note first of all that any field, however weak, has the result that the parameter η becomes different from zero at every temperature. Thus the field reduces the symmetry of the more symmetrical phase, and so the difference between the two phases disappears. Accordingly, the discrete phase transition point also disappears; the transition is "smoothed out". In particular, instead of the sharp discontinuity of the specific heat, there is an anomaly spread over a range of temperatures. The order of magnitude of this range can be estimated from the condition $\eta h V \sim a t \eta^2$; taking η from (143.6), we then find

$$t \sim h^{2/3} B^{1/3} V^{2/3}/a.$$

For a quantitative investigation of the transition, we write the equilibrium condition $(\partial \Phi/\partial \eta)_{T, h} = 0$:[†]

$$2at\eta + 4B\eta^3 = hV. \qquad (144.4)$$

The dependence of η on the field h is different for temperatures above and below T_c. It will be recalled that we agreed to take $a > 0$, so that the temperatures $t > 0$ $(T > T_c)$ correspond to the symmetrical phase ($\eta = 0$ for $h = 0$).

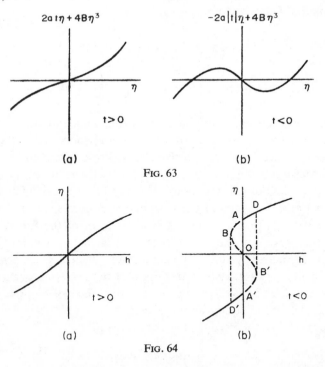

FIG. 63

FIG. 64

For $t > 0$, the left-hand side of equation (144.4) increases monotonically with η (Fig. 63a). The equation therefore has for any given value of h just one (real) root, which vanishes when $h = 0$. The function $\eta(h)$ is single-valued, and the sign of η is the same as that of h (Fig. 64a).

If $t < 0$, however, the left-hand side of (144.4) is not a monotonic function of η (Fig. 63b), and so the equation has three different real roots over a certain range of values of h, and $\eta(h)$ is no longer single-valued (Fig. 64b). This range evidently has limits given by the condition

$$\frac{\partial}{\partial \eta}(2at\eta + 4B\eta^3) = 2at + 12B\eta^2 = 0,$$

and is $-h_t < h < h_t$, where

$$h_t = \left(\frac{2}{3}\right)^{3/2} \frac{(a|t|)^{3/2}}{VB^{1/2}}. \tag{144.5}$$

It is easy to see, however, that the whole section of the curve BB' on which $(\partial \eta/\partial h)_T < 0$ corresponds to thermodynamically unstable states: differentiating equation (144.4) with respect to h, we find

$$\left(\frac{\partial \eta}{\partial h}\right)_T \left(\frac{\partial^2 \Phi}{\partial \eta^2}\right)_{T,h} = V; \tag{144.6}$$

hence we see that $(\partial^2 \Phi/\partial \eta^2)_{T,h} < 0$ when $(\partial \eta/\partial h)_T < 0$, i.e. Φ has a maximum here, not a minimum.

In the sections AB and $A'B'$, the thermodynamic potential is a minimum, but its value is greater than for the minima corresponding to the sections $A'D'$ and AD respectively, as is easily seen by direct calculation, although it is also obvious *a priori*: since the field h appears in Φ as the term $-\eta hV$, it is certainly more favourable thermodynamically for the sign of η to be the same as that of h. Thus the sections AB and $A'B'$ correspond to metastable states of the body, and the true equilibrium form of the function $\eta(h)$ is given by the continuous curve $DAA'D'$ in Fig. 64b, all points of which correspond to thermodynamically stable states. If at a given temperature $t < 0$ the field is varied, a phase transition of the first kind occurs when it passes through $h = 0$: at this point phases with values of $\eta = \pm(a|t|/2B)^{1/2}$ opposite in sign are in equilibrium together.

The susceptibility is determined as the derivative

$$\chi = (\partial \eta/\partial h)_{T; h \to 0}. \tag{144.7}$$

Differentiation of (144.4) gives

$$\partial \eta/\partial h = V/(2at + 12B\eta^2),$$

and substitution (as $h \to 0$) of $\eta^2 = 0$ for $t > 0$ and $\eta^2 = -at/2B$ for $t < 0$ leads to the result

$$\chi = V/2at \quad \text{for} \quad t > 0, \quad \chi = V/-4at \quad \text{for} \quad t < 0. \quad (144.8)$$

The infinite value of χ as $t \to 0$ is a natural consequence of the fact (already mentioned at the end of § 143) that the minimum of $\Phi(\eta)$ becomes steadily flatter as the transition point is approached. Because of this, even a slight perturbation has a great effect on the equilibrium value of η.

The quantity

$$h_t \sim (a|t|)^{3/2}/VB^{1/2}$$

gives the value of the field at which the field-induced parameter $\eta_{\text{ind}} \sim \chi h$ becomes of the same order of magnitude as the characteristic value of the spontaneous (zero-field) $\eta_{\text{sp}} \sim (a|t|/B)^{1/2}$. Fields $h \ll h_1$ are "weak" in the sense that they do not affect the thermodynamic quantities of the body in the first approximation. Fields $h \gg h_1$ are "strong" fields for which the thermodynamic quantities have values determined by the field in the first approximation; evidently, when $t = 0$, any field is strong in this sense.

In strong fields, the order parameter is

$$\eta = (hV/4B)^{1/3}. \quad (144.9)$$

It is also easily verified that in this limit the specific heat C_p is independent of the field.

§ 145. Change in symmetry in a phase transition of the second kind

In the theory given in the preceding sections we have considered a phase transition of the second kind with some definite change in symmetry of the body, assuming *a priori* that such a transition is possible. Such a theory, however, does not say whether a given change of symmetry can in fact occur by a transition of the second kind. The theory developed in the present section is designed to answer this question; it starts from a different statement of the problem, whereby a certain symmetry of the body at the transition point itself is specified, and we ask what symmetry is possible on either side of this point.

For definiteness, we shall speak of phase transitions involving a change in structure of the crystal lattice, i.e. a change in the symmetry of the configuration of atoms in it. Let $\varrho(x, y, z)$ be the density function (defined in § 128), which gives the probability distribution of various positions of the atoms in the crystal. The symmetry of the crystal lattice is the set or group of all transformations of the coordinates under which the function $\varrho(x, y, z)$ is invariant. Here we mean, of course, the complete symmetry of the lattice, including rotations, reflections and also the infinite (discrete) set of all possible parallel displacements (translations); that is, we are concerned with one of the 230 space groups.

Let G_0 be the symmetry group of the crystal at the transition point itself. As we know from group theory, an arbitrary function $\varrho(x, y, z)$ can be represented as a linear combination of several functions $\phi_1, \phi_2, \ldots$ having the property of being transformed into combinations of one another by all the transformations in the group concerned. In general the number of these functions is equal to the number of elements in the group, but when the function ϱ itself has a certain symmetry the functions ϕ_i may be fewer in number.

Bearing this in mind, we write the density function $\varrho(x, y, z)$ of the crystal as the sum

$$\varrho = \sum_i \eta_i \phi_i,$$

where the functions ϕ_i are transformed into combinations of one another by all transformations in the group G_0. The matrices of these transformations form a representation of the group G_0. The choice of the functions ϕ_i is not unique; they can obviously be replaced by any linear combinations of themselves. The functions ϕ_i can always be so chosen as to form a number of independent sets containing the minimum number of functions, the functions in each set being transformed only into combinations of one another by all transformations in the group G_0. The matrices of the transformations of the functions in each of these sets form irreducible representations of the group G_0, and the functions themselves are the basis of these representations. Thus we can write

$$\varrho = \sum_n \sum_i \eta_i^{(n)} \phi_i^{(n)}, \qquad (145.1)$$

n being the number of the irreducible representation and i the number of the function in its basis. In what follows we shall assume the functions $\phi_i^{(n)}$ to be normalised in some definite manner.

The functions $\phi_i^{(n)}$ always include one which is invariant under all the transformations in the group G_0 and gives what is called the unit representation of the group. Thus this function (which we denote by ϱ_0) has the symmetry of G_0. Denoting the remaining part of ϱ by $\delta\varrho$, we can write

$$\varrho = \varrho_0 + \delta\varrho, \qquad \delta\varrho = \sideset{}{'}\sum_n \sum_i \eta_i^{(n)} \phi_i^{(n)}, \qquad (145.2)$$

where now the unit representation is excluded from the summation (as indicated by the prime to the summation sign). The function $\delta\varrho$ has a lower symmetry than that of G_0, since $\delta\varrho$ may also remain invariant under some transformations in this group but certainly does not do so under all. We may note that the symmetry G of the function ϱ (which clearly is the symmetry of $\delta\varrho$) has, strictly speaking, been assumed from the start to be lower than G_0, since otherwise the sum (145.1) would include only one term, the function ϱ itself, which gives the unit representation.[†]

Since the physical quantity $\delta\varrho$ must be real and must remain real under all transformations, we must assume physically irreducible representations whose base functions can be taken as real (§ 135); accordingly the functions $\phi_i^{(n)}$ are taken to be real.

The thermodynamic potential Φ of a crystal whose density function ϱ is given by (145.2) is a function of temperature, pressure and the coefficients $\eta_i^{(n)}$ (and depends, of course, on the specific form of the functions $\phi_i^{(n)}$ themselves). The actual values of the $\eta_i^{(n)}$ as functions of P and T are determined thermodynamically from the conditions of equilibrium, i.e. the conditions for Φ to be a minimum. This determines also the symmetry G of the crystal, since it is clear that the symmetry of the function (145.2), with functions $\phi_i^{(n)}$ whose laws of transformation are known, is determined by the values of the coefficients in the linear combination of the $\phi_i^{(n)}$.

If the crystal is to have the symmetry G_0 at the transition point itself, it is necessary that all the $\eta_i^{(n)}$ should be zero there, i.e. $\delta\varrho = 0$, $\varrho = \varrho_0$. Since the change in state of the crystal in a phase transition of the second kind is continuous, $\delta\varrho$ must tend continuously to zero at the transition point, not discontinuously, i.e. the coefficients $\eta_i^{(n)}$ must tend to zero through arbitrarily small values near the transition point. Accordingly, we can expand the potential $\Phi(P, T, \eta_i^{(n)})$ in powers of the $\eta_i^{(n)}$ near the transition point.

First of all let us note that, since the functions $\phi_i^{(n)}$ (belonging to the basis of each irreducible representation) are transformed into combinations of one another by the transformations in the group G_0, these transformations can be regarded as transforming (in the same manner) the coefficients $\eta_i^{(n)}$ instead of the functions $\phi_i^{(n)}$. Next, since the thermodynamic potential of the body must obviously be independent of the choice of coordinates, it must be invariant under any transformation of the coordinate system, and in particular under the transformations of the group G_0. Thus the expansion of Φ in powers of the $\eta_i^{(n)}$ can contain in each term only an invariant combination of the $\eta_i^{(n)}$ that is of the appropriate power.

No linear invariant can be formed from quantities which are transformed according to a (non-unit) irreducible representation of a group, for other-

---

[†] For magnetic transitions, the density $\varrho(x, y, z)$ would have to be replaced by the current density $\mathbf{j}(x, y, z)$ in the body. In the paramagnetic phase $\mathbf{j} = 0$; beyond the transition point $\mathbf{j} = \delta\mathbf{j}$ is small.

wise that representation would contain the unit representation, i.e. would be reducible. Only one second-order invariant exists for each representation: a positive-definite quadratic form in the $\eta_i^{(n)}$, which can always be reduced to a sum of squares.

Thus the leading terms in the expansion of Φ are of the form

$$\Phi = \Phi_0 + \sum_n{}' A^{(n)} \sum_i [\eta_i^{(n)}]^2, \qquad (145.3)$$

where the $A^{(n)}$ are functions of P and T.

At the transition point itself, the crystal must have the symmetry G_0, i.e. the equilibrium values of the $\eta_i^{(n)}$ must be zero. It is evident that Φ can have a minimum when every $\eta_i^{(n)} = 0$ only if all the $A^{(n)}$ are non-negative.

If all the $A^{(n)}$ were positive at the transition point, they would also be positive near that point, so that the $\eta_i^{(n)}$ would remain zero and there would be no change of symmetry. For some $\eta_i^{(n)}$ to be non-zero, one of the coefficients $A^{(n)}$ must change sign, and this coefficient must therefore vanish at the transition point.[†] (Two coefficients $A^{(n)}$ can vanish simultaneously only at an isolated point in the PT-plane, which is the intersection of more than one line of transitions of the second kind.)

Thus on one side of the transition point all the $A^{(n)} > 0$, and on the other side one of the coefficients $A^{(n)}$ is negative. Accordingly, all the $\eta_i^{(n)}$ are always zero on one side of the transition point, and on the other side non-zero $\eta_i^{(n)}$ appear. We conclude, therefore, that on one side of the transition point the crystal has the higher symmetry G_0, which is retained at the transition point

[†] Strictly speaking, this condition should be more accurately stated as follows. The coefficients $A^{(n)}$ depend, of course, on the particular form of the functions $\phi_i^{(n)}$, being quadratic functionals of these which depend on P and T as parameters. On one side of the transition point, all these functionals $A^{(n)}\{\phi_i^{(n)}; P, T\}$ are positive-definite. The transition point is defined as that at which (as P or T varies gradually) one of the $A^{(n)}$ can vanish:

$$A^{(n)}\{\phi_i^{(n)}; P, T\} \geqslant 0.$$

This vanishing corresponds to a definite set of functions $\phi_i^{(n)}$, which may in principle be determined by solving the appropriate variational problem. These will also be the functions $\phi_i^{(n)}$ which determine the change $\delta\varrho$ at the transition point. Substituting these functions in $A^{(n)}\{\phi_i^{(n)}; P, T\}$, we obtain just the function $A^{(n)}(P, T)$ which satisfies the condition of vanishing at the transition point. The functions $\phi_i^{(n)}$ may then be regarded as given, as will be assumed below; the allowance for the variation of the $\phi_i^{(n)}$ with P and T would lead to correction terms of higher order than those of interest here.

itself, while on the other side of the transition point the symmetry is lower, and so the group G is a sub-group of the group G_0.

The change in sign of one of the $A^{(n)}$ causes the appearance of non-zero $\eta_i^{(n)}$ belonging to the nth representation. Thus the crystal with symmetry G_0 becomes one with density $\varrho = \varrho_0 + \delta\varrho$, where

$$\delta\varrho = \sum_i \eta_i^{(n)} \phi_i^{(n)} \tag{145.4}$$

is a linear combination of the base functions of any one of the irreducible representations of the group G_0 (other than the unit representation). Accordingly we shall henceforward omit the index n which gives the number of the representation, meaning always the one which corresponds to the transition considered.

We shall use the notation

$$\eta^2 = \sum_i \eta_i^2, \quad \eta_i = \eta\gamma_i, \tag{145.5}$$

(so that $\sum \gamma_i^2 = 1$) and write the expansion of Φ as

$$\Phi = \Phi_0(P, T) + \eta^2 A(P, T) + \eta^3 \sum_\alpha C_\alpha(P, T) f_\alpha^{(3)}(\gamma_i)$$
$$+ \eta^4 \sum_\alpha B_\alpha(P, T) f_\alpha^{(4)}(\gamma_i) + \ldots, \tag{145.6}$$

where $f_\alpha^{(3)}$, $f_\alpha^{(4)}$, ... are invariants of the third, fourth, etc., orders formed from the quantities γ_i; in the sums over α there are as many terms as there are independent invariants of the appropriate order which can be formed from the γ_i. In this expansion of the thermodynamic potential, the coefficient A must vanish at the transition point. In order that the transition point itself should be a stable state (i.e. in order that Φ should have a minimum at that point when $\eta_i = 0$), the third-order terms must vanish and the fourth-order terms must be positive-definite. As has been mentioned in § 143, a line of phase transitions of the second kind (in the PT-plane) can exist only if the third-order terms in the expansion of Φ vanish identically. This condition may now be formulated as requiring that it should be impossible to construct from the η_i third-order invariants which are transformed according to the corresponding irreducible representation of the group G_0.[†]

[†] In the language of the theory of representations, this signifies that the symmetric cube $[\Gamma^3]$ of the representation Γ in question must not contain the unit representation. For the (literally) irreducible representations of the space groups, there cannot be more than one third-order invariant (the proof is given by M. S. Shur, *Soviet Physics JETP* **24**, 845, 1967). When two representations are combined as one physically irreducible representation, two third-order invariants may occur.

Assuming this condition to be satisfied, we write the expansion as far as the fourth-order terms in the form

$$\Phi = \Phi_0 + A(P, T)\eta^2 + \eta^4 \sum_\alpha B_\alpha(P, T) f_\alpha^{(4)}(\gamma_i). \qquad (145.7)$$

Since the second-order term does not involve the γ_i, the latter are determined simply from the condition for a minimum of the fourth-order terms, i.e. of the coefficient of η^4 in (145.7).[†] Denoting the minimum value of this coefficient simply by $B(P, T)$ (which must be positive, as shown above), we return to the expansion of Φ in the form (143.3), η being determined from the condition that Φ is a minimum regarded as a function of η alone, as in § 143. The values of the γ_i thus found determine the symmetry of the function

$$\delta\varrho = \eta \sum_i \gamma_i \phi_i, \qquad (145.8)$$

i.e. the symmetry G of the crystal which is formed in the transition of the second kind from a crystal of symmetry G_0.[‡]

In this formalism, the set of quantities η_i acts as an order parameter describing the difference between the unsymmetrical and symmetrical phases. We see that in general this parameter has several components; the ratios $\gamma_i = \eta_i/\eta$ determine the symmetry of the unsymmetrical phase, and the common factor η gives a quantitative measure of the deviation from the specified symmetry.

The conditions derived above, however, are not yet sufficient to ensure the possibility of a phase transition of the second kind. A further essential condition is obtained if we consider a fact (hitherto deliberately ignored) relating to the classification properties of representations of space groups.[§]

[†] It may happen than there is only one fourth-order invariant, $(\Sigma\eta_i^2)^2 = \eta^4$. In this case, the fourth-order term is independent of the γ_i, and higher-order terms must be used to determine the γ_i; these terms depend on the γ_i. The use of higher-order terms may also be necessary when the minimisation of the fourth-order terms depending on the γ_i reduces them to zero.

[‡] In § 143 we have considered a transition with a given change of symmetry. Using the concepts defined here, we can say that the quantities γ_i were assumed to have given values (so that the function $\delta\varrho$ had a given symmetry). With the problem stated in these terms, the absence of the third-order term (in the expansion (143.3)) could not be a sufficient condition for the existence of a line of transition points of the second kind, since it does not exclude the possibility that there are third-order terms in the general expansion with respect to several η_i (if the irreducible representation is not one-dimensional). For example, if there are three η_i and the product $\gamma_1\gamma_2\gamma_3$ is invariant, the expansion of Φ contains a third-order term, whereas this term will vanish if the function $\delta\varrho$ has a certain symmetry which requires that one or two γ_i should be zero.

[§] The results and examples given below are due to E.M. Lifshitz (1941).

We have seen in § 134 that these representations can be classified not only by a discrete parameter (such as the number of the small representation) but also by the parameter $\mathbf{k}$, which takes a continuous series of values. The coefficients $A^{(n)}$ in the expansion (145.3) must therefore depend not only on the discrete number n but also on the continuous variable $\mathbf{k}$.

Let a phase transition correspond to the vanishing (as a function of P and T) of the coefficient $A^{(n)}(\mathbf{k})$ with a given number n and a given $\mathbf{k} = \mathbf{k}_0$. In order that the transition should actually occur, it is necessary that $A^{(n)}$ as a function of $\mathbf{k}$ should have a minimum for $\mathbf{k} = \mathbf{k}_0$ (and therefore for all vectors of the star of $\mathbf{k}_0$), i.e. the expansion of $A^{(n)}(\mathbf{k})$ in powers of $\mathbf{k} - \mathbf{k}_0$ about $\mathbf{k}_0$ should contain no linear terms. Otherwise, some coefficients $A^{(n)}(\mathbf{k})$ necessarily vanish before $A^{(n)}(\mathbf{k}_0)$ and a transition of the type in question cannot occur. A convenient formulation of this condition can be obtained on the basis of the following arguments.

The value of $\mathbf{k}_0$ determines the translational symmetry of the functions ϕ_i, and therefore that of the function $\delta\varrho$ (145.8), i.e. it determines the periodicity of the lattice of the new phase. This structure must be stable in comparison with those which correspond to values of $\mathbf{k}$ close to $\mathbf{k}_0$. But a structure with $\mathbf{k} = \mathbf{k}_0 + \varkappa$ (where $\varkappa$ is small) differs from that with $\mathbf{k} = \mathbf{k}_0$ by a spatial "modulation" in the periodicity of the latter, that is, by the appearance of non-uniformity over distances ($\sim 1/\varkappa$) which are large compared with the periods (cell dimensions) of the lattice. Such non-uniformity can be macroscopically described by regarding the parameters η_i as slowly varying functions of the coordinates (whereas the functions ϕ_i oscillate over interatomic distances). Thus we obtain the requirement that the state of the crystal should be stable with respect to loss of macroscopic homogeneity.[†]

When the quantities η_i are not constant in space, the thermodynamic potential per unit volume of the crystal will depend not only on the η_i but also on their derivatives with respect to the coordinates (in the first approximation, on the first derivatives). Accordingly Φ (for unit volume) must be expanded in powers of the η_i and of their gradients $\nabla\eta_i$ near the transition point. If the thermodynamic potential (of the whole crystal) is to be a minimum for constant η_i, it is necessary that the first-order terms in the gradients in this expansion should vanish identically. (The terms quadratic in the

---

[†] It must be emphasised, however, that these arguments assume transitions in which the symmetry of the less symmetrical phase is the same along the whole line of transition points, i.e. the value of $\mathbf{k}_0$ is independent of temperature. As well as this class of phase transitions (which alone is covered by the discussion in this section), there can also be transitions in which $\mathbf{k}_0$ is dependent on temperature and so the periodicity of the less symmetrical phase varies along the line of transition points. Such transitions will be considered elsewhere (see *Electrodynamics*) in connection with magnetic phase transitions.

derivatives must be positive-definite, but this imposes no restriction on the η_i, since such a quadratic form exists for η_i which are transformed by any of the irreducible representations.)

Among the terms linear in the derivatives, the only ones that can be of interest are those simply proportional to $\partial \eta_i/\partial x$, $\ldots$, and those containing the products $\eta_i \, \partial \eta_k/\partial x$, $\ldots$. The higher-order terms are clearly of no importance. The thermodynamic potential of the whole crystal, i.e. the integral $\int \Phi \, dV$ over the whole volume, is to be a minimum. The integration of all the total derivatives in Φ gives a constant which does not affect the determination of the minimum of the integral. We can therefore omit all terms in Φ which are simply proportional to derivatives of the η_i. Among the terms containing products $\eta_i \, \partial \eta_k/\partial x$, $\ldots$ we can omit all symmetrical combinations $\eta_k \, \partial \eta_i/\partial x + \eta_i \, \partial \eta_k/\partial x = \partial(\eta_i \eta_k)/\partial x$, $\ldots$, leaving the antisymmetrical parts

$$\eta_k \frac{\partial \eta_i}{\partial x} - \eta_i \frac{\partial \eta_k}{\partial x}, \ldots . \tag{145.9}$$

The expansion of Φ can contain only invariant linear combinations of the quantities (145.9). Hence the condition for a phase transition to be possible is that such invariants do not appear.

The components of the gradients $\nabla \eta_i$ are transformed as the products of the components of a vector and the quantities η_i. The differences (145.9) are therefore transformed as the products of the components of a vector and the antisymmetrised products of the quantities η_i. Consequently the requirement that no linear scalar can be formed from the quantities (145.9) is equivalent to the requirement that no combinations which transform as the components of a vector can be formed from the antisymmetrised products

$$\chi_{ik} = \phi_i \phi_k' - \phi_k \phi_i' ; \tag{145.10}$$

here the ϕ_i and ϕ_i' are the same base functions of the relevant irreducible representation, which we regard as taken at two different points x, y, z and x', y', z' in order that the difference shall not be identically zero.[†] Labelling the base functions by the two suffixes $k\alpha$ (as in § 134), we write the difference (145.10) in the form

$$\chi_{k\alpha, \, k'\beta} = \phi_{k\alpha} \phi_{k'\beta}' - \phi_{k\alpha}' \phi_{k'\beta}, \tag{145.11}$$

where $k, k', \ldots$ are vectors of the same star.

[†] In the language of the theory of representations, this means that the antisymmetric square $\{\Gamma^2\}$ of the representation Γ in question must not contain the irreducible representations by which the components of a vector are transformed.

Let the vector **k** occupy the most general position and have no proper symmetry. The star of **k** contains n vectors according to the number of rotational elements in the group (or $2n$ if the space group itself does not include inversion), each **k** being accompanied by the different vector $-\mathbf{k}$. The corresponding irreducible representation is given by the same number of functions $\phi_\mathbf{k}$ (one for each **k**, and so we omit the suffix α). The quantities

$$\chi_{\mathbf{k}, -\mathbf{k}} = \phi_\mathbf{k}\phi'_{-\mathbf{k}} - \phi'_\mathbf{k}\phi_{-\mathbf{k}} \tag{145.12}$$

are invariant under translations. Under the rotational elements, these n (or $2n$) quantities are transformed into combinations of one another, giving a representation of the corresponding point group (crystal class) with dimension equal to the order of the group. But this representation (called a *regular* representation) contains all the irreducible representations of the group, including those by which the components of a vector are transformed.

Similar considerations show that it is possible to form a vector from the quantities $\chi_{\mathbf{k}\alpha, -\mathbf{k}\beta}$ in cases where the group of the vector **k** contains one axis and planes of symmetry passing through that axis.

This discussion becomes inapplicable, however, if the group of the vector **k** contains axes which intersect one another or intersect planes of symmetry, or contains inversion; such groups will be said to have a central point. In such cases the question of constructing a vector from the quantities (145.11) requires separate treatment in each particular case. In particular, such a vector certainly can not be constructed if the group of **k** contains inversion, so that **k** and $-\mathbf{k}$ are equivalent, and only one function $\phi_\mathbf{k}$ corresponds to each **k** in the star; in this case there are no $\chi_{\mathbf{kk'}}$ invariant under translations (as the components of a vector must necessarily be).

Thus the requirement formulated above greatly restricts the possible changes of symmetry in a phase transition of the second kind. Of the infinity of different irreducible representations of the group G_0, we need consider only a comparatively small number for which the group of the vector **k** has a central point.

A proper symmetry of this kind can, of course, occur only for vectors **k** which occupy certain exceptional positions in the reciprocal lattice, their components being equal to certain simple fractions $\left(\frac{1}{2}, \frac{1}{3}, \frac{1}{4}\right)$ of the basic vectors of that lattice. This means that the change in the translational symmetry of the crystal (i.e. in its Bravais lattice) in a phase transition of the second kind must consist in an increase by a small factor in some of the basic lattice vectors. Investigation shows that in the majority of cases the only possible change in the Bravais lattice is a doubling of the lattice vectors. In addition, in body-centred orthorhombic, tetragonal and cubic and face-centred cubic lattices some lattice vectors can be quadrupled, and in a hexagonal lattice tripled. The volume of the unit cell can be increased by a factor of 2, 4 or 8, in a face-centred cubic lattice also by 16 or 32, and in a hexagonal lattice by 3 or 6.

Transitions are, of course, also possible without change of Bravais lattice (corresponding to irreducible representations with $\mathbf{k} = 0$). The change in symmetry then consists in a decrease in the number of rotational elements, i.e. a change in the crystal class.

We may note the following general theorem. A phase transition of the second kind can occur for any change in structure which halves the number of symmetry transformations; such a change may occur either by a doubling of the unit cell for a given crystal class or by a halving of the number of rotations and reflections for a given unit cell. The proof is based on the fact that, if the group G_0 has a sub-group G of half the order, then the irreducible representations of G_0 always include a one-dimensional representation given by a function which is invariant under all transformations of the sub-group G and changes sign under all the remaining transformations of the group G_0. It is clear that in this case there are no odd-order invariants, and no quantities of the type (145.11) can be formed from one function.

The following theorem also appears to be valid. Phase transitions of the second kind cannot occur for changes in structure which reduce to one-third the number of symmetry transformations, owing to the presence of third-order terms in the expansion of Φ.

Finally, to illustrate the practical applications of the general theory given above, let us consider the occurrence of ordering in alloys which, in the disordered state, have a body-centred cubic lattice with atoms at the vertices and centres of cubic cells, as in Fig. 61b (§ 142).[†] The problem is to determine the possible types of ordering (called *superlattices* in crystallography) which can appear in such a lattice in a phase transition of the second kind.

For a body-centred cubic lattice, the reciprocal lattice is face-centred cubic, If the edge of the body-centred cubic lattice cell is taken as the unit of length, the edge length of the cubic cell in the reciprocal lattice is $2 \cdot 2\pi$, and in this lattice the following vectors $\mathbf{k}$ have proper symmetry groups with a central point:

$$
\begin{array}{lll}
\text{(a)} & (000) & O_h \\[4pt]
\text{(b)} & (\tfrac{1}{2}\tfrac{1}{2}\tfrac{1}{2}) & O_h \\[4pt]
\text{(c)} & (\tfrac{1}{4}\tfrac{1}{4}\tfrac{1}{4}), \quad (-\tfrac{1}{4} -\tfrac{1}{4} -\tfrac{1}{4}) & T_d \\[4pt]
\text{(d)} & (0\tfrac{1}{4}\tfrac{1}{4}), \quad (\tfrac{1}{4}0\tfrac{1}{4}), \quad (\tfrac{1}{4}\tfrac{1}{4}0), & \\[4pt]
& (0\tfrac{1}{4} -\tfrac{1}{4}), \quad (-\tfrac{1}{4}0\tfrac{1}{4}), \quad (\tfrac{1}{4} -\tfrac{1}{4}0) & D_{2h}
\end{array}
\qquad (145.13)
$$

[†] Such a lattice belongs to the symmorphic space group O_h^9.

These symbols show the components of the vectors **k** along the edges of the cubic reciprocal lattice cell (x, y, z axes) as fractions of the edge lengths. In order to obtain the vectors **k** in the units specified above, these numbers must be multiplied by $2 \cdot 2\pi = 4\pi$. In (145.13) only non-equivalent vectors are shown, i.e. the vectors of each star.

The subsequent discussion is greatly simplified by the fact that not all small representations need be considered in solving the problem in question. The reason is that we are concerned only with the possible changes of symmetry that can occur by the formation of a superlattice, that is, by an ordered arrangement of atoms at existing lattice sites without relative displacement. In this case the unit cell of the disordered lattice contains only one atom. Hence the appearance of the superlattice can only mean that the lattice points in different cells become non-equivalent. The change $\delta\varrho$ in the density distribution function must therefore be invariant under all rotational transformations of the group of **k** (without simultaneous translation). Thus only the unit small representation is admissible, and accordingly u_α may be replaced by unity in the base functions (134.3).

Let us now consider in turn the stars listed in (145.13).

(a) The function with **k** $= 0$ has complete translational invariance, i.e. in this case the unit cell is unchanged, and since each cell contains only one atom no change of symmetry can occur.

(b) The function $e^{2\pi i(x+y+z)}$ corresponds to this **k**. The linear combination (of this function and the functions obtained from it by all rotations and reflections) which has the symmetry O_h of the group of **k** is

$$\phi = \cos 2\pi x \cos 2\pi y \cos 2\pi z. \qquad (145.14)$$

The symmetry of the phase formed is that of the density function $\varrho = \varrho_0 + \delta\varrho$, $\delta\varrho = \eta\phi$.[†] The function ϕ is invariant under all transformations of the class O_h and under translations along any edge of the cubic cell, but not under a translation through half the space diagonal, $\left(\frac{1}{2}\,\frac{1}{2}\,\frac{1}{2}\right)$. Hence the ordered phase has a simple cubic Bravais lattice with two non-equivalent points in the unit cell, (000) and $\left(\frac{1}{2}\,\frac{1}{2}\,\frac{1}{2}\right)$; these will be occupied by different atoms. The alloys which can be completely ordered in this way have the composition AB (e.g. the alloy CuZn mentioned in § 142).

(c) The functions corresponding to these **k** which have the symmetry T_d are

$$\begin{aligned}
\phi_1 &= \cos \pi x \cos \pi y \cos \pi z, \\
\phi_2 &= \sin \pi x \sin \pi y \sin \pi z.
\end{aligned} \qquad (145.15)$$

---

† This does not mean, of course, that the change $\delta\varrho$ in an actual crystal is given by the function (145.14). Only the symmetry of the expression (145.14) is important.

From these we can form two fourth-order invariants: $(\phi_1^2+\phi_2^2)^2$ and $(\phi_1^4+\phi_2^4)$. The expansion of Φ (145.7) therefore has the form

$$\Phi = \Phi_0 + A\eta^2 + B_1\eta^4 + B_2\eta^4(\gamma_1^4+\gamma_2^4). \tag{145.16}$$

Here two cases must be distinguished. Let $B_2 < 0$; then Φ as a function of γ_1 and γ_2, with the added condition $\gamma_1^2+\gamma_2^2 = 1$, has a minimum for $\gamma_1 = 1$, $\gamma_2 = 0$. The function $\delta\varrho = \eta\phi_1$ has the symmetry of the class O_h with a face-centred Bravais lattice, whose cubic cell has a volume 8 times that of the original cubic lattice cell. The unit cell contains 4 atoms; the cubic cell, 16 atoms. By placing like atoms at equivalent lattice sites we find that this superlattice corresponds to a ternary alloy of composition ABC_2 with atoms in the following positions:

4A (000), $\left(0\,\tfrac{1}{2}\,\tfrac{1}{2}\right)$ & cyclic,

4B $\left(\tfrac{1}{2}\,\tfrac{1}{2}\,\tfrac{1}{2}\right)$, $\left(0\,0\,\tfrac{1}{2}\right)$ & cyclic,

8C $\left(\tfrac{1}{4}\,\tfrac{1}{4}\,\tfrac{1}{4}\right)$, $\left(\tfrac{3}{4}\,\tfrac{3}{4}\,\tfrac{3}{4}\right)$, $\left(\tfrac{1}{4}\,\tfrac{3}{4}\,\tfrac{3}{4}\right)$ & cyclic, $\left(\tfrac{1}{4}\,\tfrac{1}{4}\,\tfrac{3}{4}\right)$ & cyclic;

here the coordinates of the atoms are given in units of the edge length of the new cubic lattice cell, which is twice that of the original cell (see Fig. 65a); "& cyclic" denotes cyclic interchange. If the B and C atoms are identical we obtain an ordered lattice of composition AB_3.

○ A
● B
◎ C

(a)   (b)

Fig. 65

Now let $B_2 > 0$. Then Φ has a minimum at $\gamma_1^2 = \gamma_2^2 = \tfrac{1}{2}$, so that $\delta\varrho = \eta(\phi_1+\phi_2)/\sqrt{2}$ (or $\eta(\phi_1-\phi_2)/\sqrt{2}$, which leads to the same result).[†] This function has the symmetry of the class O_h with the same face-centred Bravais lattice as in the preceding case but only two sets of equivalent points, which can be occupied by atoms of two kinds A and B:

8A (000), $\left(\tfrac{1}{4}\,\tfrac{1}{4}\,\tfrac{1}{4}\right)$, $\left(\tfrac{1}{4}\,\tfrac{3}{4}\,\tfrac{3}{4}\right)$ & cyclic, $\left(0\,\tfrac{1}{2}\,\tfrac{1}{2}\right)$ & cyclic,

8B $\left(\tfrac{1}{2}\,\tfrac{1}{2}\,\tfrac{1}{2}\right)$, $\left(\tfrac{3}{4}\,\tfrac{3}{4}\,\tfrac{3}{4}\right)$, $\left(\tfrac{1}{4}\,\tfrac{1}{4}\,\tfrac{3}{4}\right)$ & cyclic, $\left(0\,0\,\tfrac{1}{2}\right)$ & cyclic

(see Fig. 65b).

[†] The fact that γ_1 and γ_2 are in each case simply numbers is due to the presence in Φ of only one term (which depends on γ_1 and γ_2). When there are more fourth-order invariants, the sets of γ_i that minimise Φ may include some that depend on P and T.

(d) The following functions with the required symmetry D_{2h} correspond to these vectors $\mathbf{k}$:

$$\phi_1 = \cos \pi(y-z), \quad \phi_3 = \cos \pi(x-y), \quad \phi_5 = \cos \pi(x-z),$$
$$\phi_2 = \cos \pi(y+z), \quad \phi_4 = \cos \pi(x+y), \quad \phi_6 = \cos \pi(x+z).$$

From these we can form one third-order invariant and four fourth-order invariants, and so the expansion (145.6) becomes

$$\begin{aligned}
\Phi = \;&\Phi_0 + A\eta^2 + C\eta^3(\gamma_1\gamma_3\gamma_5 + \gamma_2\gamma_3\gamma_6 + \gamma_1\gamma_4\gamma_6 + \gamma_2\gamma_4\gamma_5) \\
&+ B_1\eta^4 + B_2\eta^4(\gamma_1^4 + \gamma_2^4 + \gamma_3^4 + \gamma_4^4 + \gamma_5^4 + \gamma_6^4) \\
&+ B_3\eta^4(\gamma_1^2\gamma_2^2 + \gamma_3^2\gamma_4^2 + \gamma_5^2\gamma_6^2) \\
&+ B_4\eta^4(\gamma_1\gamma_2\gamma_3\gamma_4 + \gamma_3\gamma_4\gamma_5\gamma_6 + \gamma_1\gamma_2\gamma_5\gamma_6).
\end{aligned}$$

Because cubic terms are present, a phase transition of the second kind is impossible in this case. To examine whether isolated points of continuous transition can exist and the properties of such points (see § 150) it would be necessary to investigate the behaviour of the function Φ near its minimum; we shall not pause to do so here.

The above example shows what rigid limitations are imposed by the thermodynamic theory on the possibility of phase transitions of the second kind; for example, in this case they can exist only when superlattices of three types are formed.

The following fact may also be pointed out. In case (c), when $B_2 < 0$, the actual change in the density function, $\delta\varrho = \eta\phi_1$, corresponds to only one of the two parameters γ_1, γ_2 which appear in the thermodynamic potential (145.16). This illustrates an important feature of the foregoing theory: in considering a particular change in the lattice in a phase transition of the second kind, it may be necessary to take account of other, "virtually possible", changes.

§ 146. Fluctuations of the order parameter

It has already been mentioned several times that the actual point of a phase transition of the second kind is a singularity of the thermodynamic functions for the body. The physical nature of this singularity consists in an anomalous increase in the fluctuations of the order parameter, which in turn is due to the already described flatness of the thermodynamic potential minimum near the transition point. The form of this increase in the Landau model is easily found. We shall assume that the change of symmetry in the transition is described by only one parameter η.

The minimum work needed to bring the system out of equilibrium for given constant values of the pressure and temperature is equal to the change $\Delta\Phi_t$ in the thermodynamic potential.[†] The fluctuation probability for constant P and T is therefore

$$w \propto \exp\left(-\Delta\Phi_t/T\right). \qquad (146.1)$$

In this section the equilibrium value of η will be denoted by $\bar{\eta}$. For a small departure from equilibrium,

$$\Delta\Phi_t = \tfrac{1}{2}(\eta-\bar{\eta})^2\,(\partial^2\Phi_t/\partial\eta^2)_{P,\,T}.$$

By means of (144.6) we can express the derivative $\partial^2\Phi_t/\partial\eta^2$ in terms of the susceptibility of the substance in a weak field, according to the definition (144.7). Then the fluctuation probability (at temperatures near the transition point T_c) is

$$w \propto \exp\left[-(\eta-\bar{\eta})^2\,V/2\chi T_c\right].$$

Hence the mean square fluctuation is

$$\langle(\Delta\eta)^2\rangle = T_c\chi/V. \qquad (146.2)$$

This expression can also be derived directly from the fluctuation–dissipation theorem. To do so, we need only note that, if the field h is identified with the external interaction f (with frequency $\omega = 0$) which appears in the formulation of the theorem (§ 124), then the corresponding quantity x is $\Delta\eta V$, and the generalised susceptibility $\alpha(0)$ is the product χV. Formula (146.2) follows from (124.14).

According to (144.8), the mean square (146.2) increases as $1/t$ when $T \to T_c$. For a fuller elucidation of the nature and significance of this divergence, let us determine the spatial correlation function of the fluctuations of the order parameter. We shall be concerned with long-wavelength fluctuations in which the fluctuating quantity varies slowly through the body; it is these fluctuations which, as we shall see later, increase anomalously near the transition point.

For an inhomogeneous body (such as it becomes when the inhomogeneous fluctuations are taken into account), the thermodynamic potential would have to be expressed as the integral $\Phi_t = \int\Phi\,dV$ of the potential density, a function of the coordinates of a point in the body. However, in the description of the thermodynamic state by the potential Φ, the number N of particles in the body is given, not its volume (which depends on P and T).

[†] In this section the thermodynamic potential (Φ, and later Ω) for the body as a whole is denoted by the suffix t; letters without suffixes are used for the potential values per unit volume.

It is therefore useful to change to a description by another potential which relates to some chosen and specified volume V in the medium, containing a variable number N of particles. This can be $\Omega_t(T, \mu)$, a function of temperature and chemical potential μ (for given V); the variable P is here replaced by μ, which has analogous properties; like P, it is constant throughout a system in equilibrium.

Near the transition point, the η-dependent terms in the expansion of the function $\Phi(P, T, \eta)$ (144.3) constitute a small increment to $\Phi_0(P, T)$, and after η has been determined by minimisation the remaining terms are all of the same order of magnitude. According to the theorem of small increments, we can therefore immediately write down a similar expansion for the potential $\Omega(\mu, T, \eta)$:

$$\Omega(\mu, T, \eta) = \Omega_0(\mu, T) + \alpha t \eta^2 + b \eta^4 - \eta h, \tag{146.3}$$

with the same coefficients but expressed in terms of a different variable, μ instead of T; the potential Ω is here taken per unit volume, and so the coefficients are $\alpha = a/V$, $b = B/V$.[†]

The expansion (146.3) relates to a homogeneous medium. In an inhomogeneous body, it includes not only different powers of η itself but also derivatives of various orders with respect to the coordinates. For long-wavelength fluctuations, we need take only the terms in the expansion that contain derivatives of the lowest order (and the lowest powers of these). The terms linear in the first-order derivatives, of the form $f(\eta)\, \partial\eta/\partial x_i$, give integrals over the surface on integration through the volume, which represent a surface effect unimportant here.[‡] The same is true of terms proportional to $\partial^2\eta/\partial x_i\, \partial x_k$. Thus the first terms to be taken into account in the expansion of Ω in derivatives are those proportional to $\eta\, \partial^2\eta/\partial x_i\, \partial x_k$ or $(\partial\eta/\partial x_i)(\partial\eta/\partial x_k)$. The former reduce to the latter on integration over the volume. We find ultimately that the function Ω written above has to be supplemented by terms of the form

$$g_{ik}(\mu, T)\frac{\partial\eta}{\partial x_i}\frac{\partial\eta}{\partial x_k}; \tag{146.4}$$

as always, summation over repeated vector suffixes is implied. We shall take only the simplest case (corresponding to cubic symmetry with $\eta = 0$), when

---

[†] Here, however, it has to be remembered that the expansion of the coefficient $A \cong \alpha t$ must now be made in powers of the difference $t = T - T_c(\mu)$, not $T - T_c(P)$; in this sense, the value of the coefficient $\alpha = a/V$ changes.

[‡] There are no first-order terms in the first derivatives in the expansion of Ω, even when the transition is described by several order parameters. In such cases, the proposition can be verified only by using also the conditions for the stability of the body at the transition point (§ 145).

$g_{ik} = g\delta_{ik}$; this case already shows all the characteristic features of the correlation function. Thus we write the thermodynamic potential density as

$$\Omega = \Omega_0 + \alpha t \eta^2 + b\eta^4 + g(\partial\eta/\partial\mathbf{r})^2 - \eta h. \tag{146.5}$$

It is evident that, for stability of a homogeneous body, $g > 0$; otherwise, Ω_t cannot have a minimum when $\eta = $ constant.

Considering fluctuations for given μ and T, we must write their probability as

$$w \propto \exp(-\Delta\Omega_t/T),$$

since the minimum work needed in these conditions to bring the system out of equilibrium is $R_{\min} = -\Delta\Omega_t$. Specifying the value of η in a chosen volume V does not prevent the exchange of particles (or energy) between this volume and the surrounding "medium". We can therefore consider the fluctuations of η for constant μ (and T); cf. the beginning of § 115.

Let us take the particular case of fluctuations in the symmetrical phase (in the absence of a field h); then $\bar{\eta} = 0$, so that $\Delta\eta = \eta$. As far as the second-order terms, the change in the potential Ω_t is[†]

$$\Delta\Omega_t = \int[\alpha t(\Delta\eta)^2 + g(\partial\,\Delta\eta/\partial\mathbf{r})^2]\,dV. \tag{146.6}$$

We now proceed as in § 116, expanding the fluctuation $\Delta\eta(\mathbf{r})$ as a Fourier series in the volume V:

$$\Delta\eta = \sum_{\mathbf{k}}\Delta\eta_{\mathbf{k}}e^{i\mathbf{k}\cdot\mathbf{r}}, \quad \Delta\eta_{-\mathbf{k}} = \Delta\eta_{\mathbf{k}}^*; \tag{146.7}$$

the gradient is

$$\partial\,\Delta\eta/\partial\mathbf{r} = \sum_{\mathbf{k}}i\mathbf{k}\,\Delta\eta_{\mathbf{k}}e^{i\mathbf{k}\cdot\mathbf{r}}.$$

On substitution of these expressions in (146.6), the integration over the volume causes all terms to vanish except those which contain the products $\Delta\eta_{\mathbf{k}}\Delta\eta_{-\mathbf{k}} = |\Delta\eta_{\mathbf{k}}|^2$. The result is

$$\Delta\Omega_t = V\sum_{\mathbf{k}}(gk^2 + \alpha t)|\Delta\eta_{\mathbf{k}}|^2,$$

and hence

$$\langle|\Delta\eta_{\mathbf{k}}|^2\rangle = T/2V(gk^2 + \alpha t); \tag{146.8}$$

[†] The theory of fluctuations based on an expression of this type was first developed by L. S. Ornstein and F. Zernicke (1917) for fluctuations near the critical point.

cf. the derivation of (116.12) from (116.10). We see that as $t \to 0$ it is in fact the long-wavelength fluctuations with $k \sim \sqrt{(\alpha t/g)}$ which increase.[†] It must be emphasised that the formula (146.8) itself is valid only for sufficiently long wavelengths $1/k$, which must always be large compared with interatomic distances.

The required correlation function will be denoted by

$$G(\mathbf{r}) = \langle \Delta\eta(\mathbf{r}_1)\,\Delta\eta(\mathbf{r}_2)\rangle, \quad \mathbf{r} = \mathbf{r}_1 - \mathbf{r}_2, \tag{146.9}$$

and is calculated as the sum

$$G(\mathbf{r}) = \sum_{\mathbf{k}} \langle |\Delta\eta_{\mathbf{k}}|^2\rangle e^{i\mathbf{k}\cdot\mathbf{r}}$$

or, changing to integration in **k**-space,

$$G(\mathbf{r}) = \int \langle |\Delta\eta_{\mathbf{k}}|^2\rangle e^{i\mathbf{k}\cdot\mathbf{r}} V\,d^3k/(2\pi)^3. \tag{146.10}$$

Using the Fourier transformation formula given in the footnote to § 117, Problem 3, we find (for $r \neq 0$)

$$G(r) = (T_c/8\pi g r)\exp(-r/r_c), \tag{146.11}$$

where

$$r_c = \sqrt{(g/\alpha t)} \tag{146.12}$$

is called the *correlation radius* of the fluctuations; it determines the order of magnitude of distances at which the correlation decreases significantly. As the transition point is approached, the correlation radius increases as $1/\sqrt{t}$, and at the transition point itself the correlation function decreases as $1/r$.

When $r = 0$, the integral (146.10) determines the mean square fluctuation of the parameter η in an infinitesimal volume element; it diverges for large k. This divergence, however, is simply due to the invalidity, in this region, of the expression (146.8) (which relates to long-wavelength fluctuations), and signifies only that $\langle(\Delta\eta)^2\rangle$ contains a term independent of t.

It must be emphasised, to avoid misunderstanding, that the previous expression (146.2) determines the fluctuations of the parameter η averaged

---

[†] Analogous results are, of course, obtained on the other side of the transition point, in the unsymmetrical phase. Here $\bar{\eta} = (-\alpha t/2b)^{1/2}$, and the change in the potential Ω_t (again as far as quantities $\sim (\Delta\eta)^2$) is

$$\Delta\Omega_t = \int [-2\alpha t(\Delta\eta)^2 + g(\partial\,\Delta\eta/\partial\mathbf{r})^2]\,dV,$$

instead of (146.6). It is clear that the results obtained for $\langle|\Delta\eta_{\mathbf{k}}|^2\rangle$ (and for the correlation function below) will differ from the above only in that αt is replaced by $2\alpha|t|$.

over a volume V whose linear dimension $l \gg r_c$; this quantity may be denoted by $\langle (\Delta \eta)^2 \rangle_V$. The mean value of the function $\Delta \eta (\mathbf{r})$ over the volume V is just the Fourier component $\Delta \eta_{\mathbf{k}=0}$; it is therefore natural that the expression (146.8) for $k = 0$ coincides with (146.2). The latter can also be derived from the correlation function by the obvious formula

$$\langle (\Delta \eta)^2 \rangle_V = \frac{1}{V^2} \int \langle \Delta \eta (\mathbf{r}_1) \, \Delta \eta (\mathbf{r}_2) \rangle \, dV_1 \, dV_2$$

$$= \frac{1}{V} \int G(\mathbf{r}) \, dV, \tag{146.13}$$

which is valid for any finite volume V. We may note that at the point $t = 0$ itself (where $G \propto 1/r$) this integral is proportional to $1/l$, where l is the linear dimension of the region in which the fluctuations are considered. The mean square $\langle (\Delta \eta)^2 \rangle_V$ depends on the shape of the region as well as on its volume.

We can now formulate the condition that determines the validity of this theory of fluctuations based on the expansion (146.5). This condition must be that the mean square fluctuation of the parameter η averaged over the correlation volume is small compared with the characteristic value $\bar{\eta^2} \sim \alpha |t|/b$. This quantity is given by (146.2) when $V \sim r_c^3$, and we arrive at the condition

$$T_c \chi / r_c^3 \ll \alpha |t|/b, \tag{146.14}$$

or (with χ and r_c from (144.8) and (146.12))

$$\alpha |t| \gg T_c^2 b^2 / g^3 \tag{146.15}$$

(A. P. Levanyuk, 1959; V. L. Ginzburg, 1960).[†]

The determination of the temperature dependences in the formulae derived above has also required expansions in powers of $t = T - T_c$ (in the coefficients of the expansion with respect to η). The admissibility of such an expansion implies that the condition $t \ll T_c$ is satisfied; for it to be compatible with (146.16), we must certainly have

$$T_c b^2 / \alpha g^3 \ll 1. \tag{146.16}$$

The conditions (146.14)–(146.16) ensure that the fluctuations are sufficiently small, and also that the whole Landau theory of phase transitions given in the preceding sections is applicable. We see that a temperature range in which this theory is valid exists only if the inequality (146.16) is satisfied. In such cases, the conclusions of the theory remain valid as regards the

---

[†] This condition can also be verified by a direct calculation of the fluctuation correction to the specific heat of the body near the transition point; see § 147, Problem.

selection rules for the possible changes of symmetry in transitions.[†] As regards the temperature dependence of the thermodynamic quantities, however, there is necessarily a narrow range of temperatures near T_c where the Landau theory is inapplicable. The conclusions of this theory must therefore refer only to states of the two phases outside that temperature range. For example, the expressions derived in § 143 for the discontinuities of the thermodynamic quantities are to be regarded as the differences of their values at the two ends of the range. We shall call the immediate neighbourhood of the point T_c, corresponding to the opposite of the inequality (146.15), the *fluctuation range*; in it, the fluctuations play the dominant role.

In the above calculations, no account has been taken of the elastic properties that distinguish a solid from a liquid,[‡] or the deformation of a body in consequence of ordering (which we shall call *striction*). Within the Landau theory, these effects do not alter the conclusions reached in the preceding sections. The combined action of the two effects may, however, considerably alter the fluctuations of the order parameter, and therefore the nature of the phase transition. The investigation of this problem demands an extensive use of elasticity theory, and is therefore outside the scope of the present volume. Here, we shall simply state some results.

The striction deformation may be either linear or quadratic in the order parameter, depending on the symmetry of the crystal. These two cases show a different influence of the elastic properties of the body on the phase transition.

For linear striction, let γ denote the order of magnitude of the proportionality coefficients between the components u_{ik} of the deformation tensor and the order parameter: $u_{ik} \sim \gamma\eta$. The influence of this effect on the fluctuations appears in the neighbourhood of the transition point for which $\alpha t \lesssim \gamma^2/\lambda$, where λ is the order of magnitude of the elastic moduli of the body. In many cases, striction is a weak effect, and in this sense γ is a small quantity. Then the temperature range mentioned is narrow and lies within the fluctuation range.

The long-wavelength fluctuations $(k \lesssim \sqrt{(\gamma^2/\lambda g)})$ are then suppressed, and the correlation radius does not increase beyond values $r_c \sim \sqrt{(g\lambda/\gamma^2)}$. The specific heat therefore has only a finite discontinuity at the transition point, as in the Landau theory.[§]

[†] For transitions described by more than one order parameter, the establishment of all the conditions for the Landau theory to be valid requires more detailed analysis.

[‡] What is important here is not so much the actual anisotropy of these properties but rather the fact that the deformations cannot be reduced to a single hydrostatic compression. In this sense, the following discussion would also apply to an isotropic solid with a non-zero shear modulus.

[§] See A. P. Levanyuk and A. A. Sobyanin, *JETP Letters* **11**, 371, 1970.

Quadratic striction leads to different results.[†] This effect likewise suppresses the fluctuations, but to a smaller extent. Whereas the specific heat would become infinite at the transition point without allowance for striction (see § 148), the quadratic striction leads instead to a small discontinuity of the entropy, i.e. the phase transition is of the first kind and close to the second kind; the specific heat remains finite, although it has anomalously high values.[‡]

PROBLEM

Determine the correlation radius of the fluctuations of the order parameter in an external field h at $T = T_c$.

SOLUTION. The equilibrium value $\bar{\eta}$ is given by (144.9) and the thermodynamic potential density is

$$\Omega = \Omega_0 + b\eta^4 + g(\partial\eta/\partial \mathbf{r})^2 - h\eta$$

$$= \bar{\Omega} + \frac{3b^{1/3}h^{2/3}}{2^{1/3}}(\eta-\bar{\eta})^2 + g(\partial\eta/\partial \mathbf{r})^2.$$

For the correlation function we again obtain (146.11), with the correlation radius

$$r_c = 2^{1/6}g^{1/2}/3^{1/2}b^{1/6}h^{1/3}.$$

§ 147. The effective Hamiltonian

Before going on to describe the properties of the phase transition outside the range of applicability of the Landau theory (i.e. in the immediate neighbourhood of the transition point) we shall show how the statistical problem of investigating these properties could be formulated.[§]

According to (35.3), the thermodynamic potential Ω is given by the partition function

$$\Omega = -T\log\sum_N e^{\mu N/T}\int e^{-E_N(p,\,q)T}\,d\Gamma_N, \qquad (147.1)$$

where the integration is taken over the whole phase space of a system of N particles. If the integration is taken only over the part of phase space that

---

[†] This case occurs, in particular, for transitions from the paramagnetic to the ferromagnetic state, where the order parameter is the magnetisation vector of the crystal. A linear dependence of the deformation on the magnetisation is excluded by the requirement of symmetry under time reversal (which leaves the deformation unchanged but changes the sign of the magnetic moment).

[‡] See A. I. Larkin and S. A. Pikin, *Soviet Physics JETP* **29**, 891, 1969.

[§] This formulation of the problem for phase transitions of the second kind is due to L. D. Landau (1958).

corresponds to a given distribution of the order parameter $\eta(\mathbf{r})$, the functional $\Omega[\eta(\mathbf{r})]$ defined by (147.1) may be regarded as the potential corresponding to that distribution. A continuous distribution $\eta(\mathbf{r})$ may here be conveniently replaced by a discrete set of complex variables $\eta_k = \eta'_k + i\eta''_k$, the components in the Fourier expansion (146.7). Then the definition of $\Omega[\eta]$ becomes

$$\Omega[\eta(\mathbf{r})] = -T \log \sum_N e^{\mu N/T} \int \exp\left(-\frac{E_N(p,q)}{T}\right)$$
$$\times \prod_k \delta(\eta'_k - \eta'_k(p,q;N))\, \delta(\eta''_k - \eta''_k(p,q;N))\, d\Gamma_N, \quad (147.2)$$

where $\eta_k(p,q;N)$ are the quantities η_k as functions of the point p, q in phase space. It is evident that, with this definition,

$$\Omega = -T \log \int \exp\left(-\Omega[\eta]/T\right) \prod_k (V\, d\eta'_k\, d\eta''_k). \quad (147.3)$$

In § 146 it has been shown that only fluctuations with small wave vectors $\mathbf{k}$ are subject to an anomalous increase near the transition point; these fluctuations therefore govern the nature of the singularity of the thermodynamic functions. At the same time, such quantitative characteristics of the substance as the transition temperature T_c itself are determined mainly by the short-range atomic interactions, which correspond to the short-wavelength components η_k. This physically obvious fact is represented in the integral over states by the correspondence between large values of $\mathbf{k}$ and large phase volumes.

Let k_0 (the cut-off parameter) be a value of k small compared with the reciprocal of the characteristic atomic dimensions. The long-wavelength part of the distribution $\eta(\mathbf{r})$ is given by the sum

$$\tilde{\eta}(\mathbf{r}) = \sum_{k < k_0} \eta_k e^{i\mathbf{k}\cdot\mathbf{r}}, \quad (147.4)$$

and the thermodynamic potential $\Omega[\tilde{\eta}]$ corresponding to this distribution is given by (147.2), where the product over $\mathbf{k}$ is to be extended only to values $k < k_0$. Accordingly, the relation between $\Omega[\tilde{\eta}]$ and Ω is given by (147.3) with integration only over $\tilde{\eta}_k$ with $k < k_0$.†

Near the transition point, the functional $\Omega[\tilde{\eta}]$ can be expanded in powers of the function $\tilde{\eta}(\mathbf{r})$; since this function is a slowly varying one, we need take in the expansion only the terms of lowest order in its derivatives. The expansion must also take into account the existence of the phase transition, since

---

† To simplify the treatment, we regard the physical quantity $\tilde{\eta}$ as classical. This assumption is not important, since the long-wavelength variable $\tilde{\eta}$ is certainly classical. For quantum systems, however, it is necessary to satisfy a condition of the form $\hbar k_0 u \ll T$, where u is the characteristic rate of propagation of oscillations of the order parameter.

the value of T_c is determined by the short-wavelength components excluded from $\tilde{\eta}$. This means that the expansion of $\Omega[\tilde{\eta}]$ must have the form (146.5) directly:

$$\Omega[\tilde{\eta}] = \Omega_0 + \int [\alpha t \tilde{\eta}^2 + b\tilde{\eta}^4 + g(\nabla\tilde{\eta})^2 - h\tilde{\eta}] \, \mathrm{d}V.$$

Finally, omitting the tilde, we arrive at the following expression for the thermodynamic potential Ω:

$$\Omega - \Omega_0 = -T \log \int \exp\left(-H_{\mathrm{eff}}/T_c\right) \prod_{k < k_0} (V \, \mathrm{d}\eta'_{k} \, \mathrm{d}\eta''_{k}), \quad (147.5)$$

where

$$H_{\mathrm{eff}} = \int [\alpha t \eta^2 + b\eta^4 + g(\nabla\eta)^2 - h\eta] \, \mathrm{d}V \qquad (147.6)$$

acts as the *effective Hamiltonian* of the system undergoing a phase transition.

In the range where the Landau theory is applicable, the fluctuations are small. This means that, in the integral over states (147.5), the important values of η are those lying in a narrow range near the value $\bar{\eta}$ that minimises the effective Hamiltonian. Using the saddle-point method (i.e. replacing the exponent by its expansion about the minimum), we ought to recover the thermodynamic potential as in the Landau theory; hence the coefficients in the effective Hamiltonian and in the Landau thermodynamic potential must be exactly the same. The fluctuation corrections, however, cause a certain shift of the transition temperature T_c relative to the value $T_c^{(0)}$ which occurs in (147.6) through the difference $t = T - T_c^{(0)}$.

The integral (147.5) is taken over an infinity of variables η_k (when the effective Hamiltonian has been expressed in terms of these variables by substituting $\eta(\mathbf{r})$ from (147.4)). If this "continuum integral" could be calculated, we should know the nature of the singularity of $\Omega(\mu, T)$ near the transition point. The calculation is impossible, however.

In the creation of the singularity, the significant fluctuations are those with wave numbers $k \sim 1/r_c$. When $t \to 0$, the correlation radius $r_c \to \infty$, so that very small values of k become important. It is therefore very probable that the nature of the singularity does not depend on the choice of the cut-off parameter k_0. If this singularity is assumed to consist in the presence in the thermodynamic potential of terms containing fractional powers of the temperature t and the field h, the above assertion implies that the indices of these powers (the *critical indices*) are independent of k_0.

Hence it must in turn follow that these indices are independent of the specific values of the coefficients b and g in the effective Hamiltonian, and therefore of μ or P, on which b and g depend. For a change $k_0 \to k_0/\lambda$ is equivalent to a change in the coordinate scale ($\mathbf{r} \to \lambda\mathbf{r}$), and such a change therefore cannot affect the critical indices. On the other hand, the transformation $\mathbf{r} \to \lambda\mathbf{r}$ changes the coefficient g in the effective Hamiltonian, but not b; the critical indices must thus be independent of g. Similarly, by making the simultaneous transformations $\mathbf{r} \to \lambda\mathbf{r}$, and $\eta \to \lambda\eta$ for the continuum integration variable, we change b but not g, and the critical indices are therefore also independent of b. A change in the coefficient α is unimportant, since it is cancelled by a corresponding change in the scale of t, which cannot affect the exponent.

Thus we should expect that the critical indices are the same for all systems with an effective Hamiltonian of the form (147.6). They may differ, however, if the symmetry of the system is such that (with one order parameter, as before) the term quadratic in the derivatives in the effective Hamiltionian has the more general form (146.4).

Continuing this line of argument, we may expect that in more general cases also, when the change of symmetry in the transition is described by more than one order parameter, the critical indices depend only on the structure of the effective Hamiltonian, not on the specific values of the coefficients in it. Here the "structure" of the Hamiltonian includes the number and form of the fourth-order invariants (and the signs and inequalities relating to their coefficients) and the form of the terms quadratic in the derivatives of the order parameters. The questions then arising, however, have as yet hardly been studied at all.

Lastly, we may add a few remarks on the calculation of the successive terms in the expansion of the partition function (147.5), (147.6) in powers of b. Let $h = 0$ and $t > 0$, so that $\bar{\eta} = 0$; when $b = 0$, the effective Hamiltonian is

$$H_{\text{eff}}^{(0)} = V \sum_{k < k_0} (\alpha t + gk^2) |\eta_\mathbf{k}|^2 ; \tag{147.7}$$

it separates into a sum of terms each depending only on one of the $\eta_\mathbf{k}$. The integral over states is easily calculated (see Problem). Further terms in the expansion (corresponding to the inclusion of the "interaction" between fluctuations with different $\mathbf{k}$) are products of different $\eta_\mathbf{k}$ averaged over the Gaussian distribution [$\propto \exp(-H_{\text{eff}}^{(0)}/T_c)$]. For such integrals a theorem is valid whereby the mean value of the product of several $\eta_\mathbf{k}$ is equal to the sum of the products of pairs of mean values of factors chosen in all possible ways from those concerned. Each such mean value is a correlation function of fluctuations (in the $\mathbf{k}$ representation), and hence the calculation of the successive terms in the expansion in powers of b amounts to that of various integrals of products of correlation functions.[†] As the transition point is

---

[†] The theorem stated plays a role here analogous to that of Wick's theorem in quantum electrodynamics, and the separate terms of the series may be transformed by graphs analogous to Feynman diagrams. An account of the resulting "diagram technique" for calculating the partition function is given by A. Z. Patashinskiĭ and V. L. Pokrovskiĭ, *Fluctuation Theory of Phase Transitions*, Pergamon Press, Oxford, 1979.

approached, these integrals diverge, but it is not possible to distinguish among them any set of "most strongly" divergent ones over which a summation could be carried out.[†]

In the above formulation of the problem, it is assumed that the nature of the singularity is independent of any higher-order terms in the expansion of the effective Hamiltonian in powers of η. There are good reasons for supposing that this is in fact so, since such terms lead to integrals that diverge less strongly than those resulting from the term $\sim \eta^4$.

PROBLEM

Find the first fluctuation correction to the specific heat in the range where the Landau theory is valid (A. P. Levanyuk, 1963).

SOLUTION. The calculation will be given for the symmetrical phase in the absence of a field. In the first approximation, the effective Hamiltonian is (147.7). A calculation of the integral over states from (147.5) gives

$$\Omega - \Omega_0 = -T_c \sum_{k < k_0} \log \frac{\pi T}{\eta(\alpha t + g k^2)}$$

$$= T_c V \int_0^{k_0} \log \frac{\eta(\alpha t + g k^2)}{\pi T} \cdot \frac{2\pi k^2 \, dk}{(2\pi)^3};$$

the integration is over half of k-space, since η_k and η_{-k} are not independent. Being the small correction in the potential Ω, this expression also gives the correction to the potential Φ. A twofold differentiation of this expression with respect to t gives the correction to the specific heat:

$$C_p - C_{p0} = \frac{T_c^2 V \alpha^2}{4\pi^2} \int_0^\infty \frac{k^2 \, dk}{(\alpha t + g k^2)^2} = \frac{T_c^2 V \alpha^{3/2}}{16\pi g^{3/2}} \cdot \frac{1}{\sqrt{t}}. \tag{1}$$

If this correction is to be small compared with the specific heat discontinuity (143.8), we again obtain the condition (146.15) for the Landau theory to be applicable, in the form

$$\alpha |t| \gg T_c^2 b^2 / 32\pi^2 g^3. \tag{2}$$

The large numerical factor in the denominator on the right-hand side should be noted.

[†] Such a distinguishing is possible in the formal problem of a phase transition in four-dimensional space (in which case the integrals diverge logarithmically as $t \to 0$). This is the basis of a method proposed by K. G. Wilson (1971) for estimating the critical indices, which are calculated for the case of a space of $4 - \varepsilon$ dimensions (where ε is small), the result then being extrapolated to $\varepsilon = 1$.

§ 148. Critical indices

The existing theory of phase transitions of the second kind is based on certain hypotheses that are entirely plausible though not rigorously proved. It rests also, of course, on the confirmation of these hypotheses by experimental results and by numerical calculations for certain simple models.

These results suggest that, as $T \to T_c$, the derivative $\partial C_p/\partial T$ always becomes infinite, and so in many cases does the specific heat C_p itself. From this we can already draw various conclusions about the behaviour of certain other thermodynamic quantities. This will be done here on the assumption that the specific heat itself becomes infinite (A. B. Pippard, 1956).

The tending of $C_p = T(\partial S/\partial T)_P$ to infinity means that the entropy of the body can be written as

$$S = S(T, P - P_c(T)),$$

where $P = P_c(T)$ is the equation of the curve of the points of the phase transition in the PT-plane; the derivative of S with respect to its second argument becomes infinite as $P - P_c \to 0$. Denoting differentiation with respect to that argument by a prime and retaining only the divergent terms, we have

$$(\partial S/\partial T)_P = -S'\, dP_c/dT,$$

$$-(\partial V/\partial T)_P = (\partial S/\partial P)_T = S',$$

whence

$$C_p = T_c \frac{dP_c}{dT} \left(\frac{\partial V}{\partial T}\right)_P \quad \text{as} \quad T \to T_c, \tag{148.1}$$

i.e. the thermal expansion coefficient tends to infinity in the same way as C_p.

It is easily seen that this derivation amounts to equating to zero the divergent part of the derivative of S along the curve of transition points. It is therefore natural that (148.1) has the same form as the equation (143.10) obtained by differentiating $\Delta S = 0$ along the same curve, and differs from it only in the absence of the symbol Δ. We can hence write down at once another relation by analogy with (143.9):

$$-\left(\frac{\partial V}{\partial P}\right)_T = \left(\frac{\partial V}{\partial T}\right)_P \frac{dT_c}{dP} = \frac{C_p}{T_c}\left(\frac{dT_c}{dP}\right)^2, \tag{148.2}$$

i.e. the isothermal compressibility also becomes infinite (but the adiabatic compressibility remains finite, by (16.14)). The specific heat C_v remains finite; from (143.14) we see that it is not even discontinuous at the transition point: since the right-hand side of (143.14) is zero (because $(\partial V/\partial P)_T$ is infinite), $\Delta C_v = 0.$[†] The same is true of the derivative $(\partial P/\partial T)_V$; substitution of (148.2) in (16.10) shows that on the transition line

$$(\partial P/\partial T)_V = dP_c/dT. \tag{148.3}$$

It should be emphasised that the above results essentially depend on the occupation, by the phase transition points of the second kind, of an entire line in the PT-plane, the slope of this line being finite.

Let the temperature dependence of the specific heat in the fluctuation range be written

$$C_p \propto |t|^{-\alpha}, \tag{148.4}$$

where again $t = T - T_c$. We shall see later in this section that there is reason to suppose the values of the exponent α equal on either side of the transition point (and the same is true of the other exponents introduced below). The proportionality factors in (148.4) are, of course, different on the two sides. Since the quantity of heat $\int C_p\, dT$ must always be finite, it follows that $\alpha < 1$. If only $\partial C_p/\partial T$, and not the specific heat itself, tends to infinity, then $-1 < \alpha < 0$; the expression (148.4) in that case determines only the singular part of the specific heat: $C_p = C_{p0} + C_{p1}|t|^{-\alpha}$.

The way in which the equilibrium value of the order parameter tends to zero in the unsymmetrical phase may be written as

$$\eta \propto (-t)^\beta, \quad \beta > 0. \tag{148.5}$$

By definition, the exponent β refers only to the unsymmetrical phase. We shall assume for definiteness here and below that temperatures $t < 0$ correspond to the unsymmetrical phase.

To describe the properties of the fluctuations themselves of the parameter η, an exponent ν is used which defines the temperature dependence of the correlation radius:

$$r_c \propto |t|^{-\nu}, \quad \nu > 0, \tag{148.6}$$

and an exponent ζ which gives the decrease of the correlation function with increasing distance at $t = 0$:

$$G(r) \propto r^{-(d-2+\zeta)}, \tag{148.7}$$

where d is the dimension of space ($= 3$ for ordinary bodies). The expression (148.7) is written in this form in order to have a definition suitable also for phase transitions of the second kind in two-dimensional systems ($d = 2$).

[†] The impossibility of infinite C_v on the transition line is also evident from the fact that it would lead to $C_v = T(dV_c/dT)^2 (\partial P/\partial V)_T$ (cf. (143.14)), which is certainly impossible, as C_v is positive and $(\partial P/\partial V)_T$ is negative. The specific heat C_v has, however, an infinite derivative on the transition line (see Problem).

The law (148.7) is valid also for non-zero $|t| \ll T_c$, but only for distances $r \ll r_c$.

The exponents in (148.4)–(148.7) are called *critical indices* or *critical exponents*. It must be emphasised that the degree of accuracy of the subsequent derivation of relations between the critical indices does not allow the distinguishing of logarithmic factors superimposed on power laws. In this sense, for example, a zero exponent may mean either that the quantity tends to a constant limit or that it increases logarithmically.

A further group of indices are used to describe the properties of a body in the fluctuation range when an external field h is present. Here we must distinguish the ranges of fields that are "weak" and "strong" in the sense described at the end of § 144: $h \ll h_t$ or $h \gg h_t$, where h_t is the field value for which the field-induced parameter $\eta_{ind} \sim \chi h$ becomes of the same order as the characteristic value of the spontaneous-order parameter $\eta_{sp}(t)$. The region of weak fields has an exponent γ giving the variation of the susceptibility:

$$\chi \propto |t|^{-\gamma}, \quad \gamma > 0. \tag{148.8}$$

The indices defined above may also be assigned to this region: the laws (148.4)–(148.6) derived for zero field relate, of course, also to the limiting case of weak fields.

For the opposite case of strong fields, we define critical indices which give the field dependence of the thermodynamic quantities and the correlation radius:

$$C_p \propto h^{-\epsilon}, \tag{148.9}$$
$$\eta \propto h^{1/\delta} \quad (\delta > 0), \tag{148.10}$$
$$r_c \propto h^{-\mu} \quad (\mu > 0); \tag{148.11}$$

we take for definiteness the case $h > 0$.†

The universality of the limiting laws for the behaviour of a substance in the fluctuation range near a phase transition point of the second kind, in the sense discussed in § 147, implies a similar universality of the critical indices. Thus we must expect their values to be the same for all transitions with a change of symmetry described by only one order parameter.

The critical indices are connected by various exact relations. Some of these follow almost immediately from the definitions of the indices, and we shall first derive these.

It has been shown in § 144 that the application of an external field h smooths out the phase transition over a certain temperature range. The magnitude

---

† The Landau theory corresponds to the following values of the critical indices:

$$\alpha = 0, \quad \beta = \tfrac{1}{2}, \quad \gamma = 1, \quad \delta = 3, \quad \epsilon = 0, \quad \mu = \tfrac{1}{3}, \quad \nu = \tfrac{1}{2}, \quad \zeta = 0.$$

of this range of t can be estimated from the condition $\eta_{ind}(h) \sim \eta_{sp}(t)$ mentioned above, if this is now taken as a condition on t for given h. According to the definitions (148.5) and (148.8), we have

$$\eta_{sp} \propto |t|^\beta, \quad \eta_{ind} = \chi h \propto h|t|^{-\gamma},$$

and equating the two gives

$$|t|^{\beta+\gamma} \propto h. \tag{148.12}$$

On the other hand, the same smoothing range can be estimated from the requirement that the field part of the thermodynamic potential, $-V\eta h$, is of the same order of magnitude as the thermal part, which is $\sim t^2 C_p$, since $C_p = -T \partial^2 \Phi/\partial T^2$. Hence we find $|t|^{t+2-\alpha-\beta} \propto h$, and, expressing h in terms of t from (148.12), obtain the equation

$$\alpha + 2\beta + \gamma = 2 \tag{148.13}$$

(J. W. Essam and M. E. Fisher, 1963).

We next use the obvious fact that, at the edge of the region of smoothing of the transition (i.e. with the condition (148.12)), we can equally well express each thermodynamic quantity in terms of the temperature t or the field h. Hence we find, for example,

$$\eta \propto |t|^\beta \propto h^{1/\delta},$$

and, expressing h in terms of t from (148.12), obtain the equation

$$\beta\delta = \beta + \gamma \tag{148.14}$$

(B. Widom, 1964). By a similar method starting from the two limiting forms of the specific heat C_p, we find

$$\varepsilon(\beta+\gamma) = \alpha. \tag{148.15}$$

The equations (148.14) and (148.15) relate the indices that determine the temperature dependence of the thermodynamic quantities in weak fields and their dependence on h in strong fields.

An analogous equation is found by the same method for the indices giving the behaviour of the correlation radius:

$$\mu(\beta+\gamma) = \nu. \tag{148.16}$$

Lastly, one further relation can be derived by an estimate of the expressions on either side of (146.13). According to (146.2) and the definition (148.8), the mean square fluctuation in a given volume V is

$$\langle (\Delta\eta)^2 \rangle_V = T_c \chi/V \propto |t|^{-\gamma}.$$

The integral of the correlation function is determined by the region of space $\sim r_c^d$ in which this function is considerably different from zero and, according to the definition (148.7), its order of magnitude is $\propto r_c^{-(d+2+\zeta)}$. Hence the magnitude of the integral is (in d-dimensional space)

$$\propto r_c^d \cdot r_c^{-(d-2+\zeta)} = r_c^{2-\zeta} \propto |t|^{-\nu(2-\zeta)}.$$

A comparison of the two expressions gives

$$\nu(2-\zeta) = \gamma. \tag{148.17}$$

Thus we have five relations between the eight indices, and so they can all be expressed in terms of only three independent ones.

From this we can, in particular, draw the conclusion already mentioned, that the values of the "temperature" indices α, γ, ν are the same on both sides of the transition point: for, if γ, say, were different for $t > 0$ and $t < 0$, it would follow from (148.14) that δ also depends on the sign of t. This index, however, relates to strong fields h, which satisfy only the condition $h \gg h_t$ independent of the sign of t, and it too therefore cannot depend on this sign; the same is true of the other two "field" indices, ε and μ. From the relations (148.13) and (148.16), we then find that α and ν also are independent of the sign of t.

The results obtained enable us to draw some conclusions about the thermodynamic functions of the system for any relationship between t and h. This will be demonstrated for the function $\eta(t, h)$, which may be written in the form

$$\eta = h^{1/\delta} f\left(\frac{t}{h^{1/\beta\delta}}, \; t\right)$$

(for given P). The choice of the first argument of f is determined by the condition (148.12), which separates the cases of weak and strong fields (and we have put $\beta + \gamma = \beta\delta$ in accordance with (148.14)); this argument takes all values from small to large. The argument t is always small near the transition point, and must be taken as zero in obtaining the principal term in the function $\eta(t, h)$. Thus we arrive at the expression

$$\eta(t, h) = h^{1/\delta} f(t/h^{1/\beta\delta}), \qquad h > 0, \tag{148.18}$$

where f is a function of the one argument $x = t/h^{1/\beta\delta}$. The expression (148.18) is written for $h > 0$; owing to the symmetry of the system under a simultaneous change of sign of h and η, the formula for $h < 0$ is obtained from (148.18) by simply substituting $-h$ for h and $-\eta$ for η.

In strong fields ($x \ll 1$), the limiting form (148.10) must be obtained; this means that

$$f(x) = \text{constant} \quad \text{for} \quad x \to 0. \tag{148.19}$$

Moreover, when $h \neq 0$ the order parameter is non-zero both for $t > 0$ and for $t < 0$, and $t = 0$ is not physically distinctive; this means that the function $f(x)$ has an expansion in integral powers of x.

In weak fields when $t < 0$, the order parameter follows the law (148.5), and for $t > 0$ we must have $\eta = \chi h$ with χ given by (148.8); from these requirements it follows that

$$f(x) \propto (-x)^\beta \quad \text{as} \quad x \to -\infty, \quad f(x) \propto x^{-\gamma} \quad \text{as} \quad x \to \infty. \quad (148.20)$$

see p492

The concept of a weak field presupposes that $t \neq 0$. For a given non-zero value of t, a zero field is not a singularity of the thermodynamic functions. Hence the function $\eta(t, h)$ for $t \neq 0$ can be expanded in integral powers of the variable h (and the expansion is different for $t > 0$ and $t < 0$). A natural formulation of this property would, however, require $\eta(t, h)$ to be written not in the form (148.18) but in terms of a function of the variable $h/t^{\beta\delta}$.

Similar considerations can be applied to the correlation function of fluctuations of the order parameter. For instance, in the absence of a field it depends on the parameter t as well as on the distance r. Near the transition point, however, the correlation function $G(r; t)$ can be written as

$$G(r; t) = \frac{1}{r^{d-2+\zeta}} g(rt^\nu), \quad (148.21)$$

i.e. in terms of a function of the one variable $x = rt^\nu$. As $x \to 0$, this function tends to a constant limit, in accordance with the definition (148.7), and as $x \to 0$ it decays exponentially, the correlation radius as a function of temperature being given by (148.6).

PROBLEM

Find the law of temperature dependence as $t \to 0$ for the derivative $\partial C_v / \partial T$ if C_p tends to infinity as (148.4) with $\alpha > 0$.

Solution. With greater accuracy than in (148.1), (148.2), we write for $t \to 0$

$$C_p = T_c \frac{dP_c}{dT} \left(\frac{\partial V}{\partial T} \right)_P + a,$$

$$\left(\frac{\partial V}{\partial T} \right)_P = -\left(\frac{\partial V}{\partial P} \right)_T \frac{dP_c}{dT} + \frac{b}{T_c} \frac{dT_c}{dP},$$

with a and b constants. Substituting these expressions in (16.9), we find

$$C_v \cong a - b - b^2 / C_p.$$

If C_p increases as $|t|^{-\alpha}$, the $\partial C_v / \partial T \propto |t|^{-(1-\alpha)}$. At $t = 0$ the function $C_v(t)$ has a maximum at a cusp with a vertical tangent.

§ 149. Scale invariance

The relations (148.13)–(148.17) do not involve any assumptions about the nature of the fluctuation pattern near the transition point.[†] Further conclusions about the critical indices require specific assumptions in this respect.

We may note that the theory in general involves two characteristic dimensions which determine the spatial distribution of the fluctuations: the correlation radius r_c and the dimension r_0 of the part of the body in which the mean square fluctuation of the order parameter is comparable with its characteristic equilibrium value.[‡] The inequality (146.14) which ensures the applicability of the Landau theory may be written $r_c \gg r_0$ (for, according to (146.13) and (146.11), we have, in a volume $V \sim r_0^3$, $\langle (\Delta\eta)^2 \rangle \sim T_c/g r_0$ and, equating this to $\eta^2 \sim \alpha|t|/b$, we find $r_0 \sim T_c b/g\alpha|t|$; comparison with r_c (146.12) gives the condition (146.15)). As $t \to 0$, r_0 increases more rapidly than r_c, and they become comparable at the boundary of the Landau region. The chief assumption about the fluctuation region (this being defined by the opposite inequality to (146.15)) is that the theory here contains no small parameter. In particular, we must everywhere have $r_0 \sim r_c$, so that r_c is the only dimension characterising the fluctuations. This is called the hypothesis of scale invariance (L. P. Kadanoff, 1966; A. Z. Patashinskiĭ and V. L. Pokrovskiĭ, 1966).

To estimate the fluctuations in the volume $V \sim r_c^3$, we can use formula (146.2).[§] Substituting in the condition

$$T_c \chi / V \sim \eta^2 \tag{149.1}$$

the volume $V \sim r_c^d$ and then expressing χ, r_c and η as powers of t according to the definitions of the critical indices, we obtain $\nu d - \gamma = 2\beta$ or, using (148.13),

$$\nu d = 2 - \alpha. \tag{149.2}$$

By combining this relation with those derived in § 148, we can express all the critical indices in terms of only two independent ones.[‖]

[†] It is therefore not surprising that all these relations are satisfied in the Landau theory also.

[‡] This discussion refers, of course, only to the distribution at distances large compared with atomic dimensions.

[§] In this form (i.e. expressed in terms of the susceptibility χ) this formula is general and does not depend on the assumptions made in the Landau theory, as mentioned after (146.2).

[‖] In the Landau theory there is no scale invariance, and equation (149.2) is therefore invalid.

The requirement of scale invariance enables us to derive in a uniform way all the relations between the critical indices. To do so, we shall first give a more formal statement of this requirement.

Let the scale of all spatial distances change by the same factor: $\mathbf{r} \to \mathbf{r}/u$, where u is some constant. Then the scale invariance consists in the assertion that the scales of measurement of t, h and η may be changed in such a way that all the relations in the theory remain unaltered. In other words, we can choose the exponents Δ_t, Δ_h, Δ_η (called the *scaling dimensions*) in the transformations

$$t \to tu^{\Delta_t}, \quad h \to hu^{\Delta_h}, \quad \eta \to \eta u^{\Delta_\eta} \quad \text{when} \quad \mathbf{r} \to \mathbf{r}/u \qquad (149.3)$$

in such a way that the factor u does not appear in any of the relations.

The change in the spatial scale must, in particular, bring about a similar change in the correlation radius of the fluctuations ($r_c \to r_c/u$); this ensures that the asymptotic form of the correlation function $\sim \exp(-r/r_c)$ is invariant. According to the definitions (148.6) and (148.11), with $h = 0$ the correlation radius $r_c = \text{constant} \times t^{-\nu}$, and with $t = 0$, $r_c = \text{constant} \times h^{-\mu}$. Applying the transformation (149.3) and the condition that the coefficients in these expressions should remain unchanged, we obtain

$$\Delta_t = 1/\nu, \quad \Delta_h = 1/\mu. \qquad (149.4)$$

Let us next consider the change in the thermodynamic potential due to an infinitesimal change in the field h. According to (144.2),

$$d\Phi = -V\eta \, dh$$

(with $t = \text{constant}$ and, as always, $P = \text{constant}$). In the scale transformation, the volume $V \to V/u^d$; with the condition that $d\Phi$ is unchanged, i.e. that

$$\left(Vu^{-d}\right)\left(\eta u^{\Delta_\eta}\right)\left(dh u^{\Delta_h}\right) = V\eta \, dh,$$

we find

$$\Delta_\eta = d - \Delta_h = d - 1/\mu. \qquad (149.5)$$

Thus the dimensions Δ_t, Δ_h and Δ_η are expressed in terms of the two critical indices μ and ν. The requirement of scale invariance of the other relations leads to the expression of the other critical indices in terms of these two.

Let us specify the condition for invariance of the "equation of state" of the system, i.e. the expression of the order parameter in terms of the temperature and the field, $\eta = \eta(t, h)$. This means that we must have

$$\eta(tu^{\Delta_t}, hu^{\Delta_h}) = u^{\Delta_\eta}\eta(t, h).$$

The solution of this functional equation has the form

$$\eta(t, h) = h^{\Delta_\eta/\Delta_h} f(t/h^{\Delta_t/\Delta_h})$$
$$= h^{\mu d - 1} f(t/h^{\mu/\nu}). \qquad (149.6)$$

Handwritten marginal notes:

L, L/u

$L \Rightarrow$ L/u

everything should remain the same.

independent of u

$r_c \to r_c/u$

$t^{-\nu} \Rightarrow (tu^{\delta_t})^{-\nu}$

$= t^{-\nu} u^{-\nu\Delta_t}$

must $= \dfrac{t^{-\nu}}{u}$

$\Rightarrow \Delta_t \doteq \dfrac{1}{\nu}$

but Φ may change

what happens when $d = \dfrac{1}{\mu}$?

$\mu \doteq \dfrac{2}{5} \Rightarrow d \doteq 2.5$

and d_c

choose u such that

$h u^{\Delta_h} = 1$

$u = h^{-1/\Delta_h}$

Similar arguments can be applied to the thermodynamic potential $\Phi(t, h)$, or more precisely to its singular part, which we shall use Φ below to denote. Being additive, the total thermodynamic potential of the body is proportional to its volume. Hence the requirement of its invariance under the scale transformation may be written

$$\frac{V}{u^d}\Phi(tu^{\Delta_t}, hu^{\Delta_h}) = \frac{V}{u^d}\Phi(tu^{1/\nu}, hu^{1/\mu}) = V\Phi(t, h),$$

IC this is potential density it works out

from which

$$\boxed{\Phi(t, h) = h^{d\mu}\phi(t/h^{\mu/\nu}).}$$ (149.7)

The functions f and ϕ in (149.6) and (149.7) are of course related, since $-\partial\Phi/\partial h = \eta V$. These expressions have been written for $h > 0$; in view of the symmetry of the effective Hamiltonian under the changes $h \to -h$, $\eta \to -\eta$, the formulae for $h < 0$ are obtained from those given above by making these same changes.[†]

$\phi(x) + h^{-1/2}\phi'(x) = f(x)$

Similar arguments can be based on (149.7). As already mentioned in connection with (148.18), for a given non-zero h the thermodynamic functions have no singularity with respect to t and must therefore have expansions in integral powers of this variable. This means that, when $h \neq 0$ and $t \to 0$, the function $\phi(x)$ in (149.7) is expanded in integral powers of the small variable $x = t/h^{\mu/\nu}$. The first terms in this expansion are

$$\Phi(t, h) \propto h^{\mu d}\left[1 + c_1\frac{t}{h^{\mu/\nu}} + c_2\frac{t^2}{h^{2\mu/\nu}} + \cdots\right],$$ (149.8)

where c_1 and c_2 are constant coefficients. Now, applying the condition that the order parameter and the specific heat, calculated as

$$\eta = -\frac{1}{V}\frac{\partial\Phi}{\partial h}, \quad C_p \cong -T_c\frac{\partial^2\Phi}{\partial t^2},$$

should have the forms $\eta \propto h^{1/\delta}$ and $C_p \propto h^{-\varepsilon}$ as $t \to 0$ (these correspond to the strong-field case), we get two relations between the critical indices:

$$(\mu\delta - 1)\delta = 1, \quad \mu\left(\frac{2}{\nu} - \delta\right) = \varepsilon;$$

this should specify the indices then not independent

[†] It may be mentioned once more, however, that in the effective Hamiltonian η appears as a variable over which the continuum integration is taken in the integral over states. In the thermodynamic formulae, η is the equilibrium value of the order parameter, which is given by the derivative $\partial\Phi/\partial h$ (or $\partial\Omega/\partial h$) of the thermodynamic potential determined from the integral over states. The symmetry of the effective Hamiltonian leads, of course, to an analogous symmetry in the thermodynamic relations.

it is easily verified that they in fact follow from the relations derived previously by a different method.

Next, let t be non-zero; then the thermodynamic quantities have no singularity when h passes through zero, and so the function $\Phi(t, h)$ can be
expanded in integral powers of h. This means that for $h \to 0$ and $t \neq 0$
the expansion of the function $\phi(x)$ in powers of the small variable $1/x = h^{\mu/\nu}/t$
must have the form

$$\phi(x) \propto x^{\nu d}[1 + c_1 x^{-\nu/\mu} + c_2 x^{-2\nu/\mu} + \ldots];$$

the factor $x^{\nu d}$ compensates the non-integral power $h^{d\mu}$, and the expansion
variable $x^{-\nu/\mu} \propto h$. The expansion is, however, different for $t > 0$ and for
$t < 0$. When $t > 0$, the potential $\Phi(t, h)$ contains only even powers of h,
since the derivative $-\partial\Phi/\partial h = V\eta$ must be (in the symmetrical phase) an
odd function of h:

$$\Phi \propto t^{\nu d}\left[1 + c_2 \frac{h^2}{t^{2\nu/\mu}} + \ldots\right], \quad t > 0, \quad h \to 0. \qquad (149.9)$$

As $h \to 0$, the specific heat must behave as $t^{-\alpha}$, and the order parameter
must be $\eta = \chi h \propto ht^{-\gamma}$ (corresponding to the weak-field case); it is easily
seen that the resulting relations are again equivalent to those already known.
If the temperature $t < 0$, then the expansion of $\Phi(t, h)$ as $h \to 0$ contains all
integral powers of h:

$$\Phi \propto (-t)^{\nu d}\left[1 + c_1 \frac{h}{(-t)^{\nu/\mu}} + c_2 \frac{h^2}{(-t)^{2\nu/\mu}} + \ldots\right], \quad t < 0, \quad h \to 0$$

$$(149.10)$$

(with, of course, different coefficients c_1 and c_2).[†] It is easily verified that the
required form $(-t)^\beta$ is obtained for the spontaneous (h-independent) order
parameter.

The transformation of the correlation radius has been discussed above.
We have still to consider the correlation function of the fluctuations of the
parameter η for $t \to 0$ and apply the conditions for scale invariance of the
expression

$$G(r) = \text{constant} \times r^{-(d-2+\zeta)} \qquad (t = 0).$$

Here it must be assumed that the fluctuating quantities $\eta(\mathbf{r})$ at different
points in space are transformed independently in the same way as the mean

---

[†] If (149.10) relates, say, to fields $h > 0$, the formula for $h < 0$ is obtained from
it by the substitution $h \to -h$. It may be recalled (see § 144) that for $t < 0$ the
states in fields of opposite sign refer to physically identical "phases" differing
in the sign of the order parameter (both spontaneous and field-induced); as $h \to 0$,
the two phases are in equilibrium with each other.

value η. It is important that the distances r concerned are small compared with the correlation radius, but large compared with interatomic distances. Then the correlation function is transformed by $G \to G u^{2\Delta\eta}$, and we have the condition

$$d + 2 - \frac{2}{\mu} = \zeta. \tag{149.11}$$

This also is a consequence of the relations already known.

Let us finally consider the numerical values of the critical indices. The experimental results and numerical calculations indicate that (in three-dimensional space) the indices α and ζ are quite small: $\alpha \sim 0.1$, $\zeta \sim 0.05$. The first line of the following table gives the values of the other indices found by putting $\alpha = \zeta = 0$ ($d = 3$). The second line gives the values obtained by taking for α and ζ the values estimated by Wilson's method (§ 147) for transitions described by an effective Hamiltonian (147.6) with a single order parameter:

α	β	γ	δ	ε	μ	ν	ζ	
0	$\frac{1}{3}$	$\frac{4}{3}$	5	0	$\frac{2}{5}$	$\frac{2}{3}$	0	(149.12)
0.08	0.33	1.26	4.8	0.05	0.40	0.64	0.04	

§ 150. Isolated and critical points of continuous transition

The curve of phase transitions of the second kind in the PT-plane separates phases of different symmetry, and cannot, of course, simply terminate at some point, but it may pass into a curve of phase transitions of the first kind. A point at which this happens may be called a *critical point* of a transition of the second kind; it is in some ways analogous to an ordinary critical point. See the point K in Fig. 66; in this and subsequent diagrams, continuous and broken lines represent curves of phase transition points of the first and second kinds respectively.[†]

FIG. 66

[†] The term *tricritical point* is also used in the literature for a point such as K.

In the Landau theory, the properties of a substance near such a point may be investigated by the method (§ 143) of expansion in powers of the order parameter (L. D. Landau, 1935).

In the expansion (143.3) the critical point is given by the vanishing of the two coefficients $A(P, T)$ and $B(P, T)$; if $A = 0$ but $B > 0$ we have a transition of the second kind, and so the curve of such transitions terminates only where B changes sign. For the state of the body at the critical point itself to be stable, it is necessary that the fifth-order term should be identically zero and the sixth-order term should be positive. Thus we start from the expansion

$$\Phi(P, T, \eta) = \Phi_0(P, T) + A(P, T)\eta^2 + B(P, T)\eta^4 + D(P, T)\eta^6, \quad (150.1)$$

very flat

with $A_{cr} = 0$, $B_{cr} = 0$, $D_{cr} > 0$ at the critical point.

In the unsymmetrical phase, the minimisation of the thermodynamic potential gives

$$\eta^2 = \frac{1}{3D}[-B + \sqrt{(B^2 - 3AD)}]. \quad (150.2)$$

For the entropy $S = -\partial\Phi/\partial T$ of this phase we have, omitting terms of higher order in η, $S = S_0 - a\eta^2$, where $a = \partial A/\partial T$. A further differentiation gives the specific heat

$$C_p = \frac{Ta^2}{2\sqrt{(B^2 - 3AD)}}, \quad (150.3)$$

where only the term whose denominator vanishes at the critical point is shown.

Let $T_0 = T_0(P)$ be the temperature for which $B^2 - 3AD = 0$; it is evident that $T_0 = T_{cr}$ when $P = P_{cr}$. The first term in the expansion of $B^2 - 3AD$ in powers of $T - T_0$ is

$$B^2 - 3AD = -3a_0 D_0 (T - T_0). \quad (150.4)$$

Near the critical point, the difference $T_c(P) - T_0(P)$ is a second-order small quantity: since we have $A = 0$ when $T = T_c(P)$, the difference

$$T_c(P) - T_0(P) = -B^2/3a_0 D_0, \quad (150.5)$$

and tends to zero as B^2 when $P \to P_{cr}$.

Substitution of (150.4) in (150.3) gives

$$C_p = \left(\frac{T^2 a^3}{12D}\right)_{cr}^{1/2} \frac{1}{\sqrt{(T_0 - T)}}; \quad (150.6)$$

to the same accuracy, the coefficient in this formula may be taken at T_{cr} instead of T_0. Thus the specific heat of the unsymmetrical phase increases as $1/\sqrt{(T_0 - T)}$ as the critical point is approached.

For states actually on the curve of transitions of the second kind, we find, substituting $A = 0$ in (150.3) (or (150.5) in (150.6)),

$$C_p^{(II)} = T_{cr}a_{cr}^2/2B. \tag{150.7}$$

The quantity B is zero at the critical point and is proportional to $T-T_{cr}$ (or $P-P_{cr}$) near that point.

Let us now determine the specific heat of the unsymmetrical phase, on a line of transitions of the first kind but again near the critical point. At points on this line, two different phases (symmetrical and unsymmetrical) are in equilibrium with each other. The value of the parameter η in the unsymmetrical phase is determined by the condition of equilibrium $\Phi(\eta) = \Phi_0$, and at the same time we must have $\partial\Phi/\partial\eta = 0$. Substitution of Φ in (150.1) gives the equations

$$A+B\eta^2+D\eta^4 = 0, \quad A+2B\eta^2+3D\eta^4 = 0,$$

whence

$$\eta^2 = -B/2D, \tag{150.8}$$

and substitution of this value in the equation $\Phi(\eta) = \Phi_0$ then gives

$$4AD = B^2. \tag{150.9}$$

This is the equation of the line of transitions of the first kind.

The specific heat of the unsymmetrical phase on this line is found by simply substituting (150.9) in (150.3):

$$C_p^{(I)} = T_{cr}^2 a_{cr}^2/|B|. \tag{150.10}$$

Comparison with (150.7) shows that the specific heat on the line of transitions of the first kind is twice as great as on the line of transitions of the second kind at the same distance from the critical point. The heat of transition from the unsymmetrical to the symmetrical phase is

$$q = T_{cr}(S_0-S)$$
$$= (aT/2D)_{cr}\,|B|. \tag{150.11}$$

We can also show that the curve of transitions of the first kind passes smoothly into the curve of transitions of the second kind at the critical point. On the first curve, the derivative dT/dP is determined by the condition

$$2D\,dA+2A\,dD-B\,dB = 0,$$

obtained by differentiating (150.9). The equation of the curve of transitions of the second kind is $A = 0$, so that dT/dP is determined by the condition $dA = 0$. At the critical point, $A = 0$ and $B = 0$, so that the two conditions coincide and dT/dP has no discontinuity. Similarly, it can be shown that the second derivative d^2T/dP^2 does have a discontinuity.

As the critical point is approached along the line $P = P_c$, the specific heat C_p varies as $|t|^{-1/2}$ (cf. (150.6)), i.e. the index $\alpha = \frac{1}{2}$. The order parameter in the unsymmetrical phase varies according to $\eta \cong (-A/3D)^{1/4} \propto |t|^{1/4}$, i.e. the index $\beta = \frac{1}{4}$. The index ν which determines the behaviour of the correlation radius then has the value $\frac{1}{2}$, as at any transition point of the second kind in the Landau theory; the vanishing of B does not affect the result in the approximation used to derive (146.8). The values of the remaining indices are derived from (148.13)–(148.17) as $\gamma = 1,\ \delta = 5,\ \varepsilon = \mu = \frac{2}{5},\ \zeta = 0$.

We know already that the Landau theory, on which the results given here are based, is inapplicable near a line of transitions of the second kind. It is noteworthy, however, that the conditions for this theory to be valid are more easily satisfied as the critical point is approached, as is seen from the inequality (146.15), where B appears on the right. Of course, the vanishing of B does not mean that there are no fluctuation corrections at the critical point, but the index values given above are in accordance with the scale invariance relation (149.2). It is therefore natural that the fluctuation theory leads only to further logarithmic corrections (which, it may be recalled, are not shown by the values of the indices).

Let us next consider (again in terms of the Landau theory) some properties of points of intersection of lines of phase transition of the first and second kinds.

The symmetry of the unsymmetrical phase in a phase transition of the second kind is determined (as shown in § 145) by the minimisation of the fourth-order terms in the expansion of Φ as functions of the coefficients $\gamma_i = \eta_i/\eta$. These terms also depend on P and T, however, and it may happen that the unsymmetrical phase has different symmetries on different parts of the line of transitions. In the simplest such case, we have an intersection of a line of transitions of the second kind (AC in Fig. 67) with one of tran-

Fig. 67

sitions of the first kind (BD). The region I is the symmetrical phase; the symmetry groups of phases II and III are sub-groups of that of phase I. They are not, however, in general sub-groups of each other, and therefore the curve

BD separating these phases is a line of transitions of the first kind. At *B*, all three phases are identical.[†]

Figure 68 shows a possible type of intersection of several lines of transitions of the second kind. If I is the most symmetrical phase, the symmetry groups of phases II and III are sub-groups of that of phase I, and that of phase IV is a sub-group of both those of phases II and III.[‡]

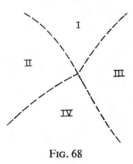

Fig. 68

Finally, it remains to consider the case where the third-order terms in the expansion of the thermodynamic potential do not vanish identically. In this case the condition for the existence of a point of continuous phase transition requires that the coefficients $C_\alpha(P, T)$ of the third-order invariants in the expansion (145.6) should vanish, as well as $A(P, T)$. It is evident that this is possible only if there is not more than one third-order invariant, since otherwise we should obtain more than two equations for the two unknowns *P* and *T*. When there is only one third-order invariant, the two equations $A(P, T) = 0, C(P, T) = 0$ determine pairs of values of *P* and *T*, i.e. there are isolated points of continuous phase transition.

Since these points are isolated, they must lie in a certain way at the intersection of curves (in the *PT*-plane) of phase transitions of the first kind. Since such isolated points of continuous transition have not yet been observed experimentally, we shall not pause to give a detailed discussion here, but simply mention the results.[§]

The simplest type is that shown in Fig. 69a. Phase I has the higher symmetry, and phases II and III the same lower symmetry, these two phases

[†] The fluctuation corrections may probably cause a singularity at *B*, the curves *AB* and *CB* forming a cusp.

[‡] A point of intersection of the type shown in Fig. 67 is called *bicritical* in the literature, and one as in Fig. 68 *tetracritical*.

[§] See L. Landau, *Zhurnal éksperimental'noĭ i teoreticheskoĭ fiziki* **7**, 19, 1937; translation in *Collected Papers of L. D. Landau*, p. 193, Pergamon, Oxford, 1965.

differing only in the sign of η. At the point of continuous transition (O in Fig. 69) all three phases become identical.

In more complex cases two or more curves of phase transition of the first kind (e.g. two in Fig. 69b) touch at the point of continuous transition. Phase I has the highest symmetry, phases II and III a lower symmetry, phases IV and V another lower symmetry, these pairs of phases differing only in the sign of η.

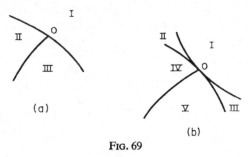

(a)

(b)

FIG. 69

§ 151. Phase transitions of the second kind in a two-dimensional lattice

The impossibility of a general theoretical determination of the critical indices lends especial interest to a simple model which allows an exact analytical solution of the problem of phase transitions of the second kind. This is a two-dimensional lattice model, for which the phase-transition problem was first solved by L. Onsager (1944).[†]

The model considered is a plane square lattice having N points, at each of which is a "dipole" with its axis perpendicular to the lattice plane. The dipole can have two opposite orientations, so that the total number of possible configurations of the dipoles in the lattice is 2^N.[‡] To describe the various configurations we proceed as follows. To each lattice point (with integral coordinates k, l) we assign a variable σ_{kl} which takes two values ± 1, corresponding to the two possible orientations of the dipole. If we take into account only the interaction between adjoining dipoles, the energy of the configuration may be written

$$E(\sigma) = -J \sum_{k,l=1}^{L} (\sigma_{kl}\sigma_{k,l+1} + \sigma_{kl}\sigma_{k+1,l}), \qquad (151.1)$$

[†] The original method used by Onsager was extremely complex. Later, various authors simplified the solution. The method described below (which in part makes use of certain ideas in the method of M. Kac and J. C. Ward (1952)) is due to N. V. Vdovichenko (1964).

[‡] This model is known as the *Ising model;* it was in fact first used by W. Lenz (1920) and was studied by E. Ising (1925) for the one-dimensional case (in which there is no phase transition).

where L is the number of points in a lattice line, the lattice being regarded as a large square, and $N = L^2$.[†] The parameter $J\,(> 0)$ determines the energy of interaction of a pair of adjoining dipoles, which is $-J$ and $+J$ for like and unlike orientations of the two dipoles respectively. Then the configuration with the least energy is the "completely polarised" (ordered) configuration, in which all the dipoles are oriented in the same direction. This configuration is reached at absolute zero; as the temperature increases, the degree of ordering decreases, becoming zero at the transition point, when the two orientations of each dipole become equally probable.

The determination of the thermodynamic quantities requires the calculation of the partition function

$$Z = \sum_{(\sigma)} e^{-E(\sigma)/T} = \sum_{(\sigma)} \exp\left\{\theta \sum_{k,l}(\sigma_{kl}\sigma_{k,\,l+1}+\sigma_{kl}\sigma_{k+1,\,l})\right\}, \qquad (151.2)$$

taken over all the 2^N possible configurations ($\theta = J/T$). The equation

$$\exp(\theta\sigma_{kl}\sigma_{k'l'}) = \cosh\theta + \sigma_{kl}\sigma_{k'l'}\sinh\theta = \cosh\theta(1 + \sigma_{kl}\sigma_{k'l'}\tanh\theta)$$

is easily verified by expanding both sides in powers of θ and using the fact that all the $\sigma_{kl}^2 = 1$. The expression (151.2) can therefore be written

$$Z = (1-x^2)^{-N}S, \qquad (151.3)$$

where

$$S = \sum_{(\sigma)} \prod_{k,l=1}^{L}(1 + x\sigma_{kl}\sigma_{k,\,l+1})(1 + x\sigma_{kl}\sigma_{k+1,\,l}) \qquad (151.4)$$

and $x = \tanh\theta$.

The summand in (151.4) is a polynomial in the variables x and σ_{kl}. Since each point (k, l) has four neighbours, each σ_{kl} can appear in the polynomial in powers from zero to four. After summation over all the $\sigma_{kl} = \pm 1$ the terms containing odd powers of σ_{kl} vanish, and so a non-zero contribution comes only from terms containing σ_{kl} in powers 0, 2 or 4. Since $\sigma_{kl}^0 = \sigma_{kl}^2 = \sigma_{kl}^4 = 1$, each term of the polynomial which contains all the variables σ_{kl} in even powers gives a contribution to the sum which is proportional to the total number of configurations, 2^N.

Each term of the polynomial can be uniquely correlated with a set of lines or "bonds" joining various pairs of adjoining lattice points. For example, the diagrams shown in Fig. 70 correspond to the terms

(a) $x^2\sigma_{kl}\sigma_{k+1,\,l}^2\sigma_{k+1,\,l-1}$,

(b) $x^8\sigma_{kl}^2\sigma_{k+1,\,l}^2\sigma_{k+1,\,l-1}^2\sigma_{k,\,l-1}^4\sigma_{k-1,\,l-2}^2\sigma_{k-1,\,l-1}^2\sigma_{k-1,\,l-2}$,

(c) $x^{10}\sigma_{kl}^2\sigma_{k+1,\,l}^2\sigma_{k+1,\,l-1}^2\sigma_{k,\,l-1}^2\sigma_{k-1,\,l-2}^2\sigma_{k-1,\,l-1}^2$
$\quad\quad \times \sigma_{k-1,\,l-2}^2\sigma_{k-1,\,l-3}^2\sigma_{k-2,\,l-3}^2\sigma_{k-2,\,l-2}^2$.

[†] The number L is, of course, assumed macroscopically large, and edge effects (due to the special properties of points near the edges of the lattice) will be neglected throughout the following discussion.

Each line in the diagram is assigned a factor x and each end of each line a factor σ_{kl}.

The fact that a non-zero contribution to the partition function comes only from terms in the polynomial which contain all the σ_{kl} in even powers signifies geometrically that either 2 or 4 bonds must end at each point in the diagram. Hence the summation is taken only over closed diagrams, which may be self-intersecting (as at the point $k, l-1$ in Fig. 70b).

Fig. 70

Thus the sum S may be expressed in the form

$$S = 2^N \sum_r x^r g_r ,$$

(151.5)

where g_r is the number of closed diagrams formed from an (even) number r of bonds, each multiple diagram (e.g. Fig. 70c) being counted as one.

The subsequent calculation is in two stages: (1) the sum over diagrams of this type is converted into one over all possible closed loops, (2) the resulting sum is calculated by reducing it to the problem of the "random walk" of a point in the lattice.

We shall regard each diagram as consisting of one or more closed loops. For non-self-intersecting diagrams this is obvious; for example, the diagram in Fig. 70c consists of two loops. For self-intersecting diagrams, however, the resolution into loops is not unique: a given diagram may consist of different numbers of loops for different ways of construction. This is illustrated by Fig. 71, which shows three ways of representing the diagram in Fig. 70b as one or two non-self-intersecting loops or as one self-intersecting loop. Any intersection may similarly be traversed in three ways on more complicated diagrams.

Fig. 71

It is easy to see that the sum (151.5) can be extended to all possible sets of loops if, in computing the number of diagrams g_r, each diagram is taken with the sign $(-1)^n$, where n is the total number of self-intersections in the loops of a given set, since when this is done all the extra terms in the sum necessarily cancel. For example, the three diagrams in Fig. 71 have signs $+$, $+$, $-$ respectively, so that two of them cancel, leaving a single contribution to the sum, as they should. The new sum will also include diagrams with "repeated bonds", of which the simplest example is shown in Fig. 72a. These diagrams are not permissible, since some points have an odd number of bonds meeting at them, namely three, but in fact they cancel from the sum, as

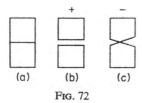

FIG. 72

they should: when the loops corresponding to such a diagram are constructed each bond in common can be traversed in two ways, without intersection (as in Fig. 72b) and with self-intersection (Fig. 72c); the resulting sets of loops appear in the sum with opposite signs, and so cancel. We can also avoid the need to take into account explicitly the number of intersections by using the geometrical result that the total angle of rotation of the tangent in going round a closed plane loop is $2\pi(l+1)$, where l is a (positive or negative) integer whose parity is the same as that of the number ν of self-intersections of the loop. Hence, if we assign a factor $e^{\frac{1}{2}i\phi}$ to each point of the loop $\left(\text{with the angle of rotation there } \phi = 0, \pm\frac{1}{2}\pi\right)$, the product of these factors after going round the whole loop will be $(-1)^{\nu+1}$, and for a set of s loops the resultant factor is $(-1)^{n+s}$, where $n = \sum\nu$.

Thus the number of intersections need not be considered if each point on the loop is taken with a factor $e^{\frac{1}{2}i\phi}$ and a further factor $(-1)^s$ is taken for the whole diagram (set of loops) in order to cancel the same factor in $(-1)^{n+s}$.

Let f_r denote the sum over single loops of length r (i.e. consisting of r bonds), each loop having a factor $e^{\frac{1}{2}i\phi}$ at each point on it. Then the sum over all pairs of loops with total number of bonds r is

$$\frac{1}{2!} \sum_{r_1+r_2=r} f_{r_1}f_{r_2};$$

the factor $1/2!$ takes into account the fact that the same pair of loops is obtained when the suffixes r_1 and r_2 are interchanged, and similarly for groups of three or more loops. Thus the sum S becomes

$$S = \sum_{s=0}^{\infty} (-1)^s \frac{1}{s!} \sum_{r_1, r_2, \ldots = 1}^{\infty} x^{r_1 + \cdots + r_s} f_{r_1} \cdots f_{r_s}.$$

Since S includes sets of loops with every total length $r_1 + r_2 + \ldots$, the numbers $r_1, r_2, \ldots$ in the inner sum take independently all values from 1 to ∞.[†]
Hence

$$\sum_{r_1, \ldots, r_s} x^{r_1 + \cdots + r_s} f_{r_1} \cdots f_{r_s} = \left(\sum_{r=1}^{\infty} x^r f_r \right)^s$$

and S becomes

$$S = \exp \left(- \sum_{r=1}^{\infty} x^r f_r \right). \tag{151.6}$$

This completes the first stage of the calculation.

It is now convenient to assign to each lattice point the four possible directions from it and to number them by a quantity $\nu = 1, 2, 3, 4$, say, as follows:

$$
\begin{array}{c}
2 \\
\uparrow \\
3 \leftarrow \cdot \rightarrow 1 \\
\downarrow \\
4
\end{array}
$$

We define as an auxiliary quantity $W_r(k, l, \nu)$ the sum over all possible paths of length r from some given point k_0, l_0, ν_0 to a point k, l, ν (each bond having as usual the factor $e^{\pm i\phi}$, where ϕ is the change of direction to the next bond); the final step to the point k, l, ν must not be from the point to which the arrow ν is directed.[‡] With this definition, $W_r(k_0, l_0, \nu_0)$ is the sum over all loops leaving the point k_0, l_0 in the direction ν_0 and returning to that point. It is evident that

$$f_r = \frac{1}{2r} \sum_{k_0, l_0, \nu_0} W_r(k_0, l_0, \nu_0): \tag{151.7}$$

both sides contain the sum over all single loops, but $\sum W_r$ contains each loop $2r$ times, since it can be traversed in two opposite directions and can be assigned to each of r starting points on it.

[†] Loops with more than N points make no contribution to the sum, since they must necessarily contain repeated bonds.

[‡] In fact $W_r(k, l, \nu)$ depends, of course, only on the differences $k - k_0$, $l - l_0$.

From the definition of $W_r(k, l, v)$ we have the recurrence relations

$$
\begin{aligned}
W_{r+1}(k, l, 1) &= W_r(k-1, l, 1)+e^{-\tfrac{1}{2}i\pi}W_r(k, l-1, 2)\\
&\quad+0+e^{\tfrac{1}{2}i\pi}W_r(k, l+1, 4),\\
W_{r+1}(k, l, 2) &= e^{\tfrac{1}{2}i\pi}W_r(k-1, l, 1)+W_r(k, l-1, 2)\\
&\quad+e^{-\tfrac{1}{2}i\pi}W_r(k+1, l, 3)+0,\\
W_{r+1}(k, l, 3) &= 0+e^{\tfrac{1}{2}i\pi}W_r(k, l-1, 2)\\
&\quad+W_r(k+1, l, 3)+e^{-\tfrac{1}{2}i\pi}W_r(k, l+1, 4),\\
W_{r+1}(k, l, 4) &= e^{-\tfrac{1}{2}i\pi}W_r(k-1, l, 1)+0\\
&\quad+e^{\tfrac{1}{2}i\pi}W_r(k+1, l, 3)+W_r(k, l+1, 4).
\end{aligned}
\tag{151.8}
$$

The method of constructing these relations is evident: for example, the point $k, l, 1$ can be reached by taking the last $(r+1)$th step from the left, from below or from above, but not from the right; the coefficients of W_r arise from the factors $e^{\pm i\phi}$.

Let A denote the matrix of the coefficients in equations (151.8) (with all k, l), written in the form

$$
W_{r+1}(k, l, v) = \sum_{k', l', v'} A(klv \mid k'l'v')\, W_r(k', l', v').
$$

The method of constructing these equations enables us to associate with this matrix an intuitive picture of a point moving step by step through the lattice with a "transition probability" per step from one point to another which is equal to the corresponding element of the matrix A; its elements are in fact zero except when either k or l changes by ± 1 and the other remains constant, i.e. the point traverses only one bond per step. It is evident that the "probability" of traversing a length r will be given by the matrix A^r. In particular the diagonal elements of this matrix give the "probability" that the poin, will return to its original position after traversing a loop of length r, i.e. they are equal to $W_r(k_0, l_0, v_0)$. Hence

$$
\operatorname{tr} A^r = \sum_{k_0, l_0, v_0} W_r(k_0, l_0, v_0).
$$

Comparison with (151.7) shows that

$$
f_r = \frac{1}{2r}\operatorname{tr} A^r = \frac{1}{2r}\sum_i \lambda_i^r,
$$

where the λ_i are the eigenvalues of the matrix A. Substituting this expression in (151.6) and interchanging the order of summation over i and r, we obtain

$$
\begin{aligned}
S &= \exp\left\{-\frac{1}{2}\sum_i\sum_{r=1}^{\infty}\frac{1}{r}x^r\lambda_i^r\right\}\\
&= \exp\left\{\tfrac{1}{2}\sum_i \log(1-x\lambda_i)\right\}\\
&= \prod_i \sqrt{(1-x\lambda_i)}.
\end{aligned}
\tag{151.9}
$$

The matrix Λ is easily diagonalised with respect to the suffixes k and l by using a Fourier transformation:

$$W_r(p, q, v) = \sum_{k, l=0}^{L} e^{-2\pi i(pk+ql)/L} W_r(k, l, v). \tag{151.10}$$

Taking Fourier components on both sides of equations (151.8), we find that each equation contains only $W_r(p, q, v)$ with the same p, q, so that the matrix Λ is diagonal with respect to p and q. For given p, q its elements are

$$\Lambda(pqv \mid pqv') = \begin{bmatrix} \varepsilon^{-p} & \alpha^{-1}\varepsilon^{-q} & 0 & \alpha\varepsilon^q \\ \alpha\varepsilon^{-p} & \varepsilon^{-q} & \alpha^{-1}\varepsilon^p & 0 \\ 0 & \alpha\varepsilon^{-q} & \varepsilon^p & \alpha^{-1}\varepsilon^q \\ \alpha^{-1}\varepsilon^{-p} & 0 & \alpha\varepsilon^p & \varepsilon^q \end{bmatrix},$$

where $\alpha = e^{\frac{1}{4}i\pi}$, $\varepsilon = e^{2\pi i/L}$.

For given p, q a simple calculation shows that

$$\prod_{i=1}^{4} (1-x\lambda_i) = \det(\delta_{vv'} - x\Lambda_{vv'})$$

$$= (1+x^2)^2 - 2x(1-x^2)\left(\cos\frac{2\pi p}{L} + \cos\frac{2\pi q}{L}\right).$$

Hence, from (151.3) and (151.9), we finally obtain the partition function

$$Z = 2^N(1-x^2)^{-N} \prod_{p, q=0}^{L} \left[(1+x^2)^2 \right.$$

$$\left. - 2x(1-x^2)\left(\cos\frac{2\pi p}{L} + \cos\frac{2\pi q}{L}\right)\right]^{1/2}. \tag{151.11}$$

The thermodynamic potential is[†]

$$\Phi = -T \log Z$$

$$= -NT \log 2 + NT \log (1-x^2)$$

$$- \frac{1}{2}T \sum_{p, q=0}^{L} \log\left[(1+x^2)^2 - 2x(1-x^2)\left(\cos\frac{2\pi p}{L} + \cos\frac{2\pi q}{L}\right)\right]$$

[†] In the model under discussion the temperature affects only the ordering of dipole orientations, not the distances between dipoles (the "thermal expansion coefficient" of the lattice is zero). It is then immaterial whether we consider the free energy or the thermodynamic potential.

or, changing from summation to integration,

$$\Phi = -NT \log 2 + NT \log (1-x^2)$$

$$-\frac{NT}{2(2\pi)^2} \int_0^{2\pi} \int_0^{2\pi} \log \left[(1+x^2)^2 - 2x(1-x^2)(\cos \omega_1 + \cos \omega_2) \right] d\omega_1 \, d\omega_2$$

$$(151.12)$$

(remembering that $x = \tanh (J/T)$).

Let us now examine this expression. The function $\Phi(T)$ has a singularity at the value of x for which the argument of the logarithm in the integrand can vanish. As a function of ω_1 and ω_2, this argument is a minimum for $\cos \omega_1 = \cos \omega_2 = 1$, when it equals $(1+x^2)^2 - 4x(1-x^2) = (x^2+2x-1)^2$. This expression has a minimum value of zero for only one (positive) value of x, $x_c = \sqrt{2}-1$; the corresponding temperature T_c ($\tanh (J/T_c) = x_c$) is the phase transition point.

The expansion of $\Phi(t)$ in powers of $t = T-T_c$ near the transition point includes a singular term as well as the regular part. Here we are interested only in the singular term, the regular part being simply replaced by its value at $t = 0$. To find the form of the singular term, we expand the argument of the logarithm in (151.12) in powers of ω_1, ω_2 and t about the minimum; the integral then becomes

$$\int_0^{2\pi} \int_0^{2\pi} \log \left[c_1 t^2 + c_2(\omega_1^2 + \omega_2^2) \right] d\omega_1 \, d\omega_2,$$

where c_1 and c_2 are constants. Carrying out the integration, we find that the thermodynamic potential near the transition point has the form

$$\Phi \cong a + \tfrac{1}{2} b(T-T_c)^2 \log |T-T_c|, \qquad (151.13)$$

where a and b are further constants (with $b > 0$). The potential itself is continuous at the transition point, but the specific heat becomes infinite in accordance with the formula

$$C \cong - bT_c \log |T-T_c|, \qquad (151.14)$$

which is symmetrical about the transition point.

In this model, the order parameter is represented by the mean dipole moment at the lattice point (the spontaneous polarisation of the lattice), which is non-zero below the transition point and zero above it. The temperature dependence of this quantity can also be ascertained; near the transition point, the order parameter tends to zero according to

$$\eta = \text{constant} \times (T_c - T)^{1/8} \qquad (151.15)$$

(L. Onsager, 1947).[†]

[†] A comparatively simple method of solving this problem is given by N. V. Vdovichenko, *Soviet Physics JETP* **21**, 350, 1965.

The correlation function is defined as the mean value of the product of the fluctuations of the dipole moment at two lattice points. The correlation radius is found to tend to infinity as $1/|T-T_c|$ when $T \to T_c$, and at the point $T = T_c$ itself the correlation function decreases with increasing distance according to

$$\langle \Delta\sigma_{kl}\, \Delta\sigma_{mn} \rangle \propto [(k-m)^2 + (l-n)^2]^{-1/8}.$$

These results, and those of solving the problem of the properties of the same model in an external field, show that the behaviour of the model near the phase transition point satisfies the requirements of the scale invariance hypothesis. The critical indices have the following values:

$$\alpha = 0, \quad \beta = \tfrac{1}{8}, \quad \gamma = \tfrac{7}{4}, \quad \delta = 15, \quad \varepsilon = 0, \quad \mu = \tfrac{8}{15},$$
$$\nu = 1, \quad \zeta = \tfrac{1}{4}; \tag{151.16}$$

ζ is found from (148.7) with $d = 2$.[†]

§ 152. Van der Waals theory of the critical point

It has been noted in § 83 that the critical point of liquid/gas phase transitions is a singularity of the thermodynamic functions of the substance. The physical nature of this singularity is similar to that which occurs at phase transition points of the second kind: in the latter case it is due to the increased fluctuations of the order parameter, and in the approach to the critical point there are increased fluctuations of the density of the substance. This analogy in the physical nature results in a certain analogy in the possible mathematical description of the two phenomena, to be discussed in § 153.

First, however, let us consider as a necessary preliminary the description of critical phenomena based on neglecting the fluctuations. In such a theory (analogous to the Landau approximation in the theory of phase transitions of the second kind) the thermodynamic quantities for the substance as functions of the variables V and T are assumed to have no singularities, so that they can be expanded in powers of the small changes in these variables. The results given in the rest of this section therefore depend only on the vanishing of the derivative $(\partial P/\partial V)_T$.

Let us first ascertain the conditions for the substance to be stable when

$$(\partial P/\partial V)_T = 0. \tag{152.1}$$

[†] As regards the critical indices, a logarithmic increase corresponds to a zero exponent (see after (148.7)).

In deriving the thermodynamic inequalities in § 21, we started from the condition (21.1), which led to the inequality (21.2), which is satisfied if the conditions (21.3), (21.4) hold. The case (152.1) of interest here corresponds to the particular case of the extremum conditions with the equality sign in (21.4):

$$\frac{\partial^2 E}{\partial S^2} \frac{\partial^2 E}{\partial V^2} - \left(\frac{\partial^2 E}{\partial V \, \partial S} \right)^2 = 0. \tag{152.2}$$

The quadratic form in (21.2) may now be either positive or zero, depending on the values of δS and δV, and so the question whether $E - T_0 S + P_0 V$ has a minimum requires further investigation.

We must obviously examine the case where in fact the equality sign occurs in (21.2):

$$\frac{\partial^2 E}{\partial S^2} (\delta S)^2 + 2 \frac{\partial^2 E}{\partial S \, \partial V} \delta S \, \delta V + \frac{\partial^2 E}{\partial V^2} (\delta V)^2 = 0. \tag{152.3}$$

Using (152.2), this equation may be written

$$\frac{1}{\partial^2 E / \partial S^2} \left(\frac{\partial^2 E}{\partial S^2} \delta S + \frac{\partial^2 E}{\partial S \, \partial V} \delta V \right)^2 = \frac{1}{\partial^2 E / \partial S^2} \left[\delta \frac{\partial E}{\partial S} \right]^2 = (C_v / T)(\delta T)^2 = 0.$$

Thus the equation (152.3) implies that we must consider deviations from equilibrium at constant temperature ($\delta T = 0$).

At constant temperature the original inequality (21.1) becomes $\delta F + P \, \delta V > 0$. Expanding δF in powers of δV and making use of the assumption that $\partial^2 F / \partial V^2 = -(\partial P / \partial V)_T = 0$, we find

$$\frac{1}{3!} \left(\frac{\partial^2 P}{\partial V^2} \right)_T \delta V^3 + \frac{1}{4!} \left(\frac{\partial^3 P}{\partial V^3} \right)_T \delta V^4 + \ldots < 0.$$

If this inequality holds for all δV, we must have[†]

$$(\partial^2 P / \partial V^2)_T = 0, \qquad (\partial^3 P / \partial V^3)_T < 0. \tag{152.4}$$

Let us now consider the equation of state of a substance near the critical point. Instead of the variables T and V, it will be more convenient to use the variables T and n, where n is the number of particles per unit volume. We shall also use the notation

$$t = T - T_c, \qquad p = P - P_c, \qquad \eta = n - n_c. \tag{152.5}$$

[†] It may be noted that the case of equality in (21.3) is impossible in this discussion, since the condition (21.4) would then be violated. The simultaneous vanishing of the two expressions (21.3) and (21.4) is also impossible: if we add a further condition to the vanishing of $(\partial P / \partial V)_T$ and $(\partial^2 P / \partial V^2)_T$, there result three equations in two unknowns, which in general have no common solution.

In these variables, the conditions (152.1) and (152.4) become

$$(\partial p/\partial \eta)_t = 0, \quad (\partial^2 p/\partial \eta^2)_t = 0, \quad (\partial^3 p/\partial \eta^3)_t > 0 \quad \text{at} \quad t = 0. \quad (152.6)$$

Taking only the first terms in the expansion in powers of small t and η, we write the dependence of pressure on temperature and density in the form

$$p = bt + 2at\eta + 4B\eta^3 \quad (152.7)$$

with constant a, b and B. There are no terms in η and η^2 in this expansion, from the first two conditions (152.6), and from the third condition $B > 0$. When $t > 0$, all states of a homogeneous body are stable (there is nowhere a separation into phases), and we must have $(\partial p/\partial \eta)_t > 0$ for all η, whence $a > 0$. The terms in $t\eta^2$ and $t^2\eta$ may be omitted, since they are certainly small in comparison with that in $t\eta$; the term in $t\eta$ itself must be retained, since it contributes to the derivative

$$(\partial p/\partial \eta)_t = 2at + 12B\eta^2 \quad (152.8)$$

needed below.

The expression (152.7) determines the isotherms of a homogeneous substance near the critical point (Fig. 73). These have a form similar to the van der Waals isotherms (Fig. 19, § 84). When $t < 0$, they have a minimum and a

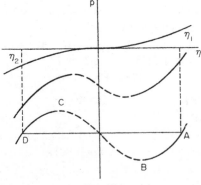

FIG. 73

maximum, and a horizontal section (AD on the bottom isotherm) corresponding to the condition (84.2) gives the equilibrium transition from liquid to gas. Taking V in this condition to be the molecular volume

$$v = \frac{1}{n} \sim \frac{1}{n_c} - \frac{\eta}{n_c^2}, \quad (152.9)$$

we can write the condition as

$$\int_A^D \eta \, dp = \int_{\eta_1}^{\eta_2} \eta \left(\frac{\partial p}{\partial \eta}\right)_t d\eta = 0. \quad (152.10)$$

After substitution of (152.8), this condition gives the following values for the densities of the two phases in equilibrium:

$$\eta_1 = -\eta_2 = \sqrt{(-at/2B)}. \tag{152.11}$$

The densities η_1' and η_2' corresponding to the limits of the metastable regions (points B and C in Fig. 73) are given by the condition $(\partial p/\partial \eta)_t$, whence[†]

$$\eta_1' = -\eta_2' = \sqrt{(-at/6B)}. \tag{152.12}$$

Substitution of (152.11) makes the sum of the last two terms in (152.7) zero. Hence

$$p = bt \qquad (t < 0) \tag{152.13}$$

is the equation of the curve of equilibrium of liquid and vapour in the *pt*-plane (and therefore $b > 0$).[‡] According to the Clapeyron–Clausius equation (82.2), the heat of evaporation near the critical point is

$$q \cong bT_c(\eta_1 - \eta_2)/n_c^2. \tag{152.14}$$

It then follows from (152.11) that as $t \to 0$ this heat tends to zero according to

$$q \propto \sqrt{-t}. \tag{152.15}$$

From (16.10) it follows that the specific heat C_p becomes infinite at the critical point while $(\partial p/\partial \eta)_t$ tends to zero. From (152.8) we find

$$C_p \propto 1/(at + 6B\eta^2). \tag{152.16}$$

In particular, for states on the equilibrium curve, $\eta \propto \sqrt{-t}$, and therefore $C_p \sim 1/(-t)$.

Lastly, let us consider in this theory the density fluctuations near the critical point. The necessary formulae have already been derived in § 116, and to apply them we need only establish the specific form of ΔF_t, the change in the total free energy of the body in a deviation from equilibrium.

We write

$$\Delta F_t = \int (F - \bar{F})\,dV,$$

where F is the free energy per unit volume and $\bar{F}$ its mean value, which is constant through the body, and expand $F - \bar{F}$ in powers of the density fluctuation $\Delta n = n - \bar{n}$ (or, equivalently, $\Delta\eta = \eta - \bar{\eta}$) at constant temperature. The first term in the expansion is proportional to Δn, and on integration

---

[†] In a theory which takes account of the singularity of the thermodynamic quantities at the boundary of the metastable states, there is no curve BC.

[‡] When $t > 0$, equation (152.13) determines the critical isochore, the curve of constant density ($\eta = 0$) passing through the critical point.

over the volume it vanishes, since the total number of particles in the body remains unchanged. The second term is[†]

$$\frac{1}{2}\left(\frac{\partial^2 F}{\partial n^2}\right)_T (\Delta n)^2 = \frac{1}{2n_c}\left(\frac{\partial P}{\partial \eta}\right)_t (\Delta n)^2.$$

Together with this term, which vanishes at the critical point itself, we must also include another term quadratic in Δn, due to the inhomogeneity of a body with fluctuating density. Without repeating here the arguments given in § 146, we can state at once that this term is quadratic in the first derivatives of Δn with respect to the coordinates; in an isotropic medium, such a term must be the square of the gradient. Thus we arrive at an expression of the form[‡]

$$\Delta F_t = \int \left\{ \frac{1}{2n_c}\left(\frac{\partial p}{\partial \eta}\right)_t (\Delta n)^2 + g\left(\frac{\partial \Delta n}{\partial r}\right)^2 \right\} dV. \qquad (152.17)$$

Now expressing Δn as a Fourier series (116.9), we can bring this to the form (116.10), with the function

$$\phi(k) = \frac{1}{n_c}\left(\frac{\partial p}{\partial \eta}\right)_t + 2gk^2$$

$$= \frac{2}{n_c}(at + 6B\bar{\eta}^2) + 2gk^2,$$

and then, using (116.14), find the Fourier transform of the required correlation function:

$$\nu(k) = \tfrac{1}{2}T[at + 6B\bar{\eta}^2 + gn_c k^2]^{-1}; \qquad (152.18)$$

since the denominator here is small, the term 1 in $\nu(k)$ may be neglected. This formula is entirely analogous to (146.8). The correlation function $\nu(r)$ in the coordinate representation therefore has the same form (146.11), with the correlation radius

$$r_c = [gn_c/(at + B\bar{\eta}^2)]^{1/2}. \qquad (152.19)$$

In particular, $r_c \propto t^{-1/2}$ on the critical isochore ($\bar{\eta} = 0$).

[†] Since the free energy F is referred to a fixed (unit) volume of substance, not to a fixed number of particles in it, $(\partial F/\partial n)_T = \mu$. The second derivative is

$$\left(\frac{\partial^2 F}{\partial n^2}\right)_T = \left(\frac{\partial \mu}{\partial n}\right)_T = \frac{1}{n}\left(\frac{\partial P}{\partial n}\right)_T,$$

since, for $T = $ constant, $d\mu = v\,dP$, where $v = 1/n$ is the molecular volume.

[‡] The fact that ΔF_t is expressed as an integral of a function of a point in the body (not of two points as in the general expression (116.8)) results from the assumption that Δn varies slowly; the long-wavelength components of the density fluctuations are considered.

§ 153. **Fluctuation theory of the critical point**

The formulae derived in § 152 enable us to establish a certain analogy between the thermodynamic description of the properties of a substance near the critical point and near phase transition points of the second kind.

To do so, we proceed in the spirit of the Landau theory by first regarding η not as a definite function of P and T but as an independent variable whose equilibrium value is given by minimising a thermodynamic potential $\Phi(P, T, \eta)$. The latter is to be chosen so that this minimisation in fact gives the correct equation (152.7). The requirement is met by the expression[†]

$$\Phi(P, T, \eta) = \Phi_0(P, T) + \frac{N}{n_c^2}[-(p - bt)\eta + at\eta^2 + B\eta^4]. \qquad (153.1)$$

Comparing (153.1) and (144.3), we now see that there is an analogy between the description of a phase transition of the second kind in an external field in the Landau theory and the description of a critical point between a liquid and a gas in the van der Waals theory. In the latter case, the order parameter is represented by the change in density of the substance, $\eta = n - n_c$, and the external field by the difference

$$h = p - bt. \qquad (153.2)$$

If $\Phi(t, h)$ is the thermodynamic potential of the body near the phase transition point of the second kind (for some fixed value of the pressure), the expression $\Phi(t, p - bt)$ gives the form of the thermodynamic potential near the critical point. The whole of the discussion in § 146 about the change from the potential Φ to the potential Ω is valid for any case, and so the analogy applies also to the potentials Ω in the two problems.

It has been shown in § 147 how we can go from the thermodynamic potential Ω in the Landau theory to the effective Hamiltonian which describes the phase transition in exact fluctuation theory. The analogy mentioned therefore leads us to expect that the laws of behaviour of the thermodynamic quantities near the critical point are the same (with the appropriate change of significance of η and h) as the limiting laws in the fluctuation region of the phase transition of the second kind in an external field (which is described by only one order parameter).

It must be emphasised immediately that this identification can only be approximate. In the theory of phase transitions based on the effective

---

[†] The (here unimportant) coefficient of the square bracket is chosen so that after minimisation the expression (153.1) gives the correct potential $\Phi(P, T)$. It may seem strange that (153.1) is not symmetrical in p and t, in that the coefficient of η^2 does not involve p. In reality, the η^2 term is significant only if the coefficient $p - bt$ of η is small, and in that case we can equally well write $at\eta^2$ or $ap\eta^2/b$.

Hamiltonian (147.6), there is an exact symmetry with respect to the transformation $h \to -h$, $\eta \to -\eta$ (because of the exact vanishing of the third-order term $\sim \eta^3$). In critical-point theory, this symmetry is only approximate; the absence from (153.1) (and therefore from the effective Hamiltonian) of terms which violate this symmetry is due only to neglecting them as small in comparison with the remaining terms. We can therefore assert only that the principal terms in the limiting relationships in the two cases must be the same.[†]

In the theory of phase transitions with $t > 0$ and $h = 0$, we have $\eta = 0$; with $t < 0$ and $h \to 0$, two phases are in equilibrium which have non-zero values η_1 and η_2 of the order parameter, where $\eta_1 = -\eta_2$ (the points A and A' in Fig. 64b, § 144); this equation is an exact consequence of the above-mentioned symmetry of the effective Hamiltonian. For a critical point, the corresponding equation is

$$p - bt = 0, \tag{153.3}$$

which determines the critical isochore ($\eta = 0$, i.e. $n = n_c$) for $t > 0$, and the liquid/vapour equilibrium line for $t < 0$. The equation $\eta_2 = -\eta_1$ here signifies the symmetry of the phase equilibrium line in the $t\eta$-plane, and an extension of the analogy shows that these values tend to zero as $t \to 0$ according to

$$\eta_1 = -\eta_2 \propto (-t)^\beta, \tag{153.4}$$

with the same exponent as in (148.5).[‡] But, since the invariance of the effective Hamiltonian under a change in the sign of η (when $h = 0$) is only approximate, the question arises of the limiting temperature dependence of the sum $\eta_1 + \eta_2$. From the foregoing discussion we can say only that it is of a higher order of smallness than η_1 and η_2 themselves; we shall return to this at the end of the section.

Figure 74 shows the phase diagram in the ηt-plane. The region of separation into two phases is hatched, and its boundary is a symmetrical curve, in accordance with (153.4).

[†] The analogy described must not, of course, conceal the physical difference between the two phenomena: for a phase transition of the second kind, we have a whole curve of transition points, which separates (in the PT-plane) the regions of existence of two phases with different symmetry. A critical point, however, is an isolated point (at the end of an equilibrium curve) in the phase diagram of two phases having the same symmetry.

[‡] Here and in the rest of § 153, when referring to the critical indices for transitions of the second kind, we mean specifically the values of these indices for transitions described by only one order parameter, with an effective Hamiltonian of the form (147.6). The van der Waals theory of the critical point corresponds to the values of the indices given for the Landau theory in the footnote to (148.11).

FIG. 74

The heat of evaporation is related to the difference $\eta_1-\eta_2$ by (152.14). It therefore tends to zero as $|t| \to 0$, by the same law

$$q \propto (-t)^\beta. \tag{153.5}$$

The general equation of state of a homogeneous substance throughout the neighbourhood of the critical point (in the ηT-plane) may be written

$$p-bt = \pm|\eta|^\delta f(t/|\eta|^{1/\beta}), \tag{153.6}$$

where the two signs correspond to $\eta > 0$ and $\eta < 0$ (B. Widom, 1965). This formula corresponds to equation (148.18) in the theory of phase transitions (when solved for h).

The same considerations of analyticity apply to $f(x)$ in (153.6) as were discussed in § 149 for transitions of the second kind.

For example, with a given non-zero value of η, a change in the sign of t nowhere causes a passage through the critical point; hence $t = 0$ is not a singularity of (153.6). This function may therefore be expanded in integral powers of t. Thus $f(x)$ is expanded in integral powers of x. The first two terms of this expansion are $1+c_1x$, so that the equation of state becomes

$$p-bt \propto \pm|\eta|^\delta \left[1+c_1\frac{t}{|\eta|^{1/\beta}}+ \dots\right] \quad \text{for} \quad |t| \ll |\eta|^{1/\beta}; \tag{153.7}$$

the first term in the expansion corresponds to the definition (148.10) for a strong field in the theory of phase transitions. In Fig. 74 the dotted lines show diagrammatically the boundaries of the region to which this equation of state applies. In this region, two further limiting cases may be distinguished. If $t \ll p$ (in particular, on the critical isotherm, i.e. on the line $t = 0$), then

$$p \propto \pm|\eta|^\delta. \tag{153.8}$$

If $t \gg p$ (in particular, on the critical isobar, i.e. on the line $p = 0$), then

$$t \propto \pm|\eta|^\delta. \tag{153.9}$$

A comparison of (153.8) and (153.9) shows that there is symmetry between p and t, as there should be.[†]

Similarly, for a given non-zero value of t, the value $\eta = 0$ is not a singularity. Hence, for $t > 0$ and $\eta \to 0$, the function (153.6) can be expanded in integral powers of η, and the expansion can contain only odd powers, again because of the symmetry of the effective Hamiltonian under a simultaneous change of sign of η and h. From this it follows that[‡]

$$f(x) \propto x^{\beta\delta}(c_1 x^{-\beta} + c_3 x^{-3\beta} + \ldots) \quad \text{for} \quad x \to \infty;$$

the factor $x^{\beta\delta}$ cancels the non-integral power η^δ, and the expansion variable $x^{-\beta} \propto \eta$. Thus the equation of state becomes

$$p - bt \propto t^\gamma[c_1 \eta + c_3 \eta^3 t^{-2\beta} + \ldots] \quad \text{for} \quad t \gg |\eta|^{1/\beta}, \quad (153.10)$$

where we have used the equation $\beta\delta = \beta + \gamma$, (148.14). The region where this equation is valid is also shown diagrammatically in Fig. 74. The first term of the expansion (153.10) corresponds to the relation $\eta = \chi h \propto h t^{-\gamma}$ in the theory of phase transitions in a weak field.

The behaviour of the derivatives of p of various orders with respect to η (with t constant) depends on the direction in the ηt-plane along which the critical point is approached. If the approach is along the critical isotherm ($t = 0$), the function $p(\eta)$ is given by (153.8). In practice, the value of δ lies between 4 and 5. Hence not only $(\partial p/\partial \eta)_t$ but the derivatives of several higher orders tend to zero along the critical isotherm.

If the critical point is approached in any other direction (lying outside the region of separation into two phases, i.e. along a radial line $t = \text{constant} \times |\eta|$ with the constant positive), the inequality $t \gg |\eta|^{1/\beta}$ holds, since in practice $1/\beta > 1$. The equation of state then gives

$$(\partial p/\partial \eta)_t \propto t^\gamma \to 0,$$

and the second derivative is

$$(\partial^2 p/\partial \eta^2)_t \propto \eta t^{\gamma - 2\beta} = t^{\gamma - \beta} \eta/t^\beta.$$

The factor $\eta/t^\beta \ll 1$, and $t^{\gamma - \beta} \to 0$, since in practice $\gamma > \beta$. Thus the derivative $(\partial^2 p/\partial \eta^2)_t$ also tends to zero.

[†] For $t \propto \eta^\delta$, the argument of $f(x)$ in (153.6) is $x \propto t/t^{1/\beta\delta} \ll 1$, since in practice $\beta\delta = \beta + \gamma > 1$. This proves that the case $t \gg p$ is in fact possible in the equation of state (153.7).

[‡] The case $x \to -\infty$ is not realistic, since values of $|\eta|^{1/\beta} \ll |t|$ with $t < 0$ are in the region of separation.

The behaviour of the specific heat of the substance in the critical region can be ascertained from the expression for the thermodynamic potential,

$$\Phi(p, t) = |h|^{2-\alpha}\, \phi(t/|h|^{1/(\beta+\gamma)}), \quad h = p - bt, \qquad (153.11)$$

which is written down immediately by analogy with (149.7) in the theory of phase transitions, with the identical substitution of indices $d\nu = 2-\alpha$, $\mu/\nu = 1/(\beta+\gamma)$. Without repeating the whole argument, we can put down at once, by analogy with (149.9) and (149.10), the necessary limiting expressions:[†]

$$\Phi(p, t) \propto t^{2-\alpha} \quad \text{for} \quad t > 0, \quad h \to 0, \qquad (153.12)$$
$$\Phi(p, t) \propto (-t)^{2-\alpha}\,[1 + c_1|h|/(-t)^{\beta+\gamma}] \quad \text{for} \quad t < 0, \quad h \to 0. \qquad (153.13)$$

By twice differentiating (153.12), we find the specific heat on the critical isochore $p - bt = 0$, $t > 0$:

$$C_v \propto t^{-\alpha}. \qquad (153.14)$$

Since differentiation with $h = 0$ and $t > 0$ signifies differentiation with $\eta = 0$, this is the specific heat at constant volume. Thus the specific heat C_v on the critical isochore behaves like C_p in a phase transition of the second kind.

According to (16.10),

$$C_p - C_v \propto \frac{(\partial p/\partial t)_\eta^2}{(\partial p/\partial \eta)_t}.$$

As the critical point is approached, the derivative $(\partial p/\partial t)_\eta$ tends to a constant limit b, as is easily seen from the equation of state (153.7) or (153.10). Hence

$$C_p \propto (\partial p/\partial \eta)_t^{-1}. \qquad (153.15)$$

The divergence of this expression as the critical point is approached is stronger than that of C_v; the term C_v has therefore been omitted in comparison with C_p.

Lastly, let us consider the asymmetry of the curve of coexistence of phases near the critical point (V. L. Pokrovskiĭ, 1972). As already mentioned, this asymmetry can occur only when terms are taken into account in the effective Hamiltonian which make it no longer symmetrical under the transformation

---

[†] It may be recalled that Φ here denotes (as in § 149) the singular part of the thermodynamic potential. While it is a small correction to the principal (non-singular) part, it also gives a similar correction to the other thermodynamic potentials. On the phase equilibrium curve, the characteristic magnitude of this correction is $\propto t^{2-\alpha}$, a point that will be made use of in § 154.

$h \to -h$, $\eta \to -\eta$. The first such term is[†] $\sim \eta^2 h$; its occurrence may be formally regarded as resulting from the replacement of t in the effective Hamiltonian by $t + \text{constant} \times h$, so that

$$a\eta^2 t \to a\eta^2(t + \text{constant} \times h).$$

This change in the effective Hamiltonian causes a similar change in the thermodynamic potential expressed as a function of h and t:

$$\Phi(h, t) \to \Phi(h, t + \text{constant} \times h).$$

Near the curve of coexistence of phases, the function $\Phi(h, t)$ is given by (153.13); the required density is found by differentiating with respect to h. The result is

$$[V\eta]_{h \to \pm 0} = (-\partial\Phi/\partial h)_{h \to \pm 0}$$
$$\propto \mp c_1(-t)^\beta + (2-\alpha) \, \text{constant} \times (-t)^{1-\alpha}.$$

The first term gives the known values (153.4) of the densities on the symmetrical coexistence curve; this term disappears in the sum $\eta_1 + \eta_2$, leaving

$$\eta_1 + \eta_2 \propto (-t)^{1-\alpha}. \tag{153.16}$$

which gives the required form. In practice $1 - \alpha > \beta$, so that the asymmetry is in fact relatively slight: $(\eta_1 + \eta_2)/\eta_1 \to 0$ as $t \to 0$. The sum $\eta_1 + \eta_2$ is in practice positive; this means that, when it is taken into account, the shape of the coexistence curve is changed in the manner shown in Fig. 75.

FIG. 75

[†] The inclusion in the effective Hamiltonian of a term $\sim \eta^3 t$ would not destroy the symmetry, since such a term could be eliminated by simply applying a transformation $\eta \to \eta + \text{constant} \times t$. In this connection it may be recalled (cf. the footnote following (149.7)) that η in the effective Hamiltonian is only a continuum integration variable, and therefore the transformation mentioned does not affect the integral over states.

CHAPTER XV

SURFACES

§ 154. Surface tension

HITHERTO we have entirely neglected effects resulting from the presence of surfaces of separation between different bodies.[†] Since, as the size of a body (i.e. the number of particles in it) increases, surface effects increase much more slowly than volume effects, the neglect of surface effects in the study of volume properties of bodies is entirely justified. There are, however, a number of phenomena which depend in fact on the properties of surfaces of separation.

The thermodynamic properties of such an interface are entirely described by one quantity, a function of the state of the bodies, defined as follows. We denote by $\mathscr{S}$ the area of the interface, and consider a process whereby this area undergoes a reversible change by an infinitesimal amount $d\mathscr{S}$. The work done in such a process is obviously proportional to $d\mathscr{S}$, and so can be written as

$$dR = \alpha \, d\mathscr{S}. \tag{154.1}$$

The quantity α thus defined is a fundamental characteristic of the interface, and is called the *surface-tension coefficient*.

Formula (154.1) is exactly analogous to the formula $dR = -P \, dV$ for the work done in a reversible change in the volume of a body. We may say that α plays the same part in relation to the surface as the pressure does in relation to the volume. In particular, we can easily show that the force on unit length of the perimeter of any part of the interface is equal in magnitude to α and is directed tangentially to the surface and along the inward normal to the perimeter.

Here we have assumed that α is positive. The fact that it must indeed always be positive is shown by the following argument. If $\alpha < 0$, the contour bounding the surface would be subject to forces along the outward normal, i.e. tending to "stretch" the surface; the interface between two phases would

---

[†] In reality, of course, phases in contact are separated by a thin transition layer, but the structure of this is of no interest here, and we may regard it as a geometrical surface.

therefore tend to increase without limit, and the phases would mix and cease to exist separately. If $\alpha > 0$, on the other hand, the interface tends to become as small as possible (for a given volume of the two phases). Hence, for example, if one isotropic phase is surrounded by another, it will take the form of a sphere (the effect of an external field, e.g. gravity, being neglected, of course).

Let us now consider in more detail the surface tension at the interface between two isotropic phases, liquid and vapour, of the same substance. If an interface between two phases in equilibrium is concerned, it must be remembered that their pressure and temperature are in a definite functional relation given by the equation of the phase equilibrium curve, and α is then essentially a function of only one independent variable, not of two.

When surface effects are neglected, the differential of the energy of a system of two phases (of the same substance), for a given total volume V of the system, is $dE = T\,dS + \mu\,dN$; in equilibrium, the temperatures T and chemical potentials μ of the two phases are equal, and this equation can therefore be written for the whole system. When the presence of surface effects is taken into account, the right-hand side of the equation must clearly include also the expression (154.1):

$$dE = T\,dS + \mu\,dN + \alpha\,d\mathfrak{s}. \tag{154.2}$$

It is, however, more convenient to take as the fundamental thermodynamic quantity not the energy but the potential Ω, the thermodynamic potential in terms of the independent variables T and μ (and the volume V). The convenience of Ω in this case arises because T and μ are quantities which have equal values in the two phases, whereas the pressures are not in general equal when surface effects are taken into account; see § 156. For the differential $d\Omega$, again with $V = $ constant, we have

$$d\Omega = -S\,dT - N\,d\mu + \alpha\,d\mathfrak{s}. \tag{154.3}$$

The thermodynamic quantities (such as E, Ω and S) for the system under consideration can be written as the sum of a "volume" part and a "surface" part. This division, however, is not unique, since the number of particles in each phase is indeterminate to the extent of the number of particles in the transition layer between the phases; the same is true of the volumes of the phases. This indeterminacy is of just the same order of magnitude as the surface effects with which we are concerned. The division can be made unique if the following reasonable condition is imposed: the volumes V_1 and V_2 of the two phases are defined so that, in addition to the equation $V_1 + V_2 = V$, where V is the total volume of the system, the equation $n_1 V_1 + n_2 V_2 = N$ is satisfied, where N is the total number of particles in the system, and $n_1 =$

$n_1(\mu, T)$ and $n_2 = n_2(\mu, T)$ are the numbers of particles per unit volume in each phase (the phases being regarded as unbounded).

These two equations determine the choice of the volumes V_1 and V_2 (and the numbers of particles $N_1 = n_1 V_1$, $N_2 = n_2 V_2$), and hence also the volume parts of all other thermodynamic quantities. We shall denote volume parts by the suffix 0, and surface parts by the suffix s; by definition, $N_s = 0$.

From (154.3) we have, for constant T and μ (and therefore constant α), $d\Omega = \alpha\, d\mathfrak{s}$; it is therefore evident that $\Omega_s = \alpha\mathfrak{s}$, and so

$$\Omega = \Omega_0 + \alpha\mathfrak{s}. \tag{154.4}$$

Since the entropy $S = -(\partial\Omega/\partial T)_{\mu, \mathfrak{s}}$, the surface part of it is[†]

$$S_s = -\partial\Omega_s/\partial T = -\mathfrak{s}\, d\alpha/dT. \tag{154.5}$$

Next, let us find the surface free energy; since $F = \Omega + N\mu$ and $N_s = 0$,

$$F_s = \alpha\mathfrak{s}. \tag{154.6}$$

The surface energy is

$$E_s = F_s + TS_s = (\alpha - T\, d\alpha/dT)\mathfrak{s}. \tag{154.7}$$

The quantity of heat absorbed in a reversible isothermal change of surface area from $\mathfrak{s}_1$ to $\mathfrak{s}_2$ is

$$Q = T(S_{s2} - S_{s1}) = -T(d\alpha/dT)(\mathfrak{s}_2 - \mathfrak{s}_1). \tag{154.8}$$

The sum of the heat Q and the work $R = \alpha(\mathfrak{s}_2 - \mathfrak{s}_1)$ in this process is equal to the change in energy $E_{s2} - E_{s1}$, as it should be.

The law of corresponding states (§ 84) can be qualitatively applied to the surface tension between a liquid and its vapour. From this law we should expect that the dimensionless ratio of α to a quantity with dimensions erg/cm² formed from the critical temperature and critical pressure will be a universal function of the reduced temperature T/T_c:

$$\alpha/(T_c P_c^2)^{1/3} = f(T/T_c); \tag{154.9}$$

at temperatures well below the critical value, this ratio is approximately 4.

At a critical point, the liquid and gas phases become identical, the interface between them ceases to exist, and the surface tension coefficient must become zero. Using the ideas of the fluctuation theory of the critical point, we can express the manner of its tending to zero in terms of the critical indices defined in § 148.

[†] The coefficient α is a function of only one independent variable; for such a function the partial derivatives with respect to μ and T have no meaning. But, by putting $N_s = -(\partial\Omega_s/\partial\mu)_T = 0$, we have formally assumed that $(\partial\alpha/\partial\mu)_T = 0$; in this case we clearly have $d\alpha/dT = (\partial\alpha/\partial T)_\mu$, and this has been used in (154.5).

As the critical point is approached, the width of the transition layer between the two phases increases and becomes of macroscopic size. Sufficiently close to the critical point, it must be of the order of the correlation radius r_c of the fluctuations. In order to find the surface tension, we need only now multiply the width r_c by the characteristic thermodynamic potential density, or rather its singular part, which determines the critical phenomena. This density is $\propto (-t)^{2-\alpha}$; see the footnote to (153.12).[†] Thus the surface tension coefficient is $\propto r_c(-t)^{2-\alpha}$; since $r_c \propto (-t)^{-\nu}$ from (148.6), and $2-\alpha = \nu d = 3\nu$ from (149.2), we have finally

$$\alpha \propto (T_c-T)^{2\nu} \tag{154.10}$$

(B. Widom, 1965). In practice, $2\nu \cong 1.3$.[‡]

§ 155. Surface tension of crystals

The surface tension of an anisotropic body, a crystal, is different at different faces; it may be said to depend on the direction of the face, i.e. on its Miller indices. The form of this dependence is somewhat unusual. Firstly, the difference in the values of α for two crystal planes with arbitrarily close directions is itself arbitrarily small, i.e. the surface tension can be represented as a continuous function of the direction of the face. It can nevertheless be shown that this function nowhere has a definite derivative. For example, if we consider a set of crystal planes intersecting along a common line, and denote by ϕ the angle of rotation around this line, which defines the direction of the plane, we find that the function $\alpha = \alpha(\phi)$ has two different derivatives for every value of ϕ, one for increasing and the other for decreasing values of the argument.[§]

Let us suppose that the surface tension is a known function of the direction of the faces. The question arises how this function can be used to determine the equilibrium form of the crystal. (It must be emphasised that the crystal shape observed under ordinary conditions is determined by the conditions of growth of the crystal and is not the equilibrium shape.) The equilib-

[†] The use of the same letter for the surface tension and the critical index α is unlikely to cause any misunderstanding.

[‡] The width of the transition layer is equal to the correlation radius in the van der Waals theory also, and therefore the form $\propto (-t)^{2-\alpha-\nu}$ for the surface tension (without the use of (149.2)) would be valid in that theory. With $\alpha = 0$, $\nu = \frac{1}{2}$, the result would be $(-t)^{3/2}$.

[§] This is discussed in more detail by L. D. Landau, *Sbornik v chest' 70-letiya A. F. Ioffe*, p. 44, Moscow, 1950; translation in *Collected Papers of L. D. Landau*, p. 540, Pergamon, Oxford, 1965.

rium form is determined by the condition for the free energy F to be a minimum (for given T, μ and volume V of the crystal) or, what is the same thing, by the condition for its surface part to be a minimum. The latter is

$$F_s = \oint \alpha \, d\mathfrak{s},$$

the integral being taken over the whole surface of the crystal; for an isotropic body $\alpha = \text{constant}$, $F_s = \alpha \mathfrak{s}$, and the equilibrium form is determined simply by the condition for the total area $\mathfrak{s}$ to be a minimum, i.e. it is a sphere.

Let $z = z(x, y)$ be the equation of the surface of the crystal, and let $p = \partial z/\partial x$, $q = \partial z/\partial y$ denote the derivatives which determine the direction of the surface at each point; α can be expressed as a function of these, $\alpha = \alpha(p, q)$. The equilibrium form is given by the condition

$$\int \alpha(p, q) \sqrt{(1+p^2+q^2)} \, dx \, dy = \text{minimum} \qquad (155.1)$$

with the added condition of constant volume

$$\int z \, dx \, dy = \text{constant}. \qquad (155.2)$$

This variational problem leads to the differential equation

$$\frac{\partial}{\partial x} \frac{\partial f}{\partial p} + \frac{\partial}{\partial y} \frac{\partial f}{\partial q} = 2\lambda, \qquad (155.3)$$

where

$$f(p, q) = \alpha(p, q) \sqrt{(1+p^2+q^2)} \qquad (155.4)$$

and λ is a constant.

Next, we have by definition $dz = p \, dx + q \, dy$; with the auxiliary function

$$\zeta = px + qy - z, \qquad (155.5)$$

we find $d\zeta = x \, dp + y \, dq$ or

$$x = \partial \zeta/\partial p, \quad y = \partial \zeta/\partial q, \qquad (155.6)$$

ζ being here regarded as a function of p and q. Writing the derivatives with respect to x and y in (155.3) as Jacobians, multiplying both sides by $\partial(x, y)/\partial(p, q)$ and using (155.6), we obtain the equation

$$\frac{\partial(\partial f/\partial p, \partial \zeta/\partial q)}{\partial(p, q)} + \frac{\partial(\partial \zeta/\partial p, \partial f/\partial q)}{\partial(p, q)} = 2\lambda \frac{\partial(\partial \zeta/\partial p, \partial \zeta/\partial q)}{\partial(p, q)}.$$

This equation has an integral $f = \lambda \zeta = \lambda(px + qy - z)$, or

$$z = \frac{1}{\lambda} \left(p \frac{\partial f}{\partial p} + q \frac{\partial f}{\partial q} - f \right). \qquad (155.7)$$

This is just the equation of the envelope of the family of planes

$$px + qy - z = \alpha(p, q)\sqrt{(1 + p^2 + q^2)}/\lambda, \tag{155.8}$$

where p and q act as parameters.

This result can be expressed in terms of the following geometrical construction. On each radius vector from the origin we mark off a segment ot length proportional to $\alpha(p, q)$, where p and q correspond to the direction of that radius vector.[†] A plane is drawn through the end of each segment af right angles to the radius vector; then the envelope of these planes gives the equilibrium form of the crystal (G. V. Vul'f).

It can be shown[‡] that the unusual behaviour of the function α mentioned at the beginning of this section may have the result that the equilibrium form of the crystal determined by this procedure will include a number of plane areas corresponding to crystal planes with small values of the Miller indices. The size of the plane areas rapidly decreases as the Miller indices increase. In practice this means that the equilibrium shape will consist of a small number of plane areas which are joined by rounded regions instead of intersecting at sharp angles.

§ 156. Surface pressure

The condition that the pressures of two phases in contact are equal has been derived (in § 12) from the equality of the forces exerted on the interface by the two phases; as elsewhere, surface effects were neglected. It is clear, however, that if the interface is not plane a displacement of it will in general change its area and therefore the surface energy. In other words, the existence of a curved interface between the phases leads to additional forces, as a result of which the pressures of the two phases will not be equal. The difference between them is called the *surface pressure*.

This quantity is determined by the condition of mechanical equilibrium: the sum of the forces acting on each phase at the interface should be zero. This sum is in turn given by the derivative of a thermodynamic potential with respect to the displacement of the interface, the other variables corresponding to this potential remaining constant; cf. (11.3) and (15.11). In particular, in differentiating the potential Ω we must regard P, μ and the total volume of the system as constants.

Let us consider two isotropic phases (two liquids, or a liquid and a vapour). We shall assume that one of the phases (phase 1) is a sphere (of ra-

---

[†] The direction cosines of the radius vector are proportonal to p, q, -1.

[‡] See the paper quoted in the first footnote to this section.

dius r) surrounded by the other phase. Then the pressure is constant within each phase, and the total thermodynamic potential Ω of the system is

$$\Omega = -P_1 V_1 - P_2 V_2 + \alpha \mathring{s}, \qquad (156.1)$$

the first two terms forming the volume part of the potential; the suffixes 1 and 2 refer to the two phases.

The pressures of the two phases satisfy the equations $\mu_1(P_1, T) = \mu_2(P_2, T) \equiv \mu$, where μ is the common value of the two chemical potentials. Hence, for constant μ and T, we must regard P_1 and P_2 as constant also, and likewise the surface-tension coefficient α. Since $V = V_1 + V_2$ is constant, we find as the condition of mechanical equilibrium

$$\left(\frac{\partial \Omega}{\partial r} \right)_{T, \mu, V} = -(P_1 - P_2) \frac{dV_1}{dr} + \alpha \frac{d\mathring{s}}{dr}.$$

Finally, substituting $V_1 = 4\pi r^3/3$, $\mathring{s} = 4\pi r^2$, we obtain the required formula:

$$P_1 - P_2 = 2\alpha/r. \qquad (156.2)$$

For a plane interface ($r \to \infty$) the two pressures are equal, as we should expect.

Formula (156.2) determines only the difference between the pressures in the two phases. We shall now calculate each of them separately.

The pressures P_1 and P_2 satisfy the equation $\mu_1(P_1, T) = \mu_2(P_2, T)$. The common pressure P_0 in the two phases when the interface is plane is determined at the same temperature by the relation $\mu_1(P_0, T) = \mu_2(P_0, T)$. Subtraction of these two equations gives

$$\mu_1(P_1, T) - \mu_1(P_0, T) = \mu_2(P_2, T) - \mu_2(P_0, T). \qquad (156.3)$$

Assuming that the differences $\delta P_1 = P_1 - P_0$, $\delta P_2 = P_2 - P_0$ are relatively small and expanding the two sides of equation (156.3) in terms of δP_1 and δP_2, we find

$$v_1 \, \delta P_1 = v_2 \, \delta P_2, \qquad (156.4)$$

where v_1 and v_2 are the molecular volumes (see (24.12)). Combining this with formula (156.2) written in the form $\delta P_1 - \delta P_2 = 2\alpha/r$, we find the required δP_1 and δP_2 as

$$\delta P_1 = \frac{2\alpha}{r} \frac{v_2}{v_2 - v_1}, \quad \delta P_2 = \frac{2\alpha}{r} \frac{v_1}{v_2 - v_1}. \qquad (156.5)$$

For a drop of liquid in a vapour, $v_1 \ll v_2$; regarding the vapour as an ideal gas, we have $v_2 = T/P_2 \cong T/P_0$, and so

$$\delta P_l = 2\alpha/r, \quad \delta P_g = 2v_l \alpha P_0/rT, \qquad (156.6)$$

where for clarity the suffixes l and g are used in place of 1 and 2. Thus we see that the vapour pressure over the drop is greater than the saturated vapour pressure over a plane liquid surface, and increases with decreasing radius of the drop.

When the drop is sufficiently small and $\delta P_g/P_0$ is no longer small, formulae (156.6) become invalid, since the large variation of the vapour volume with pressure means that the expansion used to derive (156.4) from (156.3) is no longer permissible. For a liquid, whose compressibility is small, the effect of a change of pressure is slight, and the left-hand side of (156.3) can again be replaced by $v_l \, \delta P_l$. On the right-hand side we substitute the chemical potential of the vapour in the form $\mu = T \log P_g + \chi(T)$, obtaining

$$\delta P_l = P_l - P_0 = (T/v_l) \log (P_g/P_0).$$

Since in this case $\delta P_l \gg \delta P_g$, the difference $P_l - P_0$ can be replaced by $P_l - P_g$; using formula (156.2) for the surface pressure, we then have finally

$$\log (P_g/P_0) = 2\alpha v_l/rT. \tag{156.7}$$

For a bubble of vapour in a liquid we similarly obtain the same formulae (156.6), (156.7) but with the opposite signs.

§ 157. **Surface tension of solutions**

Let us now consider an interface between a liquid solution and a gas phase (a gas and a solution of it in a liquid, a liquid solution and its vapour, etc.).

As in § 154, we divide all thermodynamic quantities for the system under consideration into volume and surface parts, the manner of division being determined by the conditions $V = V_1 + V_2$, $N = N_1 + N_2$ for the volume and number of solvent particles. That is, the total volume V of the system is divided between the two phases in such a way that, on multiplying V_1 and V_2 by the corresponding numbers of solvent particles per unit volume, and adding, we obtain just the total number N of solvent particles in the system. Thus by definition the surface part $N_s = 0$.

As well as other quantities, the number of solute particles will also be written as a sum of two parts, $n = n_0 + n_s$. We may say that n_0 is the quantity of solute which would be contained in the volumes V_1 and V_2 if it were distributed in each with a constant concentration equal to the volume concentration of the corresponding solution. The number n_0 thus defined may be either greater or less than the actual total number n of solute particles. If $n_s = n - n_0 > 0$, this means that the solute accumulates at a higher concentration in the surface layer (called *positive adsorption*). If $n_s < 0$, the concentration in the surface layer is less than in the volume (*negative adsorption*).

The surface-tension coefficient of the solution is a function of two independent variables, not one. Since the derivative of the potential Ω with respect to the chemical potential is minus the corresponding number of particles, n_s can be found by differentiating $\Omega_s = \alpha \mathscr{S}$ with respect to the chemical potential μ' of the solute:[†]

$$n_s = -\partial\Omega_s/\partial\mu' = -\mathscr{S}(\partial\alpha/\partial\mu')_T. \qquad (157.1)$$

Let us assume that the pressure of the gas phase is so small that its effect on the properties of the liquid phase may be neglected. Then the derivative of α in formula (157.1), which must be taken along the phase equilibrium curve at the temperature concerned, can be replaced by the derivative at constant (viz. zero) pressure (and constant T). Regarding α as a function of the temperature and the concentration c of the solution, we can rewrite formula (157.1) as

$$n_s = -\mathscr{S}(\partial\alpha/\partial c)_T (\partial c/\partial\mu')_{T,P}. \qquad (157.2)$$

According to the thermodynamic inequality (96.7), the derivative $(\partial\mu'/\partial c)_{T,P}$ is always positive. Hence it follows from (157.2) that n_s and $(\partial\alpha/\partial c)_T$ have opposite signs. This means that, if the solute raises the surface tension (α increases with increasing concentration of the solution), it is negatively adsorbed. Substances which lower the surface tension are positively adsorbed.

If the solution is a weak one, the chemical potential of the solute is of the form $\mu' = T \log c + \psi(P, T)$; substituting this in (157.2), we find

$$n_s = -\mathscr{S}(c/T)(\partial\alpha/\partial c)_T. \qquad (157.3)$$

A similar formula,

$$n_s = -\mathscr{S}(P/T)(\partial\alpha/\partial P)_T, \qquad (157.4)$$

is obtained for the adsorption of a gas (at pressure P) by a liquid surface.

If not only the solution but also the adsorption from it is weak, α can be expanded in powers of c; we have approximately $\alpha = \alpha_0 + \alpha_1 c$, where α_0 is the surface tension at the interface between two phases of pure solvent. Then we have from (157.3) $\alpha_1 = -n_s T/\mathscr{S}c$, and hence

$$\alpha - \alpha_0 = -n_s T/\mathscr{S}. \qquad (157.5)$$

The resemblance between this formula and van't Hoff's formula for the osmotic pressure should be noticed (the volume being here replaced by the surface area).

[†] The coefficient α is now a function of two independent variables, e.g. μ and T; the derivative $\partial\Omega_s/\partial\mu'$ must be taken at constant T and chemical potential μ of the solvent. The condition $N_s = -(\partial\Omega_s/\partial\mu)_{\mu',T} = 0$ used here implies that we formally take $(\partial\alpha/\partial\mu)_{\mu',T} = 0$ and therefore we can write equation (157.1) (cf. the second footnote to § 154).

§ 158. Surface tension of solutions of strong electrolytes

The change in surface tension of a liquid when a strong electrolyte is dissolved in it can be calculated in a general form for weak solutions (L. Onsager and N. N. T. Samaras, 1934).

Let $w_a(x)$ denote the additional energy of an ion (of the ath kind) because of the free surface at a distance x from the ion ($w_a(x)$ tending to zero as $x \to \infty$). The ion concentration near the surface differs from its value c_a within the solution by a factor $e^{-w_a/T} \cong 1 - w_a/T$. The contribution of the surface to the total number of these ions in the liquid is therefore

$$n_{as} = -\frac{3c_a}{vT} \int_0^\infty w_a \, dx, \tag{158.1}$$

where v is the molecular volume of the solvent.

To calculate the surface tension, we begin from the relation

$$3 \, d\alpha = -\sum_a n_{as} \, d\mu_a', \tag{158.2}$$

where the summation is over all the kinds of ion in the solution. For weak solutions ($\mu_a' = T \log c_a + \psi_a$),

$$3 \, d\alpha = -T \sum_a \frac{n_{as}}{c_a} \, dc_a. \tag{158.3}$$

Substitution of (158.1) gives

$$d\alpha = \frac{1}{v} \sum_a dc_a \int_0^\infty w_a \, dx. \tag{158.4}$$

It will be seen from the subsequent results that the main contribution to the integral comes from distances x which are large compared with the distances between molecules but small compared with the Debye length $1/\varkappa$.

The energy w_a consists of two parts:

$$w_a = \frac{\varepsilon - 1}{\varepsilon(\varepsilon + 1)} \frac{e^2 z_a^2}{4x} + ez_a \phi(x). \tag{158.5}$$

The first term arises from the "image force" on a charge ez_a in a medium with dielectric constant ε at a distance x from its surface. Since $x \ll 1/\varkappa$, the screening effect of the ion cloud round the charge does not alter this energy. In the second term, $\phi(x)$ denotes the change (owing to the presence of the surface) in the field potential due to all the other ions in the solution. This term is unimportant here, however, since it disappears on substituting

(158.5) in (158.4) because of the electrical neutrality of the solution ($\sum c_a z_a = 0$, and therefore $\sum z_a \, dc_a = 0$).

Thus, carrying out the integration in (158.4), we find

$$d\alpha = \frac{(\varepsilon-1)e^2}{4\varepsilon(\varepsilon+1)v} \sum_a \log \frac{1}{\varkappa a_a} \, d(z_a^2 c_a).$$

The logarithmic divergence of the integral at both limits confirms the statement made above concerning the range of integration; we have naturally taken as the upper limit the screening length $1/\varkappa$, and as the lower limit a quantity a_a of the order of atomic dimensions, but different for the different kinds of ion. Since $\varkappa^2$ is proportional to the sum $\sum z_a^2 c_a$,[†] we see that the expression obtained is a total differential and so can be integrated directly, giving

$$\alpha-\alpha_0 = \frac{(\varepsilon-1)e^2}{8\varepsilon(\varepsilon+1)v} \sum_a c_a z_a^2 \log \frac{\lambda_a z_a^2}{\sum_b c_b z_b^2}, \tag{158.6}$$

where α_0 is the surface tension of the pure solvent and the λ_a are dimensionless constants.

This formula gives the solution of the problem. It should be noticed that the dissolution of a strong electrolyte increases the surface tension of the liquid.

§ 159. Adsorption

Adsorption in the restricted sense includes cases where the solute is concentrated at the surface of a solid or liquid *adsorbent*,[‡] and hardly any of it enters the volume of the adsorbent. The adsorbed film thus formed can be described by the *surface concentration* γ, defined as the number of particles of the *adsorbate* (adsorbed substance) per unit surface area. At low pressures of the gas from which adsorption occurs, the concentration γ must be proportional to the pressure;[§] at high pressures, however, γ increases less rapidly and tends to a limiting value corresponding to the formation of a *monomolecular layer* with the adsorbate molecules closely packed together.

Let μ' be the chemical potential of the adsorbate. By the same method as was used in § 96 for ordinary solutions we can derive for adsorption the thermodynamic inequality

$$(\partial\mu'/\partial\gamma)_T > 0, \tag{159.1}$$

[†] The expression for $\varkappa^2$ in a solution differs from (78.8) only by a factor ε in the denominator.

[‡] For definiteness we shall consider adsorption from a gas phase.

[§] This rule is, however, not obeyed in practice for adsorption on a solid surface, since such a surface is never sufficiently homogeneous.

which is entirely analogous to (96.7). From (157.1) we have

$$\gamma = -(\partial \alpha / \partial \mu')_T = -(\partial \alpha / \partial \gamma)_T (\partial \gamma / \partial \mu')_T, \tag{159.2}$$

and (159.1) therefore implies that

$$(\partial \alpha / \partial \gamma)_T < 0, \tag{159.3}$$

i.e. surface tension decreases as the surface concentration increases.

The minimum work which must be done to form the adsorbed film is equal to the corresponding change in the thermodynamic potential Ω:

$$R_{\min} = \mathcal{S}(\alpha - \alpha_0), \tag{159.4}$$

where α_0 is the surface tension on the surface before adsorption. Hence, using (91.4), we find the heat of adsorption

$$Q = -\mathcal{S}T^2 \left(\frac{\partial}{\partial T} \frac{\alpha - \alpha_0}{T} \right)_P. \tag{159.5}$$

The adsorbed film may be regarded as a kind of two-dimensional thermo-dynamic system, which may be either isotropic or anisotropic, despite the isotropy of the two volume phases.[†] The question arises of the possible types of symmetry of the film.

At the end of § 137, we have already mentioned that, although the exist-ence of a two-dimensional crystal lattice of arbitrarily great extent is impos-sible (since it is smoothed out by thermal fluctuations), a film may still show solid-crystal properties if it is of relatively small size. But the smoothed-out structure is still anisotropic, as noted at the end of § 138. Such an anisotropic film can have only an axis of symmetry perpendicular to its plane, and planes of symmetry passing through this axis. The only possible types of symmetry for such systems are therefore the point groups C_n and C_{nv}.

As with three-dimensional solids, different phases can exist in two-dimensional films. The equilibrium conditions for two phases in a film require the equality of their surface tensions as well as their temperatures and chemical potentials. The condition on the surface tensions corresponds to the condition of equal pressures in three-dimensional phases, and simply expresses the necessity for the forces exerted by each phase on the other to balance.

[†] Here we are considering adsorption on a liquid surface; adsorption on a solid surface is of no interest from this point of view, since, as mentioned above, such a surface is almost always inhomogeneous.

It may be noted that anisotropy of the interface between two isotropic phases (liquid and vapour) of the same pure substance is also possible in principle.

§ 160. Wetting

Let us consider adsorption on the surface of a solid from a vapour at a pressure close to the saturation value. The equilibrium concentration γ is determined by the condition that the chemical potential of the adsorbate μ' is equal to that of the vapour μ_g. Various cases can occur according to the dependence of μ' on γ.

Let us suppose that the quantity of adsorbate gradually increases and the adsorbed layer becomes a liquid film of macroscopic thickness. The surface concentration γ then becomes a conventionally defined quantity proportional to the film thickness l: $\gamma = \varrho l/m$, where m is the mass of a molecule and ϱ the density of the liquid. As the film thickness increases, the chemical potential of the substance forming it tends to μ_l, the chemical potential of the liquid in bulk. We shall measure the value of μ' (for given P and T) from this limiting value, i.e. write $\mu' + \mu_l$ in place of μ'; thus, by definition, $\mu' \to 0$ as $\gamma \to \infty$.

The chemical potential of the vapour can be written as $\mu_g = \mu_l(T) + T \log (P/P_0)$, where $P_0(T)$ is the saturated vapour pressure; here we have used the fact that the saturated vapour is, by definition, in equilibrium with the liquid, i.e. we must have $\mu_g = \mu_l$ when $P = P_0$.[†] The surface concentration is determined by the condition $\mu' + \mu_l = \mu_g$, or

$$\mu'(\gamma) = T \log (P/P_0). \qquad (160.1)$$

If this equation is satisfied by several values of γ, the one which corresponds to a stable state is that for which the potential Ω_s is a minimum. Taking the value per unit area of the surface, we obtain a quantity which may be called (in the general case of any film thickness) the "effective surface-tension coefficient" α at the solid–vapour boundary, and which takes into account the presence of the layer between them. Integrating the relation (159.2), we can write

$$\alpha(\gamma) = \int_{\gamma}^{\infty} \gamma \frac{d\mu'}{d\gamma} \, d\gamma + \alpha_{sl} + \alpha_{lg}. \qquad (160.2)$$

The constant is so chosen that as $\gamma \to \infty$ the function $\alpha(\gamma)$ becomes the sum of the surface tensions at the "bulk"-phase (solid–liquid and liquid–gas) interfaces.

It may also be recalled that a necessary condition for the thermodynamic stability of a state is the inequality (159.1), which is valid for any γ.

---

† The liquid itself is regarded as incompressible, i.e. we neglect the dependence of its chemical potential on the pressure.

Let us now consider some typical cases which may occur, depending on the nature of the function $\mu'(\gamma)$. In the diagrams given below, the continuous curve shows the form of this function in the region of macroscopically thick films of liquid, while the broken curve is that for adsorbed films of "molecular" thickness. It is, of course, not strictly possible to represent the function in these two regions in one diagram to the same scale, and to this extent the diagrams are a convention.

In the first case shown (Fig. 76a) the function $\mu'(\gamma)$ decreases monotonically with increasing γ (i.e. with increasing film thickness) in the range of macroscopic thicknesses. For molecular dimensions the function $\mu'(\gamma)$ always tends to $-\infty$ as $T \log \gamma$ when $\gamma \to 0$, this law corresponding to a "weak solution" of the adsorbate on the surface. The equilibrium concentration is determined, according to (160.1), by the point of intersection of the curve with a horizontal line $\mu' = \text{constant} \leqslant 0$. In this case, this can occur only at molecular concentrations, i.e. ordinary molecular adsorption must occur, as discussed in § 159.

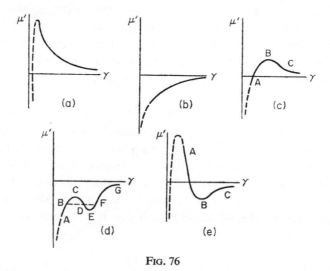

Fig. 76

If $\mu'(\gamma)$ increases monotonically but is everywhere negative (Fig. 76b), then in equilibrium a liquid film of macroscopic thickness is formed on the surface of the adsorbent. In particular, when the pressure $P = P_0$ (saturated vapour), the film formed must be so thick that the properties of the substance in it do not differ from those of the liquid in bulk, i.e. the saturated vapour must be in contact with its own liquid phase. In that case we say that the liquid *completely wets* the solid surface in question.

More complicated cases are also thoretically possible. For example, if the function $\mu'(\gamma)$ passes through zero and has a maximum (Fig. 76c) we have a case of wetting but with formation of a film stable only at thicknesses below a certain limit. The film of finite thickness corresponding to the point A is in equilibrium with the saturated vapour. This state is separated from the other stable state (equilibrium of the solid wall with the bulk liquid) by a metastable region AB and a region of complete instability BC.

The type of curve shown in Fig. 76d corresponds to a film which is unstable over a certain range of thickness. The line BF which cuts off equal areas BCD and DEF joins points B and F which have equal values of α (and equal μ'), as is easily seen from (160.2). The branches AB and FG correspond to stable films; the range CE is completely unstable, while BC and EF are metastable.

The two boundaries of the instability range (the points B and F) correspond in this case to macroscopic film thicknesses. Instability in the range from a certain macroscopic thickness to molecular thicknesses would correspond to a curve of the type shown in Fig. 76e, but such a curve would more likely lead to non-wetting, since the limit of stability would correspond to a point on BC where a horizontal line cuts off equal areas below the upper part and above the lower part of the curve. But this is usually impossible, since the latter area, which is related to the van der Waals forces (see below), is small compared with the former, which is related to the considerably greater forces at molecular distances. This means that the surface tension everywhere on BC is greater than that which would correspond to molecular adsorption on a solid surface, and the film will therefore be metastable.

The chemical potential of the liquid film (measured from μ_l) represents the difference between the energy of the substance in the film and that in the bulk liquid. It is therefore clear that μ' is determined by the interaction forces between atoms at distances large compared with atomic dimensions and $\sim l$ (van der Waals forces). The potential $\mu'(l)$ can be calculated in a general form, the result being expressed in terms of the permittivities of the solid wall and the liquid (see Part 2).

§ 161. The angle of contact

Let us consider three bodies in contact, solid, liquid and gas (or one solid and two liquid), distinguishing them by suffixes 1, 2, 3 respectively and denoting the surface-tension coefficients at the interfaces by α_{12}, α_{13}, α_{23} (Fig. 77).

Three surface-tension forces act on the line where all three bodies meet, each force being in the interface between the corresponding pair of bodies. We denote by θ the angle between the surface of the liquid and the plane surface of the solid, called the *angle of contact*. The value of this angle is determined by the condition of mechanical equilibrium: the resultant of the three surface-tension forces must have no component along the surface of the

FIG. 77

solid. Thus $\alpha_{13} = \alpha_{12} + \alpha_{23} \cos \theta$, whence

$$\cos \theta = (\alpha_{13} - \alpha_{12})/\alpha_{23}. \tag{161.1}$$

If $\alpha_{13} > \alpha_{12}$, i.e. the surface tension between the gas and the solid is greater than that between the solid and the liquid, than $\cos \theta > 0$ and the angle of contact is acute, as in Fig. 77. If $\alpha_{13} < \alpha_{12}$, however, the angle of contact is obtuse.

From the expression (161.1) we see that in any actual case of stable contact the inequality

$$|\alpha_{13} - \alpha_{12}| \leqslant \alpha_{23} \tag{161.2}$$

must necessarily hold, since otherwise the condition of equilibrium would lead to an imaginary value of the angle θ, which has no meaning. On the other hand, if α_{12}, α_{13}, α_{23} are regarded as the values of the corresponding coefficients for each pair of bodies by themselves, without the third one, then it may well happen that the condition (161.2) is not satisfied. Actually, however, it must be remembered that when three different substances are in contact there may in general be an adsorbed film of each substance on the interface between the other two, and this lowers the surface tension. The resulting coefficients α will certainly satisfy the inequality (161.2), and such adsorption will necessarily occur if the inequality would not be satisfied without it.

If the liquid completely wets the solid surface, then a macroscopically thick liquid film, not an adsorbed film, is formed on the surface. The gas will therefore be everywhere in contact with the same liquid substance, and the surface tension between the solid and the gas is not involved at all. The condition of mechanical equilibrium gives simply $\cos \theta = 1$, i.e. the angle of contact is zero.

Similar arguments are valid for contact between three bodies of which none is solid: a liquid drop (3 in Fig. 78) on the surface of another liquid (1) adjoining a gas (2). In this case the angles of contact θ_1 and θ_2 are determined

FIG. 78

by the vanishing of the resultant of the three surface-tension forces, i.e. of their vector sum:

$$\alpha_{12} + \alpha_{13} + \alpha_{23} = 0. \tag{161.3}$$

Here it is evident that none of the quantities α_{12}, α_{13}, α_{23} can be greater than the sum or less than the difference of the other two.

§ 162. Nucleation in phase transitions

If a substance is in a metastable state, it will sooner or later enter another state which is stable. For example, supercooled vapour in time condenses to a liquid; a superheated liquid is converted into vapour. This change occurs in the following manner. Owing to fluctuations, small quantities of a new phase are formed in an originally homogeneous phase; for example, droplets of liquid form in a vapour. If the vapour is the stable phase, these droplets are always unstable and eventually disappear. If the vapour is supercooled, how-ever, then when the droplets formed in it are sufficiently large they are stable and in time begin to grow and form a kind of centre of condensation for the vapour. The droplets must be sufficiently large in order to compensate the unfavourable energy change when a liquid–vapour interface is formed.[†]

Thus there is a certain minimum or critical size necessary for a *nucleus*, as it is called, of a new phase formed in a metastable phase, in order for it to become a centre for formation of the new phase. Since one phase or the other is stable for sizes less than and greater than the critical, the "critical nucleus" is in unstable equilibrium with the metastable phase. In what follows we shall discuss the probability that such nuclei occur.[‡] Because of the

---

[†] It should be borne in mind that this mechanism of formation of a new phase can actually occur only in a sufficiently pure substance. In practice, the centres of formation of the new phase are usually various kinds of "impurity" (dust particles, ions, etc.).

[‡] The calculation of the probability that nuclei of any size occur is given in Problem 2, and illustrates the relationships described.

rapid decrease in the probability of fluctuations of increasing size, the beginning of the phase transition is determined by the probability that nuclei of just this minimum necessary size occur.

Let us consider the formation of nuclei in isotropic phases: the formation of liquid droplets in supercooled vapour, or of vapour bubbles in superheated liquid. A nucleus may be regarded as spherical, since, owing to its very small size, the effect of gravity on its shape is entirely negligible. For a nucleus in equilibrium with the surrounding medium we have, from (156.2), $P' - P = 2\alpha/r$, and the radius of the nucleus is therefore

$$r_{cr} = 2\alpha/(P' - P); \tag{162.1}$$

the primed and unprimed letters everywhere refer to the nucleus and to the main (metastable) phase respectively.

According to the general formula (112.1), the probability w of a fluctuation producing a nucleus is proportional to $\exp(-R_{min}/T)$, where R_{min} is the minimum work needed to form the nucleus. Since the temperature and chemical potential of the nucleus have the same values as in the surrounding medium (the main phase), this work is given by the change in the potential Ω in the process. Before the formation of the nucleus, the volume of the metastable phase was $V + V'$ and its potential $\Omega = -P(V + V')$; after the formation of the nucleus of volume V', the potential Ω of the whole system is $-PV - P'V' + \alpha\mathscr{S}$. We therefore have

$$R_{min} = -(P' - P)V' + \alpha\mathscr{S}. \tag{162.2}$$

For a spherical nucleus $V' = 4\pi r^3/3$ and $\mathscr{S} = 4\pi r^2$, and replacing r by (162.1) we find

$$R_{min} = 16\pi\alpha^3/3(P' - P)^2. \tag{162.3}$$

As in § 156, we denote by P_0 the pressure of both phases (at a given temperature T) when the interface between them is plane; in other words, P_0 is the pressure for which the given value of T is the ordinary phase transition point, from which the superheating or supercooling is measured. If the metastable phase is only slightly superheated or supercooled, the differences $\delta P = P - P_0$, $\delta P' = P' - P_0$ are relatively small and satisfy the equation (156.4):

$$v' \, \delta P' = v \, \delta P, \tag{162.4}$$

where v' and v are the molecular volumes of the nucleus and the metastable phase. Replacing $P' - P$ by $\delta P' - \delta P$ in (162.3) and expressing $\delta P'$ in terms of δP from (162.4), we find for the probability of formation of a nucleus in a slightly superheated or supercooled phase

$$w \propto \exp\left\{-\frac{16\pi\alpha^3 v'^2}{3T(v - v')^2 (\delta P)^2}\right\}. \tag{162.5}$$

In the formation of vapour bubbles in a superheated liquid we can neglect v in this formula in comparison with v':

$$w \propto \exp\left\{-\frac{16\pi\alpha^3}{3T(\delta P)^2}\right\}.\qquad(162.6)$$

In the formation of liquid droplets in a supercooled vapour we can neglect v' in (162.5) in comparison with v, and substitute $v = T/P \cong T/P_0$. This gives

$$w \propto \exp\left\{-\frac{16\pi\alpha^3 v'^2 P_0^2}{3T^3(\delta P)^2}\right\}.\qquad(162.7)$$

The degree of metastability can be defined by the difference $\delta T = T - T_0$ between the temperature T of the metastable phase (with which the nucleus is in equilibrium) and the temperature T_0 of equilibrium of the two phases when the interface is plane, instead of by δP. According to the Clapeyron–Clausius formula, δT and δP are related by

$$\delta P = \frac{q}{T_0(v - v')}\,\delta T,$$

where q is the molecular heat of the transition from the metastable phase to the nucleus phase. Substituting δP in (162.5), we obtain for the probability of formation of a nucleus

$$w \propto \exp\left\{-\frac{16\pi\alpha^3 v'^2 T_0}{3q^2(\delta T)^2}\right\}.\qquad(162.8)$$

If saturated vapour is in contact with a solid surface (the wall of a vessel) which is completely wetted by the liquid, condensation of the vapour will occur without nucleation, directly on this surface. The formation of a liquid film on the solid surface in this case does not require work to be done to form the interface, and so the existence of a metastable phase (supercooling of the vapour) is impossible.

For the same reason, superheating of a solid with a free surface is in general impossible. This is because usually liquids completely wet the surface of a solid phase of the same substance; consequently, the formation of a liquid layer on the surface of a melting body does not require work to be done to form a new surface.

The formation of nuclei within a crystal on melting can, however, occur if the necessary conditions of heating are maintained: the body must be heated internally and its surface kept at a temperature below the melting point. The probability of formation of nuclei then depends on elastic deformations accompanying the creation of liquid droplets within the solid.

PROBLEMS

PROBLEM 1. Determine the probability of formation of a nucleus of a liquid on a solid surface for a given (non-zero) value of the angle of contact θ.

SOLUTION. The nucleus will have the shape of a segment of a sphere with base radius $r \sin \theta$, r being the radius of the sphere. Its volume is $V = \frac{1}{3}\pi r^3(1-\cos\theta)^2(2+\cos\theta)$, and the surface areas of the curved part and the base are respectively $2\pi r^2(1-\cos\theta)$ and $\pi r^2 \sin^2\theta$. Using the relation (161.1) for the angle of contact, we find that the change in Ω_s on formation of the nucleus is

$$\alpha \cdot 2\pi r^2(1-\cos\theta)-\alpha\cos\theta \cdot \pi r^2\sin^2\theta = \alpha\pi r^2(1-\cos\theta)^2(2+\cos\theta),$$

where α is the surface-tension coefficient between the liquid and the vapour. This change in Ω_s is the same as would occur in the formation, in the vapour, of a spherical nucleus of volume V and surface tension

$$\alpha_{\text{eff}} = \alpha\left(\frac{1-\cos\theta}{2}\right)^{2/3}(2+\cos\theta)^{1/3}.$$

Accordingly, the required formulae for the formation of nuclei are obtained from those derived in the text on replacing α by α_{eff}.

PROBLEM 2. Find the probability of formation of a nucleus of arbitrary dimensions.

SOLUTION. We regard the metastable phase as an external medium containing the nucleus, and calculate the work of formation of the nucleus from formula (20.2): $R_{\min} = \Delta(E-T_0S+P_0V)$ or, since in this case the process occurs at constant temperature equal to the temperature of the medium, $R_{\min} = \Delta(F+P_0V)$. To determine this quantity, it is sufficient to consider only the amount of substance which enters the other phase (since the state of the remaining substance in the metastable phase remains unchanged). Again denoting the quantities pertaining to the substance in the original and the new phase by unprimed and primed letters respectively, we have

$$R_{\min} = [F'(P')+PV'+\alpha\mathfrak{s}]-[F(P)+PV] = \Phi'(P')-\Phi(P)-(P'-P)V'+\alpha\mathfrak{s}; \quad (1)$$

for a nucleus in unstable equilibrium with the metastable phase we should have $\Phi'(P') = \Phi(P)$ and thus return to (162.2).

Assuming the degree of metastability to be small, we have $\Phi'(P') \cong \Phi'(P)+(P'-P)V'$, and (1) thus reduces to

$$R_{\min} = n[\mu'(P)-\mu(P)]+\alpha\mathfrak{s},$$

where $n = V'/v'$ is the number of particles in the nucleus. For a spherical nucleus,

$$R_{\min} = -\frac{4\pi r^3}{3v'}[\mu(P)-\mu'(P)]+4\pi r^2\alpha. \quad (2)$$

In the metastability range, $\mu(P) > \mu'(P)$ and hence the first term (the volume term) is negative. The expression (2) may be said to describe the potential barrier which has to be overcome for the formation of a stable nucleus. It has a maximum at

$$r = r_{\text{cr}} = 2\alpha v'/[\mu(P)-\mu'(P)],$$

corresponding to the critical radius of the nucleus. For $r < r_{cr}$ a decrease of r is energetically favourable and the nucleus is absorbed; for $r > r_{cr}$ an increase of r is favourable and the nucleus grows.[†]

§ 163. The impossibility of the existence of phases in one-dimensional systems

A problem of theoretical interest is that of the possibility of existence of more than one phase in one-dimensional (linear) systems, i.e. those in which the particles lie along a line. The following argument shows that in fact thermodynamic equilibrium between two homogeneous phases in contact at a single point and having arbitrarily large extent in length is not possible (L. D. Landau, 1950).

To prove this, let us imagine a linear system consisting of an alternation of sections formed by two different phases. Let Φ_0 be the thermodynamic potential of this system, without allowance for the existence of points of contact between different phases, i.e. the thermodynamic potential of the total amounts of the two phases without regard to their division into sections. To take into account the effect of the points of contact, we note that the system may be formally regarded as a "solution" of these points in the two phases. If the "solution" is weak, the thermodynamic potential Φ of the system will be

$$\Phi = \Phi_0 + nT \log (n/eL) + n\psi,$$

where n is the number of points of contact in a length L. Hence

$$\partial \Phi / \partial n = T \log (n/L) + \psi.$$

When the "concentration" n/L is sufficiently small (i.e. for a small number of sections of different phases), $\log(n/L)$ is large and negative, and therefore $\partial \Phi / \partial n < 0$. Thus Φ decreases with increasing n, and since Φ must tend to a minimum n will increase (until the derivative $\partial \Phi / \partial n$ becomes positive). That is, the two phases will tend to intermingle in shorter and shorter sections, and therefore cannot exist as separate phases.

[†] The calculation of $R_{\min}$ for $r = r_{cr}$ naturally leads to formula (162.5) if we note that under these conditions $\mu(P) - \mu'(P) \cong (v - v') \delta P$.

INDEX